MICROARRAY IMAGE
AND
DATA ANALYSIS

THEORY AND PRACTICE

Digital Imaging and Computer Vision Series

Series Editor

Rastislav Lukac

Foveon, Inc./Sigma Corporation
San Jose, California, U.S.A.

MICROARRAY IMAGE
AND
DATA ANALYSIS

THEORY AND PRACTICE

EDITED BY
Luis Rueda

CRC Press
Taylor & Francis Group
Boca Raton London New York

CRC Press is an imprint of the
Taylor & Francis Group, an **informa** business

CRC Press
Taylor & Francis Group
6000 Broken Sound Parkway NW, Suite 300
Boca Raton, FL 33487-2742

First issued in paperback 2019

© 2014 by Taylor & Francis Group, LLC
CRC Press is an imprint of Taylor & Francis Group, an Informa business

No claim to original U.S. Government works

ISBN-13: 978-1-4665-8682-6 (hbk)
ISBN-13: 978-1-138-37480-5 (pbk)

Library of Congress Cataloging-in-Publication Data

Microarray image and data analysis : theory and practice / editor, Luis Rueda.
       pages cm. -- (Digital imaging and computer vision series)
    Includes bibliographical references and index.
    ISBN 978-1-4665-8682-6 (hardback)
    1. Biotechnology--Data processing. 2. Protein microarrays. 3. Image processing. I. Rueda, Luis.

TP248.25.A96M53 2014
572.028'7--dc23   2013049255

**Visit the Taylor & Francis Web site at
http://www.taylorandfrancis.com**

**and the CRC Press Web site at
http://www.crcpress.com**

Dedication

I would like to dedicate this book to my wife, Ivanna, and my children, Maria, Gabriel and Rocio, for their constant love and support.

Contents

Preface

Since their discovery, microarrays have emerged as one of the most important biotechnological tools for studying the behavior of thousands of genes and genomic features on a genome-wide scale. The development of accurate and affordable technologies for effective and quick detection becomes paramount when considering the cost and time allowed to conduct a biological experiment. This has influenced the way in which transcriptional, proteomic and cell developmental studies are being carried out, providing a systematic way to study gene expression, elucidation of the proteome, discovering binding sites and methylation studies, all of these in a synchronized way. Transcriptional studies are one of the most important applications of microarrays, and for this, the development of gene expression arrays has helped enormously. Measuring the transcriptome or the entire repertoire of transcripts in an organism or cell lines provides a wealth of data about the relationships among DNA, transcription, translation, and phenotype. Advances in the microarray field have also allowed accurate detection that can uncover specific genetic mutations that may result in a number of human disorders. In this regard, microarrays, whether on full-length genes or short oligonucleotide assays, constitute an efficient tool for parallel and systematic analysis.

Although different shapes, resolutions, and features of microarrays may vary substantially across platforms, techniques and applications, they can be defined in general as a small glass slide with a collection of spots or elements in which biochemical reactions take place in order to reveal quantitative information about gene or protein expression, or other biological interaction of the molecules, under specific conditions. A microarray can be conceived as a small device (Greek word "mikro" for small), in which biomolecules are arranged in a systematic manner (French word "arayer" for array). In practice, a microarray is a device that has a rectangular shape and measures a few cm long – it is typically ordered, microscopic, planar, and specific.

Microarrays have been widely used by many researchers in diverse studies in biology. To mention a few facts, there have been more than 50,000 publications pertaining to microarrays, if PubMed is used as the source of information. Extending the scope to publications on "array" would list more than 120,000 abstracts in PubMed. Various databases and resources contain a wealth of microarray data for studies of gene expression, proteomics, tissue analysis, DNA methylation, genome-wide association studies, copy number variations and aberrations, and single nucleotide polymorphism, among others. The Gene Expression Omnibus database of the National Center for Integrative Biomedical Informatics of the United States contains a wealth of gene expression data, which are publicly accessible via the Internet. These databases contain millions of samples, mostly based on microarray studies.

Since the introduction of microarrays, huge amounts of data have been generated, and hence it has been imperative to create and improve specialized tools and algorithms for the acquisition, analysis, and management and interpretation of the data. In every aspect, microarray data analysis is invariably an essential component of microarray studies. Roughly speaking, data analysis tools for microarrays are derived from a variety of bioinformatics methods and techniques that bridge the gap among molecular biology, computer science, statistics, and mathematics. Microarray data analysis takes on the images that are obtained from scanning and up to the biological and functional analysis, involving a sequence of well-established steps aiming at identifying locations of spots or probe cells, background, and noise, and to subsequently quantify and normalize the signals. Numerical data resulting from the pre-processing stages are analyzed by means of different techniques for clustering, data mining, and classification, followed by visualization, and network and functional biological analysis. Data management involves storage, databases, resources, and compression, which are important steps that complement the whole process of analysis. Microarray image and data analysis is essential, since the aim of the whole experimental procedure is to obtain meaningful biological conclusions, which depends on the accuracy of the analytical steps, mainly those at the beginning of the pre-processing stage.

This book provides a comprehensive review of the main, up-to-date methods, tools, and techniques for microarray image and data analysis. Internationally recognized experts address specific research topics and challenges in their areas of expertise, providing valuable knowledge about the state-of-the-art methods in the field, and covering the main steps for image processing, gridding, segmentation, noise treatment, and normalization. Biological aspects and applications of microarrays extend to a wide range of biological studies in cancer research, and different microarray types and platforms such as DNA, various types of oligonucleotide and tissue microarrays, as well as the most recent tools for microarray data analysis in the Bioconductor suite. Machine learning methods for microarray data analysis cover the main aspects of clustering, biclustering, multi-dimensional microarray data analysis, and reconstruction of regulatory networks.

The book is intended for students as a graduate textbook and as a reference for the design of graduate courses at an advanced level, while providing sufficient level of detail for general readers as well. In addition, one of the goals of this book is to provide a useful reference for researchers and practitioners in academia and industry who develop tools and carry out critical analysis, management, and interpretation of microarray data. By covering state-of-the-art methods in microarray image and data analysis, the book provides a handy reference for professors, researchers, and graduate and advanced-level undergraduate students in multi-disciplinary fields including bioinformatics, computer science, statistics, biology, biochemistry, genomics, and biomedical engineering. While the readers will also learn about the current state of the microarray technology and how data generated by the microarray experiments are analyzed, they will understand the current landscape regarding the applications, availability, and affordability of the research utilizing this technology.

This book has been made possible through the help of many people who participated at different stages of the production process. I would like to thank all of them for their valuable help and support to make this happen. I am indebted to the book series editor, Rastislav Lukac, who specially invited me to edit this book and guided me through the main steps. My special thanks are for Adnan Ali for adding a self-contained biological insight to the book, which provides an excellent complement to the computing and statistical methods being presented. Special thanks are also for Iman Rezaeian for his tireless hours of work put into the edition and format conversions, as well as his contribution to a chapter. I would also like to thank Alioune Ngom for his contribution and help in reviewing, and for his constant logistic support, and Yifeng Li for his additional help in the review process.

Finally, I would like to thank the expert contributors for putting their efforts in carefully writing up-to-date and high quality chapters about their specific topic, while keeping the whole volume well synchronized, and for their help in the reviewing process. I would like to acknowledge the external reviewers, Michael Crawford, Meng Li, A. Sri Nagesh, Moysés Nascimento, Tuan Pham, Armando Pinho, Alexander Pozhitkov, Khedidja Seridi, Guifang Shao, Ruisheng Wang, Xiongwu Wu, Hong Yan, and Jie Zheng. With their solid expertise in the specific topics, they added the valuable feedback needed to establish these contributions as leading in the field.

Editor

Luis Rueda obtained his Ph.D. in computer science from Carleton University, Canada, in 2002. He joined the School of Computer Science at the University of Windsor in Canada, in 2002 as an Assistant Professor. After spending two years at the University of Concepción, Chile, he was appointed as an Associate Professor within the School of Computer Science and with the Pattern Recognition and Bioinformatics Lab at the University of Windsor. He has been recently promoted to Full Professor. His research interests are mainly focused on theoretical and applied machine learning and pattern recognition, mostly in the fields of transcriptomics, interactomics, and genomics. His research has mostly been centered on microarray and next generation sequence data analysis, including image and signal processing, pattern analysis and gene selection, and on finding relevant patterns in prediction of protein-protein interactions and stability of protein complexes, including domains, short-linear motifs, and complex types. Luis Rueda holds three patents on data encryption, secrecy, and stealth. He has published more than 100 publications in prestigious journals and conferences in machine learning and bioinformatics. He has participated in editorial and technical committees for conferences and journals. He is a Senior Member of the IEEE, and a Member of the Association for Computing Machinery and of the International Society for Computational Biology.

Contributors

José Adélaïde

Aix-Marseille Université; Institut
  Paoli-Calmettes, Marseille, France;
Inserm U1068; CNRS 7258, Centre de
  Recherche en Cancérologie de
  Marseille, Marseille, France
adelaidej@ipc.unicancer.fr

Adnan Ali

School of Computer Science, University
  of Windsor
Windsor, Ontario, Canada
aali@uwindsor.ca

Jose Manuel Arteaga-Salas

Cooperation Group CMA, Helmholtz
  Center Munich
Munich, Germany
jose.arteaga@helmholtz-muenchen.de

Wassim Ayadi

LaTICE, ESSTT, Université de Tunis,
  Tunisia
LERIA, Université d'Angers, France
wassim.ayadi@gmail.com

Dimitris Bariamis

Dept. of Informatics and
  Telecommunications, University of
  Athens
Athens, Greece
dimitris.bariamis@gmail.com

Matthew Bashton

Bioinformatics Support Unit, Newcastle
  University
Newcastle upon Tyne, UK
matthew.bashton@newcastle.ac.uk

Ismahane Bekhouche

Aix-Marseille Université; Institut
  Paoli-Calmettes, Marseille, France;
Inserm U1068; CNRS 7258, Centre de
  Recherche en Cancérologie de
  Marseille, Marseille, France
bekhouchei@ipc.unicancer.fr

Francis Bell

Biomedical Engineering, Science and
  Health Systems, Drexel University
Philadelphia, Pennsylvania, USA
fxb22@drexel.edu

François Bertucci

Aix-Marseille Université; Institut
  Paoli-Calmettes, Marseille, France;
Inserm U1068; CNRS 7258, Centre de
  Recherche en Cancérologie de
  Marseille, Marseille, France
bertuccif@ipc.unicancer.fr

Ghislain Bidaut

Aix-Marseille Université; Institut
  Paoli-Calmettes, Marseille, France;
Inserm U1068; CNRS 7258, Centre de
  Recherche en Cancérologie de
  Marseille, Marseille, France
ghislain.bidaut@inserm.fr

Daniel Birnbaum

Aix-Marseille Université; Institut
  Paoli-Calmettes, Marseille, France;
Inserm U1068; CNRS 7258, Centre de
  Recherche en Cancérologie de
  Marseille, Marseille, France
daniel.birnbaum@inserm.fr

Max Chaffanet
Aix-Marseille Université; Institut
  Paoli-Calmettes, Marseille, France;
Inserm U1068; CNRS 7258, Centre de
  Recherche en Cancérologie de
  Marseille, Marseille, France
chaffanetm@ipc.unicancer.fr

Tai-Been Chen
Dept. of Medical Imaging and
  Radiological Sciences, I-Shou
  University
Kaohsiung, Taiwan
ctb@isu.edu.tw

Chih-wen Cheng
Dept. of Electrical and Computer
  Engineering, Georgia Institute of
  Technology
Atlanta, Georgia, USA
cwcheng83@gatech.edu

Simon Cockell
Bioinformatics Support Unit, Newcastle
  University
Newcastle upon Tyne, UK
simon.cockell@newcastle.ac.uk

Urška Cvek
Dept. of Computer Science, Louisiana
  State University Shreveport
Center for Molecular and Tumor
  Virology, Louisiana State University
  Health Sciences Center Shreveport
Shreveport, Louisiana, USA
ucvek@lsus.edu

Mourad Elloumi
LaTICE, ESSTT, Université de Tunis
Tunisia
mourad.elloumi@fsegt.rnu.tn

Pascal Finetti
Aix-Marseille Université; Institut
  Paoli-Calmettes, Marseille, France;
Inserm U1068; CNRS 7258, Centre de
  Recherche en Cancérologie de
  Marseille, Marseille, France
finettip@ipc.unicancer.fr

Caroline C. Friedel
Institute for Informatics,
  Ludwig-Maximilians-University
  Munich
Munich, Germany
Caroline.Friedel@lmu.de

Maxime Garcia
Aix-Marseille Université; Institut
  Paoli-Calmettes, Marseille, France;
Inserm U1068; CNRS 7258, Centre de
  Recherche en Cancérologie de
  Marseille, Marseille, France
maxime.u.garcia@gmail.com

Colin S. Gillespie
School of Mathematics & Statistics,
  Newcastle University
Newcastle upon Tyne, UK
colin.gillespie@newcastle.ac.uk

Arnaud Guille
Aix-Marseille Université; Institut
  Paoli-Calmettes, Marseille, France;
Inserm U1068; CNRS 7258, Centre de
  Recherche en Cancérologie de
  Marseille, Marseille, France
arnaud.guille@inserm.fr

Jin-Kao Hao
LERIA, Université d'Angers
France
hao@info.univ-angers.fr

Miguel Hernández-Cabronero
Dept. of Information and
Communications Engineering,
Universitat Autonoma de Barcelona
Bellaterra, Barcelona, Spain
miguel.hernandez@uab.cat

Sonal Kothari
Dept. of Electrical and Computer
Engineering, Georgia Institute of
Technology
Atlanta, Georgia, USA
sonalkothari86@gmail.com

Yifeng Li
School of Computer Science, University
of Windsor
Windsor, Ontario, Canada
li11112c@uwindsor.ca

Henry Horng-Shing Lu
Institute of Statistics, National Chiao
Tung University
Hsinchu, Taiwan
hslu@stat.nctu.edu.tw

Rastislav Lukac
Foveon, Inc. / Sigma Corp.
San Jose, California, USA
lukacr@colorimageprocessing.com

Michael W. Marcellin
Dept. of Electrical and Computer
Engineering, The University of
Arizona
Tucson, Arizona, USA
mwm@email.arizona.edu

Dimitris Maroulis
Dept. of Informatics and
Telecommunications, University of
Athens
Athens, Greece
dmaroulis@di.uoa.gr

Raphaële Millat-Carus
Aix-Marseille Université; Institut
Paoli-Calmettes, Marseille, France;
Inserm U1068; CNRS 7258, Centre de
Recherche en Cancérologie de
Marseille, Marseille, France
raphaele.millatcarus@gmail.com

Alioune Ngom
School of Computer Science, University
of Windsor
Windsor, Ontario, Canada
angom@uwindsor.ca

Rehman Qureshi
Biomedical Engineering, Science, and
Health Systems, Drexel University
Philadelphia, Pennsylvania, USA
raq22@drexel.edu

Iman Rezaeian
School of Computer Science, University
of Windsor
Windsor, Ontario, Canada
rezaeia@uwindsor.ca

Luis Rueda
School of Computer Science, University
of Windsor
Windsor, Ontario, Canada
lrueda@uwindsor.ca

Renaud Sabatier
Aix-Marseille Université; Institut
Paoli-Calmettes, Marseille, France;
Inserm U1068; CNRS 7258, Centre de
Recherche en Cancérologie de
Marseille, Marseille, France
sabatierr@ipc.unicancer.fr

Ahmet Sacan
Biomedical Engineering, Science, and
Health Systems, Drexel University
Philadelphia, Pennsylvania, USA
ahmet.sacan@drexel.edu

Michalis Savelonas
Dept. of Informatics and
  Telecommunications, University of
  Athens
Athens, Greece
msavel@di.uoa.gr

Joan Serra-Sagristà
Dept. of Information and
  Communications Engineering,
  Universitat Autonoma de Barcelona
Bellaterra, Barcelona, Spain
joan.serra@uab.cat

Shahram Shirani
Dept. of Electrical and Computer
  Engineering, McMaster University
Hamilton, Ontario, Canada
shirani@mcmaster.ca

Todd H. Stokes
Dept. of Biomedical Engineering,
  Emory University and Georgia
  Institute of Technology
Atlanta, Georgia, USA
todd.h.stokes@gmail.com

Marjan Trutschl
Dept. of Computer Science, Louisiana
  State University Shreveport
Center for Molecular and Tumor
  Virology, Louisiana State University
  Health Sciences Center Shreveport
Shreveport, Louisiana, USA
mtrutsch@lsus.edu

Meng-Yuan Tsai
Institute of Statistics, National Chiao
  Tung University
Hsinchu, Taiwan
u9826804@stat.nctu.edu.tw

May D. Wang
Dept. of Biomedical Engineering,
  Emory University and Georgia
  Institute of Technology
Atlanta, Georgia, USA
maywang@bme.gatech.edu

Yiqian Zhou
Biomedical Engineering, Science, and
  Health Systems, Drexel University
Philadelphia, Pennsylvania, USA
yz86@drexel.edu

1 Introduction to Microarrays

Luis Rueda and Adnan Ali

CONTENTS

1.1 INTRODUCTION

During the last decade, microarrays have emerged as essential tools for many researchers to monitor biomolecules' behaviors for thousands of genes simultaneously. Since their discovery, diverse applications and complex issues in processing and analysis of high-throughput data have been the center of modern molecular biology research, because an accurate analysis is one of the keys to provide a meaningful interpretation of the data. Microarrays are used to explore different manifestations of the genomic information in terms of cellular processes and biochemical products that ultimately dictate the behavior of a particular organism or a specific cell type and its interactions with others. Proteins, cells, tissues, and an entire organism can be studied at different levels of detail in a massive and parallel way, offering innumerable possibilities and applications in biomedical research, medical genetics, pharmacology, disease diagnosis, staging and prognosis, pharmacotherapeutics, evolutionary studies, conservation science, and genetic polymorphic studies, to mention a few.

This introductory chapter discusses basic concepts of molecular biology, genomics and regulation of gene expression. Thereafter, an overview of the principles and fundamentals of microarrays is presented from a general perspective. In a superficial way, more emphasis is given on the means and methods for microarray production, scanning and the main steps of microarray image acquisition. Included also in this chapter are discussions, both general and specific, of the main methods for pre-processing and analyzing microarrays, including different aspects of gridding, segmentation, noise treatment, data storage and retrieval, and microarray data analysis tools available to the scientific community. The chapter also presents an overview of the existing resources for microarray data, publications, public databases and repositories, and analytical tools.

1.2 MOLECULAR BIOLOGY AND GENOMICS

This section introduces the principles of molecular biology, from the basic concepts of nucleotides, nucleic acids, amino acids, and proteins to other macromolecules of interest. A fundamental concept in molecular biology, the central dogma, defines a basic flow of genetic information from DNA to RNA and finally into a protein product. Encoded within the genomic DNA molecule is all of the genetic information in form of its specific nucleotide sequence, which is ultimately transcribed into RNA. In turn, RNA encodes the information (codons) for translation into a protein molecule that contains a specific sequence of amino acid residues. Since the discovery of nucleic acid molecules, the central dogma has evolved into its present form where other aspects, such as reverse-transcription of RNA into cDNA molecules, are also considered to be part of the basic concept (Figure 1.1). Other forms of heterogeneous RNA molecules, tRNA, rRNA, siRNA, miRNA, and lincRNA do not encode protein products but rather have specific functions in other important cellular activities, including the process of translation itself. In addition, protein folding, structural modifications and molecular interactions are required for the proper function or activity of some proteins and their domains (with different gray scales in Figure 1.1), conforming net-

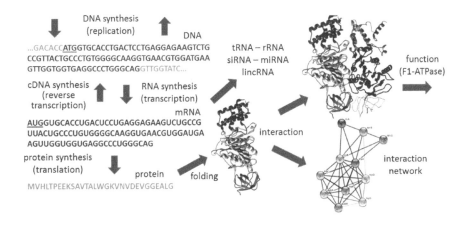

FIGURE 1.1 The basic concept of the flow of genetic information in biological systems stems from the central dogma in biology. DNA is transcribed into RNA molecules, which in turn encodes a specific amino acid sequence in a protein molecule. After folding, most proteins adopt a structure and interact with other molecules.

works of protein and domain interactions. For more details, the reader should refer to more specific textbooks in the subject [1, 2]. In the following sections, a basic understanding of the molecular structure of macromolecules and their interactions with each other are discussed with respect to the concept of microarray hybridization.

1.2.1   GENOME STRUCTURE AND FUNCTION

In eukaryotic cells, the genomic DNA is present in a discrete nucleus in the form of chromosomes wherein the double stranded DNA is complexed with a number of protein molecules, such as histones. In contrast, the genomic DNA in prokaryotic cells is a circular nucleoid structure present in the cell without any nuclear envelope. During the last two decades, genome sequencing projects of several organisms, both eukaryotic and prokaryotic, have been completed where the entire genome sequences of the genomic DNA have been identified and comparative analyses have been performed.

The genome of an organism presents a complete set of genes and is maintained generation after generation. It is important to note that the eukaryotic genomic DNA comprises coding as well as non-coding regions. Following the transcription of genes using genomic DNA as a template, the coding regions are eventually spliced together and then ultimately translated into a final protein product in the cytoplasm of the cell. The identity of a gene in the genome is specified by the presence of open reading frames (ORF) containing the coding regions, and regulatory regions that may be present upstream or downstream of the open reading frame (coding region). In the context of microarrays, for the study of gene expression, we are mainly concerned with the coding regions of the genomic DNA. Therefore, it is important to define the concept of transcriptome, which represents the entire repertoire of genes that are expressed in a cell at a given time under defined conditions. Since the RNA molecules transcribed from coding regions may be non-coding RNA molecules or messenger RNA (mRNA) coding molecules, the transcription of the genes results in mRNA molecules, which are more relevant when one considers gene expression to study the production of protein products in a cell.

Primarily, the production of proteins in the cell ultimately renders certain cell types to acquire their desired function or phenotype. This is the property of a protein molecule that gives the proper morphology and physiology of each cell type. Certainly, the role of other types of important molecules, like carbohydrates, polysaccharides, and lipids in this regard cannot be ignored either. Since higher eukaryotic organisms comprise multiple types of cells, each cell type has its own structure and function, and hence is able to form part of a specific tissue type by cellular differentiation leading to the *division of labor*. In this sense, a proteome represents a complete set of proteins being produced by specific mRNA molecules expressed in a specific transcriptome. Different proteins may function independently or in combination and coordination with their other protein or carbohydrate counterparts. Thus, a diversity of interactions of macromolecules is achieved in a very complex manner, and is responsible for the presence of multiple types of cells, and hence a variety of tissues within a single higher eukaryotic organism.

The cells in an organism are exposed to a number of external and internal stimuli that can initiate a cascade of signaling mechanisms that eventually affect the regulation of gene expression from the genomic DNA that resides in the nucleus in case of eukaryotic organisms. The process of these signaling mechanisms is commonly termed as *signal transduction*. There are numerous signaling cascades in cells that run in parallel under normal or stress conditions. Cells respond to these signals in an appropriate manner to carry out normal cellular functioning. Any malfunctioning in such signaling mechanisms causes cells to behave in an otherwise inappropriate manner. Such malfunctioning in signal transduction mechanisms has been shown to cause severe disorders, including cancer. This topic is further discussed in Chapter 2.

1.2.2 MACROMOLECULES

All living organisms are composed of cells, which present a wide range of variations from organism to organism and from cell to cell within the same organism. The molecules that form the cellular structures are generally divided into (i) small molecules, such as fatty acids, amino acids, nucleotides, and sugar molecules, and (ii) large molecules or macromolecules, such as lipids, proteins, nucleic acids, and polysaccharides [3]. Primarily, small molecules interact to make polymers that give rise to different polymeric structures of macromolecules. Here, we briefly consider the molecular structure and function of nucleic acid and protein molecules. Since these macromolecules are produced by aggregation of nucleotide and amino acid molecules, respectively, it is necessary to understand how it is possible that small molecules polymerize to form macromolecular structures that contain unique properties and functions, and contain all the genetic information.

1.2.3 NUCLEIC ACIDS

A nitrogenous base linked to a sugar is called *nucleoside*, and a nucleoside linked to one or more phosphate groups is called *nucleotide*. Nucleotides make the primary building blocks of nucleic acid molecules, which can be *deoxyribonucleic acid* (DNA) or *ribonucleic acid* (RNA) molecules. The DNA has a 2′-deoxyribose, whereas an RNA molecule has a ribose instead. In summary, a typical nucleotide molecule comprises three components: a nitrogenous base, a sugar, and a phosphate molecule. In any nucleic acid molecule, the nitrogenous base is either a purine (adenine and guanine) or a pyrimidine (cytosine and thymine). A nucleic acid molecule comprises a chain of nucleotide molecules linked together by phosphodiester linkages. The backbone of the resulting polynucleotide is composed of alternating sugar and phosphate residues wherein the 5′ end of one pentose ring is linked to the 3′ end of the next pentose ring via a phosphate group in between. Each nucleic acid molecule contains the four types of bases (A, G, C and T) in a defined order or a sequence. In both DNA and RNA molecules, the same type of purines are present. However, a difference exists with respect to the pyrimidines. In a DNA molecule, cytosine and thymine are present, whereas cytosine and uracil are present in an RNA molecule. Therefore, when writing the nucleic acid sequence of a DNA molecule with sequence ACGT, the corresponding identical sequence is written as ACGU for an RNA molecule. The terminal ends of nucleic acid molecules contain a 5′ most terminal free phosphate group and a 3′ most OH group. Traditionally, the sequence of a nucleic acid molecule is written in 5′ to 3′ direction. The terminal ends are often used to tag the nucleic acid molecules with different labels, which aid in their detection using conventional biological assays. The synthesis of RNA molecules from a genomic DNA template within the cell is called *transcription*.

By virtue of their structure, there is a specific interaction between purines and pyrimidines. So far, we have considered only a single linear polynucleotide molecule, which is produced by polymerizing individual nucleotides. When considering a double stranded DNA molecule, which occurs in nature, the specific interactions between purines and pyrimidines become crucial. In this regard, adenine can form bonding with thymine; and guanine can interact with cytosine. This hydrogen bonding between nitrogenous bases is termed as *base pairing* and is said to be complementary in nature. Therefore, the sequence of a given DNA molecule contains two polynucleotide molecules held together by *hydrogen bonding* in an antiparallel fashion, i.e., one strand or polynucleotide in 5′-3′ orientation will face its complementary strand in 3′-5′ direction.

It is also important to note that, in nature, RNA molecules exist in single stranded conformation. The same property held for nucleotides to form hydrogen bonding with their respective bases allows us to make *in vitro* assays, where one polynucleotide is used as a probe to detect its corresponding complementary strand in a given biological sample, by an assay commonly called *hybridization*. Hybridization can occur between DNA-DNA, RNA-RNA, and DNA-RNA. The ability of single stranded nucleic acid molecules to hybridize can be used as a measure of complementarity between the two nucleic acid sequences. In fact, this is the basic principle used in gene expression studies using microarray platforms. Hybridization between nucleic acid molecules can be controlled by determining the *in vitro* conditions under which they can denature, renature or hybridize. The detection of hybridization is a common routine and various detection methods are used in a biological laboratory setup. Traditionally, Southern blots have been used to detect DNA-DNA hybridization, and Northern blots for the detection of RNA molecules using a labeled DNA probe (DNA-RNA hybridization). Microarrays primarily provide a method for detection of RNA molecules using immobilized DNA probes. A variety of techniques exist for DNA sequencing in which the sequence of individual nucleotides is determined using chemical reactions for the detection of the nucleotides.

1.2.4 PROTEINS

Protein molecules form major components in a cell. Amino acid residues are linked to each other by peptide bonds and form small continuous chains called *peptides*. Within the cytoplasm of a cell, the synthesis of a polypeptide, encoded by an mRNA strand, is called *translation*. A protein molecule consists of one or more linear polymers and comprises a specific sequence or order of amino acid residues, where each residue contains one of the 20 amino acids that exist in nature—amino acid molecules are called *residues* when they are incorporated into a polypeptide. A protein molecule may contain several polypeptides, where each polypeptide is synthesized separately. Also, a polypeptide chain may be truncated into several shorter peptides, which may be part of a final structure of a protein. Following the synthesis of polypeptides, they may undergo several modifications within the cell, achieving specific properties for the proper functioning of the final protein product. The sequence of amino acids in a polypeptide can be determined *in vitro* by a chemical reaction termed as *Edman degradation*.

Each amino acid consists of an amino group that is linked to a carbon, which is ultimately linked to a carboxyl group. In addition to carbon, each residue has a distinctive side chain or R group that determines the specific properties of the amino acid contained in it. In a polypeptide, the amino acids are arranged in an amino (N-terminus) to carboxyl (C-terminus) direction. The primary structure of a protein molecule is determined by the composition of a variety of amino acid residues and their specific order in the polypeptide. Different types of amino acid residues present unique properties with respect to polarity, charge, acidity and basicity due to the nature of their form and structure. Following the synthesis of a polypeptide, further post-translational modifications provide specific properties to the protein structure and function. Each protein has a unique structure, which is achieved through the intracellular modifications of the polypeptide molecules and via its interaction(s) with a number of other proteins or carbohydrate molecules.

A proper folding of some protein molecules is crucial to their function within the cell or outside the cell when released. During the process of folding, the chain of amino acid residues can fold into secondary or tertiary structures with the help of chaperones within the cytoplasm of the cell. Proteins can be structural proteins, such as the fibrous protein keratin, or globular proteins, such as enzymes that can catalyze specific enzymatic reactions. Also, proteins can be retained within the cell or released from the cell where they perform their extracellular functions. Secreted enzymes and hormones are examples of some secreted proteins.

Historically, the identification of antibodies has had a significant impact in protein biology. Antibodies are also protein molecules that bind with specific antigenic sites present within a target protein molecule. The specificity of binding of antibodies with their specific domains in a polypeptide molecule makes them very useful in polypeptide or protein detection assays. For example, an antibody generated against a specific antigenic site (epitope) in a protein will always bind to its specific epitope when exposed to that target protein molecule. Thus, the specificity of antibody-antigen interaction has allowed the development of a number of protein detection assays, such as Western blots, immunoprecipitation, enzyme-linked immunosorbent assays (ELISA), as well as protein and antibody microarrays. Antibodies can be labeled with different radioactive or non-radioactive labeling molecules. The labeled specific antibodies are also used to isolate, purify and detect specific proteins. Western blots are commonly used to detect specific protein molecules using labeled antibodies. One of such assays is termed as *immunoprecipitation*, where a specific antibody is used to isolate and purify a specific protein molecule from a mixture of proteins in a given biological sample. The use of antibodies to detect proteins has also been widely used in microarrays, where a variety of protein molecules can be detected simultaneously.

1.2.5 TRANSCRIPTION AND TRANSLATION

As indicated earlier in this chapter, the synthesis of RNA molecules using DNA as a template is called *transcription*. There are multiple protein factors that are involved in the initiation, elongation and termination of the transcription in the cell. In eukaryotic cells, this phenomenon occurs within the nucleus of the cell wherein the genomic DNA present in the chromosomes acts as a template. The RNA produced in the cell is identical to one strand of the double-stranded genomic DNA, which is called *coding strand*. Therefore, the complementary strand or the non-coding strand of the double-stranded genomic DNA acts as the actual template for the synthesis of new RNA. Transcription reaction is catalyzed by an enzyme called *RNA polymerase*. Transcription starts with the binding of the RNA polymerase enzyme to the promoter region of a gene that is present upstream of the coding region of the genomic DNA. This binding is facilitated by several factors without which the transcription would not proceed. In the gene, a starting point marks the start site for the synthesis of RNA.

Transcription proceeds with the movement of the polymerase complex along the template strand until it reaches a terminator sequence in the template strand. The double stranded DNA is uncoiled by the action of other protein factors making the template available for accessing the polymerase complex, which moves forward. During this process, the polymerase enzyme is capable of reading the template and adding an appropriate complementary ribonucleotide in $5'$ to $3'$ direction. Every new ribonucleotide is added to the $3'$ OH end of the previously incorporated ribonucleotide.

Once the polymerase complex reaches the terminator sequence, the complex falls off and the single-stranded newly synthesized RNA is released and is then free to move into the cytoplasm of the cell.

The region containing sequences before the starting point of the gene is called upstream, whereas the region following the terminator region is termed as downstream. The first base being transcribed into RNA is numbered as $+1$, whereas the base prior to it is -1. There is no base with a number zero in a gene. There are also further specific regions upstream of a promoter or downstream of a terminator region, which are also involved in the regulation of gene expression, e.g., enhancers. However, the discussion of such regions is not intended for the present chapter. Three different types of RNA molecules are produced in the cell through the process of transcription. These are messenger RNA (mRNA), transfer RNA (tRNA), and ribosomal RNA (rRNA). The tRNA and rRNA molecules are mainly involved in the process of synthesis of protein molecules during translation, whereas the mRNA actually provides a template for the protein synthesis. The mRNA, as apparent from its name, represents the coding region of a gene and will be translated into the final polypeptide molecule. The polypeptide molecules may further become modified or interact with other polypeptide molecules to produce the final protein molecule. Here, we limit our discussion to mRNA molecules as they are the main vehicles for the active gene expression at the transcriptional level.

1.2.6 ALTERNATIVE SPLICING

Once the pre-mRNA (nascent RNA molecule) has been synthesized, it may also be subjected to further modifications. For example, one of such modifications is termed as *RNA splicing* in eukaryotic cells. During RNA splicing, the regions in the RNA molecules that may not represent coding regions are spliced out, and a mature mRNA molecule is produced. The genomic DNA contains *exons* (coding regions) and introns (non-coding regions) that are also transcribed into mRNA. The transcription mechanism by itself cannot discriminate the introns from exons. Therefore, the RNA splicing mechanism is responsible for ensuring the synthesis of mature mRNA, which does not include any non-coding sequences.

There is also another layer of complexity here, wherein the splicing mechanism is not always used only to splice out non-coding regions. Sometimes, the splicing of some exons is also observed wherein a distinct species of mRNA molecules can be generated from a single type of a newly synthesized full length RNA molecule. For example, if a gene consists of three different exons, exon 1, exon 2, and exon 3, an alternative splicing mechanism can produce a mature mRNA molecule that may contain any combination of these three exons, e.g., exon 1-exon 2-exon 3; or exon 1-exon 2; or exon 1-exon 3, and so on. The presence of alternative splicing, which is unique to eukaryotic cells, provides another level of regulation of gene expression at the post-transcriptional level. There are many genes within the cell that are regulated by alternative splicing. Such alternative splicing results in the biodiversity of protein molecules being synthesized in the cells in which one gene can code for multiple isoforms of protein molecules with different functions. In the context of alternative splicing in the regulation of cellular activities, the use of microarrays has proven to be an excellent tool. The presence of specific exons/introns as probes on a microarray enables the detection of transcribed regions of a gene by detecting the presence of the corresponding RNA transcripts in a given sample using microarray hybridization assays. This topic is further discussed in Chapter 2 with reference to the post-transcriptional regulation of gene expression.

1.2.7 REVERSE TRANSCRIPTION

In certain RNA viruses, called retroviruses, some special enzymes are present for genetic modifications such as transposition of genetic elements. These enzymes are termed as *reverse transcriptases*. These enzymes from viral particles have been isolated and purified. Under suitable conditions, reverse transcriptase enzymes can be used to reverse transcribe an RNA molecule to obtain a complementary DNA strand, termed as cDNA. The newly synthesized cDNA molecule can be further made double stranded using the same reverse transcriptase enzyme in the same reaction, since the enzyme can read RNA or DNA templates for the synthesis of a complementary DNA strand. In fact, the reverse transcription reaction has made a significant contribution towards understanding molecular mechanisms within eukaryotic cells.

The genes that are expressed in a cell are actively transcribed and mRNA from the cellular mixture can be isolated using standard techniques. The mRNA isolated from the cell can be reverse transcribed and the resulting double stranded cDNA can be cloned into DNA vectors for sequencing, amplification by polymerase chain reaction, or further genomic studies as required. In a microarray experiment, detection of gene expression is carried out using the same reverse transcription reaction during which different labeling molecules, such as Cy3 or Cy5, can be used to label the resulting cDNA, which is then used to hybridize against the immobilized DNA probes on a solid support.

1.2.8 TRANSLATION

Following the synthesis of mRNA via gene transcription, the mechanism of translation into protein takes place in the cytoplasm, which again involves a number of factors including ribosomal machinery of the cell. This process of biochemical synthesis of protein molecules is dependent upon the specific sequence of the template mRNA transcribed from a specific gene. Within the protein coding region in the mRNA molecule each triplet (three consecutive nucleotides) is called *codon*. With a few exceptions, the genetic code related to the nucleotide triplets is considered as universal code. Each amino acid is encoded by a specific triplet codon. For example, a standard codon AUG specifies the incorporation of a methionine residue in a nascent polypeptide molecule. This codon is also termed as initiation codon because usually methionine is a standard first amino acid residue in eukaryotic polypeptides. In eukaryotes, the termination codon is one of the three codons UAA, UGA and UAG, which do not encode any amino acid residue. Translation of an mRNA molecule requires an mRNA template, tRNA molecules linked to amino acid residues, and ribosomes. Following the synthesis of polypeptide molecules, a post-translational modification of the newly synthesized polypeptide molecule occurs with the help of chaperone proteins, which allow the proper folding and conformation of the final protein structure. Other processes, such as glycosylation, acetylation, and acylation may also occur with specific polypeptides as needed before they can enter into their final secondary or tertiary conformation.

As indicated earlier, antibodies are also protein molecules that are primarily host molecules produced in response to foreign molecules (antigens) and are also protein molecules themselves. Antibodies are produced by plasma cells and circulate in the body where they interact with foreign antigens and make an antibody-antigen complex, which is ultimately removed through phagocytosis. Fundamentally, antibodies are a family of glycoproteins that share structural and functional features in some aspects. The specific interaction of antibodies with their respective antigenic sites in the target protein has allowed their frequent use in the protein detection assays. In protein microarrays, the use of known specific antibodies is employed to detect the presence of their respective binding proteins in a mixture of proteins in a given biological sample.

1.3 REGULATION OF GENE EXPRESSION

Regulation of gene expression can occur at various levels within the cell. A given gene may be considered to be regulated for its expression, for example synthesis of mRNA or protein molecules, or for its associated activity. In view of the microarrays, we consider only the expression of the genes at the level of transcription or translation. In this regard, genes may be regulated at the level of genomic DNA, transcriptional level, post-transcriptional level, translational level, or even at post-translational level. Living cells are continuously exposed to internal or external stimuli, and they are well equipped with an intracellular infrastructure to respond to these signals by changing the internal cellular activities in a number of ways (Figure 1.2). For example, in response to a chemical exposure, a specific signaling mechanism is activated that can ultimately affect the regulation of gene expression by regulating the transcription, translation, or even by modifying the post-translation structure or function of the protein molecules. The study of genetic changes in the genomic DNA on a large scale is termed as *genomics*. Similarly, obtaining a global picture of the active transcriptional state of a cell, the transcriptomics, and the entire protein expression and interactive architecture, the proteomics, are studied on a genome-wide scale (Figure 1.2).

In genomic DNA, the variation in the nucleotides may occur within the member of a given biological species. The variation may be at a single nucleotide level, termed as single nucleotide polymorphism (SNP). Such variation in a single nucleotide may also be present within the paired chromosomes giving two kinds of alleles. Allelic variations due to the presence of SNPs can occur within the coding as well as in non-coding regions. Such variations may or may not have any effect on gene expression. For example, a specific SNP in a promoter region may present a mutation that impedes binding of the RNA polymerase complex, and hence inhibiting the ini-

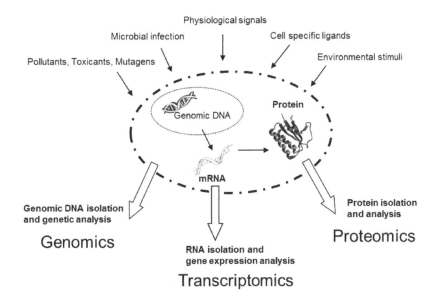

FIGURE 1.2 Analysis of complex biological processes on a genome-wide scale. The genetic changes at large scale (genomics), and the expressed genes at the RNA level (transcriptomics) or at the protein level (proteomics), and their molecular interactions can be visualized using a single microarray representative of the entire genome.

tiation of transcription. However, a greater number of SNPs appears to be present within the non-coding region in the genomic DNA, which is presumed to have some type of evolutionary significance. A number of SNPs have been associated with the presence of a range of human disorders, such as cancer, infectious diseases, and autoimmune diseases, which appear to result from particular SNPs. Even if the SNPs themselves do not cause a certain disorder, their use in the linkage analysis has allowed the identification of candidate genes that are directly associated with certain diseases. Therefore, SNPs provide important biomarkers in clinical diagnosis for certain genetic diseases. Microarrays have also provided scientists with a way to use high-throughput identification of SNP markers from a number of biological samples simultaneously. Such polymorphic analysis using either target SNP sites, or their genome-wide association studies, for certain genetic diseases using microarrays are discussed in more detail in Chapter 2.

It has been shown, however, that some modifications of the genomic DNA do not involve mutations or modifications of the DNA sequence itself. Such modifications can occur in the chromatin conformation or the biochemical modification of certain nucleotides in the DNA. One of these modifications involves methylation of cytosine nucleotides in the promoter region of some genes. Such heritable changes in the DNA, which affect gene expression or cellular phenotype, are caused by mechanisms other than changes in the DNA sequence itself, and are called *epigenetic regulation*. There are certain enzymes present within the nucleus of eukaryotic cells, which can methylate cytosine nucleotide when present in the promoter region as a dinucleotide CpG (cytosine-phosphate-guanine). Such methylation may impede the RNA polymerase complex to access the promoter to initiate the transcription of the coding region, which is downstream of the promoter region. A number of eukaryotic genes have been shown to be regulated by such methylation process. Methylation-sensitive microarrays are discussed in Chapter 2.

The next level of regulation of gene expression is at the level of transcription. Here, a number of regulatory proteins, the *trans*-acting factors (such as transcription factor), the regulatory elements within the gene, and *cis*-acting elements (such as a promoter region) interact with each other to activate or deactivate transcription. The cells can be stimulated by external factors or internal cues, which may direct the activation of transcription factors to interact with their corresponding *cis*-acting elements and allow the RNA polymerase to initiate the transcription of the gene. Certain genes are always expressed within the cell without any stimulation—such genes are called house-keeping genes or constitutively expressed genes. Examples of inducible genes are genes regulated by growth factors, hormones, other physiological signals, external stimuli, toxicants, mutagens, or high temperatures. Similarly, certain genes are expressed abnormally in cancer cells due to malfunctioning of signaling mechanisms in a cell. Such abnormal expression of certain genes can initiate a cascade of events in the cell, which ultimately modify the cellular behavior and morphology. This phenomenon may cause formation of lumps of cells due to uncontrolled growth and results in the formation of a tumor. Therefore, scientists have always been keen to understand the regulation of gene expression using traditional techniques, such as Northern blots, polymerase chain reaction, primer extension, RNase protection assays, and serial analysis of gene expression (SAGE), to name a few. The advent of microarrays has played an important role in high-throughput gene expression studies, since they can provide great detail about the transcription of a wide selection of genes in a cell at the same time under the same conditions.

One of the most important applications of microarrays is in post-transcriptional regulation studies. As discussed earlier, the molecular mechanisms, such as RNA splicing, also occur in the cell and may regulate the synthesis of specific multiple types of mRNA molecules from a single gene. It is possible to detect alternative splicing using various molecular techniques such as polymerase chain reaction and primer extension assays. Microarrays have also been employed to study the alternative splicing in the regulation of gene expression.

Translational regulation of gene expression is another area of cellular research where the application of microarrays has been widely used in recent years. Translation of mRNA templates to produce polypeptide molecules finally determines the cellular phenotype. The presence of specific proteins within the cell can be examined by isolating a mixture of the proteins of the cell employing techniques such as Western blots using specific antibodies. However, the use of microarrays has allowed scientists to take benefit of multiple antibodies, and detect a large number of protein molecules simultaneously in a single experiment. The use of protein microarrays is also discussed in Chapter 2.

1.4 MICROARRAYS

Microarray technology allows scientists to study and examine key biological questions on a genomic scale. This has influenced the way in which transcriptional, proteomic and cell developmental studies are carried out, providing a systematic way to study gene expression, elucidation of the protein expression, discovering molecular interactions and genetic studies, independently or in a synchronized fashion. Transcriptional studies are one of the most important applications of microarrays. Measuring the entire repertoire of transcripts in an organism or cell lines provides a wealth of information about the link between the DNA and the cellular phenotype. The development of gene expression arrays has helped enormously in this direction. To study the transcriptome comprising multiple transcripts expressed simultaneously in a cell or tissue, for example, about 300,000 RNA molecules in a human cell, is a challenge for biologists [4]. In this regard, microarrays, whether using full-length genes or short oligonucleotide arrays, constitute an extraordinary tool for parallel and systematic analysis.

Protein microarrays are also important in studying protein expression quantification. In this regard, tissue microarrays are one type of technology being used for protein expression patterns *in situ* (more details in Chapter 2). In addition to gene and protein expression studies, the use of microarrays in clinical studies has explored novel uses of this technology in contemporary medical sciences. Identification of single-nucleotide polymorphism (SNP), disease staging and diagnosis, drug screening, and toxicological studies are only a few to name. In particular, genes have been successfully used for detecting driver genes in cancerology (more details in Chapter 12). Other applications are discussed in detail in Chapter 2.

Although different shapes, resolutions and features of microarrays vary substantially across platforms and in respect of techniques and applications, a typical microarray can be defined as a small glass slide with a plurality of spots or elements in which biochemical reactions take place. The microarray reveals qualitative information about gene or protein expression, or other biological activity, under specific conditions of an experiment. The word microarray comes from *mikro* (small) and *arayer* (arranged). A microarray has a rectangular shape and measures a few cm long—it is typically ordered, microscopic, planar and specifically coated with a suitable substrate [5].

There are two main technologies for producing microarrays. The first one is known as *spotted arrays*, which are produced by robotic spotting or by an inkjet printer. This technology is referred to as cDNA[1] microarrays or DNA microarrays. *Oligonucleotide arrays*, on the other hand, are produced by employing mainly

---

[1]cDNA refers to "complementary" or "copy" DNA produced during the reverse transcription of RNA in the samples and which is aimed to hybridize the DNA probes printed on the slides. The term "DNA microarrays" used through this book includes cDNA, EST and genomic DNA microarrays.

one of the following approaches [6, 7, 8]: photolithographic (Affymetrix array, aka Affymetrix GeneChip), inkjet technology (Agilent), electrochemical synthesis (CombiMatrix), solid state (NimbleGen), and silica beads in microwells (Illumina arrays, aka Illumina BeadArrays). More details about these technologies are discussed in Chapter 2.

In a typical DNA microarray, the DNA molecules are selected from a library of DNA clones in the case of spotted arrays, whereas in the case of oligonucleotide arrays, the sequences are selected from known genes and then synthesized in the form of short oligonucleotides on the slides. The DNA printed on a spot of the array is known as a probe, since each spot has a known DNA sequence. The precise location of each spot on an array is known and is recorded for image processing. The detection is carried out by hybridization reaction using a labeled sample, usually by Cy3 and Cy5 fluorophores, green and red channels respectively. The labeled nucleic acid molecules (cDNA) are produced by reverse transcription from the mRNA isolated from cells, tissues or organisms. The hybridization with the probes on a microarray helps identify the presence of target complementary sequences in a given sample. Washing the array after this process eliminates any irrelevant and unhybridized molecules and prepares the slide for scanning. High resolution scanning allows devices to detect fluorescent dyes in different channels, typically green and red, which produces images for different channels or conditions, namely experiment versus control, or even normal versus diseased sample. A typical microarray scanner uses two lasers to detect Cy3 and Cy5 wavelengths to produce green and red images of the same array. In addition, the nucleic acid molecules are usually printed in duplicates or triplicates so that any differences during the hybridization procedure can be detected and normalized.

Generally speaking, microarrays are prepared by spot printing DNA fragments or cDNA clones, which are selected to represent the expressed genome of a given cell. The clones may be available as expressed sequence tags (EST) from any specific expression library. EST data is available from the widely used National Center for Biotechnology Information (NCBI). Microarrays are prepared by printing amplified PCR products using ESTs or from nucleic acid molecules isolated from the cells or tissues. The amplified products are suspended in appropriate buffers and are spot printed onto poly-L-lysine- or aminosilane-coated glass microscopic slides using a high-speed robotic system. Alternatively, microarrays are prepared by synthesizing short oligonucleotides on the coated glass slides themselves by chemical reactions on the substrate present on the slides.

Throughout the design and execution of gene expression profiling via a microarray experiment, the key element in the expression analysis is the RNA isolated from a cell, tissue or an organism (Figure 1.3). This isolated RNA is used as a template for reverse transcription to make cDNA for comparative hybridization against the probes that are present on the microarrays. Total RNA is extracted from the biological samples, experimental and control, which may further be fractionated to purify poly A+ mRNA if needed. The use of poly A+ mRNA further adds to the efficiency of reverse transcription and hence, leads to an improved quality and specificity of the hybridization reaction. Reverse transcription yields cDNA, which is labeled with two fluorescent labels, such as Cy3 and Cy5. Both of these fluorophores emit fluorescence at different wavelengths and can be detected during the scanning of the hybridized microarrays. The difference in the hybridization between the two fluorophores is representative of the prevalence of the transcripts in the hybridizing samples, one being the control and the other being the experimental. A difference in the intensity of specific fluorescence is due to relative amounts of a specific transcript in the two samples. Therefore, each probe representing a different gene is evaluated for its relative expression under control and experimental conditions in a comparative manner.

1.5 ANALYSIS OF MICROARRAY DATA

Microarray data analysis is an essential part in microarray studies, since it provides the right steps toward an interpretation of the experiments and derivation of meaningful, and hopefully the best possible, biological conclusions. Data analysis tools for microarrays are derived from a variety of bioinformatics methods and techniques that bridge the gap among molecular biology, computer science, statistics, and mathematics. It is not just as simple as running a software tool on a computer, but a well-designed, thoroughly-validated technique using algorithms and principles of the aforementioned disciplines. In particular, computer science does not study the way in which a particular software tool works, but the way in which an abstract computing model is designed, implemented and validated on a real computer to deliver the desired results as efficiently as possible. This section, and subsequently this book, provides a variety of well-studied and optimally designed methods for such analyses.

Although the approaches for analyses of different microarray technologies and types differ from each other, a generic approach for microarray data analysis is discussed here. The particulars for each type of technology are discussed for each case separately and with a sufficient level of specificity, as appropriate. The main process of microarray data analysis starts with the result of the scanning of the glass slides or chips, which produce images in different formats.

Scanning the slides at a very high resolution produces images, which depending on the type are composed of sub-grids of spots in DNA microarrays or squared probe cells in Affymetrix arrays [6]. Image processing and data analysis are the most important aspects in microarray studies. This is so, because the aim of the whole experimental procedure is to obtain meaningful biological conclusions. The analysis is carried out in a sequence of well-defined steps, which attempt to produce the most reliable results. Thus, any error in one of the stages will propagate to the subsequent stages yielding inaccuracies, and possibly misconceptions, about the answers to the relevant biological questions.

A schematic view of the process of microarray data analysis, wherein the images are obtained from scanning to obtain biological and functional analysis, is depicted in Figure 1.4. Scanning a microarray at very high resolution delivers a microarray image in a specific format. The image processing step aims at transforming the image by a series of steps into a more meaningful image with various parameters identified,

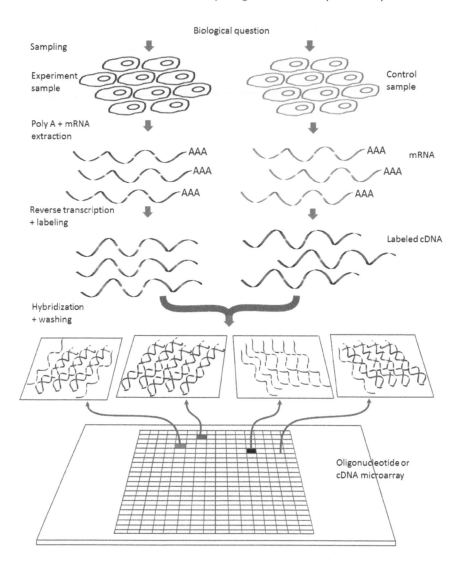

FIGURE 1.3   (SEE COLOR INSERT.) Schematic view of microarray hybridization assay and analysis for comparison of two samples (e.g., experiment versus control).

such as locations of spots or probe cells, background, signal and noise. Processed images are analyzed further in order to quantify the signals corresponding to gene expression. Noise, signal intensity, printing artifacts, shape and size of the spots, and lightning conditions may impose some bias in reading the gene expression pattern at different parts of the image, at different times or across different cell lines. The first two stages of the analysis (image processing and quantification) are different for

DNA microarrays, and oligonucleotide arrays or Affymetrix. The DNA microarray images are first processed by a segmentation algorithm in order to separate spots from background noise. For the sake of simplicity, a single DNA microarray image is shown in Figure 1.5a, and it must be noted that typical experiments generate at least two images for the green and red channels, which are then combined in the quantification and analysis steps. Oligonucleotide arrays are typically processed by proprietary algorithms associated with the technology being used. The normalization step aims at "erasing" this bias by bringing the data into "normal" conditions. Normalized data can be analyzed in many different ways, and in this sense, machine learning algorithms for clustering, data mining, or supervised classification aim at discovering relevant patterns in the data in such a way that these patterns are more easily understood by scientists when interpreting the data. An important step in the analysis is how to visualize those patterns or even the raw data—visualization techniques help obtain a better insight of the data and analyze it from different perspectives. The final step in the analysis involves functional and network analysis in terms of gene ontology, genomic features and inferring different types of biological networks.

1.5.1 MICROARRAY IMAGE PROCESSING

During the image processing and analysis, an image is considered to be a matrix of values that contains the pixel intensities. Two fundamental properties of an image are important in image processing: resolution and color depth or pixel depth [7, 9]. The resolution of an image is the number of pixels per unit of measurement. Typically, the term resolution is referred to as the number of horizontal and vertical pixels of an image, disregarding the unit of measurement. For example, a typical picture whose resolution is 640×480 may show an entire city or a single house—in the latter case the image shows much more details about the picture. In microarray images, the resolution refers to the number of pixels per spot in a DNA microarray image, or pixels per square probe cell in an oligonucleotide array. The color depth of an image is defined as the number of possible values (colors or grayscale values) the intensity of a pixel can take, and is usually given by the number of bits used to store each pixel. Thus, an 8-bit grayscale image will use 8 bits for each pixel, which contains one of 256 possible values (0 to 255), whereas a 24-bit color image uses 24 bits (8 bits for each channel of the RGB palette), which contains one of 16,777,216 possible values. The latter images are also known as true color images [9].

The format of an image is an important aspect in its representation and storage. Due to the size of the images in their primitive format, they are usually compressed before being stored. Compression can be lossless or lossy, in which case the original image may or may not be completely recovered. In the latter case, although there is some loss of information, the retrieved image has an acceptable quality. Most microarray images representing one channel are stored in tagged image file format (TIFF), which uses a lossless compression algorithm, while composite or two-channel images are stored in JPEG file interchange format (JFIF, aka JPEG), which involves lossy compression [9]. Compression of microarray images is a grow-

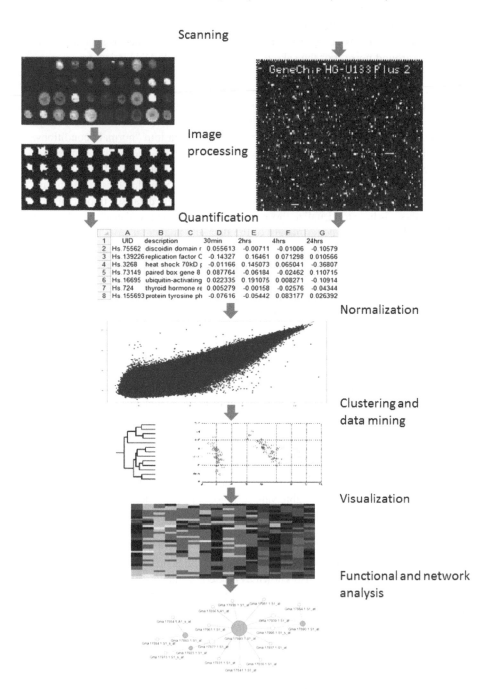

FIGURE 1.4   (SEE COLOR INSERT.) Schematic view of processing the microarray data and its analysis.

ing and exciting field that allows efficient storage and management of microarray images and data. A detailed discussion about the most recent methods is presented in Chapter 8.

The format, resolution, and steps involved in the microarray image processing stage vary, depending on the technology being used. The DNA microarray images have typically lower resolution than the Affymetrix or Illumina BeadArray images, which are produced by more powerful scanners. In terms of layout, a DNA microarray image contains rows and columns of spots separated from each other by some gap, while Affymetrix microarray images contain square probe cells adjacent to each other, each represented with a few pixels. Discussions on two types of technologies, DNA and Affymetrix, are given in the next subsections.

1.5.1.1 DNA Microarrays

DNA microarray images contain spots arranged in a grid with a certain number of rows and columns. Most images contain a certain number of sub-grids, which facilitates the location of the spots and the overall analysis. DNA array experiments are usually carried out in two channels, experimental and control, typically associated with the green and red channels. The array is scanned for each of these channels separately. In the first pass the scanner uses a laser light that detects the green dye, and in the second pass the scanner does so for the red. Since these two phases are independent from each other, two grayscale (typically 16-bit TIFF) images are produced. The example of Figure 1.5 depicts a real DNA microarray image downloaded from the Stanford Microarray Database (SMD) [10], which corresponds to a study of the global transcriptional factors for hormone treatment of *Arabidopsis thaliana* samples. The full image, Figure 1.5a, contains $12 \times 4 = 48$ sub-grids and was scanned at a $1,910 \times 5,550$ resolution. The spacing among different sub-grids allows specialized algorithms to identify each sub-grid, which contains $18 \times 18 = 324$ spots; the spot resolution is 24×24 pixels. One of the sub-grids is shown in Figure 1.5b. However, as shown later, noise present and different shapes for the spots make the image processing steps far from trivial.

The images from the two channels may be combined into a single color image, called the composite image. This can be achieved by superimposing the two grayscale images creating the composite image. More specifically, this can be done by simply taking the intensity of each pixel from the red channel image and placing it in the first component of the RGB triplet of the composite image. The same procedure can be applied for the green channel. The composite image will contain variations of green and red in different intensities and combinations. Portions of two DNA microarray images (green and red) are shown in Figure 1.6, along with the corresponding composite image. A green-like spot will indicate that the gene is more expressed in the experiment, while a red-like spot will indicate the gene is more expressed in the control. A black spot will represent no expression in either channel, while a yellow spot will indicate that the gene is expressed in both channels equally. Different variations of yellow (orange, more "greenish" or "reddish") will show variations of expressions between the two channels. It is important to know

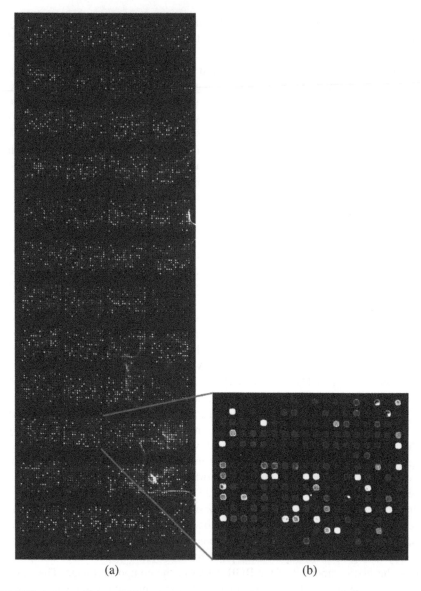

(a)   (b)

FIGURE 1.5 (a) Original DNA microarray image, 20385-ch1 (green channel), from the SMD; (b) sub-grid extracted from the 9th row and 2nd column.

that the composite image is usually used for visualization purposes only. The computational analysis is done by processing each grayscale image individually.

Image processing of DNA microarrays has its own particulars. Roughly speaking, the aim is to find the positions of the spots and then identify the pixels that represent gene expression, separating them from the background and noise. In a nutshell, the

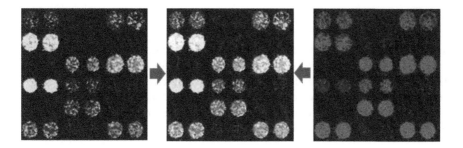

FIGURE 1.6   (SEE COLOR INSERT.) Portions of two DNA microarray images (green and red channels), along with the corresponding composite image.

main steps involved in processing a DNA microarray image are the following: spot addressing or gridding, segmentation, noise treatment and removal and background correction, which are discussed in more detail below.

1.5.1.1.1   Gridding

When producing DNA microarrays, many parameters are specified, such as the number and size of spots, number of sub-grids, and even their exact locations. However, many physicochemical factors produce noise, misalignment, and even deformations in the sub-grid template so that it is virtually impossible to know the exact locations of the spots after scanning, at least with the current technology, without performing complex procedures. Prior to applying the gridding process to find the locations of the spots, the sub-grids must be identified, a process that is also known as sub-gridding. Once the sub-grids are identified, the gridding step takes a sub-grid as input and aims at finding the exact location of each spot. Depending on how complex the mechanisms are, the gridding method may or may not require some parameters about the sub-girds, namely the number of rows and columns of spots, and the size of the spots in pixels, among others. Various methods have been proposed for solving this problem with some variations in terms of the amount of computer processing time, user intervention and parameters required. Among these, there are methods based on multi-level thresholding [11], machine learning [12], mathematical morphology [13], Bayesian models [14], and hill-climbing [15]. Chapters 3 and 4 present a detailed review of the existing methods and comparisons.

1.5.1.1.2   Segmentation

Image segmentation, in general, can be seen as the process of dividing the image into a set of non-overlapping regions. This process can also be seen as a classification problem in which each pixel is assigned to a specific group or class. In microarray images, pixels can be assigned to spot, background, noise, or possibly another group. Traditional segmentation techniques aimed to define a simple shape for the spot, namely circular or elliptical, are used to separate the spot from the background or noise. Other techniques such as seeded region growing, mathematical morphology [16], clustering and pattern recognition, and Markov random fields [13] have improved these approaches by adding more features using spatial information and intensities combined in order to define arbitrary shapes for the spots, including circular, elliptical, donut-like, half-moon-like and even free-form spots. Several techniques have been proposed to counteract this problem. A detailed description of the problem and the most up to date methods of segmentation of DNA microarray images are discussed in Chapters 5 and 6.

1.5.1.1.3 Noise and Background Correction

Noise is one of the main issues in microarray experiments. It can be present even when noise in microarrays can be due to the fluctuations in a probe, target and array preparation, in the hybridization process, and in background artifacts and lightning effects in the scanning and the resulting image processing. Most of the sources of noise present in microarray images, as described in [17], can be summarized as follows:

- Probes may vary from sample to sample during mRNA preparation.
- cDNA from reverse transcription may have different lengths.
- Fluctuations may exist in labeling of probes.
- Difficulties in quantifying PCR amplification of clones.
- Pins may be affected by surface properties.
- Amounts of transported target DNA fluctuates randomly.
- Hybridization is subject to different experimental parameters.
- Cross-hybridization or nonspecificity in hybridization.
- Hybridization may vary due to physical conditions in the slide.
- Stringency conditions of post-hybridization washes.
- The amount of probe DNA attached to the slide is unknown.
- Background, lightning and radiation conditions vary at different places in the slide.

Noise present in microarray images may affect the overall process of analysis if caution is not taken mainly in those stages at the beginning of the process. Lightning variations and artifacts present in the images can hopefully be detected and eliminated by sophisticated techniques for noise detection, treatment and removal. An overview of these problems as well as a detailed discussion of the latest techniques available to date are presented in Chapter 7. Most of the noise factors discussed above, especially variations in amounts of target DNA and experimental parameters, and lightning differences across the slide, can be detected and corrected by means of well-established normalization techniques. Some of these techniques are discussed in Chapters 9 and 10. Variations in amounts of target DNA and noisy features affecting the regularity and shape of spots are usually detected by well-known segmentation mechanisms such as those discussed in Chapters 5 and 6.

1.5.1.2 Oligonucleotide Arrays

Image processing and quantification of oligonucleotide arrays depend on the technology being used. Since Affymetrix and Illumina technologies are the most widely used at present, the underlying pre-processing stages for these two technologies are briefly discussed here. Unlike DNA microarrays, Affymetrix arrays use probes that are 25-nucleotide oligomers [6]. Probes are designed as pairs of perfect match (PM) and mismatch (MM). The probes for a gene transcript are grouped into a probe set, each of them containing between 11 and 20 PM and MM. An Affymetrix GeneChip is a very small ($2.54cm \times 2.54cm$) high-density chip. Scanning the chip produces an image that is composed of a matrix of cells, each cell represented by an array of pixels, typically 5×5 pixels. The image, stored in a DAT file, contains 4×4 checker-boards on the four corners, used for alignment purpose. The DAT file is a 16-bit grayscale image stored in TIFF format. The DAT file is also accompanied by a CDF file, which provides a description of the chip layout.

Image processing of an array proceeds by locating the 4×4 checker-boards by means of bi-linear interpolation. Once aligned, a refinement procedure allows for a more precise location of the probe cells. The segmentation proceeds by locating each probe cell, typically an array of 5×5 pixels, and estimating the intensity of the probe cell by using the 75th percentile of its 3×3 central pixels. This information and the standard deviation are stored in a CEL file. Since the Affymetrix technology is proprietary, most of the image processing and quantification algorithms are not made publicly available. Once an array is processed, the CEL file contains a summary of the actual array image. As the CEL file is normally made publicly available, many methods for post-processing of Affymetrix data have been proposed. The Bioconductor tools contain an open-source suite of tools in R [18, 19]—more details are discussed in Section 1.6.3 and in Chapter 17.

Although Affymetrix arrays are high-density and more reliable, they are not free of noisy factors and artifacts. One of the main sources of errors in these arrays is called blur, which implies that the pixels do not correspond to the actual region of hybridization. This is mainly due to the fact that the physical size of probe cells has decreased to 5×5 microns, increasing the resolution of the image, while being more prone to errors. There are also other errors that may be present in Affymetrix arrays, including dirt, dark and bright spots, dark clouds and shadowy circles [20], as well as blobs, lines and coffee rings, among others [21]. Other possible sources of errors that are typical in oligonucleotide arrays include cross-hybridization (probe to a transcript of another gene or mapping of the probe to an intron), alternative splicing, and single nucleotide polymorphisms [3]. More discussion and details about this technology, its image processing as well as dealing with the improvement of signal to noise ratio can be found in the references of this paragraph and in Chapters 9 and 10.

Illumina arrays differ from other types of technology mainly because they randomly arrange the probes in a hexagonal grid [8]—a typical array contains one million beads. Thus, the probes will occur a random number of times and at different random locations of the array. Although raw bead-level data from the array provides information on the locations of the beads, spatial artifacts make the image processing step difficult. The main process involves identifying the bead locations, followed by finding the bead types, to finally obtain the intensities. Once the images are scanned, pre-processing usually proceeds in three steps: registration of beads (using a bead detection algorithm), interpolation of the remaining beads, and centering the grid over the array. A final decoding procedure allows the user to identify the probe attached to each bead [22]. After the image processing step is completed, the foreground is well separated from the background.

1.5.1.3 Protein Microarrays

Protein microarrays are rather a new addition to the technology, and do not offer many advances that could have been made in regards to specific methods for image processing and analysis. The format and layout of the protein array images are similar to those of DNA microarrays. That is, they contain spots arranged in one or more sub-grids. Currently, the tools used for spot finding, segmentation and quantification of protein array images are essentially those used in DNA microarray image processing [23]. While it is an emerging field, new studies are being conducted so that more specific analytical techniques can be used for this type of technology.

1.5.2 QUANTIFICATION AND NORMALIZATION

The main purpose of quantification is to obtain a numeric value representing the expression of a particular gene transcript as acquired from the signal from a hybridized spot during the scanning of the microarray. In the case of DNA microarrays, each spot is usually associated with a gene, and hence the pixels identified in the segmentation procedure are combined in order to obtain a unique value for the expression. There are various methods for combining these pixel intensities, offering their own advantages and disadvantages. Mean, median, and mode of pixel intensities are the most widely used ones, the median being the most preferred one, since it is more resistant to the effect of outliers [7]. Other measures are volume of intensities and total signal intensities. But they may over/under estimate the actual expression due to the differences in image resolutions or brightness conditions.

Quantification of Affymetrix array images, and in general in oligonucleotide microarrays, differs from DNA microarrays in the sense that each gene transcript is represented by a probe set. Since each probe set contains various probes in terms of PM and MM, these quantities are combined in order to obtain a single value representing the expression of a gene. A commonly used value for this is the average of the differences between PM and MM over all probes in the probe set [6, 7]. Since there may be substantial variations among probes in the cell, more information can help quantification and hence analysis, such as the standard deviation.

As discussed earlier, noise, artifacts and lightning conditions may impose some bias in reading the gene expression levels at different parts of the image, at different times or across cell lines. Normalization is an important step in pre-processing microarray data, and its main purpose is to erase these biases by bringing the data into normal conditions. There are a large number of methods that have been applied to normalization of microarray data, some of them being specific to the type of microarray or any other technology being used. The most well-known methods for normalization include LOESS/LOWESS regression, B-spline smoothing, wavelet smoothing, kernel regression and support vector regression [24].

Normalization of oligonucleotide arrays has its own particulars. More discussion, methods and comparisons for Affymetrix arrays and Illumina arrays can be found in [8, 25, 26, 27, 28] and in Chapters 9 and 10. On the other hand, quantification normalization of protein array data has its own intricacies. One of the problems present in protein arrays is the variability in the quantity of material deposited on the chip. Then, higher concentrations in some spots would produce higher absolute signals. Using specific quantification and normalization approaches such as concentration dependent analysis helps alleviate these drawbacks [23]. The tendencies of the antibodies to interact with protein molecules non-specifically is also another major factor in the normalization process to account for any difference between the actual and the deduced results.

1.5.3 PATTERN ANALYSIS AND MACHINE LEARNING

Machine learning is one of the most exciting fields for data analysis that gathers knowledge from computer science, mathematics, statistics, and engineering. The main goal of machine learning is to learn patterns from data, observations and experience, and design efficient mechanisms for prediction, pattern analysis or decision making [29]. Once microarray data has been obtained as a result of image processing, quantification and normalization, the aim of this stage is to find relevant patterns of behavior in differentially expressed genes under different conditions, cell lines or organisms.

Machine learning for pattern analysis and recognition can be subdivided into two main areas, supervised or unsupervised learning [30]. In supervised learning, patterns, categories or classes are already known, and the aim is to find relevant features or descriptors and classification mechanisms to improve predictions, which help in disease classification, diagnosis, prognosis, staging, treatment and therapeutic responses [31].

Pattern classification and prediction have been successfully applied to microarray data analysis since their early stages. The most widely used classification techniques for microarrays include the well-known and widely used support vector machine [32], linear classifiers and dimensionality reduction [30, 33], distance-based classifiers or nearest neighbor approaches [30], and neural networks [34]. In the pattern classification context, feature selection, in particular gene selection, is a problem that has been applied to microarray data analysis. Gene expression studies typically involve thousands of genes, e.g., more than ten thousand for human transcriptome studies. Gene selection aims to find a small subset of genes that are differentially expressed and provide relevant information on disease classification and other tasks.

Feature selection methods can be grouped into two categories, namely wrapper methods and filter-based approaches [30]. Some of the recent methods include correlation-based filters [34], combination of multiple filters and wrappers [35], gene selection via sample weighting [36], and mutual information theoretic approaches [37], to name a few.

In an unsupervised learning scenario, classes and patterns are not known and have to be discovered. In the microarray analysis context, it mostly includes the application of clustering techniques for pattern finding. Clustering techniques have been used in microarray data analysis for a long time. Roughly speaking, the aim of clustering is to group the data into clusters or groups in such a way that genes or samples are similar or close to others in the same group and dissimilar to or far from samples in a different group [38]. Clustering is done in two ways, flat or hierarchical. In the latter, a hierarchy of classes is discovered and can be clearly visualized by means of a dendrogram. Various methods have been proposed for clustering microarray data, including hierarchical models [39], spectral clustering [40], k-means and expectation maximization [40, 41, 42], and graph and spectral clustering [43]. Advanced clustering approaches on more challenging problems include biclustering and co-clustering techniques for microarray data [44]. An extended coverage of existing approaches for clustering microarray data is given in Chapters 13 and 14.

1.5.4 TIME SERIES AND HIGHER DIMENSIONS

Time-series microarray data involves gene expression measurements at different time points. These points can occur at fixed or variable interval in times. The aim of these studies is to observe the trends of gene expression patterns across time, which is useful for clinical studies, embryogenesis, development, and gene-gene interactions. Clustering time-series microarray data is a challenging problem that involves its own particular constraints. Various methods have been proposed recently for clustering microarray time series, which either define a similarity measure appropriate for temporal expression data or pre-process the data in such a way that the temporal relationships are taken into account. The most well-known methods for clustering time-series microarray data include pairwise and multiple profile alignment [40], variation-based co-expression detection [45], Granger causality test analysis [43], and others.

Temporal information about gene expression data can be considered as a second dimensional feature, while the first dimension represents the genes themselves. Adding more information and data, such as different patients in a microarray study, cell lines or different conditions of analysis and treatment involves a higher dimensional representation. Thus, three (or higher) dimensional microarray data analysis is an evolving field that formalizes such a model [46]. More details and discussions about the models and techniques for analysis of time-series and high-dimensional microarray data are presented in Chapter 13.

1.5.5 VISUALIZATION

Visualization of microarray data and patterns discovered are very important in data analysis and mostly in interpretation of results. The intricacies of these problems lie in the fact that the data is typically high-dimensional and involves a large number of genes, typically in thousands. Many visualization techniques have been used and proposed for microarray data. Of these, heatmaps are the most widely used tool for visualization [47], which aims to show data in a color-based matrix using colors typically varying in a scale from red to green for up- or down-regulated genes, respectively. Other techniques for visualization used for microarray data are profile plots, mainly used for time-series microarray data, and scatter plots, which aim at plotting data points in a two or three dimensional space showing expression profiles and revealing useful visual information for clustering. Scatter plots can be obtained using different techniques for data and pattern analysis such as principal and independent component analysis (PCA) [30], linear discriminant analysis and dimensionality reduction [48], independent component analysis, multi-dimensional scaling, and self-organizing maps [30]. Other statistical data analysis and visualization tools that are useful for microarray data visualization are box plots, and pie charts, among others [7]. In addition, the integration of visualization techniques and data clustering poses various challenges and shows its own advantages, especially for three dimensional data. Points, clouds and explore are useful tools for microarray data visualization [48]. More details about tools and techniques for visualization of microarray data are discussed in Chapter 16.

1.5.6   FUNCTIONAL AND NETWORK ANALYSIS

Once analyzed and visualized, microarray data provides a wealth of information for further analysis and discovery of new relationships among genes, transcripts, proteins and metabolites. One of the next tasks involves finding pathways or a network in which the differentially expressed genes may have important roles in growth and development, embryogenesis, normal cellular processes, diseases, and other related conditions. Network analysis combined with visualization may also show how up-regulated transcripts explain the up- or down-regulation of other genes, which may or may not actually interact [47]. Detecting such relationships at a large scale involves the study of gene regulatory networks, protein-protein interaction networks, pathway analysis, and metabolic network modeling. These relationships also arise from studying gene expression relationships under different experimental conditions, such as time points or variable cell lines. In this regard, relevant patterns may reveal new links among genes at different conditions, and with specific interacting proteins, RNA, carbohydrates, and other molecules. Starting from a list of relevant genes under different conditions arranged in terms of some patterns, the challenge is to infer a network that represents these relationships. Various techniques have been proposed for inferring gene regulatory networks from microarray time-series and other data [49]. An extensive coverage of methods for discovering gene regulatory networks from microarray data is presented in Chapter 15.

1.6   RESOURCES FOR MICROARRAYS

Since the microarray technology was introduced, a large number of resources have been made available. These resources help researchers in a number of ways and forms by facilitating the availability of microarray data for further analysis and providing tools for storage, retrieval, annotation, pre-processing, analysis and visualization. This section discusses the most important resources available for microarray data, including publications, databases and repositories, Web applications, and software tools.

1.6.1 PUBLICATIONS

Microarrays have been used by many researchers in many studies within the area of natural sciences. To mention a few facts, as of January 2013, searching the word "microarray" in PubMed would bring more than 51,000 publications, while typing the key word "array" lists more than 120,000 abstracts. Various databases and resources contain a wealth of microarray data for studies of gene expression, proteomics, and tissue analysis, among others. The Gene Expression Omnibus (GEO) database of the National Center for Integrative Biomedical Informatics (NCBI) of the United States contains a wealth of gene expression data and is publicly accessible via the Internet [50]. As of January 2013, the GEO contained more than 870,000 samples, the majority of which are associated with the use of microarray technology. There are many other databases and resources for microarrays as well—a more detailed discussion follows at the end of this chapter.

Presented here are the results of some queries performed in PubMed, which comprises more than 22 million citations for biomedical literature from MEDLINE, life science journals, and online books [51], as of January 2013. Figure 1.7 shows a chart of the number of publications per year. These statistics were gathered by simply searching the database using "microarray" as a keyword. Sorting the publications by year shows a steady, small number at earlier stages and a rapidly growing trend after year 2000, with a steady growth in the past few years. This shows that although newer technologies have been recently introduced, microarray technology continues to be a growing field of research.

The list of these publications by considering the type of organism involved in the microarray study is shown in Table 1.1. These publications have also been searched in PubMed using "microarray" as a keyword and the scientific or common name for the organism. As observed in the table, microarray studies on human beings take up more than 63% of the publications. Studies on rat and mouse are reported in more than 25% of the publications. *S. cerevisiae* and *E. coli* are also two very well studied organisms, while others are studied to a minor extent, such as *A. thaliana*, *C. elegans*, *Drosophila*, and *D. rerio*.

Searching publications from another perspective shows interesting trends, too. The charts in Figure 1.8 show how publications group by microarray type and manufacturer. Since microarrays have been invented, DNA arrays were the dominant technology. However, at present, oligonucleotide arrays are dominating in the microarray studies, being employed in more than 64% of the publications, while DNA microarrays have been losing terrain and now have 23% of the share. The use of protein and tissue arrays has also grown in the past few years, accumulating 7% and 5% of the publications, respectively. Inspecting the publications by manufacturers shows that the Affymetrix platform continues to be the most widely used technology with

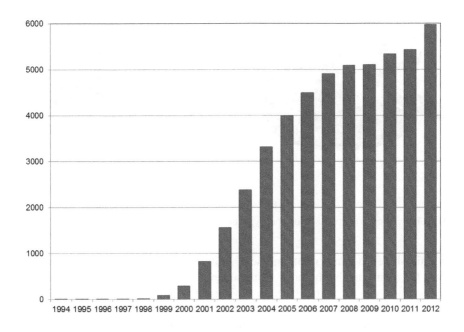

FIGURE 1.7 Microarray publications in PubMed, grouped by year of publication.

TABLE 1.1

Microarray publications in PubMed, grouped by organism.

Species	Number of publications	% of total
Homo sapiens	31,266	63.26%
Mus musculus	9,876	19.98%
Rattus norvegicus	3,618	7.32 %
Saccharomyces cerevisiae	1,930	3.91%
Escherichia coli	1,071	2.17%
Arabidopsis thaliana	669	1.35%
Caenorhabditis elegans	344	0.70%
Drosophila	326	0.66%
Danio rerio	321	0.65%
Total	49,556	

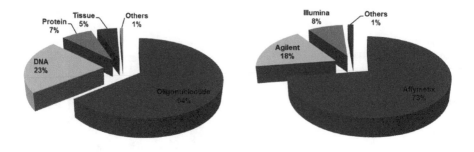

FIGURE 1.8   Microarray publications per type and technology being used.

73% of the share. To some extent, the use of Agilent arrays has also been growing in the past few years, and by the time of this survey with a share of 18%, followed by Illumina arrays with 8% of the share. To a minor extent, less than 1% of the publications involve Combimatrix and Nimblegen arrays. It is clear that Affymetrix, which is one of the pioneering technologies in microarray studies, will continue to be the dominant technology used in microarray studies.

1.6.2   DATABASES AND REPOSITORIES

With the advent of microarray technology, large amounts of data are being generated. The increase in the volume of data and the need for making these data publicly available have made possible the creation of a large number of databases and public repositories. Of the many repositories that have been created and maintained, the most widely used ones are discussed below. The corresponding Web sites and references for all these resources are listed in Table 1.2.

SMD/PUMAdb: The Stanford Microarray Database (SMD) is the most traditional databases for microarray and gene expression data [10]. It started as an effort to maintain microarray data from ongoing research at Stanford University, and to facilitate the public dissemination of that data once published, or released by the researchers. The SMD has been one of the most important public repositories for many years. As of October 2011, the SMD contained microarray data for more than 82,000 experiments. It has recently been transferred to the Princeton University Microarray Database (PUMAdb). The PUMAdb contains microarray data in raw, pre-processed and normalized format, as well as tools for data storage, retrieval, transformation, quality assessment, analysis, and visualization. As of January 2013, the PUMAdb contained microarray data for 36,592 hybridization experiments for 48 different organisms.

GEO: The Gene Expression Omnibus (GEO) is a public repository that was established by the National Center for Biotechnology Information (NCBI) of the United States, in the early 2000s, and it is still one of the most important sources for gene

TABLE 1.2

Resources and tools for analysis of microarray data.

Resource	Web site	Ref
	Databases and Repositories	
PUMAdb	`puma.princeton.edu/`	[10]
GEO	`www.ncbi.nlm.nih.gov/geo/`	[50]
ArrayExpress	`www.ebi.ac.uk/arrayexpress/`	[52]
PLEXdb	`www.plexdb.org/`	[53]
GXD	`www.informatics.jax.org/mgihome/GXD/` `aboutGXD.shtml`	[54]
PRKdb	`www2.cancer.ucl.ac.uk/Parkinson_Db2/`	[55]
GENT	`medical-genome.kribb.re.kr/GENT/`	[56]
FlyTED	`www.fly-ted.org/`	[57]
EMAGE	`www.emouseatlas.org/emage/`	[58]
FFGED	`bioinfo.townsend.yale.edu/`	[59]
	Tools for Microarray Data Analysis	
Bioconductor	`www.bioconductor.org/`	[19]
geNorm	`www.proteinatlas.org/`	[60]
GIP	`compbio.dfci.harvard.edu/tgi/`	[5]
ArrayTrack	`www.fda.gov/ArrayTrack/`	[61]
chipD	`chipd.uwbacter.org/`	[62]
cancerMA	`www.cancerma.org.uk/`	[63]
inCroMap	`www.cogsys.cs.uni-tuebingen.de/software/` `InCroMAP/`	[64]
MADTools	`www.madtools.org/`	[65]
GSEA	`www.broadinstitute.org/gsea/index.jsp/`	[66]

expression data in different forms [50]. The GEO archives and provides access to microarray, next-generation sequencing and other forms of high-throughput functional genomic data, which are submitted by the scientific community via its Web site. The GEO also provides various tools for data retrieval, analysis and means for finding gene expression patterns. Due to the increasing use and versatility of applications of microarrays for different types of research projects, the GEO also stores microarray data for other than gene-expression studies, including examination of genome copy number variation and genome-wide profiling of DNA-binding proteins. As of January 2013, the GEO contained public data for 11,018 platforms, 872,170 samples in 2,720 different datasets.

ArrayExpress is one of the main public repositories for gene expression data, and has existed since 2003 at the European Bioinformatics Institute [52]. It contains gene expression data generated by microarray or sequencing technologies. The ArrayExpress database is currently integrated with the GEO, by allowing users to import GEO-formatted data into ArrayExpress. In addition, most GEO-deposited experimental data have been imported into ArrayExpress by means of high-throughput algorithms. In addition, ArrayExpress is integrated with the Gene Expression Atlas and the sequence databases at the European Bioinformatics Institute. As of January 2013, ArrayExpress contained data for 35,682 experiments on 1,030,151 assays.

In addition to the three main repositories for microarray and gene expression data, a number of smaller databases have been created and maintained. Most of these repositories target a specific organism, disease or taxonomy. While the list of these public and private repositories is rather extensive, the most well-known and recently introduced specific repositories are include here (more information in Table 1.2). PLEXdb is a public repository that contains gene expression resources for plants and plant pathogens, mostly rice and *Arabidopsis* [53], while the gene expression database (GXD) is a resource for mouse developmental gene expression information [54]. The Parkinson's disease gene expression database [55], ParkDB, contains datasets of previously-analyzed, curated and annotated microarray data for analysis. The gene expression database of normal and tumor tissues (GENT) is a repository that contains more than 40,000 samples for gene expression patterns for human cancer and normal tissues [56]. FlyTED is the *Drosophila* testis gene expression database that contains more than 2,700 gene expression images for studies of the fruit fly, *Drosophila melanogaster* [57]. EMAGE, the mouse embryo spatial gene expression database, is a freely available online repository that contains gene expression patterns for studies of developing mouse embryo [58]. The filamentous fungal gene expression database (FFGED) is a public repository that contains gene expression data for filamentous fungal studies [59].

1.6.3 TOOLS FOR MICROARRAY DATA ANALYSIS

Since microarrays were introduced, a large number of analytical software tools for pre-processing, data analysis and visualization of gene expression profiles have been proposed. Many of these tools are freely available for downloading and/or use in Web-based servers, while some others are licensed or associated with specific types of microarray platforms. Some tools are combined with databases and repositories in the aim of facilitating the retrieval, analysis and visualization of microarray data. The most widely used tools for microarray data analysis are discussed below.

Bioconductor happens to be one of the most widely used tool suites in bioinformatics [19]. It is an open source platform of tools developed in the R statistical programming language, whose main aim is to provide solutions for analysis and understanding of high-throughput genomic data. Bioconductor has a special set of tools for analysis of microarray data including various platforms, such as Affymetrix, Illumina, Nimblegen, and Agilent, amongst the rest. It provides packages for analysis of expression arrays at different stages, including pre-processing, quality assessment, differential expression, clustering, classification, and visualization. Since its introduction, a number of new Bioconductor tools and packages for microarray data analysis have been introduced. The list would be so extensive that it would not be possible to enumerate all the tools in the present discussion. As a matter of fact, as of January 2013, there are more than 300 publications associated with Bioconductor studies on microarray or gene expression data. Discussions on specific Bioconductor packages for microarray data analysis are presented in Chapter 17.

The Protein Atlas and geNorm is another tool that uses an algorithm to compare the level of expression of housekeeping genes from a set of tested reference genes to provide optimal normalization for the microarray data [60]. The Protein Atlas, which is also a Web-based tool, permits viewing of protein expression profiles based on immunohistochemistry from a large number of human tissues, cancers, and cell lines. The enormous amount of protein data allows researchers to interpret their microarray data in view of information obtained from the subcellular localization of proteins. Further, it is possible to associate the data with certain tissue types, or cancer cells, thereby identifying the complete profile of the cells used in the study. More focused approaches can help in determining the role of subcellular components of the cell in the development and progression of specific tumor types.

The Gene Index Project (GIP) was designed to use the DNA sequence databases (EST, genomic DNA, and gene encoding sequences) to construct catalogues of genomes of reference model systems [5]. The isoforms, variants, mutants, alleles and derivatives are annotated with respect to their functional roles and with regard to their protein products. In addition, the catalogued genes are also associated with their potential cellular pathways in some model organisms. This project assists researchers in finding genes and their variants in genomic sequences with their possible functional significance, using a comparative approach.

ArrayTrack is a bioinformatics tool developed at the Food and Drug Administration (FDA) of the United States [61]. It includes an environment for microarray data processing, analysis and visualization, and incorporates statistical, pathway and gene ontology analysis. In addition, the Gene Expression Omnibus (GEO) also provides a set of tools for querying, analyzing and visualizing gene expression datasets in its database [50].

Other Web tools for analysis of microarray data include chipD, a tool for designing oligonucleotide probes in high-density tiling arrays [62], cancerMA, a Web tool for analysis of cancer microarray data [63], inCroMap, an integrated tool for analysis of microarray and pathway data [64], MADTools, a set of tools for storage, analysis and annotation of DNA microarray data [65], and GSEA, a computational method that allows the determination of statistical significance of the set of genes whose expression has been characterized on a genomic scale [66].

1.7   CONCLUSIONS AND FUTURE TRENDS

Microarrays have proven to be essential biotechnological tools for monitoring biomolecules' behaviors in cells, tissues, and organisms in a massive and parallel way. Since their introduction, many different types of technologies have been developed to introduce the versatile approaches that can be employed in biological studies at the molecular level, including gene and protein expression, genetic and epigenetic research, tissue analysis and genotyping for paternity or forensic analysis. The ability of the microarray technology to be used in diverse types of applications allows scientists to focus on many aspects of biological and medical sciences with a wide range of applications in pharmacology, medical genetics, disease diagnosis/prognosis, detection, classification, staging and treatment of diseases, drug efficacy, evolutionary studies, and genotypic characterization of selected organisms.

This chapter reviews the main aspects of microarray technologies, principles, fundamentals, production, data analysis, as well as an introduction to a wide range of resources available. Also included in this chapter are discussions of the modern concept of central dogma in view of its analysis through microarray methodology. Therefore, the first introductory part reviewed the main aspects of molecular biology and genomics and the main building blocks, nucleotides, amino acids, DNA, RNA, and proteins. The fundamental processes in the cellular processes, namely transcription and translation, in which biochemical transformations from DNA to RNA, and ultimately to protein molecules and the characterization of their molecular interactions are explored by the use of this high throughput technology. Microarray studies have had a rich and well-established history since their introduction in the early 1990s, starting from simple arrays with a few genomic markers, evolving to extremely high resolution slides with more than two million biomarkers available in today's technology.

Pre-processing and analysis of the underlying data are fundamental aspects in microarray studies, while at the same time, they offer big challenges to scientists in computational sciences. Although some of the initial stages, e.g., image processing and quantification, are highly dependent upon the type of technology being used, the main streamline for the analysis is common for most types of microarrays. Initial stages in the analysis, namely addressing, segmentation and quantification, involves spot or cell localization and identification of true signals, separating them from the slide background or noise. The efficiency of early stages in the analysis is crucial in the overall process, since any error at any of these stages is propagated to subsequent steps. Normalization, pattern analysis and machine learning, visualization, as well as functional and network analysis, provide a wealth of tools and techniques with sufficient flexibility for different types of microarray technologies.

A wide range of resources are available for microarray studies providing free online access to data, functional annotation, and data analysis. At present, oligonucleotide arrays have become the most preferred and affordable technology, while protein and tissue microarrays are still emerging tools for microarray studies. The survey presented in this chapter shows a consistent and increasing interest in microarray studies in terms of publications, databases and repositories, as well as software and Web-based tools for efficient analysis. Although there are three main repositories for microarray data, there are a large number of new databases for microarray studies, most of them focusing on specific organisms or diseases.

Overall, the chapter presents the basic aspects of microarray technology in a superficial way, targeting the general reader. The subsequent chapters cover a wide range of topics in microarray studies with sufficient level of detail for general and advanced readers, providing an up-to-date coverage outlined by worldwide experts in the respective fields. The subsequent chapters are aimed to achieve a wide-range coverage of the most important aspects in contemporary microarray studies and their applications, with detailed discussions on image processing, quantification, normalization, data management and storage, machine learning, visualization, software tools, and functional analysis, all of them independently addressing particular technologies.

ACKNOWLEDGMENTS

The authors would like to thank Iman Rezaeian for his help in preparing the LATEX version of this chapter, and Manoj Gajjarapu for his help in gathering the statistics for the plots and tables.

REFERENCES

1. H. Lodish, *Molecular Cell Biology*. New York: W. H. Freeman, 2012.
2. J. Watson, T. Baker, S. Bell, A. Gann, M. Levine, and R. Losick, *Molecular Biology of the Gene*. Books a la Carte, San Francisco: Benjamin-Cummings Publishing Company, 2013.
3. D. Nelson, A. Lehninger, and M. Cox, *Lehninger Principles of Biochemistry*. New York: W. H. Freeman, 2013.
4. J. H. Malone and B. Oliver, "Microarrays, deep sequencing and the true measure of the transcriptome," *BMC Biology*, vol. 9, no. 1, p. 34, 2011.
5. M. Schena, *Microarray Analysis*. Professional developer's guide series, Hoboken: Wiley, 2002.
6. J. M. Arteaga-Salas, H. Zuzan, W. B. Langdon, G. J. Upton, and A. P. Harrison, "An overview of image-processing methods for Affymetrix GeneChips," *Briefings in Bioinformatics*, vol. 9, no. 1, pp. 25–33, 2008.
7. S. Drăghici, *Statistics and Data Analysis for Microarrays Using R and Bioconductor*. Boca Raton: Chapman & Hall / CRC mathematical and computational biology series, CRC Press, 2012.

8. M.L. Smith, M.J. Dunning, S. Tavaré, and A.G. Lynch, "Identification and correction of previously unreported spatial phenomena using raw Illumina BeadArray data," *BMC Bioinformatics*, vol. 11, no. 1, p. 208, 2010.

9. J. Murray and W. VanRyper, *Encyclopedia of graphics file formats*. O'Reilly Series, Sebastopol: O'Reilly & Associates, 1996.

10. J. Hubble, J. Demeter, H. Jin, et al., "Implementation of GenePattern within the Stanford Microarray Database," *Nucleic Acids Research*, vol. 37, no. suppl 1, pp. D898–D901, 2009.

11. L. Rueda and I. Rezaeian, "A fully automatic gridding method for cDNA microarray images," *BMC Bioinformatics*, vol. 12, no. 1, p. 113, 2011.

12. D. Bariamis, D. K. Iakovidis, and D. Maroulis, "M3G: maximum margin microarray gridding," *BMC Bioinformatics*, vol. 11, no. 1, p. 49, 2010.

13. G. Antoniol and M. Ceccarelli, "A Markov random field approach to microarray image gridding," in *Pattern Recognition, 2004. ICPR 2004. Proceedings of the 17th International Conference on*, vol. 3, pp. 550–553, IEEE, 2004.

14. M. Ceccarelli and G. Antoniol, "A deformable grid-matching approach for microarray images," *IEEE Transactions on Image Processing*, vol. 15, no. 10, pp. 3178–3188, 2006.

15. L. Rueda and V. Vidyadharan, "A hill-climbing approach for automatic gridding of cDNA microarray images," *IEEE/ACM Transactions on Computational Biology and Bioinformatics (TCBB)*, vol. 3, no. 1, p. 72, 2006.

16. J. Angulo and J. Serra, "Automatic analysis of DNA microarray images using mathematical morphology," *Bioinformatics*, vol. 19, no. 5, pp. 553–562, 2003.

17. J. Schuchhardt, D. Beule, A. Malik, et al., "Normalization strategies for cDNA microarrays," *Nucleic Acids Research*, vol. 28, no. 10, pp. e47–e47, 2000.

18. Y. Dai, L. Guo, M. Li, and Y.-B. Chen, "Microarray R US: a user-friendly graphical interface to Bioconductor tools that enables accurate microarray data analysis and expedites comprehensive functional analysis of microarray results," *BMC Research Notes*, vol. 5, no. 1, p. 282, 2012.

19. R. C. Gentleman, V. J. Carey, D. M. Bates, et al., "Bioconductor: open software development for computational biology and bioinformatics," *Genome Biology*, vol. 5, no. 10, p. R80, 2004.

20. M. Suárez-Fariñas, M. Pellegrino, K. Wittkowski, and M. Magnasco, "Harshlight: a corrective make-up program for microarray chips," *BMC Bioinformatics*, vol. 6, 2005.

21. G. J. Upton and J. C. Lloyd, "Oligonucleotide arrays: information from replication and spatial structure," *Bioinformatics*, vol. 21, no. 22, pp. 4162–4168, 2005.

22. K. L. Gunderson, S. Kruglyak, M. S. Graige, et al., "Decoding randomly ordered DNA arrays," *Genome research*, vol. 14, no. 5, pp. 870–877, 2004.

23. D. S. DeLuca, O. Marina, S. Ray, G. L. Zhang, C. J. Wu, and V. Brusic, "Data processing and analysis for protein microarrays," in *Protein Microarray for Disease Analysis*, pp. 337–347, New York: Springer, 2011.

24. A. Fujita, J. Sato, L. Rodrigues, C. Ferreira, and M. Sogayar, "Evaluating different methods of microarray data normalization," *BMC Bioinformatics*, vol. 7, no. 1, p. 469, 2006.

25. S. Calza and Y. Pawitan, "Normalization of gene-expression microarray data," in *Computational Biology* (D. Feny, ed.), vol. 673 of *Methods in Molecular Biology*, pp. 37–52, Humana Press, 2010.

26. A. J. Hackstadt and A. M. Hess, "Filtering for increased power for microarray data analysis," *BMC Bioinformatics*, vol. 10, no. 1, p. 11, 2009.

27. R. Moffitt, Q. Yin-Goen, T. Stokes, et al., "caCORRECT2: improving the accuracy and

reliability of microarray data in the presence of artifacts," *BMC Bioinformatics*, vol. 12, no. 1, p. 383, 2011.

28. R. Schmid, P. Baum, C. Ittrich, et al., "Comparison of normalization methods for Illumina BeadChip HumanHT-12 v3," *BMC Genomics*, vol. 11, no. 1, p. 349, 2010.

29. K. Murphy, *Machine Learning: A Probabilistic Perspective*. Adaptive computation and machine learning series, Cambridge: MIT Press, 2012.

30. S. Theodoridis and K. Koutroumbas, *Pattern Recognition*. Burlington: Academic Press, 2008.

31. M. M. Li, M. Ankita Patel, and X. Hu, "Clinical applications of microarrays in cancer," in *Modern Clinical Molecular Techniques*, pp. 307–322, New York: Springer, 2012.

32. S. Abe, *Support Vector Machines for Pattern Classification*. Advances in Pattern Recognition, London: Springer, 2010.

33. P. Xu, G. N. Brock, and R. S. Parrish, "Modified linear discriminant analysis approaches for classification of high-dimensional microarray data," *Computational Statistics & Data Analysis*, vol. 53, no. 5, pp. 1674–1687, 2009.

34. F. Fernandez-Navarro, C. Hervas-Martinez, R. Ruiz, and J. C. Riquelme, "Evolutionary generalized radial basis function neural networks for improving prediction accuracy in gene classification using feature selection," *Applied Soft Computing*, vol. 12, no. 6, pp. 1787–1800, 2012.

35. Y. Leung and Y. Hung, "A multiple-filter-multiple-wrapper approach to gene selection and microarray data classification," *IEEE/ACM Transactions on Computational Biology and Bioinformatics (TCBB)*, vol. 7, no. 1, pp. 108–117, 2010.

36. L. Yu, Y. Han, and M. E. Berens, "Stable gene selection from microarray data via sample weighting," *IEEE/ACM Transactions on Computational Biology and Bioinformatics (TCBB)*, vol. 9, no. 1, pp. 262–272, 2012.

37. P. A. Mundra and J. C. Rajapakse, "SVM-RFE with MRMR filter for gene selection," *IEEE Transactions on NanoBioscience*, vol. 9, no. 1, pp. 31–37, 2010.

38. R. Xu and D. Wunsch, *Clustering*. IEEE Press Series on Computational Intelligence, Hoboken: Wiley, 2008.

39. J. J. Jay, J. D. Eblen, Y. Zhang, et al., "A systematic comparison of genome scale clustering algorithms," in *Bioinformatics Research and Applications*, pp. 416–427, Berlin: Springer, 2011.

40. N. Subhani, L. Rueda, A. Ngom, and C. J. Burden, "Multiple gene expression profile alignment for microarray time-series data clustering," *Bioinformatics*, vol. 26, no. 18, pp. 2281–2288, 2010.

41. J. Baek and G. J. McLachlan, "Mixtures of common t-factor analyzers for clustering high-dimensional microarray data," *Bioinformatics*, vol. 27, no. 9, pp. 1269–1276, 2011.

42. B. Xie, W. Pan, and X. Shen, "Penalized mixtures of factor analyzers with application to clustering high-dimensional microarray data," *Bioinformatics*, vol. 26, no. 4, pp. 501–508, 2010.

43. R. Krishna, C.-T. Li, and V. Buchanan-Wollaston, "A temporal precedence based clustering method for gene expression microarray data," *BMC Bioinformatics*, vol. 11, no. 1, p. 68, 2010.

44. K. Eren, M. Deveci, O. Küçüktunç, and U.V. Çatalyürek, "A comparative analysis of biclustering algorithms for gene expression data," *Briefings in Bioinformatics*, vol. 14, no. 3, pp. 279–292, 2013.

45. Z.-X. Yin and J.-H. Chiang, "Novel algorithm for coexpression detection in time-varying microarray datasets," *IEEE/ACM Transactions on Computational Biology and Bioinfor-

matics, vol. 5, no. 1, pp. 120–135, 2008.

46. Y. Li and A. Ngom, "Classification of clinical gene-sample-time microarray expression data via tensor decomposition methods," in *Computational Intelligence Methods for Bioinformatics and Biostatistics*, pp. 275–286, Berlin: Springer, 2011.

47. N. Gehlenborg, S. I. O'Donoghue, N. S. Baliga, et al., "Visualization of omics data for systems biology," *Nature Methods*, vol. 7, pp. S56–S68, 2010.

48. O. Rubel, G. H. Weber, M.-Y. Huang, et al., "Integrating data clustering and visualization for the analysis of 3d gene expression data," *IEEE/ACM Transactions on Computational Biology and Bioinformatics*, vol. 7, no. 1, pp. 64–79, 2010.

49. Y. Zhou, R. Qureshi, and A. Sacan, "Reconstruction of gene regulatory networks by stepwise multiple linear regression from time-series microarray data," in *7th International Symposium on Health Informatics and Bioinformatics (HIBIT)*, pp. 76–81, IEEE, 2012.

50. T. Barrett, S. E. Wilhite, P. Ledoux, et al., "NCBI GEO: archive for functional genomics data sets—update," *Nucleic Acids Research*, vol. 41, no. D1, pp. D991–D995, 2013.

51. "National Network of Libraries of Medicine [Internet]. Bethesda (MD): National Library of Medicine (US)." Available from: http://nnlm.gov/. updated: 2013 Jan 31; cited: 2013 Jan 31.

52. H. Parkinson, U. Sarkans, N. Kolesnikov, et al., "ArrayExpress updatean archive of microarray and high-throughput sequencing-based functional genomics experiments," *Nucleic Acids Research*, vol. 39, no. suppl 1, pp. D1002–D1004, 2011.

53. S. Dash, J. Van Hemert, L. Hong, R. P. Wise, and J. A. Dickerson, "PLEXdb: gene expression resources for plants and plant pathogens," *Nucleic Acids Research*, vol. 40, no. D1, pp. D1194–D1201, 2012.

54. J. H. Finger, C. M. Smith, T. F. Hayamizu, et al., "The mouse gene expression database (GXD): 2011 update," *Nucleic acids research*, vol. 39, no. suppl 1, pp. D835–D841, 2011.

55. C. Taccioli, J. Tegner, V. Maselli, et al., "ParkDB: a Parkinson's disease gene expression database," *Database*, vol. 2011, 2011.

56. G. Shin, T.-W. Kang, S. Yang, S.-J. Baek, Y.-S. Jeong, and S.-Y. Kim, "GENT: gene expression database of normal and tumor tissues," *Cancer Informatics*, vol. 10, p. 149, 2011.

57. J. Zhao, G. Klyne, E. Benson, E. Gudmannsdottir, H. White-Cooper, and D. Shotton, "FlyTED: the Drosophila testis gene expression database," *Nucleic Acids Research*, vol. 38, no. suppl 1, pp. D710–D715, 2010.

58. L. Richardson, S. Venkataraman, P. Stevenson, et al., "EMAGE mouse embryo spatial gene expression database: 2010 update," *Nucleic Acids Research*, vol. 38, no. suppl 1, pp. D703–D709, 2010.

59. Z. Zhang and J. P. Townsend, "The filamentous fungal gene expression database (FFGED)," *Fungal Genetics and Biology*, vol. 47, no. 3, pp. 199–204, 2010.

60. L. Fagerberg, P. Oksvold, M. Skogs, et al., "Contribution of antibody-based protein profiling to the human Chromosome-centric Proteome Project (C-HPP)," *Journal of Proteome Research*, vol. 12, no. 6, pp. 2439–2448, 2012.

61. J. Xu, R. Kelly, H. Fang, W. Tong, et al., "ArrayTrack: a free FDA bioinformatics tool to support emerging biomedical research—an update," *Human Genomics*, vol. 4, pp. 428–434, 2010.

62. Y. S. Dufour, G. E. Wesenberg, A. J. Tritt, et al., "chipD: a web tool to design oligonucleotide probes for high-density tiling arrays," *Nucleic Acids Research*, vol. 38, no. suppl

2, pp. W321–W325, 2010.

63. J. Feichtinger, R. J. McFarlane, and L. D. Larcombe, "CancerMA: a web-based tool for automatic meta-analysis of public cancer microarray data," *Database*, vol. 2012, 2012.

64. C. Wrzodek, J. Eichner, F. Büchel, and A. Zell, "InCroMAP: integrated analysis of cross-platform microarray and pathway data," *Bioinformatics*, vol. 29, no. 4, pp. 506–508, 2013.

65. D. Baron, A. Bihouée, R. Teusan, et al., "MADGene: retrieval and processing of gene identifier lists for the analysis of heterogeneous microarray datasets," *Bioinformatics*, vol. 27, no. 5, pp. 725–726, 2011.

66. R. A. Irizarry, C. Wang, Y. Zhou, and T. P. Speed, "Gene set enrichment analysis made simple," *Statistical Methods in Medical Research*, vol. 18, no. 6, pp. 565–575, 2009.

2 Biological Aspects: Types and Applications of Microarrays

Adnan Ali

CONTENTS

Since the production of the very first type of microarrays, there have been tremendous efforts in improving the platform to achieve higher quality arrays for reproducible high throughput data. The discovery of better substrates for the arrays, more sophisticated fabrication methods, improved production and processing of microarrays, and better data mining tools have enabled researchers to use optimal parameters to gain high quality data with a level of confidence never before achievable using conventional biotechnological tools. Primarily, researchers have used microarrays to understand cellular processes by taking a snapshot of gene expression in a multidimensional way. Basic research using microarrays has successfully engaged scientists in diverse disciplines, including biology, chemistry, toxicology, environmental studies, ecology, medical sciences, statistics, and computer science. Indeed, the discovery of microarrays has provided a common ground for collaborations among scientists across the diverse areas of research. This chapter presents an overview of the history of microarrays in general, and in particular, focuses on the fundamentals of microarrays, their types, and their use in biology, chemistry as well as in applied biomedical research. It also provides an overview of the problems and pitfalls faced by scientists using microarrays, and discusses how the recent improvements have rectified such problems. Further, the potential for this powerful technology and its possible use for researchers in the future is also discussed.

2.1 OVERVIEW

As introduced in Chapter 1, a typical microarray consists of a coated microscopic glass slide that provides a solid surface onto which nucleic acid molecules are chemically attached. Microarrays are commonly used to detect the presence of labeled nucleic acids in a biological sample by hybridizing it with the probe nucleic acid molecules present on the surface of the microarray. The detection of probes is facilitated via bound labels that emit fluorescence that can be detected by commercially available laser scanners [1]. Image acquisition is followed by the analysis and interpretation of data providing high-throughput analysis at a genomic scale. In order to study gene expression at a large scale, the labeled nucleic acids are produced by reverse transcription of mRNA expressed in a biological sample. The sample might be obtained from cells, tissues, or organisms under control or normal, and under treatment conditions [2]. The labeling step depends on the experiment and the type of microarray technology used in an experiment. In the case of the Affymetrix platform, one may construct a biotin-labeled complementary RNA target for hybridizing to the GeneChip. The use of fluorescent labeling with the two dyes Cy3 (excited by a green laser) and Cy5 (excited by a red laser) are very common. In a typical experiment, two samples are hybridized to the arrays, one labeled with each dye; this allows the simultaneous measurement of both fluorophores representing the nucleic acids expressed or present in each of the two samples. Due to the presence of a multitude of probes on the microarray, the power of this methodology permits measurement of the expression of many thousands of genes simultaneously. With respect to the nature and size of the microarray data set, it is crucial to understand the biological nature of the experiments conducted, the methods of bioinformatics employed to decipher the data, and the significance of differences revealed by mathematical, statistical, and computing tools. None of these steps would be accessible to an investigator without sufficient support and knowledge from other disciplines. Practically speaking, one needs to design the microarray experiment, acquire, analyze, and store the data using appropriate tools without changing or modifying the raw data. Only accurate, well-justified and reproducible data are useful for sharing with fellow scientists. One of the significant aspects of this high-throughput technology is that a meaningful microarray experiment is only possible with the involvement of bioinformatics tools at every stage of image acquisition and data analysis. The use of probabilistic models to identify modules of co-regulated genes, their transcriptional regulators, and conditions that influence the regulation of gene expression present a major challenge for computational scientists.

2.2 HISTORY OF MICROARRAY DEVELOPMENT

Microarrays date back to the early 1990s and the main concepts and the principles of their production were elaborated by Mark Schena and collaborators at Stanford University [1]. The prototype DNA microarrays were also presented by Schena and collaborators circa 1994, while the first Affymetrix array prototype was conceived in 1989, and the first commercial microarray was available circa 1994. The initial DNA microarray that was produced in 1995 contained 16 spots (arranged in a 4×4 grid) representing gene expression of several different lines of *Arabidopsis thaliana*. To put this into perspective, and without entering into an ironic comparison, currently an Affymetrix array can contain up to 2.6 million spots on its surface. There have been remarkable advances in the development of microarrays, as well as corresponding analytical tools during the last few years. Traditionally, standard molecular approaches have been devised to decipher the role of different genes active in a specific cellular context or phenotype. However, most of these techniques involved the analysis of one gene at a time. With the achievement of different genomic projects, including the Human Genome Project, scientists aspired to carry out genome-wide studies within the limits of experimental conditions on a feasible time scale and at an affordable cost. The microarray concept has further encouraged the use of diverse platforms to study, not only gene expression profiles, but also to compare genomes analysis using genomic microarrays [2, 3]. The repertoire has since expanded to include SNP arrays, alternative splicing arrays, methylation sensitive arrays, protein arrays, antibody arrays, as well as tissue and cell arrays. They all have played a significant role in answering some of the most interesting and fundamental questions in biology. The power of microarray technology can be easily understood when one considers the former alternative: conventional techniques like Northern blot assays, Southern blot assays, Western blot assays, polymerase chain reaction, primer extension assays, and SAGE analysis all contributed to our understanding of gene activity, but usually only a few genes at a time. The time saved by carrying out an experiment upon an entire genome via microarray methodology is an important factor, but also the information gained using such a high throughput method is vast. An emerging challenge has been to conduct and report microarray experiments so that analytical tools can be applied for comparative studies where the data from a public data repository can easily be accessed and biological connections can be made from one experiment to another. Such meta-analyses can reveal hidden connections regarding normal cell function, and the etiology of different diseases. This technology can also help us understand basic questions about evolution either by comparing the genomic sequences or SNP patterns between different organisms within a species, or between divergent animals and plants. During the last decade, microarrays have answered biological questions that would otherwise have been impossible to resolve using conventional techniques. In providing novel approaches at a large scale, microarrays have been prominent in initiating a revolution in the research and development of ancillary techniques and methods involving data analysis by the development of new algorithms. Several online resources for analysis of microarray data and for efficient public sharing of the underlying data have been established and are discussed in Chapter 1. The production of commercial microarrays from a large number of companies has introduced new and innovative approaches by which microarrays can be employed in different research and medical communities. New products have striven to keep the main advantages of the platform, namely, sensitivity and reproducibility of the data at a genomic scale. The problems encountered by scientists in the early days of the microarray era have been resolved to some extent by the production of highly consistent, reproducible, sensitive, and comprehensive arrays. It is quite interesting to note that the prototype microarrays had probe sets corresponding to only about 120 genes, which effectively detected only about 30 transcripts. However, with time and advancement in biotechnology, microarray production and use has experienced substantial improvements and delivered robust tools. Protein and tissue arrays have provided excellent tools to answer several unanswered biological questions [4]. Other protein-based microarray applications, such as epitope mapping, still present many technical challenges due to the differing properties of the bound ligands such as affinity or epitope folding patterns.

Product aspirations and development problems aside, it is clear that this is a technology that delivers: a search for microarray-based studies in PubMed reveals more than 50,000 published articles. The success of this technology and its wide range and versatility of applications has provided multiple avenues for study as well as an unprecedented means, potentially, to assemble larger scale meta-analyses the likes of which we have yet to appreciate.

2.3 CONVENTIONAL TECHNIQUES IN MOLECULAR BIOLOGY

The development of DNA arrays has provided fabulous opportunities to generate gene expression profiles and to assess the transcriptional status of genes within a given cell, tissue, or an organism. The use of this technology at the genomic level provides the analysis of thousands of genes simultaneously as well as the diversity of the available tools that can allow one to perform a comprehensive analysis of cellular processes in a focused manner [1, 4, 5]. This methodology has also allowed the isolation of novel genes, establishing protein regulatory networks, understanding of potential cross regulation of cellular signaling pathways, the down-stream regulated targets in cellular signaling, and for the identification of interacting molecules at the DNA, RNA, and protein levels. Whereas in the traditional molecular techniques, one gene at a time has been a major hurdle in the analysis at a genomic scale, the use of microarrays has provided an opportunity, both in terms of time and cost, to make such genomic analysis. In order to appreciate the advantages achieved by the use of microarrays, a basic understanding of the existing conventional techniques for the detection of macromolecules is also necessary. For the basic understanding of the concepts of molecular biology and further details on macromolecules, readers are requested to consult the biology and biochemistry textbooks [6, 7, 8]. An overview of the traditional approaches in biological research are depicted in Table 2.1.

The following is a brief description of some of the mainstream techniques in molecular biology that have been conventionally used in the art [6, 8]. Such techniques are still in practice, and in some cases their use is essential for the confirmation or validation of biological results achieved through a microarray experiment.

2.3.1 SOUTHERN BLOT ANALYSIS

The detection of DNA molecules is frequently used to identify, isolate, and quantify specific DNA molecules in a given biological sample. One of the fundamental techniques that has been in practice is termed as *Southern blot* after the name of its discoverer, Edwin Southern. Southern blotting comprises the electrophoresis of DNA molecules that may have been digested with specific restriction enzymes to generate smaller fragments. The electrophoretically separated DNA fragments are transferred and immobilized onto a solid support, usually a nylon or nitrocellulose membrane. In order to detect the presence of a particular DNA sequence on the membrane, a small DNA fragment or an oligonucleotide corresponding to the target gene sequence is labeled with radioactive or non-radioactive fluorescent molecules. Subsequent to the hybridization of labeled probe with the membrane containing the immobilized DNA

TABLE 2.1

Concepts of biological research at a genome-wide scale.

Research Projects	Traditional methods	Desired Analysis
Genomics (Genetic Analysis of Genomic DNA)	Southern blot, Dot blot, Slot blot, Macroarray, PCR, methylation-sensitive PCR, Fluorescence *in situ* hybridization (FISH)	Genotyping, SNP mapping, Mutational analysis, Sequencing, Methylation analysis, Epigenetic analysis
Transcriptomics (Gene Expression Analysis at RNA level)	Northern blot, Dot blot, Slot blot, RT-PCR, SAGE, RNase protection assay, *in situ* hybridization	Total RNA analysis, mRNA expression analysis, Gene expression profiling, Splice variants
Proteomics (Protein Expression and Molecular Interactions)	Western blot, ELISA, Immunoprecipitation, Immunostaining, Immunohistochemistry	Protein expression, Protein-Protein, Protein-DNA, Protein-carbohydrate, Protein-antibody, Enzyme-substrate, Ligand-receptor interactions

fragments, the detection of the bound fragments is carried out using standard methods. The DNA blotting method allows one to identify the presence of a specific DNA molecule that can then be isolated and cloned for further analysis. However, the main limitation for sequence detection using the Southern blot assay is that one can only use a single target gene at a time, although several samples are electrophoresed and immobilized on the solid support.

2.3.2 NORTHERN BLOT ANALYSIS

As in the Southern blot analysis, the detection of RNA molecules is carried out using the Northern blot technique. This technique is commonly used in molecular biology to study the transcriptional regulation of gene expression by detecting the specific RNA target molecules corresponding to specific genes. The RNA is isolated from the cells, tissues, or organisms and is electrophoresed and transferred to a solid support, just like the transfer of DNA in the Southern blotting procedure. The electrophoresis allows the separation of RNA molecules by size and their detection is carried out using a labeled probe by hybridization. The complementary target sequence is identified by detecting the bound labeled probe. As in the case of Southern blot analysis, the study of gene expression using Northern blotting is also limited to the detection of one gene at a time. However, multiple samples can be hybridized at the same time, and hence it is possible to study the cellular processes during various physiological or environmental conditions. The regulation of gene expression during differentiation, morphogenesis, embryogenesis, development, as well as in the cells under control, abnormal, or diseased conditions is routinely analyzed using the Northern blot procedure.

2.3.3 WESTERN BLOT ANALYSIS

Western blot or immunoblot assays are designed to detect specific proteins in a sample derived from cells, tissues, or organisms. Just like Southern or Northern blots, the transfer of protein samples is carried out following the electrophoresis of the protein sample that separates the native proteins on an acrylamide gel. The proteins may be electrophoresed using two-dimensional (2-D) or three-dimensional (3-D) separation techniques based on the structure, weight, and charge of the native or denatured proteins present in the sample. The proteins are then transferred to a membrane, usually a nitrocellulose or PVDF membrane. Following the fixation of proteins on the solid support, the presence of specific polypeptide molecules is detected by using specific antibodies, which are themselves detected by the use of secondary antibodies that are labeled with fluorescent molecules. The antibodies may be available commercially or can be raised against specific epitopes selected from a target protein. The availability of numerous antibodies (monoclonal or polyclonal antibodies) makes it feasible to detect the expression of specific protein molecules in a given control or experimental sample. Once again, Western blot assays also allow a single type of protein to be detected in one experiment. The use of antibodies in other techniques, such as immunoprecipitation and immunostaining for conducting immunoassays, is also very common in the field of biochemistry and molecular biology.

2.3.4 POLYMERASE CHAIN REACTION

For the last two decades, another technique that has gained much popularity over Southern and Northern blots is the *in vitro* amplification of nucleic acid molecules. Polymerase chain reaction (PCR) provides a very cost effective and time saving technology in molecular biology to amplify even a single copy of a target nucleic acid molecule in a sample. Primarily, PCR allows the synthesis of identical copies of DNA molecules using a single template to several orders of magnitude. PCR has frequently been used in the detection of specific DNA sequences in genomic DNA isolates using short oligonucleotide sequences termed as *primers* corresponding to the target DNA molecules. The primers are used in conjunction with four nucleotides (A, G, C, and T), a thermostable DNA polymerase, and an appropriate amplification buffer. PCR employs a heat-stable DNA polymerase (e.g., Taq polymerase), an enzyme isolated from the bacterium *Thermus aquaticus*. The genetically recombinant thermostable polymerase enzymes are also commercially available. The reaction mixture is cycled through varying temperatures for denaturing the template, annealing of the primers with the template, and extension of specific DNA molecules. A specially designed PCR machine, called a thermocycler, automatically regulates the programmed cycling of specific temperatures. The predetermined number of PCR cycles allows the amplification of a specific DNA molecule, while yielding millions of copies of one specific DNA fragment from a biological sample, which could be as small as comprising a single cell. Several modifications and diverse approaches have been introduced to the PCR methodology and have proven to be an indispensable technique in certain areas of research and clinical analysis. PCR has been used in medical and biological research for a variety of applications, including cloning and analysis of novel genes, DNA sequencing, single nucleotide polymorphism, forensic biology, functional analysis of genes, and diagnosis of hereditary and infectious diseases. A crucial manipulation of PCR for the analysis and detection of RNA molecules expressed in a sample has been employed by adding a step of reverse transcription prior to the actual PCR. During the reverse transcription reaction, the isolated RNA sample is reverse transcribed into cDNA molecules using a reverse transcriptase enzyme. The newly synthesized cDNA molecules are subjected to the PCR reaction, and thus exponentially produce a specific type of cDNA library containing a plurality of different kinds of cDNA molecules in the resulting reaction product. This technique is termed as RT-PCR and has been traditionally used to analyze the gene expression in cells, tissues, or organisms. RT-PCR generally works on a "single gene in a single experiment" basis, and therefore the whole picture of differential gene expression between cells or tissues is still difficult to obtain using this powerful approach. One of the fundamental problems in using RT-PCR for obtaining gene expression profiles is the multiplicity of the reactions that are required to carry out independent investigation of each target gene. This nature of the experimental procedure requires tedious assembly of the individual reactions, and significant care is needed to avoid any cross-contamination of primers or the templates. Therefore, PCR appears to be an essential tool for the amplification of nucleic acid molecules, which occur even in trace amounts in a given sample. However, the gene expression profiling at a genomic scale is quite challenging using PCR technique.

2.3.5 MACROARRAYS

Biologists have always been interested in searching innovative techniques that could allow them to analyze a large number of genes in a single experiment. The analysis of gene expression or identification of a specific gene sequence in a sample wherein only one gene at a time can be studied places a lot of burden on researchers, both in terms of cost and time. An earlier approach in this direction was the use of macroarrays, which provided the researchers an analysis of a plurality of genes in a less expensive way. In theory, this is similar to a microarray wherein immobilized genes or DNA fragments are present in spots on a solid support. However, the spots are much larger than the spots on a microarray (about 1 mm as compared to the spots in the range of 50μ–250μ). The hybridization and detection procedures are similar to those used in the Southern or Northern blot assays. The visualization of the signals does not require a scanner solely due to the large size of the spotted probes. The labeling may be carried out using radioisotope or non-radioactive fluorescent molecules. Although macroarrays have allowed researchers to compile gene expression profiles, there are obvious limitations with this technique as only a certain number of DNA molecules can be spotted onto a single solid support. In contrast, as we have also discussed in Chapter 1, a single microarray can allow thousands of genes to be examined simultaneously in a single hybridization assay.

2.4 BASIC CONCEPT OF ARRAY TECHNOLOGY IN BIOLOGICAL RESEARCH

As mentioned earlier, the use of microarrays is not limited to the analysis of DNA molecules nor solely to the study of gene expression profiles, *per se*. With the careful selection of appropriate surface chemistries for the microarray slide and optimal substrates, it has been possible to carry out a number of applications across various disciplines in natural sciences. Microarrays have been designed and constructed using macromolecular probes selected from a range of possibilities including: cDNAs, genomic fragments, oligonucleotides, proteins, short peptides, antibodies, living cells, and tissues, as well as low molecular weight molecules such as chemical compounds, carbohydrates, and other natural products [9].

It is quite apparent that by using a single microarray slide and two differentially labeled cDNA samples, a single experiment has the potential to reveal differences in transcriptional activity at the genomic level, which were not possible before using traditional biological assays [1, 2, 4]. In a standard microarray experiment, RNA is extracted from two samples. The two samples may represent two different treatments or even samples taken from diseased tissues or cells. Each sample containing RNA is reverse transcribed and the resulting cDNA is labeled using fluorescently tagged nucleotides (e.g., cyanine 3-dCTP or cyanine 5-dCTP). The labeled cDNAs are mixed together from the two samples and are permitted to hybridize with a microarray. Following post-hybridization washes to remove unbound labeled molecules, the microarray is scanned using a specialized laser scanner to acquire two images, one representing each fluorophore. Two separate microarray images that correspond to the emission of each of the two fluors are then analyzed using a variety of techniques, most of which are discussed in later chapters of this book. The fluorescent spots that are produced, using a laser confocal fluorescent microscope, are then quantified with regard to intensity, and ratio information is obtained following image processing. By comparing the two images' fluorescence intensities, one can tell if a particular cDNA, and hence a particular mRNA, is present in relatively higher amounts in one cell type or another. In basic biological research, microarray technology has been exploited for the identification, isolation, and characterization of novel genes, generating gene expression profiles, as well as for understanding gene regulatory networks [10, 11, 12].

It is important to note that array hybridization is not an absolute quantitative measure of gene expression. There are a number of factors involved in the production and acquisition of microarray data that can be reproduced and trusted with a greater degree of confidence. However, the efficiency of initial RNA isolation and purification, reverse transcription reaction, probe labeling, hybridization, stringency of post-hybridization washes, and the quality of the microarray slide, can greatly vary and have a significant impact on the quality of signals emitted from the hybridized array. The labeling of each sample has to be optimally standardized, and the signals must be normalized [4, 5]. In addition, the samples should be representative of fully expressed genome in the cells or tissues under study. The data obtained from the experiment must be reproducible and hence, the results obtained from different experimental samples must be compared to give an accurate estimate of the degree of change in gene expression profiles. Unlike Northern blot assays or quantitative PCR methods, microarrays do not provide absolute levels of mRNA abundance in a given sample. The result is simply the determination of the presence or absence of gene transcripts in a given sample. In essence, microarrays are useful for identifying gene expression trends. In this regard, there is a wealth of computational techniques available for analysis and clustering of differentially expressed genes, and for grouping samples with similar expression profiles to infer specific gene expression patterns or signatures [10]. Many of these techniques are discussed in the following chapters of this book.

2.5   MICROARRAY FABRICATION

There are mainly two technologies for producing microarrays that will be discussed in greater detail further below. The first one is known as spotted arrays, which are produced by robotic spotting or by ink-jet printer. The second type of production method is characterized by oligonucleotide arrays. These are produced mainly by using one of the following approaches [13, 14, 15]: photolithographic (Affymetrix array, aka Affymetrix GeneChip), ink jet technology (Agilent), electrochemical synthesis (CombiMatrix), solid state methods (NimbleGen), and by assembly upon silica beads in microwells (Illumina arrays, aka Illumina BeadArrays). The glass slides used in producing microarrays are very small in size, typically 2.54×7.62 cm for DNA microarrays (Figure 2.1) [1], and 2.54×2.54 cm for oligonucleotide arrays, such as Affymetrix arrays (Figure 2.2).

The glass slides are generally produced by printing DNA molecules at fixed positions in a grid-like pattern (Figure 2.1). In case of oligonucleotide arrays (e.g., Affymetrix arrays as shown in Figure 2.2), the microchips are produced by selection of short oligonucleotides, followed by photolithography and combinatorial chemistry. Two main technologies for constructing microarrays are further discussed below.

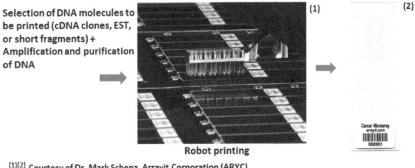

Selection of DNA molecules to be printed (cDNA clones, EST, or short fragments) + Amplification and purification of DNA

Robot printing

(1)(2) Courtesy of Dr. Mark Schena, Arrayit Corporation (ARYC).

FIGURE 2.1 Fabrication process for DNA microarrays.

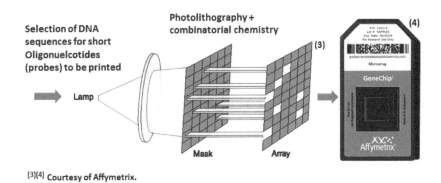

Selection of DNA sequences for short Oligonuelcotides (probes) to be printed

Photolithography + combinatorial chemistry

Lamp

Mask Array

(3)(4) Courtesy of Affymetrix.

FIGURE 2.2 Fabrication process for oligonucleotide microarrays.

2.5.1 ROBOTIC SPOTTING

Historically, microarrays have been prepared by printing selected DNA or cDNA clones that represent the genome of a given cell expressing target genes. The clones are generally obtained as expressed sequence tags (EST), which are available for the researchers from expression libraries [1, 4, 13, 16, 17, 18]. Further, the sequences of EST data are available from the National Center for Biotechnology Information (NCBI) database, and therefore, the cDNA clones can be reproduced using specific primers in a laboratory setup by employing techniques such as PCR. Regardless of the source of the DNA clone, the amplification step is vital in the construction of an array format, as it provides a pure and homogenous population of DNA molecules in each sample for the preparation of high quality probes. Similarly, commercially available oligonucleotides or independently synthesized oligonucleotide molecules can also be printed directly onto a positively charged substrate on a coated glass slide. The cDNA or oligonucleotide molecules can be deposited straight from a reagent tray (where it is suspended in a denaturing solution) onto the glass surface by a printhead containing microspotting pins or micropipettes. A liquid handling robot is commonly used for the automated amplification of DNA, and its purification and quantification prior to its delivery to the printing robot. In order to acquire a standardized consistent printing format, the probes listed above are spotted on the glass slides using a specialized automated robotic system, a commercially available microarray printer. One such method for spotting the probes on defined locations on the slide is called piezoelectric printing, which is very similar to ink-jet printing. In this method, at each spot, the electrical stimulation allows the DNA molecules to be delivered precisely onto the predefined surface via tiny jets in a gridded fashion on the slide.

2.5.2 ARRAYS PREPARED BY *IN-SITU* SYNTHESIS OF OLIGONUCLEOTIDES

Oligonucleotide arrays are also prepared by synthesizing oligonucleotides on the coated glass slides themselves by chemical reactions [1, 13]. Generally, the oligonucleotides are synthesized using photolithography. In this method, a mercury lamp is used and the projected light activates modified photosensitive DNA bases. Photolithography allows DNA synthesis to occur at predefined positions on a glass slide that is coated with the proper substrate prior to the reaction. This type of *in situ* fabrication of oligonucleotide molecules on arrays was initiated and developed by Affymetrix for producing arrays that are marketed as GeneChips. Oligonucleotide arrays have further allowed researchers to analyze even more target genes in a single experiment due to the large number of small DNA oligonucleotides that can be synthesized on a single chip. Such miniaturized DNA arrays have introduced a big revolution in genomic research and have provided an essential tool in modern medical research [19, 20, 21, 22]. In addition, GeneChips also serve to generate gene expression profiles like any other standard EST or DNA microarray. Oligonucleotide arrays are playing a prominent role in basic biological research, genotyping, gene expression profiling, genome mapping, as well as in medical and pharmacological sciences.

2.6 TYPES OF MICROARRAYS

In general the spots on microarrays, comprising DNA probe molecules, have a diameter that ranges from $50-100 \mu$m and the spots are usually separated by a distance from $200-400 \mu$m. An ordinary microscope is not able to capture the image with a high resolution for such an array [1, 4]. Therefore, as indicated earlier, the fluorescent signals from the hybridized array are captured by imaging the hybridized arrays in a specially designed scanner that can detect the fluorescence in a wide range of wavelengths. Such image acquisition in x and y directions generates a large amount of data that is further subject to analysis and interpretation. The following is a brief description of some of commonly used arrays in biological research.

2.6.1 TILING ARRAYS

One of the most popular types of microarrays is the tiling array. Tiling arrays are also produced in a similar manner to traditional microarrays [23, 24]. However, the probes present in the array are selected by choosing DNA fragments containing sequences that are dispersed in a genome at regular intervals. This is in contrast to the standard microarrays in which the probes are selected randomly across the genome. The methods of producing tiling arrays, fluorescent target, hybridization, image acquisition, and analysis of the data is almost identical to the procedures already described. The use of tiling arrays provides several advantages because the probes present in the array are accurately aligned on the solid support and because their specific location within the genome is already known. This is in contrast to EST arrays, where the identity, let alone the chromosomal position of a locus, might be unknown. Further, coding and non-coding regions selected from the genome are also included in the probe sets. As a result, probing for sequences of known or predicted genes simplifies what can be readily identified. The regulatory regions that may be present upstream or downstream of a gene can also be identified using tiling arrays, and further cloning and analysis can be pursued in a very efficient and cost effective manner. Since the arrayed sequences are known, and contiguous regions within the genome are already partially characterized, their functional significance can be confirmed by parallel biological assays. When DNA sequences are largely uncharacterized, the tiling array provides context-specific cue that can help in a functional and regulatory elucidation. For example, genomic proximity to genome regulatory blocks, or to clusters exhibiting synteny and conserved regulatory elements, may provide hints as to what to look for next. Further, these kinds of arrays have also been commonly used in transcriptome mapping (via alternative splicing-specific arrays), in the elucidation of DNA/protein interactions (ChIP-chip, DamID), in the determination of regulation of gene expression by DNA methylation (MeDIP-chip), for analyzing sensitivity to DNase (DNase Chip), and for comparative genomic studies via hybridization arrays (CGH) [7, 23, 25, 26, 27].

In addition to the characterization of previously unidentified genes and regulatory sequences, the improved format of tiling arrays also allows for the determination of regulatory networks that are involved in normal or challenged cellular processes. Although not very accurate, quantification of transcription products is also possible to some degree by adjusting the amount of overlap between the probe and target sequences. Since tiling arrays can be spotted using millions of copies of probes, it is possible that the variable pattern for spotting probes that have a range of overlap of the sequences would allow an estimate of the number of complementary copies present in the target sample. Such variable mapping resolutions can only be possible for certain regions of a large genome due to the limitations of available space on a single array.

2.6.2 CHIP-CHIP ARRAY (CHIP-ON-CHIP)

ChIP-chip assays utilize a type of tiling array in order to detect interactions between DNA and protein molecules at a genomic scale [28]. Primarily, this assay combines the interaction of proteins with specific DNA sequences and the ability of these protein molecules to be immunoprecipitated using specific antibodies. During a chromatin immunoprecipitation (ChIP) reaction, protein molecules that specifically bind to chromatin are cross-linked *in vivo* by means of a chemical reaction. The cross-linked chromatin is then fragmented using a nuclease, and specific antibodies against the target protein are used to isolate the chromatin-DNA complexes bound to the protein of interest. These specifically purified complexes are then further used to isolate and purify the DNA fragments from the complex. The purified DNA fragments are then labeled with a fluorophore and are subjected to hybridization to a genomic tiling array. Since the microarray contains specific probes that represent wide areas of the target genome, the detection of the signals on the array allows one to determine the exact identity and chromosomal location of the DNA fragments that have been immunoprecipitated. Further, by using overlapping probes or probes in close chromosomal proximity, specially designed tiling arrays can be used to screen all of the regions within the genome that may interact with the protein of interest, or even to identify unknown interacting partners.

Generally, the proteins of interest are those that interact with the chromatin under control, normal, or abnormal conditions. This methodology allows the understanding of interactions of proteins such as histones, transcription factors, DNA synthesis enzymes and ancillary proteins, replication-related proteins, as well as any modifications in these proteins that may alter their specific binding properties. Thus, the primary purpose of ChIP-on-chip assays is to accurately define the protein binding sites in a specific region of the genome. ChIP-chip experiments have successfully been used to identify binding sites of transcription factors in several organisms including yeast, *Drosophila* and mammalian systems [28, 29].

2.6.3 MEDIP-CHIP

Several gene regulation studies focus upon the epigenetic changes that are mediated by methylation patterns of genomic DNA. Such epigenetic regulation of gene expression also plays a prominent role in the cellular regulation during development, embryogenesis, and during numerous cellular activities [30, 31, 32]. Microarrays have been exploited in the analysis of methylation patterns on a genome-wide scale. Methylated DNA immunoprecipitation (MeDIP or mDIP) is commonly used to isolate and purify methylated DNA sequences from a sample. A specific antibody against the methylated cytosine nucleotide (5-methyl cytosine) is used to isolate methylated DNA fragments from a biological sample extracted from a cell or tissue. Therefore, only methylated regions of genomic DNA are immunoprecipitated and purified [30, 31]. The immunoprecipitated and purified methylated DNA sample is hybridized to a microarray following standard labeling procedures using proper fluorophores. High-throughput methylated DNA detection using DNA microarrays

(MeDIP-chip) provides a precise mapping of the methylated regions in the genome. Thus, hypermethylated regions are determined and their functional implication is assessed following sequence analysis. Understanding the methylation patterns at such a high resolution provides vital clues to the epigenetic regulation of gene expression, which may be specific for cell types. Approaches like this may provide further insight into cellular differentiation, development, and into the role of epigenetic imprinting in some disorders that are known to be affected by changes in methylation pattern of genomic DNA.

2.6.4 DNASE-CHIP

Another approach in which tiling arrays play an important role is the use of microarrays for determining hypersensitive regions within the entire genome [33]. There are certain DNA regions within the genomic DNA that are more sensitive to degradation when exposed to DNase, a DNA degrading enzyme. These hypersensitive regions are considered to represent relaxed conformations indicative of genes that are transcriptionally active. Traditional techniques, such as Southern blot assays, have restricted researchers to mapping DNase sensitivity to solely those genes already at least partially characterized. However, DNase chips permit analysis in the absence of candidate genes, and at a genome-wide scale. DNase digestion results in numerous large DNA fragments that are then labeled with a fluorophore and that are subsequently hybridized to a tiling array. Fragments that are thus identified from the microarray serve as markers for the corresponding hypersensitive sites in the genome. Such identification of the DNaseI-sensitive sites helps to accurately identify active genes as well as their associated regulatory regions including promoters, enhancers, and silencers [3, 27, 33].

2.6.5 TRANSCRIPTOME MAPPING

Using conventional biological assays, such as Northern blots or RT-PCR, one can readily determine the pattern of transcriptional expression of a particular target gene. However, taking a picture of the transcriptional activity of all the expressed genes in a cell at the same time is somewhat challenging using traditional technologies. One of the uses of microarrays, especially tiling arrays, is the simultaneous characterization of regulation of transcriptional expression genome-wide, i.e., transcriptome mapping. Transcribed eukaryotic RNA sequences are now known to encompass more that just the mRNAs that encode proteins, but also include a growing suite of miRNA, siRNA, and ncRNA. Tiling arrays provide effective means for the analysis of gene expression due to their high sensitivity, and the fact that the entire representative library of the encoding genes is present in the array. The use of arrayed genomic DNA fragments allows the detection of polyadenylated and non-polyadenylated RNA molecules. Similar approaches have been used to study the human chromosomes to identify transcriptional units [34, 35]. ESTs and oligonucleotide arrays have been frequently used in gene expression studies and have proven success in the elucidation of gene regulation on a large scale [34, 36]. Similarly, by the use of miRNA genechips, it has been possible to understand the expression of microRNAs (miRNAs) in a wide range of organisms [35]. Standard oligonucleotide array can be specifically designed to characterize and analyze miRNA expression profiles from a variety of tissues or cell lines [35].

2.6.6 COMPARATIVE GENOMIC HYBRIDIZATION

Comparative genomic hybridization (CGH) is designed to analyze DNA target samples by hybridizing them against a DNA array, usually a tiling array [37, 38]. This technique is often used in diagnostic assays to compare the genetic differences between two DNA samples. The target DNA samples may have been derived from a set of selected samples, i.e., from normal and cancer cells. Using this methodology, a tiling array comprising the entire representative genome can be employed to identify the copy numbers of specific DNA regions in question. This may also shed light on the presence of potential mutations such as deletions, insertions, and substitutions in the target DNA sequence present in the test sample. Alternatively, custom-made tiling arrays comprising specifically designed representative DNA fragments or oligonucleotides, presenting overlapping, non-overlapping, redundant, or unique sequences, may provide important clues to the genetic root cause of certain diseases or disorders by detecting chromosomal abnormalities in a relatively short period of time. Early detection of genetic disorders is a key to attain the best possible therapeutic interventions at early stages of the onset of a disease. Finally, a new type of array is just being deployed that uses unique genomic sequences to identify species and relative abundance in samples comprising complex assortments. For example, solid samples are assessed for the presence and relative abundance of different bacterial and archaea species in soil or water samples using environmental microarrays, also known as phyloarrays [39, 40, 41].

2.6.7 PROTEIN MICROARRAYS

A protein microarray, also called an immunoarray, provides a high-throughput characterization of the interactions and activities of proteins. This proteomic technology works in a similar way to the DNA arrays [42, 43, 44]. The primary advantage of this technique is that an extremely large number of individual proteins can be studied for their expression or binding abilities with their specific counterparts or specific antibodies. Typically, the study is carried out in parallel for all of the target proteins using known probes that are printed onto a solid support, usually a coated glass slide. The target protein molecules present in a given control or test sample are labeled using a fluorescent dye and are exposed to the probe molecules on the immunoarray. The fluorescence emitted from the complexes formed on the surface of the array is captured using an array scanner and the data is analyzed using a variety of computational techniques and software. The fabrication of protein arrays is based upon the same technology used for the production of DNA microarrays with modifications to provide specific substrates that can effectively bind proteins, peptides, or antibodies [9, 42]. The use of quantum dots in the detection of signals from protein microarrays has gained popularity in recent years [45]. Quantum dots are semiconductor nanocrystals that are widely used in biological assays requiring imaging. Depending on the type of reaction, quantum dots are used in labeling the probe and detecting specific molecular interactions using a wide range of wavelengths detectable by a scanner. Quantum dots are ideal for use as biological probes as they absorb and emit light through a wide spectrum of wavelengths. The combination of quantum dots and protein microarrays offer unique features that together allow detection of cancer markers in biological specimens at very low concentrations, as low as at pg/ml concentrations [45]. Quantum dots also offer remarkable advantages in providing improved signal to noise ratio and good image resolution. The problem of photobleaching, which is commonly found using standard fluorophores, is also overcome using quantum dots. The use of quantum dots together with protein microarrays enables quantitation of specific target proteins in a rapid, low-cost multiplexed assay, and hence makes it possible to analyze the biomarkers for the presence of genetic diseases.

2.6.8   TISSUE MICROARRAYS

Tissue microarrays are commonly used in the analysis of gene expression in certain tissues in multiplex assays [46]. Individual tissue samples are printed on one slide wherein each section may be composed of small 0.6–3.0mm cores of tissue arrayed at a high density on a single glass slide. The use of tissue arrays involves the transfer of multiple tissues from conventional histologic specimens to the slide. The tissue samples may have been derived from clinical biopsies and a standard microtome is used to section fixed tissue that is subsequently printed at precise locations in an arrayed pattern. The tissue array is then analyzed by any method of standard histological analysis. Each microarray tissue section can be sliced into 100–500 sections, which are then subjected to independent tests on separate arrays. *In situ* biological assays using tissue microarrays include immunohistochemistry, and fluorescent *in situ* hybridization (FISH). They are particularly useful for the analysis of tumor samples. Further, variations in the methods conducted using tissue arrays include *in situ* hybridization for RNA or DNA molecules, cytological examination, and *in situ* amplification of DNA molecules using PCR [4, 46]. Experimental uniformity and reproducibility make efficient use of a limited number of slides, and research projects can be executed in a reasonable amount of time.

2.7   CELL-BASED ARRAYS

Another revolution in microarray technology is the use of cell-based microarrays [47, 48]. Cell-based arrays may provide the use of arrays in the following types of platforms.

2.7.1 LIVING CELL ASSAYS

In this technique, expression constructs containing the desired DNA or marker molecules are spotted on an array in standard fashion. The arrays are used as a base upon which cells can be plated and transfected. The cells cultured in an appropriate medium are exposed to the array where they make contact with the substrate and many cells adhere to the nucleic acid spots thereon. Such adherence of the cells to the nucleic acid molecules on the array allows some of the DNA molecules to be transferred into the cells by the process of transfection, which is assisted by transfection reagents. The array is kept under controlled temperature and humidity conditions and cells that have retained the transfected DNA are maintained in a monolayer. This *in situ* cell transfection approach has also been coined, "reverse transfection" [44, 47, 49, 50]. The use of cellular microarrays allows for the multiplex analysis of the living cells, wherein the effect of expression of a number of different genes can be analyzed simultaneously using the same cell types under the same conditions in one experiment. Another approach by which living cells have also been analyzed by the transfected DNA is by using nucleic acid programmable protein arrays [4]. In these arrays, plasmid DNA that is to be transfected into the cells is GST-tagged and spotted on the slide together with an anti GST-polyclonal antibody. When cells are plated over the top, *in situ* expression of the transfected DNA produces protein molecules in proper environmental conditions and the nascent GST-tagged proteins are immobilized by the GST-polyclonal antibody on the spots. The array is incubated with a single fluorescently tagged probe to find protein-protein interactions, identifying expressed protein spots. The detection of GST-tag determines the presence of the expressed protein. The interaction of the GST-tagged recombinant protein with other possible partner protein molecules can then be detected by using other antibodies against target specific proteins. The effect of such interaction between protein molecules on the cellular morphology and physiology is assessed by the visualization of transfected cells on the arrays using commercially available optical tools. In an alternative method, a cell-free extract can also be used to carry out the *in situ* transcription of an immobilized GST-tagged DNA template on an array. A subsequent *in situ* translation using the same extract allows the production of a GST-tagged peptide, which is captured by a GST-polyclonal antibody that is also closely immobilized on the array. This allows the production of a peptide molecule from any DNA template at specific locations on the array. The analysis of the peptides and their interactions with other protein molecules can be easily carried out using these arrays. Since the locations and types of the corresponding DNA templates are known, further genetic analysis relies on common routine work in the laboratory.

2.7.2 REVERSE PHASE ARRAY ASSAYS

Here, protein lysates are extracted from cells and immobilized on a glass support as discrete spots. Each spot on the array contains a large number of protein molecules present in the sample. The protein array is then exposed to a specific antibody, which is eventually detected by using a secondary antibody labeled with fluorescent molecules or by using quantum dots [44, 45]. The use of a specific antibody in a single array interaction allows the analysis of solely one particular target protein in the mixture of protein samples arrayed on the solid support. Although only one protein is being analyzed in reverse phase assays, the number of target protein samples that can be simultaneously probed is potentially very large. This technique has a wide range of applications in disease diagnosis.

2.7.3 ANTIBODY ARRAYS

In this type of array, specific antibodies are immobilized on the glass slide in discrete predetermined gridded spots. Control and test target protein samples are labeled using standard techniques and exposed to antibodies on the array under standard conditions [51]. Detection of specific signals from discrete spots representing specific antibodies characterizes the expression of that particular marker probe in the given protein sample. Therefore, a number of antibodies corresponding to a variety of specific protein molecules in the given sample can be analyzed simultaneously using a single antibody microarray.

2.8   OTHER TYPES OF MICROARRAYS

Efforts are also underway to generate carbohydrate arrays or glycoarrays to detect interactions between carbohydrates and proteins [52, 53]. Carbohydrates are well known to interact with specific proteins within living cells, and such interactions are crucial to the function of some proteins, e.g., enzymes. These specific interactions between interacting partners can be detected using arrays having specific carbohydrates printed thereon in a predefined fashion. Analysis of the interaction of carbohydrates with specific protein partners is vital in certain cellular processes including growth and development. The term glycome has been commonly used to emphasize a complete set of carbohydrate population present in a particular cell type. Together with the proteomic composition of a cell, the glycome is equally important in the normal functioning of the cells. Glycoarrays can provide a comprehensive snapshot of cellular activities including post-translational modifications of the proteins and activities of the mature proteins and enzymes. The carbohydrates are normally chosen from a library and are spotted onto the slide and exposed to the mixture of proteins, which can then be detected by specific antibodies to determine the interaction of specific proteins with a known target of carbohydrate molecules. The use of glycomicroarrays has been reported by Tong et al. [52]. These glycoarrays are based on glyconanoparticles, and are utilized in the study of glycan–lectin molecular interactions. The glyconanoparticles are synthesized by linking specific carbohydrate ligand molecules to substrate present on the solid support by covalent immobilization using light activation. These glycoarrays are then exposed to target lectins, which can be labeled with fluorescent molecules. Detection of fluorescent signals entails positive interactions between glycan and lectin molecules.

2.9 APPLICATIONS OF MICROARRAYS

It is important to note that in a microarray experiment, the genes identified as being differentially expressed may be considered as candidates for subsequent studies. However, such candidate genes always need to be validated by alternative procedures. Great care has to be taken to confirm the validity of the microarray results indicating any transcriptional changes. Such confirmatory methods include Northern blot analysis, RNase protection assays, RT-PCR, *in situ* hybridization, or conducting parallel RNA interference (RNAi) assays using the same source and conditions for the samples as those used in the microarray experiment. Depending on the nature of the experimental project, further detailed studies may be carried out using high-end *in vivo* approaches, such as gene knock-out and knock-in or rescue experiments [18, 20, 21, 23]. The following is some discussion with respect to the microarray methodology and its exploitation in diverse areas of research in biological sciences.

2.9.1 MICROARRAYS IN UNDERSTANDING REGULATORY NETWORKS

Within cells, there are numerous regulatory proteins whose activity is regulated by the signaling molecules from within the cells or any external stimuli from outside the cells [6, 7, 8]. These signaling mechanisms are responsible for cellular behavior and activities. Various signaling cascades can be initiated using proper inducers or inhibitors on the cell-based arrays that may trigger cellular activities. Through the visualization of such processes, using proper tools, cells can be examined for any changes in morphology, cellular phenotypes, and any abrupt physiological change, in response to stimuli or transfection, or by exposure to toxic or other chemical compounds.

The data obtained from a typical gene expression study not only reveals the identification of a plurality of differentially expressed genes, but also sheds light on patterns of gene expression under a given set of conditions: it helps to identify when genes become active or are repressed as a cohort in response to stimuli. Gene expression data has also been used to understand interactions of different proteins and any possible regulatory relationships amongst them. Such reverse engineering methodology for the discovery of regulatory networks in a cell has been reported in some studies [54, 55]. In similar studies, gene expression data was used to infer correlations with possible interacting partners based upon their transcriptional activity in the cells [56, 57]. Variations in gene expression patterns may also reflect activity from a range of cellular processes e.g., cell signaling, metabolism, cell survival, cell death, and production of functional protein products, to name a few. A major challenge, therefore, is to organize and interpret microarray data in a way that correctly identifies the most directly affected biological or biochemical pathways within the cell under given experimental conditions. An interesting method used to differentiate indirect metabolic from specific signal transduction pathways was attempted through a meta-analysis of data from multiple experiments by generating a three dimensional expression map wherein the gene expression profiles were statistically clustered and assembled into a topomap of gene activity [58].

Over the life of microarray technology such cellular pathway analysis has gained popularity for researchers in medical sciences and pharmacology [55, 59, 60, 61]. Rational approaches for assembling statistical data regarding gene expression have been frequently used to explain the underlying molecular mechanisms in a cell. Bioinformaticians have been actively participating in the development of logical maps defining the significance of cellular pathways as indicators in specific gene expression signatures obtained from microarray experiments [62, 63, 64]. Grouping a large number of expressed genes and proteins with reference to their inherent involvement in known pathways has identified several cellular mechanisms involved in certain conditions and diseases [65, 66, 67]. The use of microarrays in understanding the development of plant *Arabidopsis thaliana* and a nematode worm *Caenorhabditis elegans* has provided interesting clues to developmental biologists [4, 5].

2.9.2 MICROARRAYS IN BIOMEDICAL RESEARCH

The popularity of the use of microarrays in medical research can be estimated using any search engine available on the Internet: the NCBI database reveals the significance of this technology in recent years. For example, over 23,000 studies have been reported in the area of biomedical research for a variety of human diseases, including cancer, diabetes, cardiovascular diseases, Alzheimer's, stroke, AIDS, cystic fibrosis, Parkinson's, autism, and anemia. For cancer research alone, there are about 10,000 published articles. Further, there are over 19,000 published studies characterizing gene expression profiles in various tissues using microarray analysis. This topic is also discussed earlier in Chapter 1 of this book.

Conducting a microarray experiment is only the first in a series of steps. The analysis and interpretation of microarray data obtained through clinical studies is complex and requires a skill set not often taught to the ordinary researcher. The computational methods essential for the correct and reliable interpretations require the collaborative involvement of bioinformaticians, computer scientists, statisticians, and molecular biologists to achieve meaningful results that are reproducible. The standardization of microarray protocols using optimal algorithms, and the availability of user-friendly software has been a major challenge in the research and development of this technology. Online databases have been developed in order to share new clinical findings [62, 63, 65]. Despite the nature of clinical data, which may not be accessible due to ethical or proprietary concerns, there is an enormous amount of microarray data that can be used to develop statistical models that best suit the needs of clinical researchers. In recent years, several clinical trials have included microarray studies to provide clinical evidence. Microarrays have widely been used in the diagnosis and staging of genetic diseases [60, 65, 68]. SNP association studies at the genomic level, for example, have been useful to predict the onset of a disease, for designing therapeutic protocols, and for determining the prognosis and susceptibility of patients to certain drugs before any prescribed therapy [37, 69].

The identification of potential gene targets using microarrays has delivered some degree of success in the molecular classification of certain cancers [32, 44, 45, 46, 69]. Microarrays have been used to analyze cases of lymphoma, leukemia, and multiple myeloma. Such studies have a significant impact on the development of modern diagnostic methods. While gene expression profiling using control and diseased samples has allowed identification of potential biomarkers and/or target genes, the use of traditional approaches retains importance for the validation of the data. It is critical that a high confidence level is achieved in the diagnosis or prognosis of the diseases. Distinct gene expression patterns in certain types of cancers, for example, can shine some light on the predisposition of relatives to the disease. However, other parameters including histological examination, blood tests, biopsies, and physiological tests are mandatory to make a firm decision for diagnosis/prognosis. Similar studies for the classification of cancers, response to certain drug treatments, ovarian cancer, and organ transplant have been performed in some independent studies [20, 70]. Differential gene expression profiles have also been generated in diabetic patients [71] and some diagnostic marker genes have been identified. Recently, the use of microarray in the genomic characterization of SARS disclosed viral gene signatures in some patients [72]. Using this approach, the virus responsible for SARS was discovered as a corona virus. Further, it can be envisioned that, in the future, the use of microarrays in the identification of the origin of metastasized cells will be possible by detecting specific markers present in the sample obtained from cancer patients where the origin of the metastasized cells is unknown.

2.9.3 MICROARRAYS FOR IDENTIFICATION OF SINGLE-NUCLEOTIDE POLYMORPHISMS

Single nucleotide polymorphism (SNP) represents a variation in a DNA sequence occurring at a single nucleotide (A, G, T, or C). The nucleic acid sequence may differ in the paired chromosomes or in the alleles. SNP may occur within the coding sequences of genes (exons), in non-coding sequences (introns), or in intergenic regions in the genome [6, 8]. SNPs are common in humans, where they occur with a frequency of about once every 1,000 bp. The presence of a specific SNP varies across different genomes. There are approximately 10 million SNPs in the human genome. SNPs are regularly characterized and submitted to a public database (dbSNP database, http://www.ncbi.nlm.nih.gov/SNP) and this compilation of SNPs allows successful mapping of the genome (e.g., Wellcome Trust; http://snp.cshl.org).

The presence of an SNP in a gene may or may not change the amino acid sequence of the protein that is being encoded by that gene. This is due to the degeneracy of the genetic code, as well as to the possibility these nucleotide changes have to reside within a gene, but outside of its coding sequences [6, 8]. The presence of SNPs in non-coding intervening sequences is presumed to have evolutionary significance. SNPs provide variations in DNA sequences that can possibly have a significant effect on the development of genetic diseases or disorders [37, 69]. Further, the presence of certain SNPs may enable the organism bearing the SNP to be more resistant or more sensitive to environmental cues including toxicants, stress, or exposure to pathogens, vaccines, antibiotics, and certain chemicals and drugs. SNP discovery has also played a significant role in the development of modern biomedical research, especially when deployed in combination with microarrays. Genome-wide SNP association studies have facilitated important research in agriculture and livestock breeding programs as well [73, 74].

Obviously, there has been extensive interest in the pharmaceutical industry to identify and employ specific SNPs that correspond to differences in sensitivity to drugs and treatment among individuals of a population. SNP associations can also be used to identify patients' profiles prior to an available therapeutic treatment. Such development of a therapeutic portfolio for identifying an optimal drug treatment plan might facilitate uncomplicated and effective treatment or recovery of the patients. Further, the use of SNP based microarrays has also allowed for identification of disease-susceptibility target genes, which may well be used as biomarkers or even as candidate genes for understanding further biochemical pathways involved in the onset or progression of a disease. Such studies combining genomic approaches in the discovery of drugs and target genes have been termed as *pharmacogenomics*, and they have extended to include research in basic sciences [59, 60, 65].

Using spotted or *in situ* synthesized oligonucleotide arrays, large-scale projects have been developed for examining genome-wide SNP associations. The analysis is based on typical hybridization assays [37, 69]. Further, validation of results is achieved by other standard molecular and biochemical methods such as mass spectroscopy and genotyping. However, a specifically designed spotted oligonucleotide gold thin film array has been reported to provide high sensitivity and reliability [75]. It is convenient that the data output in case of SNP microarray experiments is relatively simple to be interpreted, since the output only indicates the presence or absence of a specific SNP by detecting the presence or absence of fluorescence.

2.9.4 MICROARRAYS IN MICROBIOLOGY

Microarray technology has readily been exploited in genome-wide transcriptional profiling as well as in gene discovery in the field of microbiology [76]. While studying microorganisms, the use of microarrays has gained importance in medical microbiology as well and potential virulence factors and microbial pathogens have been identified using this approach. The use of microarrays in characterizing gene expression profiles in bacteria was reported in a study carried out by Dennis et al. [77]. DNA microarrays were used to analyze the expression of bacterial metabolic genes within a sample containing several microbes under control and stress conditions. This study also provided useful hints on the metabolic pathways playing a critical role in bacteria. In a similar study, using oligonucleotide arrays, Nicholson et al. studied the transcriptome of the bacterial pathogen *Bordetella bronchiseptica* [34]. Gene expression profiles at different time points during the development of biofilms suggested that the pathogen exhibits a regulated pattern of gene expression during the bacterial developmental process. Further, the different genes involved in biofilm development were also identified [34]. Similarly, studies involving host responses to pathogen infection have also employed microarrays [78, 79, 80]. Identification of differentially expressed genes in response to bacterial infection can highlight the specific mechanism of action of pathogen infection and the development of an immune response in the host. Therefore, DNA microarrays provide a platform to address multiple questions including the identification of a pathogen in case of an infection, the mechanism of infection, host responses, and the identification of potential drug targets. Allen et al. have conducted a similar analysis of viral transcriptome using both, a spotted oligonucleotide array and an Affymetrix GeneChip [72].

2.9.5 MICROARRAYS IN TOXICOLOGY

The use of microarrays in the field of toxicology has also yielded significant contributions where the merging of toxicology with genomics is also termed toxicogenomics [81]. Once again, the use of bioinformatics is critical to exploring biological processes in response to environmental contaminants or toxicants using this high throughput approach [73, 82]. However, microarray technology is still limited to only a few aquatic model systems due to the sparse availability of microarray platforms. One of the model systems used is medaka, *Oryzias latipes* [74]. Oligonucleotide arrays were used in this study to characterize gene expression in response to hypoxia. In another study, researchers employed a microarray-based approach to analyze the regulation of gene expression for known marker genes in a fish, European flounder (*Platichthys flesus*) [83]. Over 160 marker genes present on an array were analyzed using nucleic acid samples isolated from fish obtained from polluted and control regions. Among 160 genes on the array, 110 were already known biomarkers for toxic responses in this aquatic model system, as well as in other mammals. The analysis of gene expression data revealed 11 differentially expressed marker genes differentiating fish from the polluted and unpolluted regions.

Microarray applications in aquatic toxicogenomics have a great potential to facilitate the discovery of diagnostic biomarkers, and to establish stress-specific signatures and molecular signalling pathways during adaptation to stress conditions. For example, microarray results have indicated many interesting genes that are differentially expressed in response to chemicals such as sodium arsenite [84]. Sodium arsenite induces a complex cellular response in HepG2 cells and use of a microarray approach provides a global view of molecular alterations induced by sodium arsenite at relevant environmental exposure levels. The validity of the microarray data is bolstered through the provision of duplicate spots for each gene, reciprocal labeling experiments on separate chips, and extensive normalization and filtering parameters used in these experiments. The results were subsequently corroborated by RT-PCR and Northern blot analysis. Further studies are underway to explore the effect of sodium arsenite in combination with environmental contaminants to disclose potentially synergistic activity among these chemicals. This type of study is significant because the contaminants that may be found together in drinking water are likely to cause combinatorial adverse changes in cellular physiology through their toxic effects. Further research is required to infer regulatory networks by assessing gene expression patterns that may be dependent upon types of cells used in the study, as well as the duration of exposure to the chemicals. In a similar study, two herbicides, 2,4-Dichlorophenoxyacetic acid (2,4-D) and nitrate, which are found in rural ground water, were evaluated for their toxic effect on mammalian HepG2 cells [85]. The herbicides were used at environmental concentrations and the affected genes were identified using a microarray protocol. Interestingly, it was observed that the HepG2 cells indeed respond to the environmental, low-level exposure of the herbicides and produce a cellular response. The affected genes are identified as stress response, cell cycle control, and immunological and DNA repair genes. These findings also suggest important cellular pathways that may be affected by the exposure to environmental levels of these herbicidal contaminants.

2.10   ANALYTICAL TOOLS AND RESOURCES

Once the gene expression profiles have been obtained from a microarray experiment, it is necessary to derive a biologically meaningful interpretation from the statistical data of the gene expression. Merely looking at the increase or a decrease in multiple genes in a cell does not in itself provide many clues to assigning the response to a specific cellular pathway. Therefore, specifically designed tools are required for identifying differentially expressed genes and associating them with the cellular activities that have previously been linked with these genes based on knowledge in the public databases [3, 86]. The following are some of these tools and resources, which are publicly accessible through the Internet.

2.10.1   TOOLS FOR CELLULAR PROCESSES AND PATHWAYS

- GenMAPP: http://www.genmapp.org/. GenMAPP is an open source online free application that is specifically designed to analyze gene expression and correlate it with the genomic data and known biological pathways. This application allows the clustering of genes and enables researchers to correlate them with particular pathways. This integration allows a global analysis of gene expression while correlating expression with hundreds of known biological pathways. Such association of expressed genes to specific pathways also allows the elucidation of cellular processes as well as revealing new potential mechanisms of cellular response under experimental conditions.
- Database for Annotation, Visualization and Integrated Discovery (DAVID): http://david.abcc.ncifcrf.gov/. DAVID is an important resource for annotation. The online open source application, DAVID, is a Web-based application that provides a comprehensive set of functional annotation tools for translating statistical data into biologically meaningful results using gene expression profiles at a genomic scale.
- GoMiner: http://discover.nci.nih.gov/gominer/index.jsp. GoMiner is a useful tool for the manipulation and analysis of genomic data obtained through transcriptomics, proteomics, or even from metabolomics assays. The data from gene expression studies using microarrays is organized in a logical manner to help produce valuable biological associations through which inferences can be made.

- WebGestalt: `http://bioinfo.vanderbilt.edu/webgestalt/`. WebGestalt is an analytic tool specifically designed for conducting functional genomic studies, including proteomic and large-scale genetic research projects. Exhaustive lists of differentially expressed genes are continuously curated and assembled by incorporating data sets retrieved from available public resources. This tool provides a comprehensive overview of the transcriptional or proteomic state of a cell and facilitates comparative studies.
- Ingenuity: `http://www.ingenuity.com/` is also a Web-based tool for the analysis and interpretation of genomic data to improve interpretation of acquired data. The software allows researchers to reveal complex networks of differentially expressed genes, as well as to develop possible pathway connections that help in understanding normal and abnormal cellular activities with reference to particular pathways.

2.10.2 TOOLS FOR FUNCTIONAL ANNOTATION

It is essential to accurately and precisely design microarray platforms for large-scale production runs. Since thousands of genes can be accommodated on a single slide, the challenge comes to eliminate any sort of batch-to-batch discrepancies while co-relating the data acquired following the microarray experiment with the identity of a specific gene at a particular location on the array. Similarly, all known functions of the probes are significantly important for consideration following microarray data analysis [86, 87]. Thus, keeping all the accurate information regarding the exact location of a specific probe in a particular spot and its associated known function is crucial to the accuracy of the interpretation. The following are some of the resources that are commonly used to link biological information with genetic elements or probes used on a microarray platform.

- NetAffx: `http://www.affymetrix.com/analysis/index.affx` The NetAffx Analysis tool allows researchers to correlate array data with array design as well as to assist in gene annotation.
- Anni 2.1 Biosemantics: `http://www.biosemantics.org/index.php?page=anni-2-0`. This software assembles biomedical literature and published documents through an ontology-based interface. The documents and the published data retrieval and assembly are carried out in using biological concepts, including gene expression, and information regarding different drugs and diseases. In essence, this software provides a text-mining utility.
- GeneCards: `http://www.genecards.org/`. GeneCards is a searchable database of all known human genes that attempts to link all available genomic information. It integrates the available transcriptomic, genetic, proteomic, and disease information, retrieved from public resources. GeneCards uses standard nomenclature that attempts to overcome the barriers of variations in data formats amongst different platforms.
- Transcriptome map: `http://bioinfo.amc.uva.nl/HTMseq/controller.`

The quantitative mRNA expression data provided by SAGE libraries and Affymetrix GeneChips is integrated with the available genome sequence data. The HTMseq application aligns the expression of target genes with their specific position on particular chromosomes, and in a range of tissues and cell types.

- Oncomine: https://www.oncomine.org/resource/login.html. Oncomine is a Web-based application that retrieves, rationally manipulates and integrates high-throughput cancer-specific gene expression profiling data. Such expression data manipulation permits identification and assessment and classification of a large number of cancer types. The Web site also provides a searchable database of user-defined gene expression signatures, which further enables researchers to conduct basic and applied research freely.

- Ensembl: http://www.ensembl.org/index.html. The Ensembl project enables researchers to access genome databases and transcription/gene models for vertebrates and other model eukaryotic species.

- UCSC: http://genome.ucsc.edu/. This repository provides a reference database for genome sequences of different model systems from lower to higher eukaryotes.

2.11 CHALLENGES IN MICROARRAY TECHNOLOGY

The power of high-throughput genomic scale analysis of experimental data presents several challenges. One of the biggest difficulties is the variability in gene expression that is inherent due to factors that are beyond a researcher's control. Physiological or systemic differences can be those that are introduced by cells or tissues that are in different metabolic or developmental states, or by other features such as age, sex, genetic variants, and the number of samples tested in a given geographical population. Another challenge arises in the interpretation of the data: although uniformity of approach is ultimately desirable within the research community, experiment-specific conditions might require different statistical cut-off thresholds, filtering parameters, and others. Such factors have to be acknowledged and counted while extracting biological meaning so that an accurate and reproducible interpretation can be made, reported, and if necessary, reproduced.

Another limiting factor in the array technology has been the availability of highly efficient and user-friendly equipment. Such equipment is not only required to generate high-quality miniature chips, but also for achieving efficacy and reproducibility that meet the quality standards. The robotics used for printing the arrays, equipment for isolation, purification, and quantification of nucleic acids, automated thermal cyclers, liquid handling robots, and array scanners as essential components for fabrication and utilization of microarrays need to be seamlessly integrated and easy enough to run that different operators will generate a highly consistent product.

Some researchers find the associated cost as a major limitation in low- to medium-scale research laboratories. The costs associated with the entire infrastructure does not allow every researcher to generate their own customized arrays. Array-printing centers established to produce specific customized arrays have been able to maintain high standards of quality and some have achieved commercial success. The higher the number of DNA molecules to be printed on the array, the higher the cost of the whole experiment. Furthermore, batch-to-batch consistency has to be maintained to ensure the quality and comparative utility of microarrays. Cost can be kept under control if high quality, high-density, standardized arrays can be fabricated on a large scale. Since it is ideal that the arrays should not be reused even after a single use, the requirement of repeating an experiment for obtaining statistically significant results becomes even more expensive. Limited availability of types of DNA arrays for selected model systems or even having a desired platform for necessary marker probes on a single array sometimes requires researchers to use customized arrays. For example, in the case of a non-model organism, the selected gene markers for disease or environmental stressors may provide a full scope of research opportunities for some researchers, but not for others in the discipline at large.

The number of fully sequenced genomes has grown in the past few years and it is becoming increasingly important that microarrays for such genomes become cheaply and widely available. Also, in parallel, more robust analytical tools are encouraging a growing interest and need for comparative genomic studies of newly sequenced model or non-model organisms genomes. On the researcher's part, focused research goals and a cognizance of the broader field are imperative to generate well designed standard microarrays containing marker probes that will be equally useful for researchers in other laboratories.

Despite a great advancement in the production of microarrays and the versatility of approaches with which this technology endows scientists to conduct research in diverse areas of science, several challenges remain that have to be circumvented in order to obtain the maximum benefit out of this technology. It is in fact quite difficult to manipulate extremely small volumes of solutions containing the probes to be printed on a slide. It is crucial to produce microarrays that have consistent spot sizes, shapes, and densities, ordered locations, and concentrations delivered onto each point on the array. It is very important to reduce artifacts during the production of arrays, whether they are produced by spot printing or by *in situ* synthesis of oligonucleotides. Several technologies have played a role and a number of solutions for increased sensitivity of DNA microarrays are currently being developed, including the electrical polarization of the array spots [88]. The use of specialized substrates on the slides to reduce the background signals, and the use of specific surface tension reagents also aid in the precision of array printing. Similarly, even if the microarrays are consistent, the production of target requires optimal RNA isolation methods to yield reproducibly good results. Consistent methodology during the labeling of target samples, consistency in the incorporation of fluorophores during a labeling reaction, and hybridization and post-hybridization protocols are crucial. Any variations in the protocols may have significant effects on the quality of data and its reproducibility. A higher density of the labeled samples along with the smaller size of the arrays appears to provide data that can be trusted with a higher degree of confidence. Using nanolithography, researchers have been able to develop fully functional protein arrays containing spots of about 1μm diameter. Furthermore, the signal to noise ratio (SNR) was reported to be above 10 using these immunochips, even while providing a detection limit as low as 1.3ng/ml of target molecule in a given sample [42]. The resolution that can be achieved using available array scanners also delivers a level of quality that can be used to protect the data acquired, but when target expression differences are slight and subtle there is always room for improvement. The development of more advanced systems, tools and equipment, which are more user-friendly, precise, and accurate in their functionality, is needed.

Another big caveat in the successful execution of microarray experiments is the availability of appropriate computational approaches and user-friendly software for data processing and statistical analysis. The development of computational methods for microarray data analysis has a rich history. Software tools, whether open source or available commercially, are continuously under development. In addition to the software required for the microarray data analysis, there is also a great need for developing accurate algorithms for ancillary tasks such as: DNA clone selection; collection and annotation of clones; designing microarrays; and for validating the quality of microarrays after they have been produced. Also, there is a need for developing computational methods and software tools to deal with very large datasets that are produced. Here, the bioinformatics research community plays an important role. In the forthcoming chapters, this will be discussed in detail with reference to data quality, data analysis, visualization, and interpretation of data from microarray studies.

2.12 CONCLUSIONS

The development of microarray technology has revolutionized molecular biology and biomedical research. The integration of information from microarray experiments using genomics, transcriptomics, and proteomics has helped us to develop a real-time understanding of complex biological processes. The availability of exceptionally efficient platforms and methods for high-throughput and robust approaches at a competitive cost has allowed scientists to probe cellular events efficiently and in a manner that is unbiased with regard to the initial state of knowledge. One advantage of this genomic approach is the possibility to detect and compare any transcript of an organism in a comprehensive way, and hence to identify any malfunction, change, or perturbation of the genome. Similarly, the availability of faster and reliable technology for data analysis and processing will eliminate some of the problems and pitfalls associated with current microarray methodologies.

Microarrays provide an exceptional tool for defining gene expression and nucleic acid behavior in routine experiments. The availability of more comprehensive platforms and highly efficient and trustworthy analytical tools will allow accurate, fast, robust, and reproducible data output, which can best be integrated with the genomic studies conducted across numerous laboratories around the world. This feature, in combination with better specified, accessible, and comprehensive open source databases, will benefit research communities across disciplines including biology, medical sciences, computer science and statistics. Such centralized database accessibility will also permit larger scale meta-analysis where new and profound insights might be developed. Internet access to data and analytical tools for performing inter-laboratory collaborations have already proven fruitful as indicated by the number and wealth of collaborative studies published during the last decade. Furthermore, international collaborations will facilitate and encourage playing an important role in addressing urgent situations such as environmental pollution, ecotoxicology, epidemic diseases, and conservation biology for endangered species [89]. In the future, geographical comparative studies of gene expression and genetic data at the genome level could also be performed by regulatory agencies to help develop, monitor, and assess health and environmental policies.

2.13   THE FUTURE OF MICROARRAYS

While microarray methodology can occasionally burden one with technical challenges and an overwhelming amount of data, the technology provides a powerful and robust approach for high-throughput analysis at a genomic scale. Microarrays became increasingly popular in a very short period of time. Despite the considerable amount of effort that is required to acquire optimal parameters for printing, hybridization, and computational analysis of the data, the value of interdisciplinary collaborations that necessarily form among biology, biochemistry, physics, engineering, computer science, and statistics cannot be underestimated. Admittedly, significant improvements are required to make this technology accessible on a routine basis. Similarly, the amount of data generated through a single experiment is enormous and better tools to gather, analyze, and translate those data into biological sense is as important as the production of actual microarray targets, probes, and hybridization assays. The use of nanotechnology can play an important role in the precise construction of high-density microarrays. Improvements in the use of best construction materials as substrates and proper solid supports using precision tools to fabricate the arrays is also necessary for increasing productivity and accuracy of a microarray experiment. There is much more to be done for the establishment of proper tools and optimal assays in the use of protein and antibody arrays, and this field of proteomics promises to grow and empower the basic and applied sciences in the future.

Inevitably, the potential for modularization and mass production of this technology will result in its adoption for many diagnostic/prognostic assays in medical sciences. Conservation science, toxicology, microbiology, and environmental biology can certainly benefit from this technology. Some areas of science, including evolutionary biology, aquatic ecology, and conservation biology, have not yet fully exploited the promise of this technology. Although there have been several reports in which the use of microarrays have shown some interesting results, there remains a big gap in understanding complex questions such as species diversity, preservation, evolution, and a comparative analysis of genomes across the animal and plant kingdoms. Whether the technology is used for establishing the "fingerprints" or "signatures" of pertinent gene expression, or for the detection of genetic changes in the genomic DNA in cells, tissues, or organisms, it provides an insight into normal or defective intracellular processes in a very quick and elegant manner. Fingerprints or gene signatures can also help diagnosis and staging of particular epidemic diseases, across animal and plant species. The applications are limited only by imagination, cost, and accessibility. As the latter two improve, adoption of microarrays will no doubt play a role in the emergence of new paradigms, innovations, and technologies.

ACKNOWLEDGMENTS

The author would like to thank Iman Rezaeian for his help in preparing the LATEXversion of this chapter, and the anonymous reviewers for their positive input that helped enhance the chapter. The author would also like to thank Dr. Mark Schena, Arrayit Corporation (ARYC), and Affymetrix for providing the images for Figures 2.1 and 2.2.

REFERENCES

1. M. Schena, *Microarray analysis*. Hoboken: Wiley-Liss, 2003.
2. P. Hegde, R. Qi, K. Abernathy, et al., "A concise guide to cDNA microarray analysis," *Biotechniques*, vol. 29, no. 3, pp. 548–563, 2000.
3. J. D. Hoheisel, "Microarray technology: beyond transcript profiling and genotype analysis," *Nature reviews genetics*, vol. 7, no. 3, pp. 200–210, 2006.
4. A. Kozarova, S. Petrinac, A. Ali, and J. W. Hudson, "Array of informatics: Applications in modern research," *Journal of proteome research*, vol. 5, no. 5, pp. 1051–1059, 2006.
5. A. Ali and M. J. Crawford, "Developmental biology: an array of new possibilities," *Biotechnology advances*, vol. 20, no. 5, pp. 363–378, 2002.
6. H. Lodish, *Molecular Cell Biology*. New York: W. H. Freeman, 2012.
7. D. Nelson, A. Lehninger, and M. Cox, *Lehninger Principles of Biochemistry*. New York: W.H. Freeman, 2013.
8. J. Watson, T. Baker, S. Bell, A. Gann, M. Levine, and R. Losick, *Molecular Biology of the Gene*. New York: Benjamin-Cummings Publishing Company, 2013.
9. Y. M. Foong, J. Fu, S. Q. Yao, and M. Uttamchandani, "Current advances in peptide and small molecule microarray technologies," *Current opinion in chemical biology*, vol. 16, no. 1, pp. 234–242, 2012.
10. J. K. Peeters and P. J. Van der Spek, "Growing applications and advancements in microarray technology and analysis tools," *Cell biochemistry and biophysics*, vol. 43, no. 1, pp. 149–166, 2005.
11. D. W. Huang, B. T. Sherman, and R. A. Lempicki, "Bioinformatics enrichment tools: paths toward the comprehensive functional analysis of large gene lists.," *Nucleic Acids Res*, vol. 37, no. 1, pp. 1–13, 2009.
12. P. Khatri, M. Sirota, and A. J. Butte, "Ten years of pathway analysis: current approaches and outstanding challenges," *PLoS computational biology*, vol. 8, no. 2, p. e1002375, 2012.
13. J. M. Arteaga-Salas, H. Zuzan, W. B. Langdon, G. J. Upton, and A. P. Harrison, "An overview of image-processing methods for Affymetrix GeneChips," *Briefings in Bioinformatics*, vol. 9, no. 1, pp. 25–33, 2008.

14. M. L. Smith, M. J. Dunning, S. Tavaré, and A. G. Lynch, "Identification and correction of previously unreported spatial phenomena using raw Illumina BeadArray data," *BMC bioinformatics*, vol. 11, no. 1, p. 208, 2010.

15. S. Draghici, *Statistics and Data Analysis for Microarrays Using R and Bioconductor, Second Edition*. Chapman & Hall / CRC mathematical and computational biology series, New York: Taylor & Francis, 2012.

16. A. Rasooly and K. E. Herold, "Food microbial pathogen detection and analysis using DNA microarray technologies," *Foodborne pathogens and disease*, vol. 5, no. 4, pp. 531–550, 2008.

17. C.-H. Lin, J. K. Lee, M. A. LaBarge, et al., "Fabrication and use of microenvironment microarrays (MEArrays)," *Journal of visualized experiments: JoVE*, no. 68, 2012.

18. T. G. Fernandes, M. M. Diogo, D. S. Clark, J. S. Dordick, and J. Cabral, "High-throughput cellular microarray platforms: applications in drug discovery, toxicology and stem cell research," *Trends in biotechnology*, vol. 27, no. 6, pp. 342–349, 2009.

19. A. Tzouvelekis, G. Patlakas, and D. Bouros, "Application of microarray technology in pulmonary diseases," *Respir Res*, vol. 5, no. 1, p. 26, 2004.

20. C.-C. Liu, J. Hu, M. Kalakrishnan, H. Huang, and X. Zhou, "Integrative disease classification based on cross-platform microarray data," *BMC bioinformatics*, vol. 10, no. Suppl 1, p. S25, 2009.

21. S. Suthram, J. T. Dudley, A. P. Chiang, R. Chen, T. J. Hastie, and A. J. Butte, "Network-based elucidation of human disease similarities reveals common functional modules enriched for pluripotent drug targets," *PLoS computational biology*, vol. 6, no. 2, p. e1000662, 2010.

22. M. A. Hernandez, R. Schulz, T. Chaplin, et al., "The diagnosis of inherited metabolic diseases by microarray gene expression profiling," *Orphanet journal of rare diseases*, vol. 5, no. 1, p. 34, 2010.

23. T. C. Mockler and J. R. Ecker, "Applications of DNA tiling arrays for whole-genome analysis," *Genomics*, vol. 85, no. 1, pp. 1–15, 2005.

24. J. Yazaki, B. D. Gregory, and J. R. Ecker, "Mapping the genome landscape using tiling array technology," *Current opinion in plant biology*, vol. 10, no. 5, pp. 534–542, 2007.

25. T. A. Clark, C. W. Sugnet, and M. Ares, "Genomewide analysis of mRNA processing in yeast using splicing-specific microarrays," *Science*, vol. 296, no. 5569, pp. 907–910, 2002.

26. Q.-R. Chen, S. Bilke, and J. Khan, "High-resolution cDNA microarray-based comparative genomic hybridization analysis in neuroblastoma," *Cancer letters*, vol. 228, no. 1, pp. 71–81, 2005.

27. J. H. Malone and B. Oliver, "Microarrays, deep sequencing and the true measure of the transcriptome," *BMC biology*, vol. 9, no. 1, p. 34, 2011.

28. M. J. Buck and J. D. Lieb, "Chip-chip: considerations for the design, analysis, and application of genome-wide chromatin immunoprecipitation experiments," *Genomics*, vol. 83, no. 3, pp. 349–360, 2004.

29. X.-y. Li, S. MacArthur, R. Bourgon, D. Nix, D. A. Pollard, V. N. Iyer, A. Hechmer, L. Simirenko, M. Stapleton, C. L. L. Hendriks, et al., "Transcription factors bind thousands of active and inactive regions in the *Drosophila* blastoderm," *PLoS biology*, vol. 6, no. 2, p. e27, 2008.

30. M. Weber, J. J. Davies, D. Wittig, et al., "Chromosome-wide and promoter-specific analyses identify sites of differential dna methylation in normal and transformed human cells," *Nature genetics*, vol. 37, no. 8, pp. 853–862, 2005.

31. I. Laczmanska, P. Karpinski, M. Bebenek, et al., "Protein tyrosine phosphatase receptor-like genes are frequently hypermethylated in sporadic colorectal cancer," *Journal of human genetics*, vol. 58, no. 1, pp. 11–15, 2012.

32. G. Heller, V. N. Babinsky, B. Ziegler, et al., "Genome-wide CpG island methylation analyses in non-small cell lung cancer patients," *Carcinogenesis*, vol. 34, no. 3, pp. 513–521, 2013.

33. G. E. Crawford, S. Davis, P. C. Scacheri, et al., "DNase-chip: a high-resolution method to identify DNase I hypersensitive sites using tiled microarrays," *Nature methods*, vol. 3, no. 7, pp. 503–509, 2006.

34. T. L. Nicholson, M. S. Conover, and R. Deora, "Transcriptome profiling reveals stage-specific production and requirement of flagella during biofilm development in *Bordetella bronchiseptica*," *PloS one*, vol. 7, no. 11, p. e49166, 2012.

35. A. M. Krichevsky, K. S. King, C. P. Donahue, K. Khrapko, and K. S. Kosik, "A microRNA array reveals extensive regulation of microRNAs during brain development," *Rna*, vol. 9, no. 10, pp. 1274–1281, 2003.

36. D. Peršoh, A. R. Weig, and G. Rambold, "A transcriptome targeting EcoChip for assessing functional mycodiversity," *Microarrays*, vol. 1, no. 1, pp. 25–41, 2011.

37. T.-P. Yang, T.-Y. Chang, C.-H. Lin, M.-T. Hsu, and H.-W. Wang, "ArrayFusion: a web application for multi-dimensional analysis of CGH, SNP and microarray data," *Bioinformatics*, vol. 22, no. 21, pp. 2697–2698, 2006.

38. D. G. Albertson and D. Pinkel, "Genomic microarrays in human genetic disease and cancer," *Human molecular genetics*, vol. 12, no. suppl 2, pp. R145–R152, 2003.

39. E. Yergeau, S. A. Schoondermark-Stolk, E. L. Brodie, et al., "Environmental microarray analyses of antarctic soil microbial communities," *The ISME journal*, vol. 3, no. 3, pp. 340–351, 2008.

40. R. Chapman, H. L. Hayden, T. Webster, D. E. Crowley, and P. M. Mele, "Development of an environmental microarray to study bacterial and archaeal functional genes in australian soil agroecosystems," *Pedobiologia*, vol. 55, no. 1, pp. 41–49, 2012.

41. J. L. Sebat, F. S. Colwell, and R. L. Crawford, "Metagenomic profiling: microarray analysis of an environmental genomic library," *Applied and environmental microbiology*, vol. 69, no. 8, pp. 4927–4934, 2003.

42. I. Tsarfati-BarAd, U. Sauer, C. Preininger, and L. A. Gheber, "Miniaturized protein arrays: model and experiment," *Biosensors and Bioelectronics*, vol. 26, no. 9, pp. 3774–3781, 2011.

43. P. Diez, N. Dasilva, M. Gonzalez, et al., "Data analysis strategies for protein microarrays," *Microarrays*, vol. 1, no. 2, pp. 64–83, 2012.

44. C. P. Paweletz, L. Charboneau, V. E. Bichsel, N. L. Simone, T. Chen, J. W. Gillespie, M. R. Emmert-Buck, M. J. Roth, E. Petricoin, and L. A. Liotta, "Reverse phase protein microarrays which capture disease progression show activation of pro-survival pathways at the cancer invasion front," *Oncogene*, vol. 20, no. 16, pp. 1981–1989, 2001.

45. A. Zajac, D. Song, W. Qian, and T. Zhukov, "Protein microarrays and quantum dot probes for early cancer detection," *Colloids and Surfaces B: Biointerfaces*, vol. 58, no. 2, pp. 309–314, 2007.

46. D. A. Notterman, U. Alon, A. J. Sierk, and A. J. Levine, "Transcriptional gene expression profiles of colorectal adenoma, adenocarcinoma, and normal tissue examined by oligonucleotide arrays," *Cancer Research*, vol. 61, no. 7, pp. 3124–3130, 2001.

47. J. Ziauddin and D. M. Sabatini, "Microarrays of cells expressing defined cDNAs," *Nature*, vol. 411, no. 6833, pp. 107–110, 2001.

48. B. S. O. Palsson, "Methods, compositions and apparatus for cell transfection," Sept. 22 1998. US Patent 5,811,274.

49. P. Jaluria, C. Chu, M. Betenbaugh, and J. Shiloach, "Cells by design: a mini-review of targeting cell engineering using DNA microarrays," *Molecular biotechnology*, vol. 39, no. 2, pp. 105–111, 2008.

50. C. H. Kwon, I. Wheeldon, N. N. Kachouie, et al., "Drug-eluting microarrays for cell-based screening of chemical-induced apoptosis," *Analytical chemistry*, vol. 83, no. 11, pp. 4118–4125, 2011.

51. S. Milgram, R. Bombera, T. Livache, and Y. Roupioz, "Antibody microarrays for label-free cell-based applications," *Methods*, vol. 56, no. 2, pp. 326–333, 2012.

52. Q. Tong, X. Wang, H. Wang, T. Kubo, and M. Yan, "Fabrication of glyconanoparticle microarrays," *Analytical chemistry*, vol. 84, no. 7, pp. 3049–3052, 2012.

53. Y. Ma, I. Sobkiv, V. Gruzdys, H. Zhang, and X.-L. Sun, "Liposomal glyco-microarray for studying glycolipid–protein interactions," *Analytical and bioanalytical chemistry*, vol. 404, no. 1, pp. 51–58, 2012.

54. E. Segal, M. Shapira, A. Regev, et al., "Module networks: identifying regulatory modules and their condition-specific regulators from gene expression data," *Nature genetics*, vol. 34, no. 2, pp. 166–176, 2003.

55. J. Le Merrer, K. Befort, O. Gardon, et al., "Protracted abstinence from distinct drugs of abuse shows regulation of a common gene network," *Addiction biology*, vol. 17, no. 1, pp. 1–12, 2012.

56. L. Hooker, C. Smoczer, F. KhosrowShahian, M. Wolanski, and M. J. Crawford, "Microarray-based identification of Pitx3 targets during *Xenopus* embryogenesis," *Developmental Dynamics*, vol. 241, no. 9, pp. 1487–1505, 2012.

57. S. Gonsalves, S. Neal, A. Kehoe, and J. Westwood, "Genome-wide examination of the transcriptional response to ecdysteroids 20-hydroxyecdysone and ponasterone A in *Drosophila melanogaster*," *BMC genomics*, vol. 12, no. 1, p. 475, 2011.

58. S. K. Kim, J. Lund, M. Kiraly, et al., "A gene expression map for *Caenorhabditis elegans*," *Science Signaling*, vol. 293, no. 5537, p. 2087, 2001.

59. G. Hardiman, "Applications of microarrays and biochips in pharmacogenomics," in *Pharmacogenomics in Drug Discovery and Development*, pp. 21–30, Springer (Online), 2008.

60. D. Blohm, "Microarray development beyond the current state of drug discovery," *Business Briefing: Future Drug Discovery*, pp. 101–103, 2003.

61. T. Efferth and H. J. Greten, "*In Silico* analysis of microarray-based gene expression profiles predicts tumor cell response to withanolides," *Microarrays*, vol. 1, no. 1, pp. 44–63, 2012.

62. T. Barrett, D. B. Troup, S. E. Wilhite, et al., "NCBI GEO: mining tens of millions of expression profiles-database and tools update," *Nucleic acids research*, vol. 35, no. suppl 1, pp. D760–D765, 2007.

63. G. C. Tseng, D. Ghosh, and E. Feingold, "Comprehensive literature review and statistical considerations for microarray meta-analysis," *Nucleic acids research*, vol. 40, no. 9, pp. 3785–3799, 2012.

64. L. Zhao and M. J. Zaki, "TRICLUSTER: an effective algorithm for mining coherent clusters in 3D microarray data," *Proceedings of the 2005 ACM SIGMOD international conference on Management of data*, pp. 694–705, ACM, 2005.

65. J. L. Lapp, "Pharmacogenetics and pharmacogenomics: Advancing the field of personalized medicine," *MMG 445 Basic Biotechnology eJournal*, vol. 5, no. 1, pp. 25–30,

2009.

66. N. Vasli, J. Böhm, S. Le Gras, et al., "Next generation sequencing for molecular diagnosis of neuromuscular diseases," *Acta neuropathologica*, vol. 124, no. 2, pp. 273–283, 2012.

67. A. Allam, R. S. Gumpeny, et al., "Analyzing microarray data of alzheimer's using cluster analysis to identify the biomarker genes," *International Journal of Alzheimer's Disease*, vol. 2012, 2012.

68. B. Gautam, P. Katara, S. Singh, and R. Farmer, "Drug target identification using gene expression microarray data of *Toxoplasma gondii*," *International Journal of Biometrics and Bioinformatics (IJBB)*, vol. 4, no. 3, pp. 113–124, 2010.

69. X. Mao, B. D. Young, and Y.-J. Lu, "The application of single nucleotide polymorphism microarrays in cancer research," *Current genomics*, vol. 8, no. 4, p. 219, 2007.

70. M. Rudnicki, S. Eder, P. Perco, et al., "Gene expression profiles of human proximal tubular epithelial cells in proteinuric nephropathies," *Kidney international*, vol. 71, no. 4, pp. 325–335, 2006.

71. C. D. Cohen, M. T. Lindenmeyer, F. Eichinger, et al., "Improved elucidation of biological processes linked to diabetic nephropathy by single probe-based microarray data analysis," *PLoS One*, vol. 3, no. 8, p. e2937, 2008.

72. M. J. Allen, B. Tiwari, M. E. Futschik, and D. Lindell, "Construction of microarrays and their application to virus analysis," *Manual of Aquatic Viral Ecology (American Society of Limnology and Oceanography)*, pp. 34–56, 2010.

73. Z. Zhang, P. Li, X. Hu, Q. Zhang, X. Ding, and W. Zhang, "Microarray technology for major chemical contaminants analysis in food: Current status and prospects," *Sensors*, vol. 12, no. 7, pp. 9234–9252, 2012.

74. Z. Zhang, M. C. Wells, M. G. Boswell, et al., "Identification of robust hypoxia biomarker candidates from fin of medaka (*Oryzias latipes*)," *Comparative Biochemistry and Physiology Part C: Toxicology & Pharmacology*, vol. 155, no. 1, pp. 11–17, 2012.

75. H. J. Lee, T. Goodrich, and R. M. Corn, "SPR Imaging Measurements of DNA Microarrays in PDMS Microfluidic Channels on Gold Thin Films," in *Micro Total Analysis Systems 2001*, pp. 441–443, Springer (Online), 2001.

76. T. Majtan, G. Bukovska, and J. Timko, "DNA microarrays-techniques and applications in microbial systems," *Folia microbiologica*, vol. 49, no. 6, pp. 635–664, 2004.

77. P. Dennis, E. A. Edwards, S. N. Liss, and R. Fulthorpe, "Monitoring gene expression in mixed microbial communities by using DNA microarrays," *Applied and environmental microbiology*, vol. 69, no. 2, pp. 769–778, 2003.

78. A. K. Bansal, "Bioinformatics in microbial biotechnology–a mini review," *Microbial cell factories*, vol. 4, no. 19, pp. 1–11, 2005.

79. I. Bricchi, C. M. Bertea, A. Occhipinti, I. A. Paponov, and M. E. Maffei, "Dynamics of membrane potential variation and gene expression induced by *Spodoptera littoralis*, Myzus persicae, and Pseudomonas syringae in Arabidopsis," *PloS one*, vol. 7, no. 10, pp. 1–20, 2012.

80. E. S. Li and W.-T. Liu, "DNA microarray technology in microbial ecology studies-principle, applications and current limitations," *Microbes and environments*, vol. 18, no. 4, pp. 175–187, 2003.

81. R. D. Irwin, G. A. Boorman, M. L. Cunningham, A. N. Heinloth, D. E. Malarkey, and R. S. Paules, "Application of toxicogenomics to toxicology: basic concepts in the analysis of microarray data," *Toxicologic pathology*, vol. 32, no. 1 suppl, pp. 72–83, 2004.

82. T. Lettieri, "Recent applications of dna microarray technology to toxicology and eco-

toxicology," *Environmental health perspectives*, vol. 114, no. 1, pp. 1–9, 2006.

83. T. D. Williams, A. M. Diab, S. G. George, et al., "Development of the GENIPOL European flounder (*Platichthys flesus*) microarray and determination of temporal transcriptional responses to cadmium at low dose," *Environmental science & technology*, vol. 40, no. 20, pp. 6479–6488, 2006.

84. J. Nordlee, S. Petrinac, A. Ali, and L. Bharadwaj, "Altered gene expression in human hepatoma (HepG2) cells exposed to low-level Sodium Arsenite," *Clinical Medicine Reviews in Oncology*, no. 2, pp. 1–11, 2010.

85. L. Bharadwaj, K. Dhami, D. Schneberger, M. Stevens, C. Renaud, and A. Ali, "Altered gene expression in human hepatoma HepG2 cells exposed to low-level 2, 4-dichlorophenoxyacetic acid and potassium nitrate," *Toxicology in vitro*, vol. 19, no. 5, pp. 603–619, 2005.

86. E. Suarez, A. Burguete, and G. J. Mclachlan, "Microarray data analysis for differential expression: a tutorial," *PR Health Sciences Journal*, vol. 28, no. 2, pp. 89–104, 2009.

87. J. H. Kim, "Biological knowledge assembly and interpretation," *PLoS computational biology*, vol. 8, no. 12, pp. 1–12, 2012.

88. P. C. Mathias, S. I. Jones, H.-Y. Wu, et al., "Improved sensitivity of DNA microarrays using photonic crystal enhanced fluorescence," *Analytical chemistry*, vol. 82, no. 16, pp. 6854–6861, 2010.

89. W. Pennie, S. D. Pettit, and P. G. Lord, "Toxicogenomics in risk assessment: an overview of an HESI collaborative research program," *Environmental Health Perspectives*, vol. 112, no. 4, pp. 417–419, 2004.

3 Gridding Methods for DNA Microarray Images

Iman Rezaeian and Luis Rueda

CONTENTS

Image processing and analysis are two important aspects of DNA microarrays, since the aim of the whole experimental procedure is to obtain meaningful biological conclusions, which depend on the accuracy of the different stages, mainly those at the beginning of the process. DNA microarray images are obtained by scanning DNA chips at high resolution and are composed of sub-grids of spots. Although in many cases, based on the layout of the printer pins, the number of sub-grids or spots are known, due to misalignments, deformations, artifacts, or noise while producing the microarray images, these data may not be accurate or available. In this chapter, the gridding problem in DNA microarray images is addressed. In this direction, different methods for separating sub-grids and identifying spot centers in DNA microarray images are reviewed. Conceptual and experimental comparisons of some of these methods are also discussed.

3.1 DNA MICROARRAY IMAGE GRIDDING

Gridding microarray images is one of the most important stages of microarray image analysis, since any error in this step is propagated to further steps and may reduce the integrity and accuracy of the analysis dramatically. In principle, images from microarray experiments are highly structured since they are composed of high intensity spots arranged in a regular grid-like pattern. The spots are expected to be roughly circular, although in practice different shapes are possible. The first task in DNA microarray image processing is gridding [2, 5, 6, 7, 8], which, if done correctly, substantially improves the efficiency of the subsequent tasks that include segmentation, quantification, normalization, and data mining [9].

A typical DNA microarray image contains a number of sub-grids and each sub-grid contains a number of spots arranged in rows and columns. The aim of the gridding step is to perform a two-stage process in such a way that (i) the sub-grid locations are found in the first stage, and (ii) spot locations within each sub-grid can be found in the second stage. Consider an image (matrix) $A = \{a_{ij}\}, i = 1, ..., n_a$ and $j = 1, ..., m_a$, where $a_{ij} \in \mathbb{Z}^+$ and represents the intensity of pixel (i, j). In most DNA microarray images, a_{ij} is in the range $[0..65,535]$ in a TIFF image. The aim of gridding is to obtain a matrix (grid) $G = \{g_{ij}\}, i = 1, ..., n_g$ and $j = 1, ..., m_g$, where $g_{ij} = (x_i, y_j)$ denotes the center of the $(i, j)^{th}$ spot. An alternative definition of a microarray grid is to define the boundaries between rows and columns of spots, namely horizontal and vertical vectors, $v_i = [v_{i1}, ... v_{im_a}]^t$ and $h_j = [h_{j1}, ... h_{jn_a}]^t$, where $v_{ik} \in [1, m_a]$, $h_{jl} \in [1, n_a]$, $i = 1, ..., n_{g+1}$ and $j = 1, ..., m_{g+1}$, under the following conditions:

$$\forall k = 1, ..., m_a, \quad v_{ik} < v_{(i+1)+k} \tag{3.1}$$

$$\forall l = 1, ..., n_a, \quad v_{jl} < v_{(j+1)+l} \tag{3.2}$$

Each vertical and horizontal vector can be used to separate sub-grids and to find the centers of the spots, i.e., g_{ij}.

When producing DNA microarrays, many parameters are specified, such as the number and size of spots, number of sub-grids, and even their exact locations. However, so many physicochemical factors produce noise, misalignment, and even deformations in the sub-grid template that it is extremely difficult to find the exact location of the spots after scanning, at least with the current technology. Thus, to make the gridding procedure more robust, one needs to deal with the following issues [10, 11, 12, 13]:

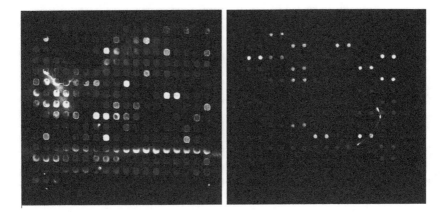

FIGURE 3.1 Irregular spots and noise-contaminated sub-grids in a typical microarray image, which make the gridding process difficult.

1. The exact locations of the sub-grids in each image can vary from slide to slide, even in the same biological experiment.
2. The relative position of the sub-grids and the distance between them can be different.
3. The sub-grids in a DNA microarray image can be of different sizes.
4. There may be some noise present in the image after scanning, or the background intensity may be too high.
5. The signal intensities may not be strong enough or uniform across the same image.
6. The size and shape of each spot may not be the same for all spots (e.g., not perfectly circular).

Figure 3.1 shows examples of two microarray image sub-grids with irregularly-shaped spots. The spots have different sizes, which makes the spot detection procedure more complicated. Moreover, the images contain a high level of noise, which affects the sub-grids during the slide scanning phase. As another example, Figure 3.2 shows a sub-grid of a DNA microarray image before and after gridding. As one can observe in this figure, the spots have been detected properly regardless of their intensity variations and noise present in the image.

3.2 OVERVIEW OF GRIDDING METHODS

The gridding process requires knowledge of the sub-girds in advance in order to proceed (i.e., the sub-gridding step). Many approaches have been proposed for sub-gridding and spot detection. Each of these approaches tries to solve the above-mentioned issues using different techniques and methods. This section provides a survey of the most recent techniques for sub-gridding and gridding. Among these

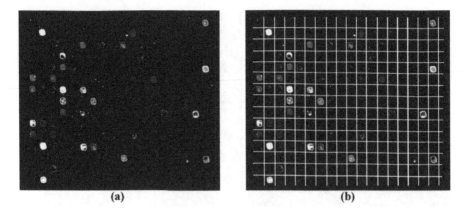

FIGURE 3.2   A sub-grid of a typical microarray image: (a) before and (b) after gridding.

methods are maximum margin and machine learning-based approaches, which are summarized at the end of this section and discussed in detail in Chapter 4. Another method that we have recently proposed is based on multi-level thresholding and is discussed in detail in Section 3.3.

3.2.1   SUB-GRIDDING

A Radon-transform-based method that separates the sub-grids in a DNA microarray image has been proposed in [14]. This method applies the Radon transform to find possible rotations of the image and then finds the sub-grids by smoothing the row or column sums of pixel intensities. However, the method does not automatically find the correct number of sub-grids, and the process is subject to data-dependent parameters.

Another approach for DNA microarray sub-gridding is a method that performs a series of steps including rotation detection and compares the row or column sums of the top-most and bottom-most parts of the image [15, 16]. This method, which detects rotation angles with respect to one of the axes, either x or y, has not been tested on images having regions with high noise (e.g., when the bottom-most $\frac{1}{3}$ of the image is very noisy).

3.2.2   THE MARKOV RANDOM FIELD

The Markov random field (MRF) is a well-known approach that involves different constraints and heuristic criteria [17]. MRFs are a class of statistical models that describe contextual constraints. As a generalization of the Markov chain model, the MRF is a set of random variables that satisfy a Markov property described by an undirected graph. The MRF shares some similarities with the Bayesian network in its representation of dependencies. There are also some differences between MRFs and Bayesian networks. The MRF has a directed and acyclic graph representation, while the Bayesian network is represented as an undirected graph. Thus, the MRF can represent some dependencies that the Bayesian network cannot (e.g., cyclic dependencies) and vice versa (e.g., induced dependencies) [18]. In [17], the MRF has been used to perform high level microarray image gridding. The input parameters of the model are the numbers of sub-grids in each row and column, as well as the number of spots in each row and column of sub-grids. The first step consists of removing contaminated noise that involves positive intensities only [17]. Thus, the thresholds already collected from locally weighted histograms at different quantile levels are used to obtain connected sets of pixels (regions), whose intensities are above some threshold. Then, the nearest neighbor distances of the regions are computed. In this way, a sharp peak corresponding to each region will be present in the distance histogram, which is used to select suitable quantile levels for thresholding.

Antoniol et al. [6] used an MRF for detecting the spots in DNA microarray images. The first step of the algorithm consists of locating the spots in the image via orientation matching transform (OMT). The second step defines a grid that interpolates the identified spots using a stochastic hill-climbing search combined with a genetic algorithm.

3.2.3 MATHEMATICAL MORPHOLOGY

Mathematical morphology is a technique used for analysis and processing geometric structures. It is based on set theory, topology, and random functions. The aim of this technique is to help remove peaks and ridges from the topological surface of the image. In [1], it has been used for gridding DNA microarray images. A gray scale image can be represented as a function $f : D_f \to G$ in such a way that $D_f \subset \mathbb{Z}^2$ and $G = g_{min}, ..., g_{max}$ is an ordered set of gray levels where $G \subset \mathbb{Z}$. From a morphological point of view, the image can be seen as a topological surface in which each pixel in the image is translated into a location and the gray level of that pixel shows the altitude of the point in the surface. Thus, the brighter the pixel (or the lower the gray level) is, the higher the altitude of that point in the surface is. In this case, applying a sequence of morphological operators (i.e., erosion and dilation [19]) helps locate the positions of the spots. Figure 3.3 shows a typical sub-grid obtained after gridding and segmentation using mathematical morphology. Some of the spots with non-uniform intensities are segmented and isolated from the background very well, while some other spots, especially those with lower intensity, have been left unidentified.

Jain's [20], Katzer's [21], and Stienfath's [22] models are integrated systems for microarray gridding and quantitative analysis. In [20], a five step procedure was applied to determine the intensity of each spot by:

1. Estimating the space between sub-grids and spots.
2. Locating the positions of the sub-grids.
3. Locating the position of each spot within the sub-grid.
4. Identifying foreground and background pixels corresponding to each spot.
5. Computing the ratio and other statistical measures.

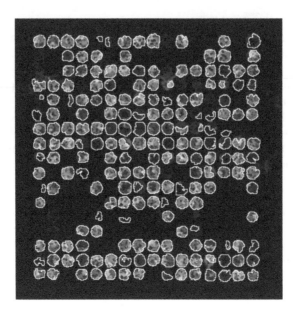

FIGURE 3.3 A segmented sub-grid for spot detection using mathematical morphology (From [1], Copyright 2003 Oxford University Press).

Locating the position of the sub-grids and spots is carried out by using the running sums of signal intensities in x and y directions. Also, the foreground and background pixels in each spot are distinguished using a threshold defined by the user.

3.2.4 THE BAYESIAN MODEL

A method for detecting spot locations based on a Bayesian model has been proposed in [2]. It uses a deformable template to fit the grid of spots using a posterior probability model in which the parameters are learned by a simulated-annealing-based algorithm. The method involves two main steps:

1. Use orientation matching and Radon transform to generate a preliminary grid.
2. Adjust the grid nodes based on local spot deformations.

Orientation matching transform [23] is an extension of the Hough transform [24], which has several advantages over the Hough transform for detecting circular shapes, such as applying it on a wide range of radii. The Radon transform is an integral-based transformation that maps the original source onto its projection space [25, 26].

Figure 3.4 shows an example of detected spots in a sub-grid using the grid-matching approach. There may be some misalignments between the actual sub-grid and the matched grid, which should be resolved before measuring the intensity level of each spot.

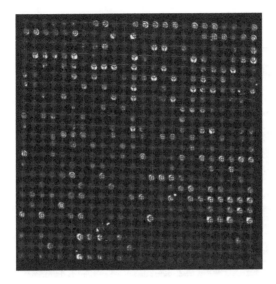

FIGURE 3.4 Detected spots in a typical sub-grid using the grid matching approach (From [2], Copyright 2006 IEEE) .

3.2.5 THE HILL-CLIMBING APPROACH

Another method for finding spot locations uses a hill-climbing approach to maximize the energy, seen as the intensities of the spots, which are fit to different probabilistic models [5]. The method finds the spots in a two-step procedure. First, the sizes of the spots are estimated by computing local energies of the pixels using a parametric probabilistic distribution. This distribution is selected from one of the three suggested: uniform, exponential, and normal distributions. Then, based on the estimated size of the spots from the previous step, the locations of the spots are found using a hill-climbing algorithm applied to the projection of the image in x and y directions and finding the peaks within the projected curve. Figure 3.5 shows an example of the spots found in a sub-grid using the hill-climbing approach. The spots within the sub-grid have been identified accurately despite the variations in the intensity levels.

3.2.6 GAUSSIAN MIXTURES

Fitting the image to a mixture of Gaussians is another technique that has been applied to grid microarray images by considering radial and perspective distortions [8]. In this technique, a grid recognition step is followed by a procedure used to eliminate nonlinear distortions caused by a possible perspective transformation.

A grid can be defined as a weighted combination of two linear vectors as follows:

$$G = \{av_1, bv_2 \mid a, b \in \mathbb{Z}, v_1, v_2 \in \mathbb{R}^2\}. \tag{3.3}$$

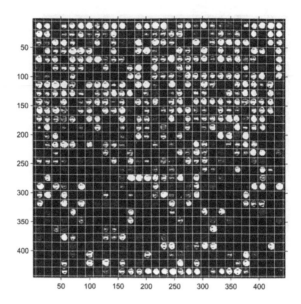

FIGURE 3.5   Detected spots in a sub-grid of a DNA microarray image using the hill-climbing approach.

A Voronoi cell, V_G, corresponding to grid G, is the set of points that are closer to the origin than any other point in the grid. More formally:

$$V_G = \{x \mid \| x \| \leq \| x - p \|, x \in \mathbb{R}^2, \forall p \in G\}. \tag{3.4}$$

In the two-dimensional space, the Voronoi cell of a grid is determined using the equidistant lines between the specific point and its eight nearest neighbors. The corresponding polygon, which is formed using all of such equidistant lines around the origin, is the Voronoi cell. Using these cells, the minimum generating matrix (MGM) is defined as a matrix that contains the elements corresponding to the x and y axes that have minimum norm. The MGM is estimated using the Gaussian mixture model (GMM) clustering algorithm. In the second step, the proposed median of infinite lines from local structures method is used to eliminate nonlinear distortions caused by perspective transformations.

3.2.7   GENETIC ALGORITHMS

Another method for gridding DNA microarray images uses an evolutionary algorithm to separate sub-grids and detect the positions of the spots [3]. The approach is based on a genetic algorithm that discovers parallel and equidistant line segments, which constitute the grid structure. In this approach, finding the separating lines between the spots is considered as an optimization problem.

The chromosomes are encoded as triplets containing three real values: the first two values are the y coordinates, which correspond to the end points of one line segment, and the third value is the distance between two adjacent line segments.

A separating line should be contained in an empty area between adjacent spots. Considering that the pixels of that area are counted as part of the background and their intensities are usually lower than the intensities of the pixels corresponding to the spots, the probability of line L_i, $P(L_i)$, is the portion of the grid given by the following equation:

$$P(L_i) = f_B^{R_{L_i}} L_i - f_S^{R_{L_i}} L_i. \tag{3.5}$$

Here, R_{L_i} denotes the region of image G in such a way that the pixels in that region have a distance less than a margin w from the line-segment L_i. Function $f_B^{R_{L_i}} L_i$ returns the percentage of pixels within region R_{L_i} in such a way that their intensity is lower than a value I_B. Similarly, function $f_S^{R_{L_i}} L_i$ returns the percentage of pixels within region R_{L_i} in such a way that their intensities are higher than a threshold I_B. I_B is defined as the most frequent value present in any pixel of image G. All pixels whose intensities are below I_B belong to the background.

Then, the fitness function for this optimization problem, $F(m)$, is defined as follows:

$$F(m) = \begin{cases} S_p(m) \times N(m), & \text{if } f_{LS}(m) \leq f_{max} \\ S_p(m), & \text{otherwise}, \end{cases} \tag{3.6}$$

where $S_p(m) = \Sigma_{i=1}^{N(m)} P(L_i)$. Also, function $f_{LS}(m)$ is the percentage of line segments $L_i, i = 1, \ldots, N(m)$, which are represented by chromosome m and have a low probability, $P(L_i)$, to be part of the grid.

Thereafter, a refinement procedure is applied to further improve the grid structure, by slightly modifying the line segments. Figure 3.6 shows an example of the grids found in a sub-grid using the genetic algorithm approach. One of the drawbacks in this method is that the accuracy of the method depends on the parameters of the genetic algorithm, which have to be set manually by the user.

3.2.8 MAXIMUM MARGIN

Maximum margin is another method for automatic gridding of DNA microarray images based on maximizing the margin between rows and columns of spots [27]. Initially, a set of grid lines is placed on the image in order to separate each pair of consecutive rows and columns of the selected spots. Then, the optimal positions of the lines are obtained by maximizing the margin between these rows and columns using a maximum margin linear classifier. For this, a gridding method that uses a support vector machine (SVM) was used in [28]. In this method, the positions of the spots on a DNA microarray image are first detected using image analysis operations. Then, a set of soft-margin linear SVM classifiers is used to find the optimal layout of the grid lines in the image. Each grid line corresponds to the separating line produced

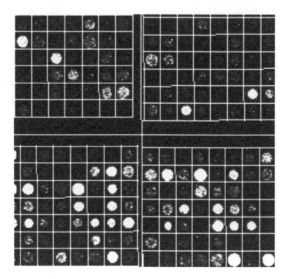

FIGURE 3.6 Detected sub-grids and spots using the genetic algorithm approach (From [3], Copyright 2008 IEEE).

by one of the SVM classifiers, which maximizes the margin between two consecutive rows or columns of spots. In [29], a series of correlation and convolution operations are used to identify sub-grids and spots in a given microarray image. This method, which has been designed based on the *ImageJ* image processing program [30], uses a two-step approach for locating the sub-grids in a microarray image in the first step, and identifying spots within each sub-grid in the next step. More details about the maximum margin method are given in Chapter 4.

3.3 OPTIMAL MULTILEVEL THRESHOLDING GRIDDING

Multi-level thresholding is another technique that is widely used in image processing for segmentation, and has been successfully applied to sub-gridding and gridding of DNA microarray images. There are a few methods that can use thresholding without input parameters. One of these methods is optimal multi-level thresholding gridding (OMTG) [4], which is an iterative approach to find the gridding for every possible number of thresholds, and then evaluate the objective function with an index of validity to find the best number of thresholds. This method has been shown to work very well for a variety of DNA microarray images with different configurations, while being almost free of parameters. A variant of the OMTG approach has been recently proposed in [31]. In this section, we discuss the original OMTG method as reported in [4].

To find the thresholds in a typical DNA microarray image, OMTG works as follows. The Radon transform is used as a pre-processing phase. Then, using *optimal*

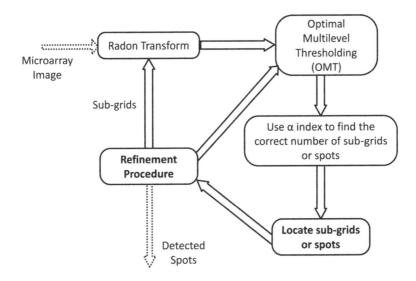

FIGURE 3.7 Schematic visualization of the OMTG process for finding sub-grids or spots in a DNA microarray image or a sub-grid.

multilevel thresholding (OMT) and an index of validity, the correct number of thresholds (or sub-grids) can be found. The same process is used to find the correct number of spots in each sub-grid. Figure 3.7 depicts the process of finding the sub-grids in a microarray image and the spots in a sub-grid. The input to the Radon transform is a DNA microarray image and the output of the whole process is the location (and partitioning) of the sub-grids. Similarly, the locations of the spots in each sub-grid are found by using OMT combined with the index to find the best number of rows and columns of spots. The input to this process is a sub-grid (already extracted from the sub-gridding step) and the output is the partitioning of the sub-grids into spots (or spot regions).

3.3.1 ROTATION ADJUSTMENT

In OMTG, rotations of the images are seen in two different directions, with respect to the x and y axes. To find two independent angles of rotation for an affine transformation, the Radon transform is applied first. Given an image $A = \{a_{x,y}\}$, the Radon transform performs the following transformation:

$$R(p,t) = \int_{-\infty}^{\infty} a_{x,t+px} dx,$$ (3.7)

where p is the slope and t its intercept. The rotation angle of the image with respect to the slope p is given by $\phi = \arctan p$. $R(\phi,t)$ denotes the Radon transform of image

A. Each rotation angle ϕ gives a different one-dimensional function, and the aim is to obtain the angle that gives the best alignment with the lines, which will occur when the lines are *parallel* to the y-axis. The best alignment will occur then at the angle ϕ_{min} that minimizes the *entropy* as follows [6]:

$$H(\phi) = -\sum_{t=-\infty}^{\infty} R'(\phi,t)\log R'(\phi,t)dt. \tag{3.8}$$

$R(\phi,t)$ is normalized into $R'(\phi,t)$, such that $\sum_t R'(\phi,t) = 1$. The positions of the pixels in the new image, $[u\ v]$, are obtained as follows:

$$[u\ v] = [x\ y]\begin{bmatrix} \cos\phi_{min_x} & \sin\phi_{min_y} \\ -\sin\phi_{min_y} & \cos\phi_{min_x} \end{bmatrix}, \tag{3.9}$$

where ϕ_{min_x} and ϕ_{min_y} are the best angles of rotation found by $R'(\phi,t)$.

3.3.2 OPTIMAL MULTILEVEL THRESHOLDING

Given a histogram that represents the frequencies of the different levels of intensities a pixel can take, the aim of thresholding is to find meaningful groups of intensity levels that can represent the image in a simplified way. When more than two groups are being considered, the problem can be solved by OMT, which aims to find an optimal partitioning of the histogram into a number of well-defined groups. In the microarray image context, the underlying histograms are obtained by computing the row and column sums of pixel intensities, obtaining discrete one dimensional functions, where the domain is given by the positions of the rows/columns of pixels. The number of bins in a histogram is given by the number of different intensity levels. The values in the histogram can be normalized to be considered as probabilities of the corresponding bins.

Before we show some examples, we note that in the rest of this chapter we use sample images from three datasets. The first dataset consists of a set of images drawn from the Stanford Microarray Database (SMD) [32], and corresponds to a study of the global transcriptional factors for hormone treatment of *Arabidopsis thaliana* samples. The second dataset consists of a set of images from Gene Expression Omnibus (GEO) [33], and corresponds to an atlantic salmon head kidney study. The third dataset consists of two images, obtained from a dilution experiment (DILN) [34]. This experiment involves a set of serially diluted oligonucleotide samples arrayed on a slide of 8×8 patches. Each column within the patch is a replicate of serial dilutions at 8 different levels.

Figure 3.8 shows a typical DNA microarray image (AT-20387-ch2) from the SMD dataset that contains 12×4 sub-grids, along with the corresponding row or column sums (histograms). These histograms are used to detect the sub-grids in the whole image. Similarly, Figure 3.9 shows one of the sub-grids obtained from AT-20387-ch2 along with the corresponding histograms for the sums of rows and columns. Each row or column sum is then processed as discussed in this section in order to obtain the optimal multilevel thresholding that is used to determine the sub-grids and the locations of the spots in each sub-grid.

More formally, OMT works as follows. Consider a histogram H, an ordered set $\{1, 2, \ldots, n-1, n\}$, where the ith value corresponds to the ith bin and has a probability, p_i. Given a microarray image (or a sub-grid), $A = \{a_{ij}\}$, H can be obtained by means of the horizontal (vertical) sum as follows: $p_i = \sum_{j=1}^{mR} a_{ij}$ ($p_j = \sum_{i=1}^{n} a_{ij}$). A threshold set T is defined as an ordered set $T = \{t_0, t_1, \ldots, t_k, t_{k+1}\}$, where $0 = t_0 < t_1 < \ldots < t_k < t_{k+1} = n$ and $t_i \in \{0\} \cup H$. The problem of multilevel thresholding consists of finding a threshold set, T^*, in such a way that a function $f : H^k \times [0, 1]^n \to R^+$ is maximized/minimized. Using this threshold set, H is divided into $k+1$ classes: $\zeta_1 = \{1, 2, \ldots, t_1\}$, $\zeta_2 = \{t_1 + 1, t_1 + 2, \ldots, t_2\}$, $\ldots$, $\zeta_k = \{t_{k-1} + 1, t_{k-1} + 2, \ldots, t_k\}$, $\zeta_{k+1} = \{t_k + 1, t_k + 2, \ldots, n\}$.

Various parametric and non-parametric thresholding methods and criteria have been proposed. The three most important criteria are Otsu's method, which aims to maximize the separability of the classes measured by means of the sum of between-class variances [35], the criterion that uses information theoretic measures in order to maximize the separability of the classes [36], and the minimum error criterion [37].

The between-class variance criterion consists of maximizing the following function of T:

$$\Psi_{BC}(T) = \sum_{j=1}^{k+1} \omega_j \mu_j^2, \tag{3.10}$$

where $\omega_j = \sum_{i=t_{j-1}+1}^{t_j} p_i$, $\mu_j = \frac{1}{\omega_j} \sum_{i=t_{j-1}+1}^{t_j} i p_i$.

To implement the between-class variance criterion, $\Psi_{BC}(T)$ is expressed as follows:

$$\Psi_{BC}(T) = \sum_{j=1}^{k+1} \omega_j \mu_j^2 = \sum_{j=1}^{k+1} \Psi_{t_{j-1}+1, t_j}, \tag{3.11}$$

where $\Psi_{t_j+1, t_{j+1}} = \omega_j \mu_j^2$. Temporary variables a and b can be computed as follows:

$$a \leftarrow p_{t_{j-1}+1} + \sum_{i=t_{j-1}+2}^{t_j} p_i, \qquad \text{and} \tag{3.12}$$

$$b \leftarrow (t_{j-1}+1) p_{t_{j-1}+1} + \sum_{i=t_{j-1}+2}^{t_j} i p_i. \tag{3.13}$$

Since from (3.12) and (3.13), a and b are known, then for the next step, $\Psi_{t_{j-1}+2, t_j}$ can be re-computed as follows in $\Theta(1)$ time:

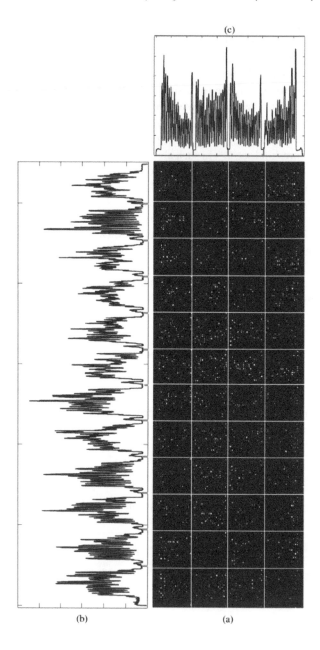

FIGURE 3.8 Sub-grid detection in microarray image AT-20387-ch2 from the SMD dataset: (a) detected sub-grids, (b) horizontal histogram and detected valleys corresponding to horizontal lines in the sub-gridding, and (c) vertical histogram and detected valleys corresponding to vertical lines in the sub-gridding.

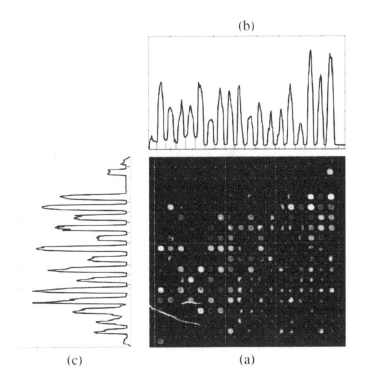

FIGURE 3.9 Spot detection in the sub-grid located in row two and column four of image AT-20387-ch2 from the SMD dataset: (a) detected spots in the sub-grid, (b) horizontal histogram and detected valleys corresponding to horizontal lines in the grid, and (c) vertical histogram and detected valleys corresponding to vertical lines in the grid.

$$a \quad \leftarrow \quad a - p_{t_{j-1}+1}, \tag{3.14}$$

$$b \quad \leftarrow \quad b - (t_{j-1} + 1)p_{t_{j-1}+1}, \text{ and} \tag{3.15}$$

$$\psi_{t_{j-1}+2, t_j} \quad \leftarrow \quad \frac{b^2}{a}. \tag{3.16}$$

The implementation of the OMT algorithm for the between-class criterion is given in Algorithm 3.1. The main algorithm invokes procedure findThresholdRanges(k) to obtain the maximum and minimum ranges for each threshold. The algorithm uses table C to store the solution to smaller values of k. The table is incrementally filled by column, for $k = 1, 2, 3$, and so on. Similarly, table D stores the thresholds for all the optimal solutions. Thus, $D(t_j, j)$ contains the value of t_{j-1} for which $\Psi^*(T_{0,j})$ is optimal. The variable "psi" represents the function $\psi_{t_{j-1}+2, t_j}$, computed as in (3.16). For this, the values of a and b are recomputed as in (3.14) and (3.15). The index i in the main for loop is used to represent t_{j-1}. Three statements are marked with "(*),", which are also used in the entropy and minimum error criteria. By following the complexity analysis for Algorithm 3.1, it is not difficult to see that it runs in $\Theta(kn^2)$ worst-case time. Note that computing the initial values of a and b requires $\Theta(n)$ time; however, these computations are outside the innermost loop.

Another criterion that can be used in OMT for microarray image gridding is the entropy-based criterion, which aims to maximize the following function of T:

$$\Psi_H(T) = \sum_{j=1}^{k+1} H_j, \tag{3.17}$$

where $H_j = \psi_{t_{j-1}+1,t_j} = \sum_{i=t_{j-1}+1}^{t_j} \frac{p_i}{w_j} \log \frac{p_i}{w_j}$.

For the implementation, the entropy-based criterion, $\Psi_H(T)$, can be expressed as follows:

$$\Psi_H(T) = \sum_{j=1}^{k+1} H_j = \psi_{t_{j-1}+1,t_j} = \sum_{i=t_{j-1}+1}^{t_j} \frac{p_i}{w_j} \log \frac{p_i}{w_j}. \tag{3.18}$$

Then, by calculating a, b, and c as follows:

$$a \leftarrow p_{t_{j-1}+1} + \sum_{i=t_{j-1}+2}^{t_j} p_i, \tag{3.19}$$

$$b \leftarrow p_{t_{j-1}+1} \log p_{t_{j-1}+1} + \sum_{i=t_{j-1}+2}^{t_j} p_i \log p_i, \quad \text{and} \tag{3.20}$$

$$c \leftarrow (\log a)p_{t_{j-1}+1} + \log a \sum_{i=t_{j-1}+2}^{t_j} p_i. \tag{3.21}$$

It is noticeable that from (3.19), (3.20), and (3.21), a, b, and c are known, and hence $\psi_{t_{j-1}+2,t_j}$ can be computed as follows in $\Theta(1)$ time:

$$a \leftarrow a - p_{t_{j-1}+1}, \tag{3.22}$$

$$b \leftarrow b - (t_{j-1}+1) \log p_{t_{j-1}+1}, \tag{3.23}$$

$$c \leftarrow c - (\log a)p_{t_{j-1}+1}, \text{and} \tag{3.24}$$

$$\psi_{t_{j-1}+2,t_j} \leftarrow -\frac{1}{a}(b-c). \tag{3.25}$$

The multi-level thresholding algorithm for the entropy-based criterion can be implemented by modifying the algorithm for the between-class criterion. This is achieved by changing the three statements marked with "(*)" in which the computations of a and b marked with $(*^1)$ are substituted for those of a, b, and c computed as in (3.19), (3.20), and (3.21). In addition, the computation of a and b marked with $(*^4)$ are substituted by those of a, b, and c computed as in (3.22), (3.23), and (3.24),

Algorithm 3.1 Multilevel Thresholding between Class Variance

Input: Probabilities, $P = \{p_1, p_2, \ldots, p_n\}$. Number of thresholds, k.
Output: A threshold set, $T = \{t_0, t_1, t_2, \ldots, t_k, t_{k+1}\}$.

  minTj, maxTj $\leftarrow$ findThresholdRanges(k)
  // Fill columns 1 to $k+1$
  $C(0,0) \leftarrow 0; D(0,0) \leftarrow 0$
  for $j \leftarrow 1$ **to** $k+1$ **do**
    for $t_j \leftarrow$ minTj(j) **to** maxTj(j) **do**
      $C(t_j, j) \leftarrow 0; \quad a \leftarrow \sum_{i=j}^{t_j} p_i; \quad b \leftarrow \sum_{i=j}^{t_j} i p_i$ $(*^1)$
      for $i \leftarrow$ minTj$(j-1)$ **to** $\min\{$maxTj$(j-1), t_j - 1\}$ **do**
        psi $\leftarrow \frac{b^2}{a}$ $(*^2)$
        if $C(i, j-1) +$ psi $> C(t_j, j)$ **then**
          $C(t_j, j) \leftarrow C(i, j-1) +$ psi
          $D(t_j, j) \leftarrow i$
        end if
        $a \leftarrow a - p_{i+1}; \quad b \leftarrow b - (i+1) p_{i+1}$ $(*^4)$
      end for
    end for
  end for
  return findThresholds(D)
  procedure findThresholdRanges(k: integer)
  for $j \leftarrow 0$ **to** $k+1$ **do**
    if $j = k+1$ **then**
      minTj$(j) \leftarrow n$
    else
      minTj$(j) \leftarrow j$
    end if
    if $j = 0$ **then**
      maxTj$(j) \leftarrow 0$
    else
      maxTj$(j) \leftarrow n - k + j - 1$
    end if
  end for
  return minTj, maxTj
  end procedure
  procedure findThresholds(D: table)
  $T(k+1) \leftarrow n$
  for $j \leftarrow k$ **downto** 0 **do**
    $T(j) \leftarrow D(T(j+1), j+1)$
  end for
  return T
  end procedure

while the computation of ψ marked with $(*^2)$ has to be substituted by that of Formula (3.25). Since computing ψ at each step can be done in $\Theta(1)$ time, the worst-case time complexity of the algorithm is $\Theta(kn^2)$.

The minimum error criterion aims to minimize the following function of T:

$$\Psi_{\text{ME}}(T) = \psi_{t_{j-1}+1,t_j}, \tag{3.26}$$

where $\psi_{t_{j-1}+1,t_j} = \sum_{j=1}^{k+1} \omega_j(\log \sigma_j + \log \omega_j)$.

As in the other criteria, by setting a, b and c to:

$$a \;\leftarrow\; p_{t_{j-1}+1} + \sum_{i=t_{j-1}+2}^{t_j} p_i, \tag{3.27}$$

$$b \;\leftarrow\; (t_{j-1}+1)p_{t_{j-1}+1} + \sum_{i=t_{j-1}+2}^{t_j} ip_i, \qquad \text{and} \tag{3.28}$$

$$c \;\leftarrow\; (t_{j-1}+1)^2 p_{t_{j-1}+1} + \sum_{i=t_{j-1}+2}^{t_j} i^2 p_i, \tag{3.29}$$

$$\tag{3.30}$$

and since from (3.27), (3.28), and (3.29), a, b, and c are known, $\psi_{t_{j-1}+2,t_j}$ can be computed as follows in $\Theta(1)$ time:

$$a \;\leftarrow\; a - p_{t_{j-1}+1}, \tag{3.31}$$

$$b \;\leftarrow\; b - (t_{j-1}+1)p_{t_{j-1}+1}, \tag{3.32}$$

$$c \;\leftarrow\; c - (t_{j-1}+1)^2 p_{t_{j-1}+1}, \text{and} \tag{3.33}$$

$$\psi_{t_{j-1}+2,t_j} \;\leftarrow\; a\frac{1}{2}\log\left[\frac{c}{a} - (\frac{b}{a})^2\right] + \log a. \tag{3.34}$$

Again, the OMT algorithm for the minimum error criterion can be implemented by introducing slight modifications to the algorithm for the between-class criterion. This is achieved by substituting the computations of a and b marked with $(*^1)$ for those of a, b, and c computed as in (3.27), (3.28), and (3.29) and initializing $C(t_j, j) \leftarrow \infty$. The computation of a and b marked with $(*^4)$ are substituted by those of a, b, and c computed as in (3.31), (3.32), and (3.33), while the computation of ψ marked with $(*^2)$ has to be substituted for that of Formula (3.34). Also, in the statement marked with $(*^3)$, the comparison symbol ">" has to be replaced by "<". To conclude this sub-section, we note that for the between class criterion, the optimal can be found in $O(kn)$, by following the algorithm introduced in [38]. That algorithm has not been implemented for the entropy-based and minimum error criterion.

3.3.3 AUTOMATIC DETECTION OF THE NUMBER OF SUB-GRIDS AND SPOTS

One of the difficulties in sub-grid and spot detection is detecting the correct number of sub-grids in a DNA microarray image and the correct number of spots within each sub-grid consequently. This stage is crucial since the whole sub-gridding and gridding processes can be done automatically, if the number of sub-grids and spots are detected without user intervention. As seen in Section 3.3.2, OMT uses the number of sub-grids (spots) as a single parameter. Thus, the correct number of sub-grids (spots) has to be determined before using the method. For this, validity indices for clustering can be used. By analyzing the traditional indices for clustering validity and their suitability to be combined with our measure, an index of validity for this specific problem can be defined.

There are different types of indices of validity for clustering (cf. [39, 40]). One of the most widely used indices for clustering is the I index, which is defined as follows [39]:

$$I(K) = \left(\frac{1}{K} \times \frac{E_1}{E_K} \times D_K \right)^2, \tag{3.35}$$

where $E_K = \Sigma_{i=1}^{K} \Sigma_{k=1}^{n_i} p_k \|k - z_i\|$, $D_K = \max_{i,j=1}^{K} \|z_i - z_j\|$, n is the total number of points in the dataset (bins in the histogram), and z_k is the center of the kth cluster. We also consider the average frequency value of the thresholds in the histogram, which is computed as follows:

$$A(K) = \frac{1}{K} \sum_{i=1}^{K} p(t_i), \tag{3.36}$$

where t_i is the ith threshold found by OMT and $p(t_i)$ is the corresponding probability value in the histogram.

As a result, the α validity index is obtained from the combination of $A(K)$ and the I index as follows [4]:

$$\alpha(K) = \sqrt{K} \frac{I(K)}{A(K)} = \frac{\left(\frac{E_1}{E_K} \times D_K \right)^2}{\sqrt{K} \Sigma_{i=1}^{K} p(t_i)}. \tag{3.37}$$

For maximizing $I(K)$ and minimizing $A(K)$, the value of $\alpha(K)$ must be maximized. Thus, the best number of thresholds K^* based on the α index is given by:

$$K^* = \operatorname*{argmax}_{1 \leq K \leq \delta} \alpha(K) = \operatorname*{argmax}_{1 \leq K \leq \delta} \frac{\left(\frac{E_1}{E_K} \times D_K \right)^2}{\sqrt{K} \Sigma_{i=1}^{K} p(t_i)}. \tag{3.38}$$

To find the best number of thresholds, K^*, an exhaustive search is performed on all positive values of K from 1 to δ, obtaining the value K^* that maximizes the α index. The value of δ can be set to $\sqrt{n}$ (cf. [41]). Figure 3.10 shows plots of the index functions for AT-20387-ch2 from the SMD dataset. As can be observed in the

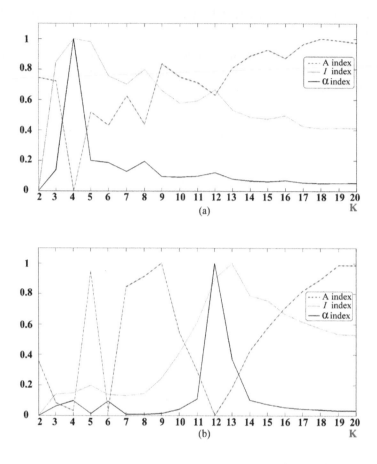

FIGURE 3.10 Plots of the index functions for AT-20387-ch2: (a) the values of the I, A, and α indices for horizontal separating lines in the sub-gridding, and (b) the values of the I, A, and α indices for vertical separating lines in the sub-gridding.

figure, finding the correct number of sub-grids using I or A index is not an easy task, while the α index clearly reveals the correct number of horizontal and vertical sub-grids by producing a semi-flat curve with a sharp peak at 4 and 12, respectively.

3.3.4   GRID REFINEMENT

As mentioned earlier, due to the alignment of blocks inside the microarray image and the arrangement of spots inside the blocks, the line segments, which have the same direction and constitute the borders of blocks (or spots), are ideally equidistant. However, this property may not always be satisfied when rotations, misalignments, and local deformations of the ideal rectangular grid exist. As a result, the obtained

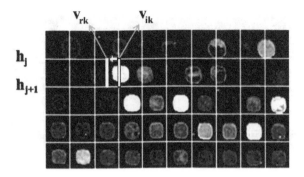

FIGURE 3.11   Example of a portion of a microarray image after applying OMTG before the refinement procedure.

line segments may slightly vary from the ideal ones.

In some cases, due to rotations, misalignments, or local deformations, some of the spots on the DNA microarray image may not be aligned with each other. Consequently, the detected grid or sub-grid may not separate those spots completely or may separate them marginally. In this case, a refinement procedure can be used to boost the performance of method. For this, each segment of horizontal or vertical lines can be moved to left or right (for vertical lines) and up or down (for horizontal lines) to find the best location separating the spots. Consider two horizontal lines h_j and h_{j+1} where $j \in [1, K^*]$ and a vertical line v_i where $i \in [1, K^*]$, and v_{ij} is bounded between h_j and h_{j+1}. Given $A = \{a_{ij}\}$, line v_{ik} can be moved left and right in such a way that $\Sigma_{i=h_j}^{h_{j+1}} a_{ik}$ is minimized. In other words, the vertical line v_{ik} can be replaced with a new vertical line, v_{rk}, in such a way that:

$$r = \operatorname*{argmin}_{v_{i-1} \leq k \leq v_{i+1}} \Sigma_{i=h_j}^{h_{j+1}} a_{ik}. \qquad (3.39)$$

Analogously, this procedure can be applied to each horizontal line. Figure 3.11 shows an example in which a vertical line is replaced by a new one during the refinement procedure. As shown in the figure, the vertical line v_{ik} is originally placed on top of a spot and does not separate two adjacent spots correctly. By moving it left using the refinement procedure, the new line v_{rk} is found in such a way that those adjacent spots are separated correctly.

As a second example of applying the refinement procedure, Figure 3.12 shows the detected spots in one of the sub-grids of 20387-ch2 of SMD before and after applying (3.39). It is clear that there are some misalignments in separating adjacent spots in the top part of the sub-grid before using the refinement procedure. After applying the refinement procedure, all the spots are separated precisely as shown in the figure.

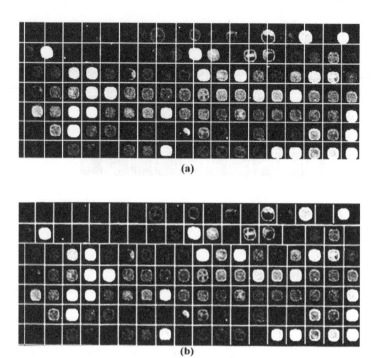

(a)

(b)

FIGURE 3.12 Result of the refinement procedure used to increase the accuracy of the gridding method: (a) detected spots in one of the sub-grids of AT-20387-ch1 from the SMD dataset before using the refinement procedure, and (b) detected spots in the same part of the sub-grid after applying the refinement procedure.

3.4 ANALYSIS OF GRIDDING METHODS

Each of the gridding methods presented in Sections 3.2 and 3.3 has its own features and uses its own motivation to detect the sub-grids and spots in a DNA microarray image. In addition, each method has its own parameters to handle various images with different characteristics. Some of these parameters need to be adjusted by the user prior to running the algorithm.

3.4.1 CONCEPTUAL COMPARISON

A conceptual comparison among micoarray image gridding methods based on their features is shown in Table 3.1. The methods included in the comparison are the following: (i) Radon transform sub-gridding (RTSG) [14], (ii) Bayesian simulated annealing gridding (BSAG) [2], (iii) genetic-algorithm-based gridding (GABG) [3], (iv) hill-climbing gridding (HCG) [5], (v) maximum margin microarray gridding (M^3G) [27], and optimal multi-level thresholding gridding (OMTG) [4].

TABLE 3.1

Conceptual comparison of gridding and sub-gridding methods based on the required number and type of input parameters and features.

Method	Parameters	Sub-grid Detection	Spot Detection	Automatic Detection No. of Spots	Rotation
RTSG	n: Number of sub-grids	√	×	×	√
BSAG	α, β: Parameters for balancing prior and posterior probability rates	×	√	√	√
GABG	μ, c: Mutation and Crossover rate, p_{max}: probability of maximum threshold, p_{low}: probability of minimum threshold, f_{max}: percentage of line with low probability to be a part of grid, T_p: Refinement threshold	√	√	√	√
HCG	λ, σ: Distribution parameters	×	√	√	×
M^3G	c: Cost parameter	×	√	√	√
OMTG	δ: maximum number of thresholds	√	√	√	√

As shown in the table, all these methods use one or more parameters to adjust their algorithms to the input image. Using more parameters can decrease the flexibility of the method, since these parameters are needed to be adjusted carefully based on the features of each microarray image before running the gridding algorithm. For example, GABG needs to set several parameters such as the mutation rate, μ, the crossover rate, c, the maximum threshold probability, p_{max}, the minimum threshold probability, p_{low}, the percentage of lines with low probability to be part of the grid, f_{max}, and the refinement threshold, T_p. Also, HCG needs to set some parameters such as λ and σ. OMTG only needs one parameter, λ, to be adjusted by the user.

TABLE 3.2

Results of the comparison of OMTG, GABG, and HCG proposed in [4], [5], and [3] respectively.

Dataset	Method	Incorrectly	Marginally	Perfectly
SMD	OMTG	1.72%	0.22%	98.06%
	GABG	5.37%	0.51%	94.12%
	HCG	2.12%	1.23%	96.65%
GEO	OMTG	0.58%	0.16%	99.26%
	GABG	4.49%	0.32%	95.19%
	HCG	2.55%	0.74%	96.71%
DILN	OMTG	1.97%	0.08%	97.95%
	GABG	4.35%	0.34%	95.31%
	HCG	3.78%	0.65%	95.57%

3.4.2   EXPERIMENTAL COMPARISON

To show how gridding methods work from another perspective, three examples of using *OMTG* are shown in Figure 3.13. The figure shows the detected sub-grids in image AT-20387-ch2 (a) and the detected spots in one of the sub-grids (b). Also, Figure 3.14 shows the sub-grids detected in image Diln4-3.3942B (a) and the detected spots in one of the sub-grids (b). As shown in the figures, in the first step, the sub-grids within the DNA microarray image are located. Then, in the second step, the spots within each sub-grid are found using the same algorithm. Although the size and number of sub-grids and spots are different, OMTG effectively finds the boundaries between the sub-grids and spots.

An experimental comparison of OMTG, GABG, and HCG is shown in Table 3.2. Since GABG and HCG use several parameters, to obtain best results for the SMD, GEO, and DILN datasets, the parameters have been adjusted manually. Without this, good results may not be achieved for a particular dataset. This makes these methods not fully independent from datasets that they process, and some user intervention is required to adjust the parameters. As shown in Table 3.2, in all three datasets, OMTG obtains the highest accuracy and HCG and GABG are placed in the second and third positions, respectively. After adjusting the parameters, HCG and GABG work very well, though.

3.4.3   BIOLOGICAL ANALYSIS

One of the most important aspects in evaluating the performance of a microarray image gridding method, and in general, for a microarray data analysis technique, is to observe the biological meaning of the results. For this, a microarray dilution

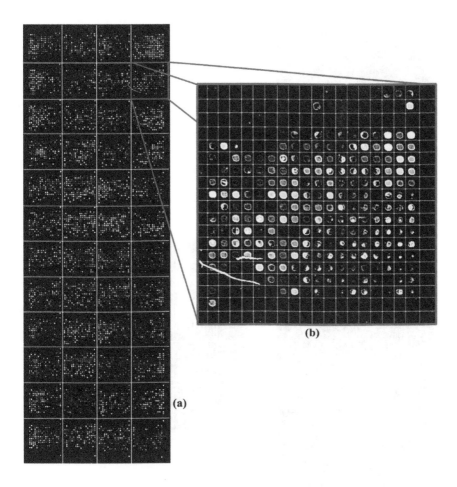

FIGURE 3.13 (a) Result of applying OMTG on image AT-20387-ch2 from the SMD dataset, and (b) detected spots in one of the sub-grids of AT-20387-ch2.

experiment that was done previously [34], the DILN dataset, is considered. The two images in the DILN dataset are considered by applying OMTG, followed by typical segmentation and quantification procedures. The two sub-grids and spots for the two images were detected using OMTG. In the next step, the *Sobel* method, which is a discrete differentiation operator, is used to detect the edge of each spot and then the region within the edge is defined as the region of each spot. As the next step, a set of opening and closing operations (using morphological dilation and erosion operators) are used to decrease the noise and artifacts in the region corresponding to each spot. Finally, the summation of all pixel intensities in the spot are considered as the *volume* of that spot which represents the expression level of the gene associated

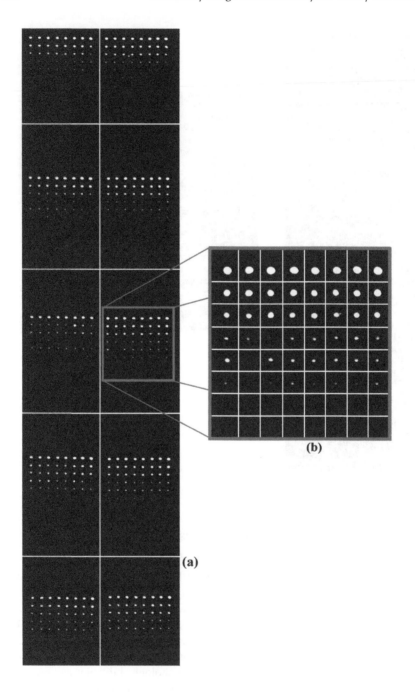

FIGURE 3.14 Gridding and sub-gridding of image Diln4-3.3942B from the Diln dataset: (a) detected sub-grids, and (b) detected spots in one of the sub-grids of Diln4-3.3942B.

TABLE 3.3

Logs of volume intensities of each dilution step for images A and B from the DILN dataset.

Dilution steps	Diln4-3.3942.01A	Diln4-3.3942.01B
1	22.02	21.75
2	20.63	20.78
3	19.75	19.94
4	18.12	18.05
5	17.98	18.25
6	16.98	17.03
7	16.18	16.17
8	15.07	15.46

with that spot.

Table 3.3 shows the volume intensity of each dilution step for images Diln4-3.3942.01A and Diln4-3.3942.01B respectively. As shown in the table, the OMTG method estimates the intensities of the dilution steps very well with near linearly decreasing steps. Figure 3.15 shows log plots of the dilution steps. The mean volume for all intensities is shown as a thick line. Also, the dashed line with slope −1 is shown as a reference line. An important aspect to note is the source of irregularity in the linearity of the thick curve as shown in Table 3.3 and Figure 3.15. As pointed out by the authors of these experiments [34], the intensities of the spots in step 4 are smaller than those of step 5. An example of this can be observed in the third and last rows of the sub-grids in Figure 3.16. As shown in Figure 3.16(b), this decrease in the intensities of the spots causes a slight nonlinearity in step 4 of the dilution steps. However, the gridding method is able to capture the nonlinear relationships present in the dilution experiments. This behavior is observed in the log plots of Figure 3.15, since the dashed line follows the thick line, which is the mean volume of all intensities in the log scale.

3.5 CONCLUSION AND DISCUSSION

This chapter presents a discussion of the DNA microarray image gridding problem. Different methods for separating sub-grids and spot centers in DNA microarray images are reviewed. These methods use different techniques to find the location of each sub-grid in the DNA microarray image and find spot locations within each sub-grid using a two step procedure. Dealing with different microarray images with various numbers of sub-grids and different spot sizes makes the gridding problem rather complex in microarray image analysis. Some of the methods use mathematical and geometrical models such as the Bayesian model or the Markov random field to gener-

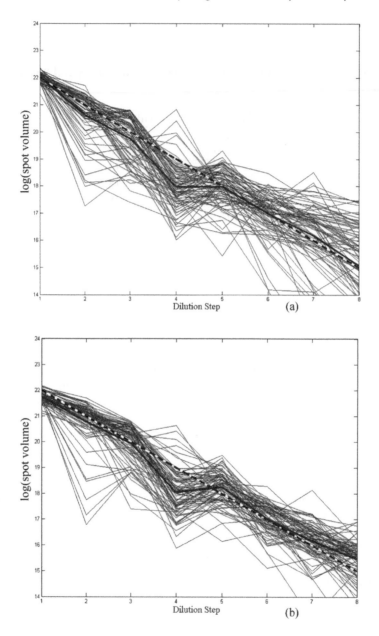

FIGURE 3.15 Relationship between logs of spot volumes and dilution steps. The y axis represents the logs of spot volumes that correspond to the dilution steps in (a) Diln4-3.3942.01A and (b) Diln4-3.3942.01B. The thick lines show the average of logs of spot volumes in different dilution steps. The dashed line corresponds to the reference line with slope $= -1$.

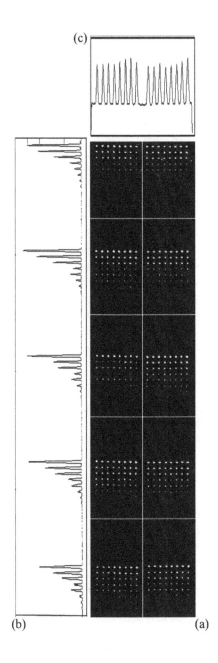

FIGURE 3.16 Result of applying OMTG on image Diln4-3.3942.01A from the Diln dataset: (a) detected sub-grids, (b) vertical histogram and detected valleys corresponding to horizontal lines, and (c) horizontal histogram and detected valleys corresponding to vertical lines.

ate the corresponding grid for each sub-grid, while other methods use sophisticated mechanisms such as a genetic algorithm or the hill-climbing approach to identify spot locations.

Most of these methods use a few parameters that the user can adjust to obtain the best results for each dataset. While this flexibility may be a benefit in some cases, generally, most methods depend on a particular dataset. This makes these models rather difficult to be used as a general method for gridding microarray images regardless of the specific properties of each microarray image. OMTG is among the few methods that uses almost no parameters and is able to perform gridding on various DNA microarray images, achieving a high accuracy and without setting parameters individually for each image or dataset. The M^3G method also performs the gridding accurately even in the presence of noise and artifacts, while taking into account rotations in the input image.

While most of the methods presented in this chapter perform very well, there are several aspects that can be improved, and which open various avenues for future research. Using more sophisticated methods with a minimum number of parameters can make the gridding phase even more accurate and flexible. Some of the modules in these methods can be replaced with more sophisticated and efficient algorithms that can boost the overall performance and accuracy of the gridding step and, in turn, the biological analysis.

ACKNOWLEDGMENTS

The authors would like to thank the support from NSERC, the Natural Sciences and Engineering Research Council of Canada, and the anonymous reviewers for their valuable feedback on the chapter. The authors would also like to thank Dr. L. Ramdas for providing the cDNA microarray images for the dilution experiments, the DILN dataset.

REFERENCES

1. J. Angulo and J. Serra, "Automatic analysis of DNA microarray images using mathematical morphology," *Bioinformatics*, vol. 19, no. 5, pp. 553–562, 2003.
2. M. Ceccarelli and G. Antoniol, "A deformable grid-matching approach for microarray images," *IEEE Transactions on Image Processing*, vol. 15, no. 10, pp. 3178–3188, 2006.
3. E. Zacharia and D. Maroulis, "Microarray image gridding via an evolutionary algorithm," *Proceedings of the IEEE International Conference on Image Processing*, pp. 1444–1447, 2008.
4. L. Rueda and I. Rezaeian, "A fully automatic gridding method for cDNA microarray images," *BMC Bioinformatics*, vol. 12, p. 113, 2011.
5. L. Rueda and V. Vidyadharan, "A hill-climbing approach for automatic gridding of cDNA microarray images," *IEEE Transactions on Computational Biology and Bioinformatics*, vol. 3, no. 1, pp. 72–83, 2006.
6. G. Antoniol and M. Ceccarelli, "A Markov random field approach to microarray image gridding," *Proceedings of the 17th International Conference on Pattern Recognition*, pp. 550–553, 2004.

7. N. Brandle, H. Bischof, and H. Lapp, "Robust DNA microarray image analysis," *Machine Vision and Applications*, vol. 15, pp. 11–28, 2003.

8. F. Qi, Y. Luo, and D. Hu, "Recognition of perspectively distorted planar grids," *Pattern Recognition Letters*, vol. 27, no. 14, pp. 1725–1731, 2006.

9. L. Qin, L. Rueda, A. Ali, and A. Ngom, "Spot detection and image segmentation in DNA microarray data," *Applied Bioinformatics*, vol. 4, no. 1, pp. 1–12, 2005.

10. M. Katzer, F. Kummert, and G. Sagerer, "Methods for automatic microarray image segmentation," *IEEE Transactions on NanoBioscience*, vol. 2, no. 4, pp. 202–214, 2003.

11. X. Wang, R. Istepanian, and Y. Song, "Application of wavelet modulus maxima in microarray spots recognition," *IEEE Transactions on NanoBioscience*, vol. 2, no. 4, pp. 190–192, 2003.

12. Y. Yang, M. Buckley, S. Dudoit, and T. Speed, "Comparison of methods for image analysis on cDNA microarray data," *Journal of Computational and Graphical Statistics*, vol. 11, no. 1, pp. 108–136, 2002.

13. A. Kuklin, S. Shams, and S. Shah, "Automation in microarray image analysis with AutoGene," *Journal of the Association for Laboratory Automation*, vol. 5, no. 5, pp. 67–70, 2000.

14. L. Rueda, "Sub-grid detection in DNA microarray images," *Proceedings of the IEEE Pacific-RIM Symposium on Image and Video Technology*, pp. 248–259, 2007.

15. Y. Wang, M. Ma, K. Zhang, and F. Shih, "A hierarchical refinement algorithm for fully automatic gridding in spotted DNA microarray image processing," *Information Sciences*, vol. 177, no. 4, pp. 1123–1135, 2007.

16. Y. Wang, F. Shih, and M. Ma, "Precise gridding of microarray images by detecting and correcting rotations in subarrays," *Proceedings of the 8th Joint Conference on Information Sciences*, pp. 1195–1198, 2005.

17. M. Katzer, F. Kummer, and G. Sagerer, "A Markov random field model of microarray gridding," *Proceedings of the 2003 ACM Symposium on Applied Computing*, pp. 72–77, 2003.

18. S. Li, *Markov Random Field Modeling in Image Analysis*. Springer, 2009.

19. L. Vincent, "Morphological grayscale reconstruction in image analysis: applications and efficient algorithms," *IEEE Transactions on Image Processing*, vol. 2, no. 2, pp. 176–201, 1993.

20. A. Jain, T. Tokuyasu, A. Snijders, et al., "Fully automatic quantification of microarray data," *Genome Research*, vol. 12, no. 2, pp. 325–332, 2002.

21. M. Katzer, and F. Kummert, and G. Sagerer, "Automatische auswertung von mikroarraybildern," in *Bildverarbeitung für die Medizin 2002*, pp. 251–254, Springer, 2002.

22. M. Steinfath, W. Wruck, and H. Seidel, "Automated image analysis for array hybridization experiments," *Bioinformatics*, vol. 17, no. 7, pp. 634–641, 2001.

23. M. Ceccarelli and A. Petrosino, "The orientation matching approach to circular object detection," in *Proceedings of the International Conference on Image Processing*, vol. 3, pp. 712–715, IEEE, 2001.

24. R. O. Duda and P. E. Hart, "Use of the Hough transformation to detect lines and curves in pictures," *Communications of the ACM*, vol. 15, no. 1, pp. 11–15, 1972.

25. S. Deans, *The Radon Transform and Some of its Applications*. Dover Publications, 2007.

26. P. Toft, *The Radon transform, theory and implementation*. PhD thesis, Danmarks Tekniske Univ., Lyngby (Denmark). Inst. for Matematisk Modellering; Danmarks Tekniske Univ., Lyngby (Denmark), 1996.

27. D. Bariamis, D. Maroulis, and D. Iakovidis, "M^3G: maximum margin microarray grid-

ding," *BMC Bioinformatics*, vol. 11, p. 49, 2010.

28. D. Bariamis, D. Maroulis, and D. Iakovidis, "Unsupervised SVM-based gridding for DNA microarray images," *Computerized Medical Imaging and Graphics*, vol. 34, no. 6, pp. 418–425, 2010.

29. E. Wu, Y. A. Su, E. Billings, et al., "Automatic spot identification for high Throughput microarray analysis," *Journal of Bioengineering and Biomedical Science*, pp. 1–9, 2012.

30. M. D. Abramoff, P. J. Magalhaes, and S. J. Ram, "Image processing with ImageJ," *Biophotonics international*, vol. 11, no. 7, pp. 36–42, 2004.

31. G.-F. Shao, F. Yang, Q. Zhang, Q.-F. Zhou, and L.-K. Luo, "Using the maximum between-class variance for automatic gridding of cDNA microarray images," *IEEE/ACM Transactions on Computational Biology and Bioinformatics (TCBB)*, vol. 10, no. 1, pp. 181–192, 2013.

32. J. Hubble, J. Demeter, H. Jin, et al., "Implementation of GenePattern within the Stanford Microarray Database," *Nucleic Acids Research*, vol. 37, no. suppl 1, pp. D898–D901, 2009.

33. T. Barrett, S. E. Wilhite, P. Ledoux, et al., "NCBI GEO: archive for functional genomics data sets—update," *Nucleic Acids Research*, vol. 41, no. D1, pp. D991–D995, 2013.

34. L. Ramdas, K. R. Coombes, K. Baggerly, et al., "Sources of nonlinearity in cDNA microarray expression measurements," *Genome Biology*, vol. 2, no. 11, 2001.

35. N. Otsu, "A threshold selection method from gray-level histograms," *IEEE Transactions on Systems, Man and Cybernetics*, vol. SMC-9, pp. 62–66, 1979.

36. J. Kapur, P. Sahoo, and A. Wong, "A new method for gray-level picture thresholding using the entropy of the histogram," *Computer Vision Graphics and Image Processing*, vol. 29, pp. 273–285, 1985.

37. J. Kittler and J. Illingworth, "Minimum error thresholding," *Pattern Recognition*, vol. 19, no. 1, pp. 41–47, 1986.

38. M. Luessi, M. Eichmann, G. M. Schuster, and A. K. Katsaggelos, "Framework for efficient optimal multilevel image thresholding," *Journal of Electronic Imaging*, vol. 18, no. 1, pp. 013004–013004, 2009.

39. U. Maulik and S. Bandyopadhyay, "Performance evaluation of some clustering algorithms and validity indices," *IEEE Transactions on Pattern Analysis and Machine Intelligence*, vol. 24, no. 12, pp. 1650–1655, 2002.

40. S. Theodoridis and K. Koutroumbas, *Pattern Recognition*. New York: Academic Press, 4th ed., 2008.

41. R. Duda, P. Hart, and D. Stork, *Pattern Classification*. New York: John Wiley and Sons, Inc., 2nd ed., 2000.

4 Machine Learning-Based DNA Microarray Image Gridding

Dimitris Bariamis, Michalis Savelonas, and Dimitris Maroulis

CONTENTS

4.1 INTRODUCTION

Most of the state-of-the-art methods for DNA microarray gridding rely on empirically estimated parameters, with strong implications on the objectivity, robustness, and overall quality of gridding results. Empirical estimation can be a laborious, time-consuming, and error-prone process, which asks of technically unskilled end-users to understand the inner algorithmic mechanisms. Machine learning provides an ideal tool towards more intelligent and less parameterized gridding methods. However, this potential of machine learning had not been fully realized in this domain until recently.

Manual tuning and user intervention are not avoided in several well-known gridding methods, such as ImaGene [3], ScanAlyze [4], and SpotFinder [5]. On the other hand, the method of Brandle et al. [6] is only semi-automatic, still requiring empirical estimation of multiple parameters. A few state-of-the-art methods have been proposed for automated gridding. The morphology-based method of Angulo and Serra [7] is limited in the sense that it requires grid rows and columns to be strictly aligned with the x and y axes of the microarray. The same limitation holds for the hill-climbing approach of Rueda and Vidyadharan [8]. An automated region segmentation method based on Markov random fields has been introduced by Katzer [9]. However, the quality of the obtained gridding is diminishing in cases of weakly expressed spots. The Bayesian grid matching method proposed by Hartelius and Carstensen [10] employs an iterative algorithm to solve a complex deformable model for accurate microarray image gridding. Blekas et al. [11] proposed a method based on Gaussian mixture models, whereas Giannakeas and Fotiadis [12] employed Voronoi diagrams for gridding. This last method requires that artificial spots are introduced in place of those spots that are very weakly expressed. It is worth noting that the use of Voronoi diagrams is equivalent to the use of a 1-nearest neighbor (1-NN) classifier.

Stochastic optimization was the first branch of machine learning to be effectively incorporated in the context of DNA microarray gridding. Antoniol and Ceccarelli [13] combined a stochastic search approach and a Markov Chain Monte Carlo method so as to account for local image deformations. Still this method is not fully automated, since it requires prior knowledge on the number of rows and columns of the spots in the microarray image. A heuristic gridding approach based on a genetic algorithm was proposed by Zacharia and Maroulis [14]. This algorithm achieves nearly optimal gridding, outperforming the method proposed by Blekas et al. [11], while being noise and rotation invariant. Still, genetic optimization is associated with long convergence times, since a multitude of possible solutions has to be created and evaluated.

This chapter presents a state-of-the-art machine learning-based gridding method, named Maximum Margin Microarray Gridding (M^3G). M^3G uses soft-margin linear Support Vector Machine (SVM) classifiers in order to achieve automated and computationally efficient gridding. The algorithm enables a robust solution, where the optimal parameter values are determined in an unsupervised fashion. The core idea in M^3G comes from the observation that gridding lines in DNA microarray images are naturally determined from the position of microarray spots, similarly to the SVM hypersurface, which is determined by the position of support the vectors. Accordingly, the maximization of the margin between consecutive rows and columns of the microarray spots is implemented by training a linear SVM with an automatically detected subset of spot positions on the microarray image. Only spots with specific properties are selected by the spot detection component, so as to filter out irregularities and artifacts. The SVM determines the optimal positioning of each grid line, whereas the use of the soft-margin variant introduces robustness to outliers. M^3G is supported by a non-parametric Radon-based rotation estimator of general applica-

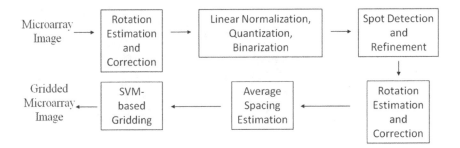

FIGURE 4.1 Pipeline of the proposed method.

bility for DNA microarray images. M^{3}G is fully unsupervised in the sense that the optimal parameter values are determined automatically. The results obtained are supported by a thorough exploration of the parameter space, the use of an extensive data set, and the comparison of the gridding results to the ground truth gridding of the reference images. All algorithms have been implemented under a GNU/Linux environment and the M^{3}G software is publicly available [15]. Aspects of this method have been originally introduced in [1, 2].

4.2 METHODS

In M^{3}G, the overall microarray image rotation is estimated by means of Radon transform. The image is counter-rotated, so as to achieve rotation invariance, and preprocessed by means of linear normalization, quantization, and binarization. Next, spots are detected as groups of consecutive white pixels that reside on the same spot edge. The set of detected spots is refined by appropriately comparing spot size distribution with normal distribution, excluding negative values. An initial estimation of average spacing between rows and columns is performed by means of the approach proposed by Ceccarelli et al. [16]. Selected spots along with the estimations of grid rows and columns are used as input to an ensemble of SVM classifiers. Each SVM classifier is used to identify one line of the final microarray image grid. In short, M^{3}G consists of a pipeline that is summarized in Figure 4.1.

4.2.1 ROTATION ESTIMATION

The Radon transform (Equation 4.1) is applied on the original DNA microarray image (Figure 4.2(a)) in order to estimate the image rotation angle:

$$R(a,r) = \int\limits_{-\infty}^{+\infty} \int\limits_{-\infty}^{+\infty} I(x,y)\delta(r - x\cos a - y\sin a)dxdy \tag{4.1}$$

where $I(x,y)$ denotes the gray level image intensity in (x,y). A similar approach for microarray rotation estimation can be found in [6]. In the transformed image, which

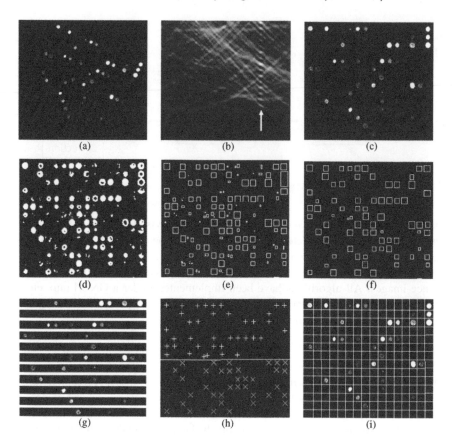

FIGURE 4.2 Outputs of M³G components [1]: (a) input microarray image, (b) result of the Radon transform (c) counter-rotated input image, (d) binarized image, (e) detected spots, (f) selected spots, (g) distance estimation between rows of spots, (h) determination of grid line, (i) gridded microarray image.

is illustrated in Figure 4.2(b), the intensity of each pixel with coordinates (a, r) is equal to the integral of the image brightness over a straight line, with an angle a to the x axis and a distance r to the origin. The rotation angle θ of the microarray image is estimated by locating the column with the highest mean brightness in the transformed image, which is denoted by the arrow. The image is subsequently counter-rotated by an angle θ, as illustrated in Figure 4.2(c).

4.2.2 LINEAR NORMALIZATION, QUANTIZATION AND BINARIZATION

As a next step, the microarray image is linearly normalized, so as to extend the intensity histogram to the full dynamic range of a 16-bit image. The linearly normalized image is quantized to 256 gray levels in order to reduce the computational complexity of the subsequent steps. Spot edges are detected by the application of the Sobel operator [17] on the normalized image. A threshold is determined automatically using the Otsu's method [18], in order to binarize the image and isolate the sharpest edges that correspond to spots, as illustrated in Figure 4.2(d). In [2], the threshold was manually set after experimentation, which leads to slightly less accurate gridding results.

Binarization is naturally associated with the issue of quantization error. However, considering that the image has been normalized to the full dynamic range, the quantization error affects a very small number of pixels. This is confirmed in the data set used for the evaluation of M^3G (see Results): the percentage of the affected pixels in each binarized image is negligible, having no effect on the subsequent grid placement.

4.2.3 SPOT DETECTION

The next component of the M^3G pipeline aims at identifying groups of consecutive white pixels residing on the same spot edge in the binarized image. Each group can be described by the location of the center pixel and its size. Figure 4.2(e) illustrates detected groups by circumscribing rectangles. Ideally, each rectangle should contain the edge pixels of a single microarray spot. However, as a result of the noise present in the image and the inter-spot proximity, it might also include artifacts or multiple merged spots, depending on the threshold used. These artifacts raise the need for a spot refinement step in order to obtain detection results as those shown in Figure 4.2(f).

4.2.4 SPOT REFINEMENT

The spot refinement process aims at removing false spots introduced by noise and artifacts. This component involves an initial evaluation of the aspect ratios of the detected pixel groups. Taking into account that the ideal spot shape is circular, the rectangles (Figure 4.2(e)) should not deviate much from being square, so that each rectangle contains only one microarray spot. Accordingly, the aspect ratio of each spot must be close to unity. Then, a lower bound s_{min} and an upper bound s_{max} of the spot sizes are calculated in order to maximize the similarity of the spot size distribution to the normal distribution. The spots with sizes exceeding the calculated bounds are considered false and are being discarded.

The similarity of spot size distribution to normal distribution is quantified by considering that spot sizes can only be positive, in contrast to the normal distribution $N(x; \mu, \sigma)$ (Equation 4.2) that also spans into negative values. The comparison should therefore be made to a variation of normal distribution for which the negative values are explicitly set to zero. Such a variation $N_m(x; \mu, \sigma)$ (Equation 4.3) can be derived by nullifying the probability of $N(x; \mu, \sigma)$ for $x < 0$ and scaling it accordingly, so that the total probability remains equal to unity. The corresponding cumulative distributions $C(x; \mu, \sigma)$ and $C_m(x; \mu, \sigma)$ are expressed by Equations 4.4 and 4.5 respectively.

$$N(x;\mu,\sigma) = \frac{1}{\sigma\sqrt{2\pi}}e^{-\frac{(x-\mu)^2}{2\sigma^2}} \qquad (4.2)$$

$$N_m(x;\mu,\sigma) = \begin{cases} 0 & x < 0 \\ \dfrac{N(x;\mu,\sigma)}{1-C(x;\mu,\sigma)} & x \le 0 \end{cases} \qquad (4.3)$$

$$C(x;\mu,\sigma) = \frac{1}{2}\left(1+erf\left(\frac{x-\mu}{\sigma\sqrt{2}}\right)\right) \qquad (4.4)$$

$$C_m(x;\mu,\sigma) = \begin{cases} 0 & x < 0 \\ \dfrac{C(x;\mu,\sigma)-C(0;\mu,\sigma)}{1-C(x;\mu,\sigma)} & x \le 0 \end{cases} \qquad (4.5)$$

The dissimilarity E between the discrete probability distribution of the spot sizes and the continuous probability distribution $N_m(x;\mu,\sigma)$ can be quantified by means of their respective cumulative distribution functions. The cumulative histogram of spot sizes $C_h(x)$ is defined as a function of the histogram $h(x)$, as shown in Equation 4.6. The dissimilarity E is defined as the total area between $C_h(x)$ and $C_m(x;\mu,\sigma)$, as shown in Equation 4.7.

$$C_h(x) = \sum_{i=0}^{x} h(i) \qquad (4.6)$$

$$E = \int_{0}^{+\infty} |C_h(x) - C_m(x;\mu,\sigma)|dx \qquad (4.7)$$

The calculated optimal bounds s_{min} and s_{max} minimize the dissimilarity E defined above. By filtering out the spots with sizes exceeding the range defined by these bounds, the resulting cumulative spot size distribution closely resembles the normal distribution, as illustrated in the example of Figure 4.3. In this case, any spot that is smaller than $s_{min} = 6.4$ pixels or larger than $s_{max}=17.1$ pixels is considered false and discarded. It is evident that the cumulative histogram of the selected spots almost coincides with the cumulative normal distribution (Figure 4.3(d)), whereas the original cumulative distribution (Figure 4.3(c)) differs substantially from the respective cumulative normal distribution.

4.2.5 ESTIMATION OF SPACING BETWEEN CONSECUTIVE ROWS AND COLUMNS

Spacing between spot rows is calculated by segmenting the input microarray image into horizontal stripes with a height of d_r pixels, as shown in Figure 4.2(g), which are then averaged. Figure 4.4 illustrates the main idea behind this process, with two examples of spacing for optimal and suboptimal d_r. A stripe is isolated for each of these two cases and a detail of each stripe is magnified. If d_r is selected so that it

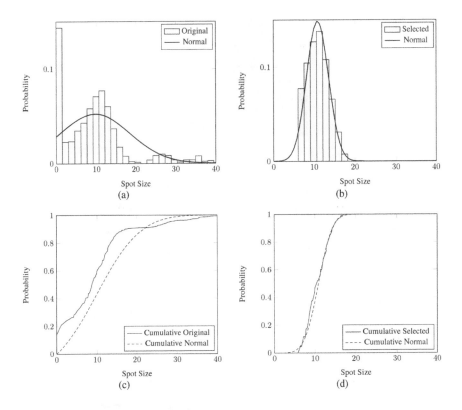

FIGURE 4.3 The (a) original, (b) selected, (c) cumulative original, and (d) cumulative selected spot size distributions compared to their respective modified normal distributions [1].

is equal to the distance between the rows, the spots of all rows will be in the same relative positions in the horizontal stripes, so that they will be highly overlapping in the resulting average stripe. Thus, the average stripe will contain well-defined spot areas. If a suboptimal value of d_r is selected, the spots will reside in different relative positions in the horizontal stripes and will thus blend with the background in the average stripe. The optimal value of d_r is selected by maximizing the standard deviation of the pixel intensities of the average stripe. The standard deviation can be used as an effective measure of spot overlap, since high values of the standard deviation indicate distinct dark and bright areas, whereas low values of the standard deviation indicate abundant gray areas. Therefore, the standard deviation should be maximized with respect to d_r in order to obtain the optimal value of d_r. The optimal column width d_c is likewise estimated using vertical stripes.

User intervention is not required in order to determine the optimal value for d_r, since a wide range of values is tested. Figure 4.5 illustrates several average stripes obtained for such a wide range of values for d_r. The standard deviation σ_{d_r} of the average stripes is calculated for all values of d_r within that range, using a small real-

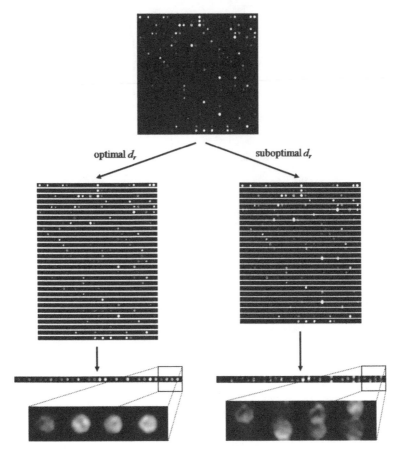

FIGURE 4.4 Generation of horizontal subimages, averaged subimage for optimal d_r and suboptimal d_r, and detail of averaged subimages [2].

FIGURE 4.5 The averaged row subimages produced for various values of d_r [2].

valued step. From all the tested values of d_r, those that result in local maxima of the standard deviation are selected (Figure 4.6). These local maxima are most often

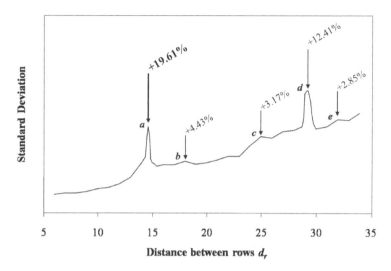

FIGURE 4.6   Standard deviation of pixel intensity as a function of distance between rows d_r. The selected point a is indicated in bold [2].

located on multiples of the optimal d_r, since such an estimation also results in highly overlapping spots. Sometimes, other local maxima may be present, depending on the rotation of the image. For each of the selected d_r values, the mean of the resulting standard deviation $\overline{\sigma}_{d_r}$ in its neighborhood is calculated. The value of d_r that results in the highest value of the $\sigma_{d_r}/\overline{\sigma}_{d_r}$ ratio is selected as optimal.

Ceccarelli et al. [16] have proposed an alternative for estimation of spacing between rows and columns of spots. They employ the orientation matching (OM) and Radon transforms in order to extract the spot positions and grid rotation, respectively. Subsequently, they project spot locations on the axes of the grid and estimate the distance between rows and columns. This method requires prior knowledge about the radii of the spots and uses a filter for noise reduction. As opposed to this approach, M^{3}G performs distance estimation without any parameters or arbitrarily selected filters, based on the maximization of the standard deviation of the average stripe. This maximization over a wide range of d_r values allows successful estimation in an unsupervised fashion, whereas the use of the average stripe acts as a low pass filter, allowing high tolerance to noise. An evaluation of the distance estimation for noisy images has been included in the Results section.

4.2.6   MAXIMUM MARGIN MICROARRAY GRIDDING

Gridding is performed by representing each spot that has been selected (Figure 4.2(e)) with a vector $\bar{x}_i$, $i = 1 \ldots N$, where N is the total number of selected spots in the image and the components of each vector $\bar{x}_i$ are the coordinates of the spot center. These vectors are assigned to distinct rows and columns, based on the previously estimated distances d_r and d_c. Accordingly, each pair of consecutive rows or

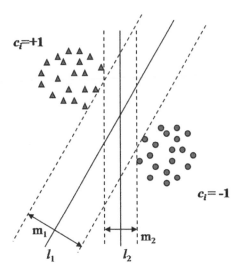

FIGURE 4.7   Separating hyperplanes and their respective margins [2].

columns of spots can be separated by a single separating line. The optimal such line is positioned in order to maximize the margin between the rows or columns of the spots. For a pair of rows numbered k and $k+1$, the vectors that belong to row k or to any row above it are assigned a class label $c_i = +1$ and the vectors that belong to row $k+1$ or to any row below it are assigned a class label $c_i = -1$. These vectors $\bar{x}_i$, along with their respective class labels c_i are provided as a training set to a linear SVM [19], which identifies the maximum margin grid line.

The SVM is provided with the training set $D = \left\{ (\bar{x}_i, c_i) | \bar{x}_i \in \Re^2, c_i \in \{-1, +1\} \right\}$, which was introduced earlier in this chapter. By solving a quadratic programming optimization problem, it determines the normal vector $\bar{w}$ and the parameter b of the separating line $\bar{w} \cdot \bar{x} - b = 0$, which maximizes the margin between vectors $\bar{x}_i$ of different classes, i.e., the margin between spots of distinct rows or columns. The width of the margin is equal to $2/\|w\|$, and hence the widest margin is found by minimizing $\|w\|$ under the constraints $c_i(\bar{w} \cdot \bar{x}_i - b) \geq 1$, i.e., requiring that all the spots lie on the correct side of the resulting grid line. Figure 4.7 presents an example of two possible lines for the separation of two classes of vectors. Although line l_2 is a valid separating line, line l_1 maximizes the margin ($m_1 > m_2$) and would therefore be chosen by the SVM.

The SVM described above is the *hard-margin* SVM. One of its properties is that the separating line is solely determined by the vectors that lie on the edges of the margin, called *support vectors*. In a linear SVM, a very small number of support vectors determine the separating line and the margin. In the case of outliers present inside the margin, the positioning of the separating line will be exclusively determined by the outliers and will thus be suboptimal for gridding. Figure 4.8 presents a gridding example in the presence of an obvious outlier, denoted by the arrow, which forces

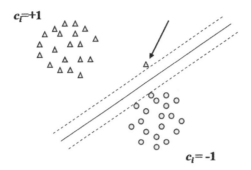

FIGURE 4.8 Reduction of margin width due to an outlier, in the case of hard-margin linear SVM [2].

the SVM to position the separating hyperplane significantly closer to vectors with a class label of +1, reducing the width of the margin. The *soft-margin* SVM copes with this issue by introducing a slack variable ξ_i for each vector $\bar{x}_i$. The constraints are then relaxed, taking the form $c_i(\bar{w} \cdot \bar{x}_i - b) \geq 1 - \xi_i$ and the separating hyperplane is determined by minimizing

$$\min \left[\frac{1}{2}\|w\|^2 + C\sum_{i=1}^{N} \xi_i \right] \tag{4.8}$$

where C is a cost parameter controlling the effect of outliers on the positioning of the resulting grid line. Large values of C result in a grid line that is mostly determined by outliers, whereas smaller values of C result in a grid line that follows the general trend of the spot locations, virtually ignoring any outliers. The hard-margin classifier can be perceived as an extreme case of a soft-margin classifier, where C is infinitely large [20].

The soft-margin SVM is used in the context of gridding so as to diminish the effects of misdetected spots that result from artifacts or noise. A small fraction of these outliers might have a shape and size similar to valid spots and could therefore pass through the selection component without being discarded. The soft-margin SVM ensures that such outliers will not have an impact on the generated grid lines. For ideal microarray images, where all spots could be successfully detected and no outliers are present, a hard-margin SVM could be used as well, but a gridding application for real microarray images requires robustness against outliers. Furthermore, in ideal noiseless images, the training set for the SVM classifier would consist only of the necessary spots, i.e., those residing on rows k and $k+1$. However, in real microarray images, there are cases where several consecutive spots might be very weakly expressed and therefore not detected. Aiming to cope with this problem, spots from rows above k and below $k+1$ are included in the training set, providing redundant data to the classifier. Using an algorithm based on the sequential minimal optimization (SMO) to solve the SVM optimization problem [21], the additional data introduces only a small computational overhead, since such algorithms evaluate vectors that are far from the separating line in only the first few iterations of their outer loops [21, 22]. The SVM has been selected over similar methods for the determination of the grid lines, such as a least squares fit, because the soft margin SVM is adjustable with regards to its tolerance to outliers through the cost parameter C [23]. In the case that row k contains less than two detected spots, the two grid lines that separate this row from rows $k-1$ and $k+1$ cannot be determined by the use of the SVM. This is a rather rare case considering that the image is normalized during the preprocessing step. To cope with this limitation, the endpoints of the two grid lines are positioned equidistantly between the endpoints of the first neighboring grid lines above and below them. In the case where the top or bottom rows of spots contain less than two spots, the endpoints of the grid lines that cannot be determined are positioned d_r pixels further from the nearest grid lines.

Figure 4.9 presents a gridding example in the presence of an obvious outlier. It is evident that for a small value of the cost parameter C (Figure 4.9(a)) the margin is determined by the rest of the spots in the row and the outlier is ignored, whereas for larger values of C (Figure 4.9(b)) the outlier affects the positioning of the separating line by moving it significantly closer to the vectors of the top row, reducing the margin and rendering it suboptimal for gridding.

4.3 RESULTS

4.3.1 EXPERIMENTAL SETTING

M^3G has been experimentally evaluated on two microarray data sets. The first data set consists of 54 DNA microarray images from the well-known Stanford Microarray Database [24]. The images are of TIFF format with a resolution of 1900×5500 pixels and 16-bit gray level depth. Each image includes 48 sub-grids of 870 spots each, resulting in a total of 2,255,040 spots in the entire data set. These microarray images have been produced for the study of the gene expression profiles of 54 specimens of BCR-ABL-positive and -negative acute lymphoblastic leukemia [25]. A subset of these images has been used in the *preliminary version* of M^3G [26] and the heuristic approach proposed in [14]. The data set is accompanied by ground truth annotations regarding the positions and the sizes of the spots. Figure 4.10 visually validates the resemblance of the distribution of the sizes of the spots in the data set to the $N_m(x; \mu, \sigma)$ distribution.

Another microarray data set was employed in the experimental evaluation, consisting of 10 sub-grids extracted from distinct microarray images that were either artificially created or obtained from public microarray databases. The sub-grids are stored in TIFF files with 16-bit gray level depth. This data set has been used by Blekas et al. [11] for the evaluation of their method and has been obtained from the authors.

Aiming to allow direct comparisons with the aforementioned methods, the same statistical analysis is performed. Each spot was evaluated as being perfectly, marginally, or incorrectly gridded when the percentage of its pixels contained within

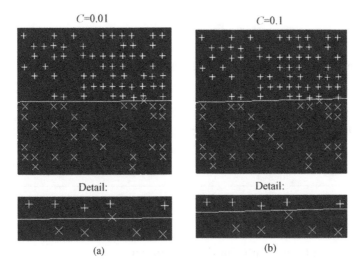

FIGURE 4.9   The effect of an outlier as a function of the SVM cost parameter C for: (a) small value of C and (b) large value of C [2].

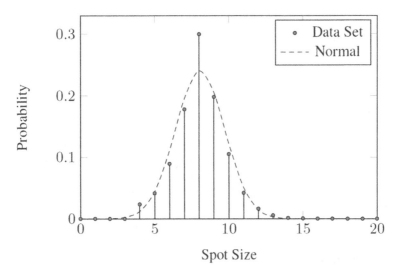

FIGURE 4.10   The distribution of spot diameters in the data set compared to the modified normal distribution $N_m(x; \mu, \sigma)$ with $\mu = 8.07$ and $\sigma = 1.66$ [1].

its grid cell is 100%, more than 80%, or less than 80%, respectively.

The evaluation platform is based on an Athlon 64 X2 3800+ processor with 3GB of RAM, whereas the sub-grids used belong to the first data set and their dimensions are roughly 450×450 pixels. For the first sub-grid in each microarray image, M³G requires 18 seconds of processing time, most of which is used for estimating the dis-

TABLE 4.1

Comparisons of gridding results for the first data set.

	Perfect	Marginal	Incorrect
Heuristic approach [14]	94.6	4.8	0.6
Preliminary method [26]	95.1	4.5	0.4
M^3G (all spots)	98.3	1.5	0.2
M^3G (50% of spots)	98.1	1.6	0.3

TABLE 4.2

Comparison of gridding results for the second data set.

	Perfect	Marginal	Incorrect
ScanAlyze [4]	48.7	22.6	28.7
SpotFinder [5]	72.8	14.3	12.9
Heuristic approach [14]	94.4	5.1	0.5
Blekas et al. [11]	89.6	9.2	1.2
M^3G	98.0	1.7	0.3

tance between consecutive rows and columns of spots. For any subsequent sub-grid of the same image, the range of possible values for d_r and d_c is reduced, resulting in 10 seconds of processing time. It is worth noting that the processing times mentioned above can be significantly reduced by optimizing the implementation and by using multiple cores of the processor, as the most time consuming parts of the gridding process can be efficiently parallelized. The heuristic approach [14] requires a processing time of 92 seconds for each microarray image sub-grid, which is nearly one order of magnitude larger than the time required by M^3G.

4.3.2 COMPARISONS

Table 4.1 presents the evaluation results of M^3G for the first data set, the results of the heuristic approach [14], and the results of the preliminary version of M^3G [26]. Two experiments were conducted in order to assess the sensitivity of M^3G to missing spots: in the first experiment, all selected spots were used for SVM training, whereas in the second experiment, half of the selected spots were randomly discarded. Both experiments resulted in perfect gridding for more than 98% of the spots in the data set, whereas only up to 1.6% and 0.3% of the spots were marginally and incorrectly gridded, respectively. These results illustrate that M^3G is robust in the case of significantly fewer detected spots.

TABLE 4.3

Euclidean distances between spot centers and grid cell centers.

	Mean	Standard Deviation
Giannakeas et al. (Red)	2.73 pixels	1.10 pixels
Giannakeas et al. (Green)	2.34 pixels	1.07 pixels
M^3G (all spots)	1.17 pixels	0.71 pixels
M^3G (50% of spots)	1.21 pixels	1.04 pixels

Table 4.2 presents the evaluation results for the second data set. M^3G achieves a significantly lower percentage of marginally and incorrectly gridded spots, compared to supervised ([5, 4]) and unsupervised ([11, 14]) approaches.

M^3G features several enhancements as compared to the preliminary version [26]: 1) unsupervised selection of spots based on the distribution of their sizes, 2) unsupervised parameter adjustment, 3) inclusion of all selected spots into the training set of each SVM, instead of only pairs of rows or columns of selected spots, and 4) enhanced rotation detection by means of the Radon transform.

Gridding accuracy is evaluated according to the evaluation process of [12], which involves the measurement of the distance between each spot center and the center of its respective grid cell. The mean value and standard deviation of these distances are used to evaluate the localization of the spots by the grid. The results of M^3G as compared to the ones presented in [12] are shown in Table 4.3.

M^3G outperforms the method presented in [12] with respect to spot localization, achieving a 50% smaller mean distance and a 33% smaller standard deviation. Even when half of the selected spots are randomly discarded to simulate the case of significantly fewer detected spots, M^3G still achieves a better localization.

The gridding performance of M^3G was evaluated by following a grid search approach for the determination of the optimal value for the SVM cost parameter C, which determines the effect that outliers or noise might have on the positioning of the separating lines generated by the SVM. Accordingly, a small value of C should be selected for successful gridding. The percentage of correctly gridded spots as a function of C is presented in Figure 4.11. The choice of $C = 10^{-2}$ is supported by this analysis, as it results in the most accurately gridded spots. Lower values of C do not increase the obtained accuracy, but the choice of a larger value reduces accuracy, due to the accompanying increase of the significance of outliers.

Figure 4.12 illustrates four gridding examples, as obtained by the application of M^3G on microarray image areas that include large and bright artifacts. Even in the vicinity of such artifacts, the gridding is not affected.

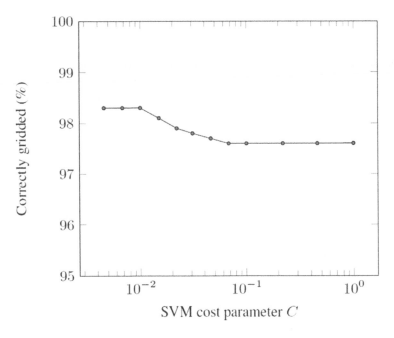

FIGURE 4.11   Percentage of correctly gridded spots as a function of the SVM cost parameter C [1].

4.3.3   EVALUATION OF ROTATION DETECTION ROBUSTNESS AND DISTANCE ESTIMATION

The robustness of the rotation estimation component employed in M^3G has been evaluated by randomly rotating the microarray images of the first data set by angle θ_{real}, ranging from $-25°$ to $+25°$. The rotation detection method was then employed to compute an estimate θ_{est} of each image rotation. The images were counter-rotated by θ_{est} and subsequently gridded. The evaluation was based on the mean and standard deviation of the difference between the real and estimated rotation of the images $\Delta\theta = \theta_{real} - \theta_{est}$. Both of these measures had values below $0.1°$. The accuracy achieved by gridding the counter-rotated images was within a range of 0.3% of the accuracy obtained using the original images. Therefore, the effects of image rotation in gridding accuracy are negligible. An example of an image rotated by $15°$, as well as the counter-rotated image and the gridding results, are illustrated in Figure 4.13. In this case, $\Delta\theta$ was equal to $0.9°$.

The distance estimation component of M^3G was evaluated with regards to its robustness to noise. Fifty sub-grids from distinct real microarray images of the first data set were randomly selected and additive Gaussian noise of $\sigma = 250$, 500 and 1000 was introduced. The resulting mean signal-to-noise ratios were 9 dB, 5.5 dB and 1 dB, respectively, whereas in several cases the introduced noise was stronger than the actual signal in the image. For all the images and noise levels tested, the

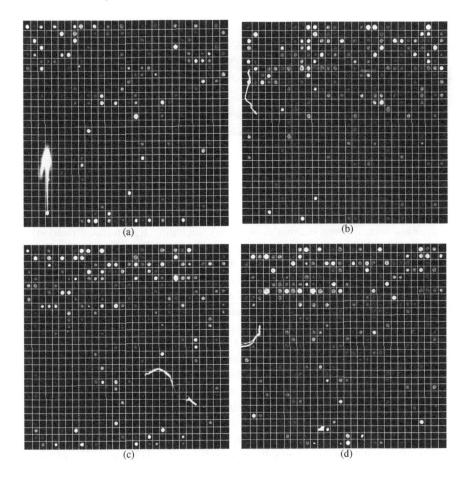

FIGURE 4.12 Four examples of successful gridding in the presence of large and bright artifacts.

distance estimation component displayed a negligible variance of up to 0.02 pixels, illustrating that the use of the average stripe for distance estimation is highly robust to noise.

4.4 CONCLUSIONS

Machine learning provides an ideal tool towards more intelligent and less parameterized methods for gridding of DNA microarray images. Despite that stochastic optimization methods had been previously employed for gridding, the potential of machine learning had not been fully realized until the introduction of M^3G. This unsupervised gridding method is based on the maximization of the margin between consecutive rows and columns of spots by means of a maximum margin classifier.

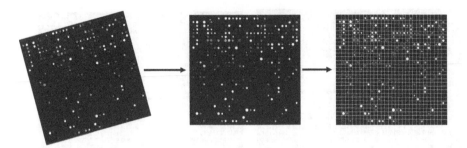

FIGURE 4.13   (a) A microarray image rotated by 15°, (b) the counter-rotated image ($\Delta\theta = 0.9°$), and (c) the resulting gridding for this image [2].

M^3G involves several preprocessing components, including Radon-based rotation estimation for the microarray image, as well as spot detection and selection. The distance between rows and columns of spots is then estimated and the positions of the selected spots are used to train a set of linear soft-margin SVMs.

The use of soft-margin SVMs reinforces robustness to outliers that result from artifacts and noise, whereas the use of redundant vectors in the SVM training set and the automatic determination of the operating parameters facilitate a substantial increase in gridding accuracy. In addition, M^3G copes with the presence of irregular and weakly expressed spots, as well as rotation. A potential weakness of M^3G is that the SVM classifiers require several detected spots in each row and column of spots. Rarely, most of the spots in a row or column might be weakly expressed and not detected. In such cases, which account for less than 0.1% of the rows and columns in the data set, the grid line positioning is determined by the nearest grid lines. The experimental evaluation on standard DNA microarray images showed that M^3G outperforms state of the art gridding methods, such as [11] and [14], providing the potential of achieving nearly optimum gridding. A future perspective is the integration of M^3G with subsequent segmentation and quantification methods in the context of a user-friendly GUI, in order to facilitate everyday use by biologists. Another research path is to study the use of alternative filters for spot detection, aiming to enhance detection accuracy. Furthermore, the effect of alternative SVM kernels is worth investigating.

REFERENCES

1. D. Bariamis, D. Iakovidis, and D. Maroulis, "M^3G: maximum margin microarray gridding," *BMC Bioniformatics*, vol. 11, no. 49, 2010.
2. D. Bariamis, D. Maroulis, and D. Iakovidis, "Unsupervised SVM-based gridding for DNA microarray images," *Computerized Medical Imaging and Graphics*, vol. 34, pp. 418–425, 2010.
3. Biodiscovery Inc., "ImaGene, http://www.test.org/doe/," 2007.
4. M. B. Eisen, "ScanAlyze, http://rana.lbl.gov/eisensoftware.htm," 2002.
5. P. Hegde, R. Qi, K. Abernathy, et al., "A concise guide to cDNA microarray analysis," *Biotechniques*, vol. 29, no. 3, pp. 548–550, 2000.

6. N. Brandle, H. Bischof, and H. Lapp, "A concise guide to cDNA microarray analysis," *Machine Vision and Applications*, vol. 15, pp. 11–28, 2003.

7. J. Angulo and J. Serra, "Automatic analysis of DNA microarray images using mathematical morphology," *Bioinformatics*, vol. 19, no. 5, pp. 553–562, 2003.

8. L. Rueda and V. Vidyadharan, "A hill-climbing approach for automatic gridding of cDNA microarray images," *IEEE/ACM Transactions Computational Biology Bioinformatics*, vol. 3, pp. 72–83, 2006.

9. M. Katzer, F. Kummert, and G. Sagerer, "A Markov random field model of microarray gridding," in *Proceedings of the ACM Symposium on Applied Computing*, (Melbourne, FL, USA), pp. 72–77, 2003.

10. K. Hartelius and J. Carstensen, "Bayesian grid matching," *IEEE Transactions on Pattern Analysis and Machine Intelligence*, vol. 25, no. 2, pp. 162–173, 2003.

11. K. Blekas, N. Galatsanos, A. Likas, et al., "Mixture model analysis of DNA microarray images," *IEEE Transactions on Medical Imaging*, vol. 24, no. 7, pp. 901–909, 2005.

12. N. Giannakeas and D. Fotiadis, "An automated method for gridding and clustering-based segmentation of cDNA microarray images," *Computerized Medical Imaging and Graphics*, vol. 33, pp. 40–49, 2009.

13. G. Antoniol and M. Ceccarelli, "Microarray image gridding with stochastic search based approaches," *Image and Vision Computing*, vol. 25, no. 2, pp. 155–163, 2007.

14. E. Zacharia and D. Maroulis, "An original genetic approach to the fully automatic gridding of microarray images," *IEEE Transactions on Medical Imaging*, vol. 27, no. 6, pp. 805–813, 2008.

15. D. Bariamis, "M3G: maximum margin microarray gridding," http://rtsimage.di.uoa.gr/downloads/m3g-0.01.tgz, 2010.

16. M. Ceccarelli and G. Antoniol, "A deformable grid-matching approach for microarray images," *IEEE Transacitons on Image Processing*, vol. 15, no. 10, pp. 3178–3188, 2006.

17. R. Gonzalez and R. Woods, *Digital Image Processing*. Saddle River: Prentice-Hall, 2006.

18. N. Otsu, "A threshold selection method from gray-level histograms," *IEEE Transactions on Systems, Man and Cybernetics*, vol. 9, pp. 62–66, 1979.

19. C. Cortes and V. Vapnik, "Support vector networks," *Machine Learning*, vol. 20, no. 3, pp. 273–297, 1995.

20. S. Theodoridis and K. Koutroumbas, *Pattern Recognition*. San Diego: Academic Press, fourth ed., 2008.

21. J. Platt, "Sequential minimal optimization: A fast algorithm for training support vector machines," tech. rep., Microsoft, Inc., 1998.

22. R. Fan, P. Chen, and C. Lin, "Working set selection using second order information for training support vector machines," *Machine Learning Research*, pp. 1889–1918, 2005.

23. J. Burges, "A tutorial on support vector machines for pattern recognition," *Data Mining and Knowledge Discovery*, vol. 2, no. 2, pp. 121–167, 1998.

24. J. Hubble, J. Demeter, H. Jin, et al., "Implementation of GenePattern within the Stanford Microarray Database," *Nucleic Acids Research, Database issue*, vol. 37, pp. D898–D901, 2009.

25. D. Juric, N. Lacayo, M. Ramsey, et al., "Differential gene expression patterns and interaction networks in bcr-abl-positive and -negative adult acute lymphoblastic leukemias," *Journal of Clinical Oncology*, pp. 1341–1349, 2007.

26. D. Bariamis, D. Maroulis, and D. Iakovidis, "Automatic DNA microarray gridding based on support vector machines," in *Proceedings of the IEEE International Conference on Bioinformatics and Biotechnology*, Athens, Greece, pp. 1–5, 2008.

5 Non-Statistical Segmentation Methods for DNA Microarray Images

Shahram Shirani

CONTENTS

5.1 INTRODUCTION

Image segmentation is defined as the process of dividing an image into constituent regions or objects. In the context of DNA microarray images, segmentation's goal is to divide each grid cell into regions corresponding to the spot (foreground) and background. Segmentation is typically performed after the microarray image has undergone a gridding step and it has been divided into cells (also known as sub-grids). After segmentation, gene expression levels are estimated from the foreground area. A number of factors make the segmentation of microarray images a challenging task. To name a few, the background is typically contaminated by noise. Spots' shapes and sizes might differ within an image from one spot to another. The intensity of a spot is not necessarily uniform. Also since the hybridization process is not homogeneous

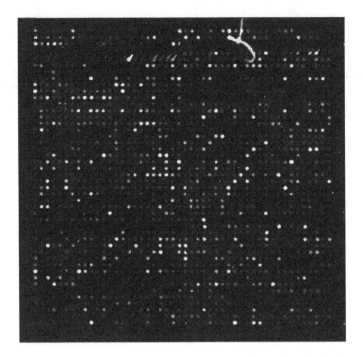

FIGURE 5.1 Part of a DNA microarray image showing the noise in the background and variation in spot shapes.

the spot regions could be broken. Moreover, the quality of DNA microarray images might vary. Figure 5.1 shows a part of a typical microarray image. The background noise and variation in the shape of the spots can be clearly seen in the picture. Figure 5.2 displays some sample spots of different DNA microarray images. As can be seen there is a wide range of variations in terms of shape and quality of the spots. As a result, segmentation techniques typically require human intervention in order to specify the required parameters or to correct their results. However, this lack of automation can significantly affect biological conclusions. In fact, it has been shown that the choice of the segmentation method has significant effects on the provision of the outcome of an experiment [2].

The main process for measuring spot intensity includes three stages [3]. In the first stage, called gridding, the position of each spot is assigned; compartments of the image containing one individual spot and background are therefore defined. In the second stage, called spot segmentation, the contour of each spot is determined. In the third stage, called intensity extraction (quantification), the intensity of each spot is calculated. Gene expression data analysis is the subsequent step of microarray analysis and is accomplished with different statistical and/or unsupervised learning methods. Robust segmentation is significant for accurate classification of gene expression and to extrapolate an assortment of spot quality measures. From the aforementioned stages, spot segmentation is the most challenging task.

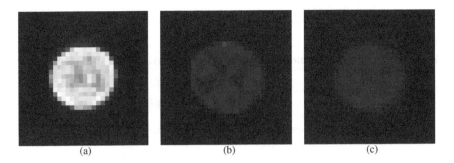

FIGURE 5.2 Examples of different spots in a DNA microarray image.

DNA microarray image segmentation methods can be divided into several groups based on the feature(s) they use to separate the spots from the background. Over the last few years, many sophisticated DNA microarray image segmentation algorithms have been developed. Some of these algorithms have been incorporated into software packages. In this chapter, ten different *non-statistical* DNA microarray segmentation algorithms are reviewed. Table 5.1 provides a summary of the segmentation approaches that are discussed in this chapter.

5.2 CIRCULAR-SHAPE-BASED SEGMENTATION

Any prior knowledge regarding the shape of spots can be incorporated into a shape-based separation of spots from the background. A simple yet effective segmentation method is based on the assumption that spots are circular with fixed, known radius. Then a circular mask of fixed radius is displaced within each cell to locate the position of the best match. Pixels inside the mask are labeled foreground and everything else is background. One example of application of this algorithm is in the ScanAlyze [4] software tool. This method has also been used as an option in many microarray analysis software tools such as GenePix [5], ScanArray Express [6], and Dapple [7]. Figure 5.3 shows part of a DNA microarray image segmented using the circular-shape-based approach.

Assuming that all the spots are circular with a fixed radius is not always a valid hypothesis. The adaptive circle segmentation technique was proposed to overcome the drawback of fixed circle segmentation. In adaptive circle segmentation the radius and center of each spot is adjusted individually for every cell. The variable circle method fits a circle with variable size onto the region containing the spot. Adaptive circle segmentation involves two steps. First, the center of each spot needs to be estimated. Second, the diameter of the circle has to be adjusted. This method is able to resolve circular spots of different sizes, but is less effective on irregularly shaped spots. The same concept can be extended to any other shape besides circles. Elliptic methods assume an elliptical shape for the spots, and can adapt to a more general shape than the adaptive circle method, but cannot recognize irregularly shaped spots [8]. Figure 5.4 shows a DNA microarray segmented using the adaptive circular shape approach.

TABLE 5.1

DNA microarray image segmentation approaches reviewed in this chapter.

Method	Description
Circular-shape-based	A circular mask of fixed radius is employed to locate the position of the spot
Region-growing	Groups foreground and background pixels into larger regions based on predefined criteria for growth
Thresholding	Iteratively computes the foreground/background threshold using the MannWhitney test
Clustering-based	Unsupervised classification algorithms are employed to divide the image into regions of foreground and background
Active-contour (Snake)-based	Spot's contour is obtained by minimizing the sum of the snake's energy and the region's energy
Shape-based	A microarray segmentation technique based on modeling spot and cell's shape with 3D models
Graph-based	A cell is mapped to a graph. The weight assigned to an edge is the dissimilarity of the two pixels corresponding to two vertices of that edge. A minimum spanning tree is used to segment the cell into spot and background
Morphology	A shape detector based on grayscale hit-or-miss transform is used to located the spot and segment the DNA microarray image
Watersheds	Segmentation by morphological watersheds (a classic image segmentation approach)
Supervised learning-based segmentation	A support vector machine is employed to classify the pixels of the DNA image into spot and background

5.3   REGION-GROWING-BASED SEGMENTATION

Region growing is a classic image segmentation approach that has been used for DNA microarray images [9]. Generally speaking, region growing is a procedure that groups pixels into larger regions based on predefined criteria for growth [10]. The algorithm segments each cell by iteratively growing separate regions with respect to predefined seed points providing starting points for the segmentation. Often, the seeds consist of a single pixel in the spot (foreground) and a pixel in the background. In each iteration, all the pixels that are not yet labeled but where at least one of their neighboring pixels is labeled are considered. From these pixels, the algorithm adds the most similar pixel (in terms of grey-level) to the segmented regions. The algorithm aims at ensuring that the final segmented regions are as homogeneous as possible given the connectivity constraint. Finally, the region originating from the foreground seed is considered as the spot, and the region originating from the background seed as the background. For microarray segmentation, the foreground and background seeds are chosen based on the outcome of the gridding step. Seeded region growing is employed in a software package named SPOT [9]. Figure 5.5 shows the result of DNA microarray image segmentation using the SPOT software.

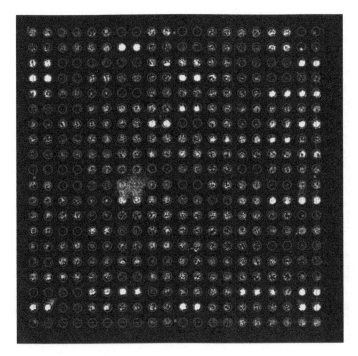

FIGURE 5.3 (SEE COLOR INSERT.) Part of a DNA microarray image segmented using the circular-shape-based approach.

5.4 THRESHOLDING

The Mann-Whitney test (MWT) is a distribution-free statistical test aimed at finding a significant difference between two distributions X and Y with mean of μ_X and μ_Y, respectively. In statistical terms, it is a test of the null hypotheses $H0 : \mu_X = \mu_Y$. This is the hypothesis that two independent samples of observations $X_1, ..., X_n$ and $Y_1, ..., Y_m$ have come from populations having the same distribution.

The principle of the MWT is as follows. To compare the distributions X and Y, the MWT picks a sample of observations of size n in X and a sample of observations of size m in Y. The MWT then orders the $(n + m)$ observations. The statistic U_X is the sum of the counts of observations of X that precedes each observation of Y. U_X is small when the population distribution of the X measurements is to the right of the population distribution of Y. Rejection of $H0$ occurs when $U_X > w_{\alpha,m,n}$ where $w_{\alpha,m,n}$ is the critical value corresponding to the significance level α (please refer to [11] for a detailed discussion on the Mann-Whitney hypothesis test).

A segmentation algorithm based on the statistical Mann-Whitney test was first introduced by Chen et al., in 1997 [12]. The algorithm iteratively computes the threshold between foreground and background using the Mann-Whitney test. First, a predefined mask is used to identify a portion of the cell that contains the foreground. The mask is based on the geometry of the potential foreground areas and

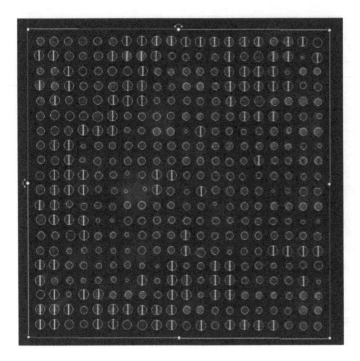

FIGURE 5.4 Part of a DNA microarray image segmented using the adaptive circular-shape-based approach.

can be constructed from specially tagged spots or other strong spots (e.g., the mask is obtained by finding all strong spots, aligning them together, averaging, and then thresholding). Eight sample pixels from the known background (outside the mask) are randomly picked as $Y_1, Y_2, \ldots, Y_8$, and the lowest eight samples from within the mask are selected as $X_1, X_2, \ldots, X_8$. The statistic U_X is calculated and, for a given confidence level α, compared with a threshold $w_{\alpha,8,8}$. Rejection of $H0$ occurs when $U_X > w_{\alpha,8,8}$. If the null hypothesis is not rejected, then a predetermined number (perhaps only one) of the eight samples from the potential spot region are discarded and the lowest eight remaining samples from the spot region are selected. The Mann-Whitney test is repeated until the null hypothesis is rejected. When H0 is rejected, the spot site is taken to be the eight pixels causing the rejection, together with all pixels in the mask whose values are greater than or equal to the minimum value of the eight. The resulting site is said to be a spot site of significance level α. If the null hypothesis is never rejected, then it is concluded that there is no appreciable probe at the spot site. Furthermore, one can require that the Mann-Whitney spot site contain a minimum number of pixels for the spot to be considered valid.

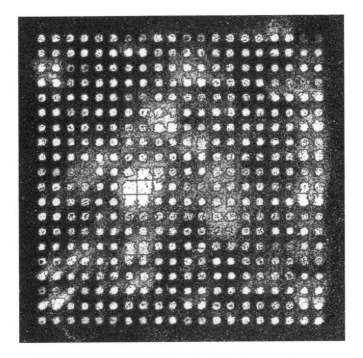

FIGURE 5.5   (SEE COLOR INSERT.) Part of a DNA microarray image segmented using the region growing approach.

5.5   CLUSTERING-BASED SEGMENTATION

Classic unsupervised classification algorithms such as the k-means can be employed for microarray image segmentation [13, 14, 15]. Since segmentation is used for dividing the image into the regions of foreground and background, the number of cluster centers k is set to two. As the initial cluster centers, the pixels with minimum and maximum intensities are selected. All data points are then assigned to the nearest cluster centers according to a distance measure (e.g., Euclidean distance). Thereafter, new cluster centers are set to the mean of the pixel values in each cluster. Finally, the algorithm is iteratively repeated until the cluster centers stay unaltered. The approach is interesting since it is independent of the shape of the spots.

The hybrid k-means microarray segmentation algorithm [16] is an extended version of the original k-means segmentation approach. Two main features of the new approach are repeated clustering and mask matching. Repeated clustering is applied, if the clustering selects only very few pixels as foreground region (mainly in the case of low foreground intensity and small bright artifacts). In such a case, the outlier pixels are removed and clustering is repeated on the background pixels, until at least m pixels (e.g., $m = 50$) are labeled as foreground area. The other feature of the hybrid approach is mask matching that integrates the spot shape into the algorithm. After the first clustering, a bivalence mask is constructed to estimate the average spot shape (see [16] for the details of how the mask is constructed). For every spot all pixels that are assigned to foreground in the spot and to background in the mask and vice versa are deleted. A convenient consequence of the mask matching step is that pixels are divided into three groups instead of two: foreground, background, and deletions. This is desirable, since it allows a separate treatment of artifacts and their elimination from further analysis. In fact, the mask is used as a template to filter out low-quality parts of single microarray spots. This yields a genuine combination of the two central features "histogram" and "shape," and thus a favorable hybrid image analysis solution.

In [17], a fuzzy clustering approach (fuzzy C-means) is proposed for segmentation of DNA microarray images. Two clusters (foreground and background) are considered. Each pixel of a microarray cell is initialized with a membership value $u_k(i)$, which is the membership of pixel i in cluster k. Clearly $\sum_{k=1}^{2} u_k(i) = 1$. Next, the centroid of each cluster is computed as:

$$v_k = \frac{\sum_{i=1}^{N} u_k(i)x_i}{\sum_{i=1}^{N} u_k} \tag{5.1}$$

where N is the number of pixels in a cell and x_i is the intensity of pixel i. After calculation of the cluster centroids, the updated membership values $u_k(i)$ are computed using the following equation:

$$u_k(i) = \frac{1}{\sum_{j=1}^{2} \left(\frac{\|v_k - x(i)\|}{\|v_j - x(i)\|} \right)^{\frac{2}{m-1}}} \tag{5.2}$$

where m is the fuzziness parameter. The above two steps are repeated until convergence is achieved.

5.6   ACTIVE-CONTOUR (SNAKE) BASED SEGMENTATION

The active-contour method used for segmentation of DNA microarray images is based on the active-contour scheme originally proposed by Kass et al. [18] also known as snake. In the active-contour approach the boundaries of the objects in an image are described by parametric curves $\Gamma(s)$, which minimize an energy functional consisting of internal components comprising prior smoothness and shape assumptions and an external component that incorporates model curve and image intensities. An object modeled this way is segmented by setting an initial contour and deforming it to minimize the energy functional. Generally the snake approach cannot be directly applied to segmentation of DNA microarray images. The main issue is the fact the spots might be very small and therefore there would be only a small number of pixels along the contour of the spot. This violates the assumption of large contour size compared to a single image pixel used in the design of the discrete active-contour algorithms.

In boundary-based segmentation, cell area R is assumed to consist of two regions, spot R_1 and background R_2. It is also assumed that a closed curve divides the cell into

R_1, R_2 where $R = R_1 \cup R_2$. We denote M_1 and M_2 as the expected values of pixels in R_1 and R_2, respectively. Let us consider a parametric description of the contour $\Gamma(s)$ where s is the parameter. The total energy induced by contour $\Gamma(s)$ is defined as the sum of the snake energies (external) and the region's energy (internal energy). The former measures the properties along the contour, while the latter measures the statistical differences between the regions separated by the contour. The total energy is written as

$$E_{total} = E_{snake} + \bar{\gamma} E_{region} \tag{5.3}$$

in which

$$E_{snake} = \int_\Gamma (\frac{\alpha}{2}|\Gamma_s|^2 + \frac{\beta}{2}|\Gamma_{ss}|^2 - ||\nabla I||^2)ds \tag{5.4}$$

$$E_{region} = \frac{\int \int_{R_1}(I - M_1)^2 dxdy + \int \int_{R_2}(I - M_2)^2 dxdy}{(M_1 - M_2)^2} \tag{5.5}$$

where α and β are constants, $\Gamma_s = [dx/ds, dy/ds]$, $\Gamma_{ss} = [d^2x/ds^2, d^2y/ds^2]$ and ∇I is the gradient of the image pixel values. In [19] the authors use an iterative algorithm to find the solution contour of the above equation. The authors also compare their spot segmentation results with those obtained by other algorithms for three sets of microarray images. Two of the sets contain some poor quality images from the Stanford Microarray Database (SMD) [20], while the third contains Agilent 60-mer oligonucleotide microarrays [21]. The Agilent microarrays are some of the best quality oligonucleotide chips available commercially. In the experiments, each cell is partitioned from adjacent cells manually, and apply the segmentation algorithm inside each region. Because the spots in the tested microarray images are small (10 to 15 pixels), the snake algorithm cannot be directly applied to segment the foreground and background of a cell. Therefore, in [19], the authors enlarge each cell by interpolation so that its size is four times that of the original cell. They then apply their algorithm to detect the contour of the enlarged spot image. Finally, the spot boundaries are obtained by down-sampling the contours detected in the enlarged image to their original size.

To evaluate the performance of the proposed algorithm, the authors of [19] compare it with the representative image analysis methods and software in GenePix Pro 5.0, which detects spots by circular boundary adjustment, and Spot 2.0, which detects spot regions by seed region growing. For the different segmentation results, they calculate the t-test value between the gray level pixels in the foreground and background, and use it to assess the performance of a segmentation algorithm. The t-test assesses whether the means of two groups are statistically different. The larger the t-test value, the better the segmentation result obtained. The authors show that the distributions of the t-values of the snake segmentation method are statistically larger than those of the other methods, which indicates that the contours derived by this method generally yield better segmentation results

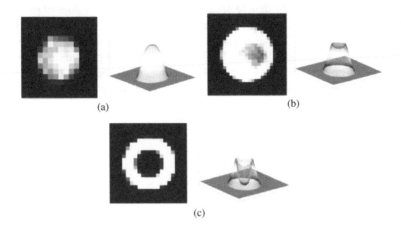

(a)   (b)

(c)

FIGURE 5.6   Different types of microarray spots in 2D and 3D dimensions: (a) a peak-shaped spot, (b) a volcano-shaped spot, and (c) a doughnut-shaped (figure from [1] with permission).

5.7   SHAPE-BASED SEGMENTATION

The authors of [1] proposes a microarray segmentation technique based on modeling spot and cell's shape. The authors argue that spots can be classified into three categories based on their shapes: peak shaped, volcano-shaped, and doughnut-shaped spots. Figure 5.6 shows these spot types. All the aforementioned spots categories can be represented using a 3D-curve representing the main body S_{MB} and a 3D-curve representing the inner-dip S_{ID} of the spot-model. Both the main-body and the inner-dip 3D curves resemble 3D Gaussian curves. Moreover, their orientation is opposite; the base of the main-body of the spot-model is down and its peak is up, while the base of the inner-dip of the spot-model is up and its peak is down. Figure 5.7 shows the S_{MB} and S_{ID} components of the 3D shape models of a peak-shaped spot a volcano-shaped spot and a doughnut-shaped spot. The spot-model $S_{Model}(x,y)$ is constructed by combining the $S_{MB}(x,y)$ and $S_{ID}(x,y)$ 3D-curves according to the following equation:

$$S_{MODEL}(x,y) = \min(S_{MB}(x,y), S_{ID}(x,y)) \tag{5.6}$$

Likewise, a model for the cell can be constructed as:

$$I_{MODEL}(x,y) = \max(B_{AV}, S_{MODEL}(x,y)) \tag{5.7}$$

where B_{AV} denotes the average background intensity of the cell. Pixels whose values are lower than B_{AV} belong to the background and their values are set equal to B_{AV}. In the next step, the parameters of the model should be set in a way to match the real image. The parameters consist of the value of the average background intensity B_{AV}, the values of the variables of the main-body S_{MB}, and the values of the variables of the inner-dip S_{ID}. In [1], a genetic algorithm is used to solve this non-linear optimization problem.

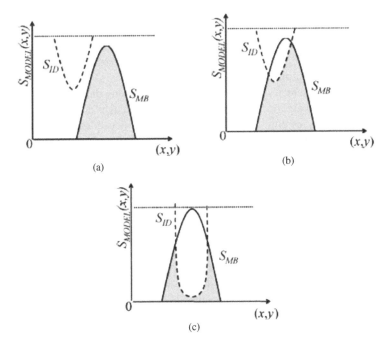

FIGURE 5.7 S_{MB} and S_{ID} components of the 3D models of: (a) a peak-shaped spot, (b) a volcano-shaped spot, and (c) a doughnut-shaped spot. The overall morphological models are the gray areas (Figure from [1] with permission)

5.8 GRAPH-BASED SEGMENTATION

In [22] a graph-based DNA microarray image segmentation is proposed. To explain the graph-based algorithm let us define P as the set of all pixels of a cell with the height and width of H_U and W_U. Let C be the set of possible grayscale values. Let I_B, be a function that assigns greyscale values to elements of P: $I_B : P \rightarrow C$. Let us consider an undirected weighted graph $G_U(V, E)$ corresponding to a cell. The vertices of the graph are the elements of P, $V(G_U) = P$. An edge exists between two vertices of this graph if their corresponding pixels are 4-neighbors. The weight $W(E)$ of an edge is defined as the absolute value of the difference between the greyscale intensities of the two pixels corresponding to the vertices of that edge. The weighted graph G_U would have small weights for smooth regions of an image grid-unit because of the similarities of neighboring pixels in that region. At the boundaries between regions the dissimilarities cause larger weights in the graph. The first step of the graph-based algorithm is to form a minimum spanning tree T_U from the weighted graph of G_U. The goal is to choose an edge in T_U where by cutting it the minimum spanning tree is divided into two subtrees corresponding to the foreground and background regions of the image. The selection process consists of two steps. In the first step, a group of candidate edges are selected. A number of tests are run on these candidate edges and only those that have certain conditions will go to the second step of the selection process. In the second step, another criterion is applied on the refined group of candidate edges. Eventually, from the pool of candidate edges, one edge is selected. By elimination of this selected edge the two desired subtrees are formed and the segmentation process is finished.

The first step in the selection process starts with decreasing the order of the edges of G_U based on their weights. Then, the first k (e.g., 10) edges are placed in set Q. Each edge in set Q is once eliminated (in a separate and independent test) from T_U to divide G_U into two subtrees. Let us assume edge q_i is a member of Q. By eliminating q_i two subtrees are produced. Each one of these subtrees represents a distinct region in the cell. A decision has to be made about which of these two regions corresponds to the spot and which one is the background. One of the two mentioned subtrees that has fewer pixels on the boundary of the cell is chosen as the spot component and denoted by $Spot(q_i)$. To see if the centroid of the spot region is close to the center of the cell or not the Euclidean distance between them is calculated as $\bar{d}$. In order to distinguish a spot from possible noise specks, the $Spot(q_i)$ component has to be larger than a certain threshold and it should also be close to the center of the cell. This means that the $\bar{d}$ distance should be smaller than a threshold d_U and the number of pixels of $Spot(q_i)$ has to be larger than a threshold n. Under these circumstances q_i becomes a member of Q_0. Threshold n indicates the minimum acceptable spot size.

The first step of the selection process is performed by eliminating an edge q_i from the tree to form two subtrees. Each subtree is studied as explained above. Then the eliminated edge is placed back in the tree and the next edge from the set Q is eliminated. This is done independently for all of the edges in set Q. Members of the set Q, which pass the above criteria, will be placed in set Q_0 and go to the second step of the selection process. In the second step of the selection process all of the candidate spot components are tested for compactness. From a pool of candidate spot components the intention is to choose the one that has higher resemblance to a circle. This is due to the fact that microarray technology is designed to generate circular spots and the spots actually turn out to have roughly circular shape. For each q_i member of Q_0 the number of pixels that lay on the boundary of the region represented by $Spot(q_i)$ is computed as $\alpha(Spot(q_i))$. Dividing $\alpha(.)^2$ by the area of the region would give a measure of the compactness of that region. Out of all candidate regions that may be formed by eliminating q_i the one that has a more circular shape, based on its compactness, is chosen as representing the spot. In the case that no edge can pass the tests performed in the first step of the selection process, it means no spot exists in that cell. For the implementation of the algorithm, Kruskal's method was used to generate the minimum spanning tree. Figure 5.8(a) shows a DNA microarray image and Figure 5.8(b) shows the segmentation results using the graph-based approach.

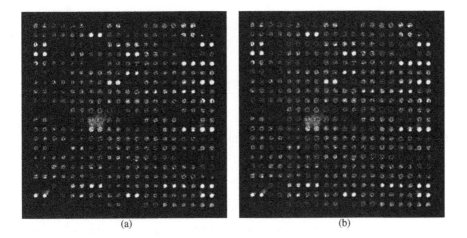

(a)   (b)

FIGURE 5.8 (SEE COLOR INSERT.) (a) A DNA microarray image. (b) Segmentation results using the graph-based method.

5.9 SEGMENTATION OF MICROARRAY IMAGES USING MATHEMATICAL MORPHOLOGY

In [23], a shape detector similar to binary hit-or-miss transform and a DNA microarray segmentation method based on it is proposed. This method is invariant to intensity transformation. Because the method is of local nature, local background changes can also be compensated. These properties make the method insensitive to changes in imaging and calibration.

The method is based on a gray-scale generalization of hit-or-miss transform. The detector has a hit window represented by points $H_1, \ldots H_h \in Z^2$ and a miss window represented by points $M_1 \ldots M_m \in Z^2$, which results in $n = h + m$ points in a window. There is also a weight vector $w = (w_0 \ldots w_n)^T \in R^{n+1}$ associated with the detector. The output of the detector is determined by a rank-filter $f_{h,m,w}(\mathbf{x})$

$$f_{h,m,w}(x) = w_0 + \sum_{i=1}^{h} w_i \left[x_i > x^{(m)} \right] + \sum_{i=h+1}^{h+m} w_i \left[x_i < x^{(m+1)} \right]$$

where $\mathbf{x} = (x_1 \ldots x_n) \in R^2$, $x^{(k)}$ is the kth order statistic of the sequence $x_1, \ldots x_n$ and $[a < b]$ is defined by

$$[a < b] = \begin{cases} 1 & \text{if} \quad a < b \\ 0 & \text{if} \quad a \geq b \end{cases}$$

If $\mathbf{H} = (H_1, \ldots, H_h)$ and $\mathbf{M} = (M_1, \ldots, M_m)$, the detector output $F_{H,M,w}(I)$ for the image I is defined as

$$(F_{H,M,w}(I))(i) = f_{h,m,w}(I(H_1 + i), \ldots I(H_h + i), I(M_1 + i), \ldots I(M_m + i))$$

First the output of the detector is compared to a threshold to determine the "hit" positions. All the points for which the detector's output is above a threshold form

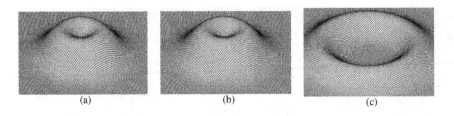

(a)   (b)   (c)

FIGURE 5.9 Volume normalized spot models with different shape variables.

the set $G_{\mathbf{H},\mathbf{M},\mathbf{w}}(I)$. Let us assume that the set $N_H(i) = \{H_1 + i, \ldots H_h + i\}$ is the hit neighborhood and the set $N_M(i) = \{M_1 + i, \ldots, M_m + i\}$ the miss neighborhood of point $i \in Z^2$. The first segmentation output $S^H_{H,M,w}(I)$ is the set of such points $j \in Z^2$ that for some $i \in G_{\mathbf{H},\mathbf{M},\mathbf{w}}(I)$, j belongs to $N_H(i)$ and $I(j)$ is greater than the mth order statistic of the sequence $I(H_1 + i), \ldots, I(H_h + i), I(M_1 + i), \ldots, I(M_m + i)$. The second segmentation output $S^M_{H,M,w}(I)$ is the set of such points $j \in Z^2$ that for some $i \in G_{\mathbf{H},\mathbf{M},\mathbf{w}}(\mathbf{I})$, j belongs to $N_M(i)$ and $I(j)$ is less than the $(m+1)^{th}$ order statistic of the sequence $I(H_1 + i), \ldots, I(H_h + i), I(M_1 + i), \ldots, I(M_m + i)$. These segmentation outputs are sets of candidate points as spot pixels and background pixels, respectively.

The hit and miss windows have to be selected based on the spot sizes in the imaging process. Usually the hit window would be chosen a little smaller than the area inside a common spot and the miss window would be chosen as an area strictly outside the spot. To reach acceptable segmentation results, the weights of the detector need to be adjusted. A training process can be used to find proper values for the weights [23]. In [23], the hit-or-miss-based segmentation method is compared with the Mann-Whitney scheme and it was observed that on the tested DNA microarray images the performances are close.

5.10   MODEL-BASED SEGMENTATION

The process of producing microarrays involves wetting a printing pin with the solution containing many copies of a DNA sequence and ejecting the droplet onto a solid substrate [24]. A spot made in such a way has a predictable shape, which is a function of droplet volume, velocity, and density. Trying many possible mathematical models, a good model that could simulate the shape of the spot is

$$f(r, r_{in}, t) = e^{-(r - r_{in})^2 / t^2} + e^{-(r + r_{in})^2 / t^2} \qquad (5.8)$$

where r is the distance from the center of the spot, and r_{in} and t are shape parameters that define the diameter of the spot and the thickness of its lobe. Figure 5.9 shows how we can have different spot shapes by adjusting these parameters. In any microarray manufacturing process, the volume of the droplet is constant. Thus the model we proposed should have a constant volume. For our model, the volume can be formulated as follows:

$$V(r_{in},t) = \int (e^{-(r-r_{in})^2/t^2} + e^{-(r+r_{in})^2/t^2}) 2r\pi dr. \tag{5.9}$$

Then, the normalized model based on r_{in} and t will be:

$$f_{norm}(r,r_{in},t) = f(r,r_{in},t)/V(r_{in},t)$$

In order to segment the image, every spot in the input image is matched with the proposed model. This model has five variables, x and y, which are the coordinates of the center of the model, r_{in} and t, which are the shape parameters, and c, which is the amplitude factor of the normalized model spot. To find the best matching, the least square image difference of the original spot and the model spot is minimized [25]. Figure 5.10 shows the result of segmenting a DNA microarray image using this method.

5.11 SEGMENTATION USING WATERSHEDS

Segmentation by morphological watersheds is a classic image segmentation approach that can be used for segmentation of DNA microarray images. The concept of watersheds is based on visualizing an image in three dimensions: two spatial coordinates and intensity [10]. In this context, a monochrome image is considered to be an altitude surface in which high-amplitude pixels correspond to ridge points and low amplitude pixels correspond to valley points. If a drop of water were to fall on any point of the surface, it would move to a lower altitude until it reached a local minimum. The accumulation of water in the vicinity of a local minimum is called catchment basin. Suppose that this three-dimensional topography is flooded from below and conceptual single pixel holes are pierced at each local minima. When the rising water in distinct catchment basins is about to merge, a dam is built to prevent the merging. The flooding ultimately reaches a level such that only the tops of the dams are visible. The dams form the boundaries of regions segmented by the watershed algorithm [10, 26].

5.12 SUPERVISED LEARNING-BASED SEGMENTATION

Giannakeas et al. [27] introduce a segmentation method that is based on classifying the pixels of a DNA microarray image into foreground (spot), background, and artifact pixels using support vector machines (SVM). Training the data required for the SVM is extracted using the fixed circle technique for real microarray images, while for simulated images the training data are extracted during the production of the images. The proposed method consists of the following steps. An automated gridding method is applied to isolate each spot of the image into a single cell. In this cell, vector filters are employed to remove the noise. Next, a pixel-by-pixel classification technique based on SVM is applied. A set of features for each pixel is then extracted. These features (14 in total) are related to the intensity of the pixel, the location of the pixel in the cell area, and the similarity of the pixel and its neighbors with a

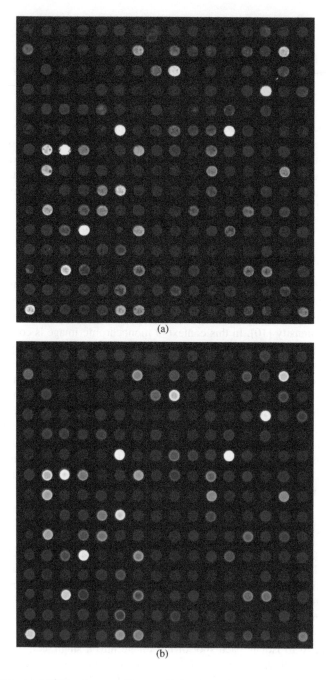

FIGURE 5.10   (a) A DNA microarray image. (b) Segmentation results using the model-based method.

theoretical spot (which is simulated by a 2D Gaussian template). The experimental results show that the obtained accuracy for the real images is about 98%, while the accuracies for the good, normal, and bad-quality simulated images are 96, 93, and 71%, respectively.

5.13   COMPARISONS, CONCLUSIONS, AND FUTURE DIRECTIONS

Since different DNA microarray segmentation algorithms have been applied to different microarray image sets, it is almost impossible to do a fair comparison of different segmentation algorithms. Moreover, many of the proposed segmentation algorithms have parameters that have to be set properly for achieving the best possible performance. This makes the comparison even more difficult. Additionally, in practice, it is not possible to obtain ground-truth information about the segmentation results. The authors of [28] also point to the fact that performance evaluation of microarray segmentation algorithms is not a straightforward task. In [28], a set of synthetic microarray images produced by a microarray simulator are used as test images. In [28], nine different DNA microarray image segmentation methods including fixed circle, adaptive circle, region growing, Mann-Whitney, k-means, and hybrid k-means have been compared. The results indicate that the k-means algorithm gave nearly error-free segmentation for the good quality images, whereas the Mann-Whitney algorithm produced clearly more erroneous segmentation for the same images. For the low quality DNA microarray images the hybrid k-means algorithm gave the best results.

Another approach that the researchers have been using is to compare DNA segmentation algorithms images from replicated experiments. There are two possible methods of replication. One method uses different microarray substrates for the same experiment. Another option is to use spot replicated microarray images where one spot is replicated a number of times on a single substrate. Replicate spots that are side-by-side are likely to be very highly correlated since they are not only printed with the same gene but are also spatially close together and therefore likely to share many common causes including local effects on the array surfaces as well as hybridization and labeling effects. Application of different algorithms on images from replicated spot experiments could result in suitable comparison between the segmentation algorithms in terms of consistency. In [22], the authors compare six DNA segmentation methods in terms of mean absolute error (MAE) of replicated spots. Their study shows that the graph-based segmentation method has a smaller MAE compared to circle-shape based and region growing-based segmentation schemes.

In this chapter, we briefly reviewed the main advantages and disadvantages of some DNA microarray segmentation algorithms. The fixed circle method is simple to implement, and works well if all the spots are circular and approximately the same size. However, the shapes and sizes of spots vary in practice, and hence circular segmentation clearly cannot satisfy the needs. An important advantage of Mann-Whitney segmentation is its lack of assumptions on spot shape; a drawback is the requirement of a background sample, which is difficult to provide reliably in an automated system. Clustering techniques have the same intrinsic initialization prob-

TABLE 5.2

Comparison of DNA microarray image segmentation approaches.

Method	Segmentation Quality	Level of automation	Features/Limitations
Circular-Shape	Low	High	Limited to circular spots
Region-growing	High	Low	Requires foreground and background seed points
Thresholding	Low	Requires a mask	Independent of the shape of a spot
Clustering	High	High	Pixels are divided into three groups: foreground, background, and deletions
Snake	High	Low	Computationally expensive
Shape	High	High	Assumes that the spots are symmetric
Graph	High	High	Computationally expensive
Morphology	High	High	Insensitive to changes in imaging and calibration
Model	High	High	Assumes that the spots are symmetric
Watersheds	High	High	Might over-segment the image
Supervised Learning-based	High	High	Requires already segmented images for training

lem as the Mann-Whitney segmentation. Watersheds and region growing methods require specification of an initial seed point.

In Table 5.2, we summarize important information for each method, including: 1) segmentation quality reported (might be based on different measures and datasets), 2) the level of automation and the need of parameter adjustment, and 3) associated features and/or limitations.

With the availability of more powerful computational resources, more advanced techniques will be developed for DNA image segmentation. Learning-based segmentation methods should also be considered for DNA image segmentation. In [29] a preliminary work based on neural networks and used for segmentation of DNA microarray images was proposed. However, more research is necessary in the field to assess the efficacy of the method.

ACKNOWLEDGMENTS

The author would like to thank Dr. Nader Karimi from Isfahan University of Technology for contributing some of the images.

REFERENCES

1. E. Zacharia and D. Maroulis, "3D spot-modeling for automatic segmentation of cDNA microarray images," *IEEE Transactions on NanoBioscience*, vol. 9, pp. 181–192, November 2010.

2. A. A. Ahmed, "Microarray segmentation methods significantly influence data precision," *Nucleic Acids Research*, vol. 32, no. 5, pp. 50–50, 2004.

3. M. Katzer, F. Kummert, and G. Sagerer, "Methods for automatic microarray image segmentation," *IEEE Transactions on NanoBioscience*, vol. 2, no. 4, pp. 202–214, 2003.

4. M. Eisen, "ScanAlyze, user manual," tech. rep., Stanford University, California, USA, March 1999.

5. *GenePix Pro 5.0 user's guide and tutorial.* Axon Instruments Inc, 2003.

6. *ScanArray Express Microarray Analysis System User Manual.* PerkinElmer Life Sciences, 2002.

7. J. Buhler, T. Ideker, and D. Haynor, "Dapple: improved techniques for finding spots on DNA microarrays.," Tech. Rep. UWTR 2000-08-05, Department of Computer Science and Engineering, University of Washington, Seattle, WA, August 2000.

8. L. Rueda and J. Rojas, "A pattern classification approach to DNA microarray image segmentation," *Pattern Recognition in Bioinformatics, Lecture Notes in Computer Science*, vol. 5780, pp. 319–330, 2009.

9. Y. H. Yang, M. Buckley, S. Dudoit, et. al, "Comparison of methods for image analysis on cDNA microarray data," *Journal of Computational and Graphical Statistics*, vol. 11, no. 1, pp. 108–136, 2002.

10. R. Gonzalez and R. Woods, *Digital Image Processing, Third Edition.* Prentice Hall, 2008.

11. E. R. Dougherty, *Probability and Statistics for the Engineering, Computing, and Physical Sciences.* Prentice-Hall, 1990.

12. Y. Chen, E. R. Dougherty, and M. L. Bittner, "Ratio-based decisions and the quantitative analysis of cDNA microarray images," *Journal of Biomedical Optics*, pp. 364–370, 1997.

13. D. Bozinov and J. Rahnenfuhrer, "Unsupervised technique for robust target separation and analysis of DNA microarray spots through adaptive pixel clustering," *Bioinformatics*, vol. 18, pp. 747–756, 2002.

14. L. Rueda and L. Qin, "A new method for DNA microarray image segmentation," *Image Analysis and Recognition, Lecture Notes in Computer Science*, vol. 3656, pp. 886–893, 2005.

15. G. Weng, Y. Hu, and Z. Li, "cDNA microarray image segmentation using shape-adaptive DCT and k-means clustering," in *2011 International Conference in Electrics, Communication and Automatic Control Proceedings*, pp. 317–324, 2012.

16. J. Rahnenfuhrer and D. Bozinov, "Hybrid clustering for microarray image analysis combining intensity and shape features," *BMC Bioinformatics*, vol. 5, no. 47, pp. 1–11, 2004.

17. J. Harikiran, D. RamaKrishna, M. Phanendra, et al., "Fuzzy C-means with bi-dimensional empirical mode decomposition for segmentation of microarray image," *International Journal of Computer Science Issues*, vol. 9, pp. 189–198, September 2012.

18. M. Kass, A. Witkin, and D. Terzopoulos, "Snakes: Active contour models," *International Journal of Computer Vision*, vol. 1, no. 4, pp. 321–331, 1988.

19. J. Ho and W.-L. Hwang, "Automatic microarray spot segmentation using a snake-fisher model," *IEEE Transactions on Medical Imaging*, vol. 27, no. 6, pp. 847–857, 2008.

20. J. Gollub, C. A. Ball, G. Binkley, et al., "The Stanford microarray database: Data access and quality assessment tools," *Nucleic Acids Research*, vol. 31, pp. 94–96, Jan. 2003.

21. Agilent, "Agilent ADS circuit components manual," tech. rep., Available at: http://www.agilent.com/find/eesof-knowledgecenter.

22. N. Karimi, S. Samavi, S. Shirani, et al., "Segmentation of DNA microarray images using

an adaptive graph-based method," *IET Image Processing*, vol. 4, no. 1, pp. 19–27, 2010.

23. P. Vesanen, M. Tiainen, and O. Yli-Harja, "Calibration-free methods in segmentation of cDNA microarray images," in *Proc. SPIE Vol. 4667, p. 291-302, Image Processing: Algorithms and Systems*, vol. 4667, pp. 291–302, 2002.

24. M. Bonner, K. McWeeny, P. Gwynne, et al., "BioChip SNP analysis assay: Development of a 3-D microarray system," *American Journal of Human Genetics*, vol. 67, no. 4, pp. 266, 2000.
M. Bonner, "BioChip SNP analysis assay: Development of a 3-d microarray system," tech. rep., Motorola BioChip Systems, Arizona, USA.

25. N. Faramarzpour, J. Bondy, and S. Shirani, "Lossless DNA image compression," in *Proceedings of the Thirty-Seventh IEEE Asilomar Conference on Signal, Systems and Computers.*, vol. 2, (Pacific Grove, CA, USA), pp. 1501–1504, 2003.

26. W. Pratt, *Digital Image Processing, Third Edition*. Wiley, 2001.

27. N. Giannakeas, P. S. Karvelis, T. P. Exarchos, et al., "Segmentation of microarray images using pixel classification—comparison with clustering based methods," *Computers in Biology and Medicine*, vol. 43, pp. 705–716, July 2013.

28. A. Lehmussola, P. Ruusuvuori, and O. Yli-Harja, "Evaluating the performance of microarray segmentation algorithms," *Bioinformatics.*, vol. 22, no. 23, pp. 2910–2917, 2006.

29. Z. Wang, B. Zineddin, J. Liang, et al., "A novel neural network approach to cDNA microarray image segmentation," *Computer Methods and Programs in Biomedicine*, vol. 11, pp. 189–198, July 2013.

6 Statistical Segmentation Methods for DNA Microarray Images

Meng-Yuan Tsai, Tai-Been Chen, and Henry Horng-Shing Lu

CONTENTS

6.1 INTRODUCTION

Microarrays allow researchers to study tens of thousands of genes at once [1]. However, due to the increase in data size, microarray data analysis has become far more difficult. The key issues lie in its complexity and computation rate. Therefore, statistical methods become crucial in these large datasets. Since DNA microarrays are typically noisy, various approaches have been proposed to improve efficiency and accuracy or the corresponding data analysis. Segmentation is one of these standard processes, which is essential in analyzing the intensities of microarray images [2, 3]. It is a process used to classify the foreground pixels from the background pixels in a spot, where the former correspond to pixels of interest and the latter simply represent some noise or just background. The main goal of segmentation is to extract important features from images and use them for quantification. Segmentation in microarray images is tricky in computation, since a poor-quality image can substantially affect the spot morphology, and hence it would be difficult to differentiate foreground from background. Thus, different methods have been proposed for simplifying the underlying computations.

To be more specific, the segmentation problem is stated mathematically as follows. Suppose $\mathbf{y}^{Cy5} = \{y_{ij}^{Cy5} | i = 1, 2, ..., m; j = 1, 2, ..., n\}$ is an integer-valued matrix representing the Cy5 image, and $\mathbf{y}^{Cy3} = \{y_{ij}^{Cy3} | i = 1, 2, ..., m; j = 1, 2, ..., n\}$ denote an integer-valued matrix representing the Cy3 image. Each element of an image is a pixel. Assume that k clusters $\mathbf{c} = \{c_i | i = 1, 2, ..., k\}$ are considered, where each cluster represents one category of pixel intensities. In practice, we assume that there are only two clusters, foreground and background, denoted as c_1 and c_2, respectively. The segmentation problem aims to assign each pixel of $\mathbf{y}^{Cy5}$ or $\mathbf{y}^{Cy3}$ to one of the classes, $\mathbf{c}$. For example, considering the two-class problem, the result of the segmentation problem will be a binary image $\mathbf{z}$, where the values of the elements in $\mathbf{z}$ are either 0 or 1. The binary value represents where each pixel belongs either background or foreground [4, 5].

Segmentation of DNA microarray images is an interesting problem both in non-statistics, i.e., machine learning, and statistics. Several methods have been presented, including fixed circle segmentation, adaptive segmentation, histogram-based segmentation, and clustering-based segmentation [6]. Most of the above approaches have been adopted in different image analysis software packages [7, 8, 9, 10, 11, 12]. Table 6.1 lists all the software packages and algorithms related to different methods, respectively. Fixed circle segmentation utilizes a circular mask with fixed diameter to confine the region of spots. The restriction in shape and diameter leads to less flexibility and poor accuracy in experimental results [13]. Adaptive circle segmentation further improves the previous method. It allows a circular mask to have a diameter that varies with the data [14]. Adaptive shape segmentation lifts the restriction in shape, which yields more flexibility and accuracy [15, 16]. Histogram-based segmentation considers the problem diversely, which aims to find the intensities of foreground and background directly [17]. Clustering-based segmentation is a pioneering undergoing method, differentiating foreground from background based on the information obtained from the data [18].

While Chapter 5 focuses on non-statistical methods such as the ones mentioned above, this chapter focuses primarily on statistical methods for segmentation of DNA microarray images. Typically, statistical methods can be divided into two groups: parametric and nonparametric methods. Three methods are introduced in this chapter: the Gaussian mixture model (GMM), kernel density estimation (KDE), and the Gaussian mixture model incorporated with kernel density estimation (GKDE) [19]. The Gaussian mixture model relies on the assumption of normality for the application to segmentation problems. This method fits both foreground and background as different normal distributions, and the overall intensities of the spot is the mixture of these two normal distributions. Although the GMM is simple, the normality assumption is a strong restriction in statistics. Therefore, it limits the flexibility of

TABLE 6.1

Segmentation methods and the corresponding algorithms and software packages.

Methods	Software/Algorithms
Fixed circle	ScanAlyze [8], GenePix [7], QuantArray [9]
Adaptive circle	GenePix [7], ScanAlyze [8], ScanArray Express [12], ImaGene [11], Dapple [14]
Adaptive shape	Spot [10], Seeded Region Growing (SRG) [16]
Histogram-based	QuantArray [9] and adaptive thresholding [17]
Clustering-based	k-means single-feature clustering microarray image segmentation (KSCMIS) [6]

the method. Kernel density estimation is a nonparametric approach in order to relax the parametric assumption, such as the normality assumption. Since the selection of all parameters is data-driven, KDE gives more flexibility in density estimation. However, higher flexibility leads to higher complexity. Hence, the Gaussian mixture model incorporated with kernel density estimation (GKDE) was developed in order to combine the advantages of the GMM and KDE.

We state the segmentation problem mathematically, and discuss some prior knowledge for better understanding in Section 2. Section 3 describes basic ideas about the mixture model, which are then extended to the GMM at the end of the section. The concept of KDE is introduced in Section 4, and is followed by details on the applications of microarray image segmentation. In addition, a new method that combines the advantages of the GMM with those of kernel density estimation, GKDE, are discussed in Section 5. All these sections contain the relevant algorithms. At the end of the chapter, sixteen microarray images are used as an example to compare the performance of different segmentation methods.

6.2   THE GAUSSIAN MIXTURE MODEL

6.2.1   INTRODUCTION TO MIXTURE MODELS

Mixture model development is attributed to practical problems where variables are measured under two or more conditions [20]. For example, the distribution of color images reflects the mixture of three primary colors, red (R), green (G), and blue (B), and the expression of one gene might be a mixture of the upstream genes in which they do or do not express. In such cases, it is more difficult or less intuitive to construct the model directly from the measurements. Therefore, mixture models were developed for these kinds of situations. For instance, the value of each pixel in the image can be interpreted from the values of three primary colors. One can consider each color feature as a separate univariate distribution, such as normal, rather than model the color map of the images directly. Thus, the color map of the images can be referred to as the combination of these color features. In another example, it is less realistic if we fit the expression of the interested gene directly. Instead, it is more informative to fit the expression of upstream genes separately, such as the binomial distribution.

The development of the mixture model started in the 1970's, and it has been extensively studied and widely applied to various topics, such as classification, clustering, density estimation and pattern recognition. More details regarding the introduction to mixture model can be obtained from Titterington, Smith, and Markov [21], McLachlan and Basord [22], and McLachlan and Peel [23]. With the thriving development of computational methods, especially Markov chain Monte Carlo (MCMC) techniques, works on mixture model have been dedicated to Bayesian approaches, including West [24], Diebolt and Robert [25], Escobar and West [26], Richardson and Green [27], Mullar and Rosner [28], and Stephens [29].

A mixture model describes the combination of subpopulations. Each subpopulation illustrates one or two characteristics of the latent features. Considering all the latent features together, one may acquire some insight regarding the overall measurements. Suppose $\{\mathbf{Y}_1, \mathbf{Y}_2, ..., \mathbf{Y}_n\}$ denote a random sample of size n, where $\mathbf{Y}_j$ is a p-dimensional random vector with probability density function $f(\mathbf{Y}_j)$ on R^p. In other words, $\mathbf{Y}_j$ contains the random variables corresponding to p measurements on the j-th recording. In addition, let $\mathbf{Y} = (\mathbf{Y}_1, \mathbf{Y}_2, ..., \mathbf{Y}_n)^T$ represent the collection of all these random vectors, where the superscript T denotes vector transpose. That is, $\mathbf{Y}$ stands for the entire sample. With the random sample $\mathbf{Y}_j$, the density $f(\mathbf{Y}_j)$ can be written in the form:

$$f(\mathbf{Y}_j) = \sum_{i=1}^{k} \pi_i f_i(\mathbf{Y}_j), \tag{6.1}$$

where $f_i(\mathbf{Y}_j)$ are densities and the π_i's are non-negative and add up to 1; that is, $0 \le \pi_i \le 1, i = 1, 2, ..., k$ and $\sum_{i=1}^{k} \pi_i = 1$. The quantities $\pi_1, \pi_2, ..., \pi_k$ are called mixing proportions or simply weights. The functions $f_1(\mathbf{Y}_j), f_2(\mathbf{Y}_j), ..., f_k(\mathbf{Y}_j)$ are the densities and are also named as component densities of the mixture, or mixture components. Since all the components in (6.1) are densities, it is apparent that $f(\mathbf{Y}_j)$ is a density as well. $f(\mathbf{Y}_j)$ can be referred to as a k-component mixture density and its relevant distribution function can be considered as a k-component mixture distribution.

Another point of view to describe a mixture model is to use the Bayesian approach [30, 31]. The Bayesian approach started to become popular due to the enhanced computation power. Its key idea lies in conditional probability:

$$p(\mathbf{Y}|\Theta) = \frac{p(\mathbf{Y}, \Theta)}{p(\Theta)}. \tag{6.2}$$

The probability of the model can be considered as the joint probability distribution of parameters and random sample. Thus, the joint probability distribution can be written in another form related to (6.2):

$$p(\mathbf{Y}, \Theta) = p(\Theta) \times p(\mathbf{Y}|\Theta). \tag{6.3}$$

Here, the parameters are no longer constants; instead, they are referred to as another random sample. Therefore, the marginal probability distribution of random sample $\mathbf{Y}$ is:

$$p(\mathbf{Y}) = \int p(\Theta) \times p(\mathbf{Y}|\Theta)d\Theta. \tag{6.4}$$

where $p(\Theta)$ is called the prior density of the parameters Θ.

Similarly to the former description, suppose $\{\mathbf{Y}_1, \mathbf{Y}_2, ..., \mathbf{Y}_n\}$ denote a random sample of size n, where $\mathbf{Y}_j$ is a p-dimensional random vector with probability density function $f(\mathbf{Y}_j)$ on R^p. Let $f(\mathbf{Y}_j|\theta_i), i = 1, 2, ..., k$ denote the component densities with parameter θ_i, and the weight π_i be the proportion of the population from component i with $\sum_{i=1}^{k} \pi_i = 1$. Usually, the mixture components are assumed to lie in the same distribution family, such as the exponential family, but with different parameters. Hence, the sampling density of $\mathbf{Y}_j$ in the above case is:

$$f(\mathbf{Y}_j|\Theta, \pi) = \pi_1 f(\mathbf{Y}_j|\theta_1) + \pi_2 f(\mathbf{Y}_j|\theta_2) + ... + \pi_k f(\mathbf{Y}_j|\theta_k). \tag{6.5}$$

The form of the sampling density of $\mathbf{Y}_j$ is definitely a mixture density as described above. The mixing proportion $\pi = (\pi_1, \pi_2, ..., \pi_k)$ can be considered as a discrete prior distribution. However, in the standard Bayesian setup, it is more appropriate to consider the prior distribution as a description of the variance in Θ over the population of interest. From this point of view, the mixture model is more adequate to resemble a hierarchical model with an unobserved indicator τ_{ji}:

$$\tau_{ji} = \begin{cases} 1 & \text{if the } j\text{-th unit is drawn from the } i\text{-th mixture component,} \\ 0 & \text{otherwise.} \end{cases} \tag{6.6}$$

Thus, the entire model now can be written as follows:

$$p(\mathbf{Y}, \tau|\Theta, \pi) = p(\tau|\pi) \times p(\mathbf{Y}|\tau, \Theta) = \prod_{j=1}^{n} \prod_{i=1}^{k} (\pi_i f(\mathbf{Y}_j|\theta_i))^{\tau_{ji}}, \tag{6.7}$$

where $\tau_{ji} \sim multinomial(1; \pi_1, \pi_2, ..., \pi_k)$. Equation (6.7) is equivalent to (6.1).

A mixture model describes a complex distribution with the combination of simple sub-distributions. Thus, it provides a satisfactory model that is more informative. For example, the expression of one gene can be considered as the co-effect of distinct upstream genes using a mixture model. In addition, it is more flexible and less complicated in modeling, providing wide applications in many research fields, such as neural networks, heterogeneity, and image processing.

6.2.2 THE GAUSSIAN MIXTURE MODELS FOR DNA MICROARRAY IMAGES

The GMM is widely used under the assumption of normality. The GMM for segmentation of DNA microarray images assumes that the distribution of foreground intensities is a Gaussian distribution $f_1(\mu_1, \sigma_1^2)$ with mean μ_1 and variance σ_1^2; while the distribution of background intensities is another Gaussian distribution $f_2(\mu_2, \sigma_2^2)$ with mean μ_2 and variance σ_2^2. Suppose y_j denotes the intensity at the j-th pixel in a spot. Then, the distribution of y_j can be modeled as (6.8):

$$f(y_j; \theta) = \pi_1 f_1(y_j; \mu_1, \sigma_1^2) + \pi_2 f_2(y_j; \mu_2, \sigma_2^2), j = 1, ..., n, \tag{6.8}$$

where

$$f_i(y_j; \theta_i) = \frac{1}{\sqrt{2\pi\sigma_i^2}} exp(\frac{-(y_j - \mu_i)^2}{2\sigma_i^2}), \theta_i = \{\mu_i, \sigma_i^2\}, i = 1, 2 \tag{6.9}$$

and π_i is the mixing (or prior) probability constrained by $0 \leq \pi_i \leq 1$ and $\pi_1 + \pi_2 = 1$.

Foreground intensities typically include most of the signal and some noise; while the background intensities mainly consist of noise. Thus, the mean of foreground intensities should exceed the mean of background intensities. In other words, the constraint $\mu_1 \geq \mu_2$ should be considered in the study, which is commonly used in the literature [19, 32]. The expectation-maximization (EM) algorithm is applied to estimate the above parameters [33]. The purpose of EM algorithm is to find maximum likelihood estimates for the parameters, where the model contains unobserved latent variables. In the expectation (E) step, the expected values (or posterior mean) of the mixing proportions π and the parameters θ are first estimated under the given initial values. Then, in the maximization (M) step, the log-likelihood of the observed data in the two-component mixture model is computed as follows:

$$\log(L(\theta|\mathbf{y})) = \sum_{j=1}^{n} \log(\sum_{i=1}^{2} \pi_i f_i(y_j; \mu_i, \sigma_i^2)). \tag{6.10}$$

If the value of the log-likelihood given the observed data is approximately equal to that of the previous step, the iteration will be terminated. Otherwise, the values of π and θ calculated in the E step are the new values for the next iteration.

In this study, the initial parameters θ and mixing proportions π are set as follows. Initially, μ_1 and μ_2 are set to the first and third quartiles of pixel intensities in one spot, while σ_1^2 and σ_2^2 are the standard deviations of the pixel intensities below the first quartile and above the third quartile, respectively. The initial values of π_1 and π_2 are set to 0.5. The segmentation algorithm of one spot using the GMM is given in Algorithm 6.1 [19].

6.3   KERNEL DENSITY ESTIMATION

6.3.1   INTRODUCTION TO KERNEL DENSITY ESTIMATION

Research about density estimation can be traced back to the 1950's. Apart from histogram, there are many other methods that estimate the densities from different points of view, such as spline, wavelet and Fourier series. Rosenblatt first introduced the kernel method for density estimation [34]. Since then, more statisticians have been

---

Algorithm 6.1 Segmenting one spot by the GMM

Input: Load the microarray image, denoted by $\mathbf{y} = \{y_1, y_2, ..., y_n\}$

$k \leftarrow 0,$

$\pi^{(0)} = \{\pi_i^{(0)}, i = 1, 2\} \leftarrow \{0.5, 0.5\},$

$\theta^{(0)} \leftarrow \{\mu_i^{(0)}, \sigma_i^{2(0)}, i = 1, 2\},$

$tol \leftarrow 10^{-2}$

repeat

$$\omega_{ji}^{(k)} = \frac{\pi_i^{(k)} f_i(y_j; \mu_i^{(k)}, \sigma_i^{2(k)})}{\sum_{m=1}^{2} \pi_m^{(k)} f_m(y_j; \mu_m^{(k)}, \sigma_m^{2(k)})}$$

$$\theta^{(k+1)} = \{\mu_i^{(k+1)}, \sigma_i^{2(k+1)}, i = 1, 2\}$$

$$= \{\frac{\sum_{j=1}^{n} \omega_{ji}^{(k)} y_j}{\sum_{j=1}^{n} \omega_{ji}^{(k)}}, \frac{\sum_{j=1}^{n} \omega_{ji}^{(k)} (y_j - \mu_i^{(k+1)})^2}{\sum_{j=1}^{n} \omega_{ji}^{(k)}}, i = 1, 2\}$$

$$\pi_i^{(k+1)} = \frac{1}{n} \sum_{j=1}^{n} \frac{\pi_i^{(k)} f_i(y_j; \mu_m^{(k)}, \sigma_i^{2(k)})}{\sum_{m=1}^{2} \pi_m^{(k)} f_m(y_j; \mu_m^{(k)}, \sigma_m^{2(k)})}$$

until

$$\log(L(\theta^{(k+1)} | \mathbf{y})) - \log(L(\theta^{(k)} | \mathbf{y})) < tol$$

$$\omega_{ji}^{(k+1)} = \frac{\pi_i^{(k+1)} f_i(y_j; \mu_i^{(k+1)}, \sigma_i^{2(k+1)})}{\sum_{m=1}^{2} \pi_m^{(k+1)} f_m(y_j; \mu_m^{(k+1)}, \sigma_m^{2(k+1)})}$$

$$y_j \in \begin{cases} foreground & \text{if } \omega_{j1}^{(k+1)} > \omega_{j2}^{(k+1)}, \\ background & \text{otherwise.} \end{cases}$$

return y

---

working on this concept. Scott and Thompson extended the kernel method to high dimensional density estimation and Sheather and Jones provided a reliable data-based bandwidth selection method for kernel density estimation [35, 36]. Silverman offered more details on kernel density estimation and another method for density estimation [37].

The probability density function $(p.d.f.)$, provides a good representation of the population of interest, and describes the shape and features of the measurements. Suppose a random variable Y that has probability density function f. Then, by definition, the distribution, and the associated probability of Y can be described as (6.11):

$$P(a < Y < b) = \int_a^b f(y) dy \quad \text{for all } a < b. \tag{6.11}$$

The methods to density estimation can be classified into two main branches: parametric and nonparametric. Parametric approaches need more rigid assumptions while nonparametric approaches do not. The first step of parametric approaches is to determine the distribution family. Based on the assumption of distribution, the parameters will then be estimated from the observed data. Taking the normal distribution with two parameters, mean μ and variance σ^2, as an example, the density f can be determined simply by estimating the parameters, μ and σ^2 from the data, and substituting the estimates into the normal density. Nonparametric approaches avoid those rigid assumptions made about the distribution. Instead, nonparametric approaches allow the data to present themselves. For example, nonparametric approaches only assume that the observed data have some density f with mean μ and variance σ^2. Then, the constraints that data provides will determine the density f. Nonparametric approaches do not need the assumption of a specific form of distribution with parameters, which provides more flexibility to the model. However, the nonparametric estimate typically needs higher computational cost than the parametric estimate.

Kernel density estimation is a nonparametric approach. It is generalized from the histogram method, after improving some of its statistical properties. Suppose a random variable Y has some unknown density f. To estimate the unknown probability density function $f(Y)$, we first consider a small region R. The probability p that Y falls in R is given by:

$$p = \int_R f(Y)dY. \tag{6.12}$$

After collecting n observations of the data, the total number k of observed data that lie in region R follows the binomial distribution with probability p:

$$binomial(k|n,p) = \frac{n!}{k!(n-k)!}p^k(1-p)^{n-k}. \tag{6.13}$$

Hence, the fraction of data points falling in R is a distribution sharply peaked around p if n is large enough, since:

$$E(k/n) = p, \tag{6.14}$$

$$Var(k/n) = \frac{p(1-p)}{n} \rightarrow 0, \text{if } n \rightarrow \infty. \tag{6.15}$$

If R is sufficiently small, the value of the integration can be approximated into:

$$p \cong f(Y)h, \tag{6.16}$$

where h stands for the length of R. Since $\frac{k}{n} \cong p$ and the density $f(Y)$ is roughly constant within the region R, (6.16) can be written as:

$$f(y) \cong \frac{k}{nh}. \tag{6.17}$$

Based on (6.17), the main idea of kernel density estimation is to find the optimal value of k under some fixed h from the data. To count the number of data points falling inside R, it is useful to define the kernel function, K. A kernel function is a density centered at the origin. The value of $K\left(\frac{y-y_j}{h}\right)$ will be one if the data point y_j is inside the region centered at y with length h. In practice, the kernel function K is usually selected to be symmetric, such as uniform or normal. These kernel functions decide the characteristics of the density estimate. Therefore, before adopting the kernel function, the purpose of the study and the property of the data should be considered [38]. The kernel density estimator with kernel function K is defined as:

$$\hat{f}(y) = \frac{1}{nh}\sum_{j=1}^{n} K(\frac{y-y_j}{h}). \tag{6.18}$$

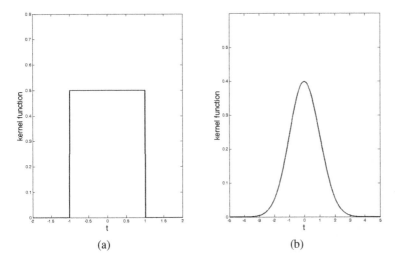

(a)   (b)

FIGURE 6.1   Kernel functions: (a) Uniform kernel function; (b) Normal kernel function.

where h is called the window width, bandwidth, or smoothing parameter, and the kernel function should satisfy the following conditions:

$$K(t) \geq 0, \tag{6.19}$$

$$\int_{-\infty}^{\infty} K(t)dt = 1. \tag{6.20}$$

The uniform kernel function, as shown in Figure 6.1(a), is much simpler in computation:

$$K(t) = \begin{cases} \frac{1}{2} & \text{if } |t| \leq 1, \\ 0 & \text{otherwise.} \end{cases} \tag{6.21}$$

However, due to its stepwise form, the uniform kernel will cause the density estimates to be ragged. The normal kernel function or Gaussian kernel function, on the other hand, is smoother and continuous, and has the form:

$$K(t) = \frac{1}{\sqrt{2\pi}} \exp\left(-\frac{t^2}{2}\right), \quad |t| < \infty. \tag{6.22}$$

Due to the special properties of the normal distribution, such as continuity and symmetry, it is the most useful kernel function in applications. The normal kernel function not only provides the advantages of bump hunting, but also decreases the complexity of the computations.

Another issue in kernel density estimation is the selection of the bandwidth h. The bandwidth h determines the spread of a kernel function. In addition, it determines the details of the observed data included in the estimated density. In other words, h decides the smoothness of the estimated density. To illustrate this point, a histogram and two normal kernel density estimates with different bandwidths are

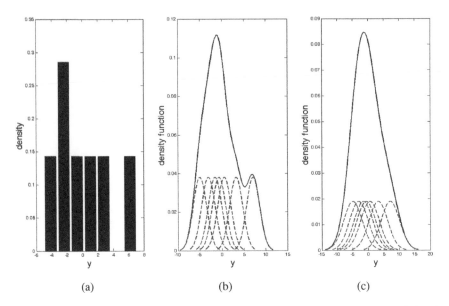

FIGURE 6.2 Simulated results: (a) histogram; (b) with bandwidth $h = 1.5$; (c) with bandwidth $h = 3.0$.

compared using simulated data: $y = (-5, -3.1, -1.6, -0.7, 0.5, 3.2, 7.1)$. The horizontal axis of the histogram is divided into 7 bins of width 2, shown in Figure 6.2(a). A normal kernel with a bandwidth of 1.5 is adopted, which has a variance of 2.25 according to (6.22), indicated by the dashed lines in Figure 6.2(b). In Figure 6.2(c), a normal kernel with variance 9 is used to estimate the density. The kernel density estimates are displayed with solid lines in both Figure 6.2(b) and Figure 6.2(c). The histogram partitions the entire region into disjoint intervals and then counts the number of observations in every interval. Thus, the histogram is not smooth. In addition, the shape of the histogram is very sensitive to the selection of the bandwidth and interval points. However, the kernel density estimates consider the joint contribution of the data around the target. Hence, the kernel density estimates evidently obtains much smoother density estimates than the histogram. Different bandwidths h give different density estimates. A smaller bandwidth leads the estimate to raggedness since it contains more spurious data artifacts. A larger bandwidth allows the kernel to cover a larger space, and thus, it results in an over-smooth density model. Therefore, finding a satisfactory bandwidth h is crucial in kernel density estimation.

Ideally, h tends to be as small as the data allow. However, it is actually a trade-off between the bias and the variance of the estimator. A very intuitive way to select the bandwidth is to consider the mean integrated squared error (MISE), or the expected L_2 risk function [39]:

$$MISE(h) = E\left[\int (\hat{f}_h(y) - f(y))^2 dy\right]. \qquad (6.23)$$

If certain criteria are satisfied: (a) the kernel function is a symmetric probability density; (b) $h(n) \rightarrow 0$ as $n \rightarrow \infty$, the MISE could be reduced to the asymptotic mean integrated squared error (AMISE) by taking the first two terms of its asymptotic expansion:

$$AMISE(h) = n^{-1}h^{-1} \int [K(t)]^2 dt + \frac{1}{4}h^4 K_2^2 \int [f''(y)]^2 dy, \qquad (6.24)$$

where $K_2^2 = \int y^2 K(y) dy$ and f'' is the second derivative of f. The AMISE can be calculated by using integrals to obtain h_{AMISE}:

$$h_{AMISE} = K_2^{-\frac{2}{5}} \left\{ \int [K(t)]^2 dt \right\}^{\frac{1}{5}} \left\{ \int [f''(y)]^2 dy \right\}^{-\frac{1}{5}} n^{-\frac{1}{5}}. \qquad (6.25)$$

An ideal bandwidth of a Gaussian kernel with variance σ^2 and the size of sample n via h_{AMISE} can be mathematically computed as follows:

$$h_{AMISE} \approx (\frac{4}{3})^{\frac{1}{5}} \sigma n^{-\frac{1}{5}}. \qquad (6.26)$$

The detailed proof of the AMISE can be found in [37]. Although either the MISE or the AMISE has an explicit mathematical form, these two are not feasible since they contains the second derivative of an unknown density function f''. Thus, more computational, data-based methods have been proposed for the selection of the bandwidth, including plug-in selectors and cross-validation selectors.

6.3.2 KERNEL DENSITY ESTIMATION FOR DNA MICROARRAY IMAGES

A Gaussian kernel has the form of (6.22). Due to its symmetry and computability, the Gaussian kernel is widely used in density estimation. Hence, we apply the Gaussian kernel to the illustrated example in the later section. Here, we introduced the algorithm about the KDE using the Gaussian kernel with automatic bandwidth selection. 128 grid points are used to obtain the density estimate. Figure 6.3 shows an example that uses the Gaussian kernel to estimate the density of Cy5 and Cy3:

$$\hat{f}(x_i) = \frac{1}{n} \sum_{j=1}^{n} \frac{1}{\sqrt{2\pi}h} \exp \left(-\frac{1}{2} \left(\frac{x_i - y_j}{h} \right)^2 \right), \qquad (6.27)$$

where y_j is the j-th sample in a spot, x_i is the i-th grid point, h is the bandwidth used in the Gaussian kernel to estimate a spot probability density function, n is the number of pixels in a spot, and $i = 1, 2, ..., 128$. The details are given in Algorithm 6.2 [19].

6.4 GAUSSIAN MIXTURE MODEL INCORPORATED WITH KERNEL DENSITY ESTIMATION

Both the GMM and KDE have their own advantages and weaknesses. The GMM can provide the initial segmentation, while the KDE method can further improve

---

Algorithm 6.2 Segmenting one spot by the KDE

Input: Load the microarray image, denoted by $\mathbf{y} = \{y_1, y_2, ..., y_n\}$
1. Find 128 grid points that are equally spaced:
$$x_i = \min(\mathbf{y}) + i \times \left(\frac{\max(\mathbf{y}) - \min(\mathbf{y})}{m}\right), \, for \, i = 1, 2, ..., m.$$
2. Calculate the data-driven bandwidth for KDE:
$$h = 0.9 \times \min\{std, \frac{IQR}{1.34}\} \times n^{-\frac{1}{5}}$$
3. Calculate the KDE using (6.27).
4. Find the first local minimum of the KDE at $x_i^* = CP$ as a cut-off point.
5. $y_j \in \begin{cases} foreground & \text{if } y_j > CP, \\ background & \text{otherwise.} \end{cases}$

return $\mathbf{y}$

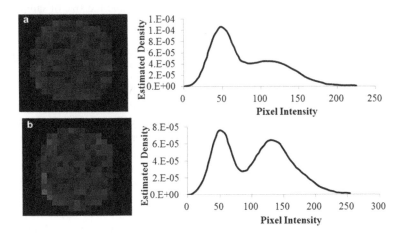

FIGURE 6.3 (**SEE COLOR INSERT.**) Two estimated density curves for spots of Cy5 (a) and Cy3 (b) dyes. Both Cy5 and Cy3 images have two intensity distributions $\hat{f}(y_j)$, which consist of background and foreground pixels. The intensity distribution of Cy5 (a) is computed with sample size $n = 289$ and bandwidth $h = 35$, and the distribution of Cy3 (b) is computed with sample size $n = 289$ and bandwidth $h = 30$. The local minimum is used as the cut-off point for segmenting the spot.

the segmentation by relaxing the assumption of normality. Although KDE has more flexibility than the GMM, its computational cost is higher than that of the GMM. Therefore, the Gaussian mixture model incorporated with kernel density estimation (GKDE) is developed by combining the advantages of the two existing methods. Once the foreground and background are found using the GMM, the KDE can be applied to find their estimated densities. Then, a cut-off point for segmenting a spot into two clusters is determined by the equality of two estimated densities. The details are given in Algorithm 6.3 [19].

Algorithm 6.3 Segmenting one spot by the GKDE

Input: Load the microarray image, denoted by $\mathbf{y} = \{y_1, y_2, ..., y_n\}$
1. Segment a spot initially using the GMM.
2. Estimate the kernel densities for foreground, $\hat{f}_f$, and background, $\hat{f}_b$.
3. Find a cut-off point CP that is close to $\hat{f}_f = \hat{f}_b$.
4. $y_j \in \begin{cases} foreground & if\ y_j > CP, \\ background & otherwise. \end{cases}$

return y

6.5   EXAMPLE

6.5.1   THE SIXTEEN MICROARRAY DATASET

This chapter is concluded with a real example using the methods discussed in the previous sections. Sixteen microarrays used herein are an attempt to understand the molecular mechanisms for the different clinical features between adenocarcinoma/adenosquamous carcinoma (AC/ASC) and squamous carcinoma (SC) of the uterine cervix [40]. The microarray images consist of 32 sub-grids with 484 spots in each sub-grid and 7,744 genes in the entire array. Each experiment is conducted twice, as shown in Figure 6.4. The microarray image for one sub-grid consists of 22 columns and 22 rows, marked from 1 to 8. In addition, for each sub-grid, one swapped array is obtained by exchanging the Cy5 and Cy3 dyes, marked from 1s to 8s. Meanwhile, eight spike genes are placed in each sub-grid to evaluate the performance and accuracy of segmentation. The arrays from (1,1s) to (4,4s) have eight designed spikes with Cy5-Cy3 ratios of $0.1, 0.1, 0.2, 0.2, 0.4, 0.4, 1.0, 1.0$. The designed spikes are placed at the 22nd column and from the 3rd to 10th rows in all sub-grids. In addition, the arrays from (5,5s) to (8,8s) have eight designed constant Cy5-Cy3 ratios of 0.2.

Different segmentation methods are used in this example for better comparison and illustration: fixed circle segmentation (ScanAlyze), adaptive segmentation (GenePix), GMM, KDE, and GKDE. Three different modes are used in adaptive segmentation: irregular, circular, and rectangular. Figure 6.5 shows the segmentation of a Cy3 spot by using different methods with the corresponding ratio of the mean. In addition, a spike gene with a known Cy5-Cy3 ratio of 1.0 is placed inside. The adaptive irregular method has the estimated ratio of mean that is much closer to the target ratio. However, the segmentation region using the adaptive irregular method may give an over- or under-estimate on spot intensities.

6.5.2   EVALUATION OF PERFORMANCE AND ACCURACY

The evaluation of performance and accuracy allows scientists to assess the performance of a method. Spike genes with known contents on microarrays represent the gold standard for evaluating the accuracy of segmentation methods. The sum of squared relative error (SSRE) and the sum of squared error (SSE) are used to

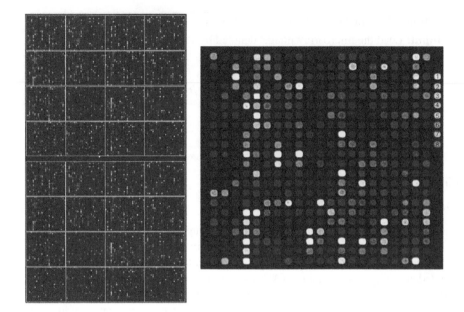

FIGURE 6.4 A microarray image (left) that contains 32 sub-grids. Two replicated spots are designed in one array, of which the upper 16 sub-grids are duplicated as the lower 16 sub-grids. The microarray image for one sub-grid (right) consists of 22 columns and 22 rows, marked from 1 to 8. For each sub-grid, one swapped array is obtained by exchanging Cy5 and Cy3 dyes, marked from 1s to 8s. In addition, eight spike genes are placed in each sub-grid (marked from 1 to 8 in the 22nd column) to evaluate the performance and accuracy of segmentation.

evaluate accuracy and are defined as follows:

$$SSRE = \sum_{j=1}^{M} \sum_{b=1}^{B} \{\frac{\hat{T}_{j,b} - T_j}{T_j}\}^2, \tag{6.28}$$

$$SSE = \sum_{j=1}^{M} \sum_{b=1}^{B} (\hat{T}_{j,b} - T_j)^2, \tag{6.29}$$

where $\hat{T}_{j,b}$ denotes the estimated ratio of means between Cy5 and Cy3 for the j-th spike gene in the b-th block, and T_j denotes the target ratio of the j-th spike gene. The number of sub-grids, B is equal to 32 and the totality of spike genes, M is 8. Smaller values of SSRE and SSE indicate closeness to the target ratio, which yields higher accuracy.

Tables 6.2 and 6.3 list the SSEs and SSREs for different methods. It is clear that all the SSEs and SSREs of KDE are smaller than those obtained by the irregular segmentation method. The relative improvement ratio (*RP*) of two segmentation methods is

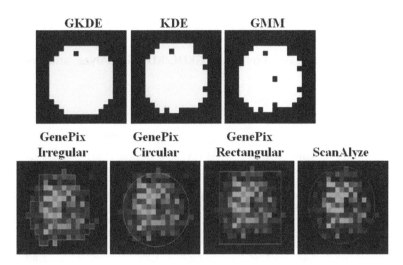

FIGURE 6.5 The results of segmentation by using fixed circle segmentation (ScanAlyze), different modes of adaptive segmentation (GenePix): irregular, circular, and rectangular compared to the methods presented in this chapter: GKDE, KDE, and GMM on a Cy3 spot that has a known spike ratio of 1.0. The estimated ratios of the mean for the above methods are 0.965, 1.013, 0.901, 1.086, 0.813, 0.866, 0.685 for GKDE, KDE, GMM, Genepix (irregular, circular, and rectangular) and fixed circle segmentation (ScanAlyze), respectively.

defined as follows:

$$RP(SSE) = \frac{min\{SSE_{GenePix}, SSE_{ScanAlyze}\} - SSE}{min\{SSE_{GenePix}, SSE_{ScanAlyze}\}} \times 100\%, \qquad (6.30)$$

$$RP(SSRE) = \frac{min\{SSRE_{GenePix}, SSRE_{ScanAlyze}\} - SSRE}{min\{SSRE_{GenePix}, SSRE_{ScanAlyze}\}} \times 100\%. \qquad (6.31)$$

Since the arrays from (1,1s) to (4,4s) are produced based on varying target ratios, the relative improvements measured by SSRE and SSE are different. The arrays from (5,5s) to (8,8s) are produced based on a constant ratio, and the relative improvements measured by the SSRE and SSE are the same. Arrays from 1s to 8s are obtained by swapping Cy3 and Cy5 dies with the same target genes in arrays from 1 to 8. Comparing the SSEs and SSREs of (6,6s) in Tables 6.2 and 6.3, although the dies are swapped, the SSEs and SSREs of GKDE, KDE, and GMM in array 6 are still close to those obtained in array 6s. However, the SSEs and SSREs of the segmentation methods in GenePix and ScanAlyze in array 6 are different from those obtained in array 6s. This indicates that GKDE, KDE, and GMM are more robust and are less sensitive to the swapping of dies than the segmentation methods in GenePix and ScanAlyze. In addition, the average relative improvement ratios of GKDE, KDE, and GMM compared to segmentation methods of GenePix and ScanAlyze for SSRE and SSE are at levels of $(50.55\%, 45.36\%)$, $(50.16\%, 48.59\%)$, and

TABLE 6.2

Comparison of SSEs obtained by GMM, GKDE, KDE, ScanAlyze, and GenePix for spike genes.

Array	Sum of Squared Errors						
	GKDE	KDE	GMM	ScanAlyze	GenePix irregular	GenePix circular	GenePix rectangular
1	2.868	2.781	2.869	25.696	22.523	27.509	29.040
2	3.024	3.019	3.027	21.155	12.082	17.091	18.741
3	5.432	5.408	5.439	42.612	33.806	39.260	41.237
4	9.391	9.290	9.700	9.944	10.446	9.915	10.198
5	0.412	0.316	0.416	0.610	0.804	0.643	0.789
6	0.305	0.309	0.306	2.136	2.304	2.203	2.340
7	2.436	2.375	2.437	3.605	4.621	176.431	4.581
8	4.439	4.076	4.440	6.549	8.792	7.877	8.293
1s	4.414	3.464	4.398	17.577	13.413	16.293	16.882
2s	2.062	2.675	2.265	3.261	3.201	2.966	3.401
3s	12.308	14.516	12.309	44.033	30.953	40.024	39.269
4s	88.786	99.959	86.532	151.779	106.721	132.938	131.203
5s	0.488	0.484	0.489	0.521	0.929	0.582	0.690
6s	0.270	0.262	0.271	3.794	4.295	4.078	4.192
7s	0.509	0.497	0.510	1.195	2.142	1.765	1.803
8s	0.399	0.401	0.400	0.703	1.020	0.859	0.989

Array	Relative Performance (%)		
	GKDE	KDE	GMM
1	87.266	87.654	87.263
2	74.971	75.013	74.950
3	83.932	84.004	83.910
4	5.286	6.301	2.165
5	32.544	48.161	31.765
6	85.710	85.525	85.678
7	32.418	34.106	32.405
8	32.220	37.761	32.213
1s	67.094	74.172	67.211
2s	30.489	9.811	23.614
3s	60.236	52.135	60.233
4s	16.805	6.335	18.917
5s	6.295	7.049	6.162
6s	92.879	93.093	92.845
7s	57.400	58.419	57.371
8s	43.201	42.942	43.105

(49.98%, 45.23%). These results reveal that GKDE, KDE, and GMM perform better than the segmentation methods in GenePix and ScanAlyze. Among these methods, the segmentation results by GKDE yield the greatest improvement. SSE and SSRE can only measure the performance when the ground truth can be obtained. However, in most cases the ground truth is difficult to acquire. Therefore, some methods are

TABLE 6.3

Comparison of SSREs obtained by GMM, GKDE, KDE, ScanAlyze, and GenePix for spike genes.

Array	Sum of Squared Errors						
	GKDE	KDE	GMM	ScanAlyze	GenePix irregular	GenePix circular	GenePix rectangular
1	85.482	82.495	85.482	243.383	301.408	320.264	258.886
2	55.009	45.899	55.025	117.817	152.598	123.584	128.127
3	80.421	77.148	80.421	286.48	317.267	303.845	317.147
4	29.861	28.021	30.170	36.042	31.277	34.409	35.664
5	10.401	7.908	10.410	15.256	20.094	16.070	19.737
6	7.605	7.729	7.647	53.392	57.603	55.068	58.491
7	60.911	59.383	60.916	90.118	115.513	115.779	114.534
8	110.991	101.908	110.992	163.735	219.798	196.928	207.335
1s	33.005	31.740	32.980	130.681	147.429	132.150	132.289
2s	26.900	27.211	26.905	32.285	31.157	33.630	37.984
3s	149.047	130.196	149.739	244.790	286.494	261.726	272.590
4s	675.010	648.212	674.388	769.75	826.411	761.916	767.239
5s	12.202	12.106	12.222	16.525	23.215	14.541	17.244
6s	6.781	6.550	6.786	94.839	107.370	101.951	104.794
7s	12.705	12.425	12.739	29.883	53.558	44.113	45.079
8s	9.910	10.025	9.997	17.570	25.496	21.485	24.715

Array	Relative Performance (%)		
	GKDE	KDE	GMM
1	64.878	66.105	64.878
2	53.310	61.042	53.296
3	71.928	73.070	71.928
4	4.528	10.409	3.539
5	31.823	48.161	31.765
6	85.757	85.525	85.678
7	32.411	34.106	32.405
8	32.213	37.761	32.213
1s	74.744	75.712	74.763
2s	13.662	12.664	13.648
3s	39.101	46.813	38.830
4s	11.406	14.923	11.488
5s	16.086	16.745	15.95
6s	92.850	93.093	92.845
7s	57.484	58.419	57.371
8s	43.599	42.942	43.105

constructed under this situation, such as Pearson's correlation, and the concordance correlation. Pearson's correlation only measures whether a linear relation exists between two random variables, while the concordance correlation coefficient allows scientists to see whether the values of two random variables is actually equal. The concordance correlation coefficient of two random variables X and Y is defined as

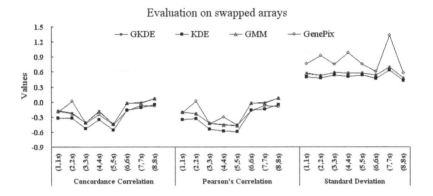

FIGURE 6.6 Concordance correlations, Pearson's correlations, and standard deviations computed between swapped arrays using GKDE, KDE, GMM, ScanAlyze, and GenePix.

follows:

$$\rho_c = \frac{2Cov(X,Y)}{Var(X) + Var(Y) + [E(X) - E(Y)]^2}. \tag{6.32}$$

The concordance correlation coefficient can be adopted to determine the degree of similarity, agreement, and reproducibility in expressions between duplicated spots of all genes in a microarray, which is expected to be close to 1. The concordance correlation coefficients of the swapped microarrays are adopted to evaluate the performance with reference to selected features with high log ratios of means in Cy5-Cy3 dyes. Since the dyes of Cy3 and Cy5 in the swapped arrays are exchanged, the negative value of the concordance correlation coefficient is obtained from the features of the swapped arrays and is expected to be close to -1.

Figure 6.6 shows the concordance correlation coefficients, Pearson's correlations, and standard deviations of eight swapped arrays, while Figure 6.7 shows those between replicates of gene expressions of the sixteen arrays. The KDE method gives higher correlation and lower standard deviation in the sixteen arrays with duplicated genes. In addition, the KDE method produces lower standard deviation and higher correlation in swapped arrays, which indicates that KDE performs well among all the methods.

The accuracy and performance can also be evaluated in terms of the numbers of spots (excluding spike genes and bad spots) in each array. A bad spot appears when the subtraction of foreground and background means is negative.

6.6 CONCLUSION

The parametric method, GMM, has the advantage of computational efficiency and effective segmentation performance when the normality assumption holds. The nonparametric method, KDE lifts the restriction of density assumption, and gives more flexibility in data density estimation. However, the complexity of computation and

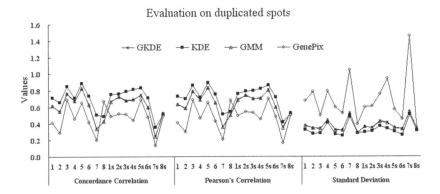

FIGURE 6.7   Concordance correlations, Pearson's correlations and standard deviations computed between duplicated spots using GKDE, KDE, GMM, ScanAlyze, and GenePix.

proper initial values need to be considered carefully in KDE. GKDE collects the advantages of both the GMM and KDE, and can resolve the selection of initial values from the KDE approach. The stopping criteria of convergence is the key issue in GKDE, and needs to be studied and improved.

In general, all the methods for segmentation can be differentiated into two classes, statistical and non-statistical methods. Statistical methods consider randomness for segmentation problems, and alleviate the effect of random noise. Consequently, statistical methods usually generate segmentation with high accuracy. Statistical methods typically take more computation time since more estimation procedures are used in the model. On the other hand, non-statistical methods are often fast. But the resulting segmentation is likely to have the noise artifacts that reduce the accuracy.

Developing a fast method that attains high accuracy remains challenging. For instance, Athanasiadis et al. further use the fuzzy Gaussian mixture model to segment microarray images, which extends the capacity of the Gaussian mixture model [41]. They also combine the traditional image processing method, wavelets with the Markov random field model [42, 43]. In addition, statistical methods combined with other research fields, such as information theory and graph models, became the pioneering research in the segmentation problem [44, 45, 46, 47]. These interdisciplinary studies give scientists different views in microarray image segmentation, and lead to highly accurate results with lower computational complexity.

ACKNOWLEDGMENTS

The authors would like to acknowledge the support of the National Science Council, National Center for Theoretical Sciences, Shing-Tung Yau Center, and the Center of Mathematical Modeling and Scientific Computing at National Chiao Tung University in Taiwan.

REFERENCES

1. M. B. Eisen, "DNA arrays for analysis of gene expression," *Methods in Enzymology*, vol. 303, pp. 179–205, 1999.

2. P. Soille, *Morphological Image Analysis: Principles and Applications*. New York: Springer, 1999.

3. Y. F. Leung and D. Cavalieri, "Fundamentals of cDNA microarray data analysis," *Trends in Genetics*, vol. 19, no. 11, pp. 649–659, 2003.

4. Y. Chen, E. R. Dougherty, and M. L. Bittner, "Ratio-based decisions and the quantitative analysis of cDNA microarray images," *Journal of Biomedical Optics*, vol. 2, pp. 364–374, 1995.

5. Y. Yang, M. Buckley, S. Dudoit, and T. Speed, "Comparison of methods for image analysis on cDNA microarray data," *Journal of Computational and Graphical Statistics*, vol. 11, pp. 108–136, 2002.

6. H. Wu and H. Yan, "Microarray image processing based on clustering and morphological analysis," in *Proceedings of First Asia Pacific Bioinformatics Conference*, Adelaide, Australia, pp. 111–118, 2003.

7. I. Axon Instruments, *GenePix 4000A User's Guide*. Axon Instruments, Inc., 1999.

8. M. B. Eisen, *ScanAlyze*. 1999.

9. GSI Lumonics, *QuantArray Analysis Software*, Operator's Manual. GSI Lumonics, 1999.

10. M. Buckley, *The Spot User's Guide*. CSIRO Mathematical and Information Sciences. 2000.

11. BioDiscovery, *ImaGene Guide for Agilent Microarrays Scanned on the NimbleGen MS200 Microarray Scanner*, BioDiscovery, 2013.

12. PerkinElmer Life Science, *ScanArray Express Microarray Analysis System, User's Manual*. PerkinElmer Life Science. 2002.

13. L. Vincent and P. Soille, "Watersheds in digital spaces: an efficient algorithm based on immersion simulations," *IEEE Transactions on Pattern Analysis and Machine Intelligence*, vol. 13, pp. 583–598, 1991.

14. S. Beucher and F. Meyer, *Mathematical Morphology in Image Processing, volume 34 of Optical Engineering*, ch. The morphological approach to segmentation: the watershed transformation, pp. 433–481. New York: Marcel Dekker, 1992.

15. J. Buhler, T. Ideker, and D. Haynor, "Dapple: Improved techniques for finding spots on DNA microarrays," tech. rep., University of Washington, August 2000.

16. R. Adams and L. Bischof, "Seeded region growing," *IEEE Transactions on Pattern Analysis and Machine Intelligence*, vol. 16, pp. 641–647, 1994.

17. A. W. C. Liew, H. Yan, and M. Yang, "Robust adaptive spot segmentation of DNA microarray images," *Pattern Recognition*, vol. 36, no. 5, pp. 1251–1254, 2003.

18. N. Giannakeasa, P. S. Karvelisc, T. P. Exarchosb, F. G. Kalatzisb, and D. I. Fotiadisb, "Segmentation of microarray images using pixel classification-comparison with clustering-based methods," *Computers in Biology and Medicine*, vol. 43, no. 6, pp. 705–716, 2013.

19. T. B. Chen, H. H. S. Lu, Y. S. Lee, and H. J. Lan, "Segmentation of cDNA microarray images by kernel density estimation," *Journal of Biomedical Informatics*, vol. 41, no. 6, pp. 1021–1027, 2008.

20. T. I. Lin, J. C. Lee, and S. Y. Yen, "Finite mixture modeling using the skew normal distribution," *Statistica Sinica*, vol. 17, pp. 909–927, 2007.

21. D. M. Titterington, A. F. M. Smith, and U. E. Markov, *Statistical Analysis of Finite*

Mixture Distributions. New York: Marcel Dekker, 1988.

22. G. J. McLachlan and K. E. Basford, *Mixture Models: Inference and Applications to Clustering*. New York: Addison-Wesley, 1985.

23. G. McLachlan and D. Peel, *Finite Mixture Models*. New York: Wiley, 2000.

24. M. West, *Bayesian Statistics*, ch. Modeling with mixtures, pp. 502–504. Oxford University Press, 4th ed., 1992.

25. J. Diebolt and C. P. Robert, "Estimation of finite mixture distributions through Bayesian sampling," *IEEE Transactions on Pattern Analysis and Machine Intelligence*, vol. 56, no. 2, pp. 363–375, 1994.

26. M. D. Escobar and M. West, "Bayesian density estimation and inference using mixtures," *Journal of the American Statistical Association*, vol. 90, no. 430, pp. 577–588, 1995.

27. S. Richardson and P. J. Green, "On Bayesian analysis of mixtures with an unknown number of components," *Journal of the Royal Statistical Society. Series B (Methodological)*, vol. 59, no. 4, pp. 731–792, 1997.

28. P. Muller and G. L. Rosner, "A Bayesian population model with hierarchical mixture priors applied to blood count data," *Journal of the American Statistical Association*, vol. 92, no. 440, pp. 1279–1292, 1997.

29. M. Stephens, "Dealing with label switching in mixture models," *Journal of the Royal Statistical Society. Series B (Statistical Methodology)*, vol. 62, no. 4, pp. 795–809, 2000.

30. A. Gelman, J. B. Carlin, H. S. Stern, and D. B. Rubin, *Bayesian Data Analysis*. London: Chapman and Hall, 2003.

31. T. Elguebaly and N. Bouguila, "Bayesian learning of generalized Gaussian mixture models on biomedical images," *Artificial Neural Networks in Pattern Recognition*, vol. 5998, pp. 207–218, 2010.

32. K. Blekas, N. P. Galatsanos, A. Likas, and I. E. Lagaris, "Mixture model analysis of DNA microarray images.," *IEEE Transactions on Medical Imaging*, vol. 24, no. 7, pp. 901–909, 2005.

33. A. P. Dempster, N. M. Laird, and D. B. Rubin, "Maximum likelihood from incomplete data via the EM algorithm.," *Journal of the Royal Statistical Society, Series B*, vol. 39, no. 1, pp. 1–38, 1977.

34. M. Rosenblatt, "Remarks on some nonparametric estimates of a density function," *Annals of Mathematical Statistics*, vol. 27, pp. 832–837, 1956.

35. D. W. Scott and J. R. Thompson, *Computer Science and Statistics: Proceedings of the Fifteenth Symposium on the Interface*, ch. Probability density estimation in higher dimensions, pp. 173–179. Amsterdam: North-Holland, 1st ed., 1983.

36. S. J. Sheather and M. C. Jones, "A reliable data-based bandwidth selection method for kernel density estimation," *Journal of the Royal Statistical Society. Series B (Methodological)*, vol. 53, no. 3, pp. 683–690, 1991.

37. B. W. Silverman, *Density Estimation for Statistics and Data Analysis*. London: Chapman and Hall, 1994.

38. M. Rosenblatt, "Curve estimates," *Annals of Mathematical Statististics*, vol. 42, pp. 1815–1842, 1971.

39. J. Engel, E. Herrmann, and T. Gasser, "An iterative bandwidth selector for kernel estimation of densities and their derivatives," *Journal of Nonparametric Statistics*, vol. 4, pp. 21–34, 1994.

40. A. Chao, T. H. Wang, Y. S. Lee, et al., "Molecular characterization of adenocarcinoma and squamous carcinoma of the uterine cervix using microarray analysis of gene expres-

sion.," *International Journal of Cancer*, vol. 119, no. 1, pp. 91–98, 2006.

41. E. I. Athanasiadis, D. A. Cavouras, P. P. Spyridonos, D. Glotsos, I. K. Kalatzis, and G. C. Nikiforidis, "Complementary DNA microarray image processing based on the fuzzy Gaussian mixture model," *IEEE Transactions on Information Technology in Biomedicine*, vol. 13, no. 4, pp. 419–425, 2009.

42. E. I. Athanasiadis, D. A. Cavouras, D. Glotsos, P. V. Georgiadis, I. K. Kalatzis, and G. C. Nikiforidis, "Segmentation of complementary DNA microarray images by wavelet-based Markov random field model," *IEEE Transactions on Information Technology in Biomedicine*, vol. 13, no. 6, pp. 1068–1074, 2009.

43. E. Athanasiadisa, D. Cavourasb, S. Kostopoulosb, D. Glotsosb, I. Kalatzisb, and G. Nikiforidisa, "A wavelet-based Markov random field segmentation model in segmenting microarray experiments," *Computer Methods and Programs in Biomedicine*, vol. 104, no. 3, pp. 307–315, 2011.

44. N. Karimi, S. Samavi, S. Shirani, and P. Behnamfar, "Segmentation of DNA microarray images using an adaptive graph-based method," *IET Image Processing*, vol. 4, no. 1, pp. 19–27, 2010.

45. C. Argyropoulosa, A. Daskalakisa, G. C. Nikiforidisa, and G. C. Sakellaropoulosa, "Background adjustment of cDNA microarray images by maximum entropy distributions," *Journal of Biomedical Informatics*, vol. 43, no. 4, pp. 496–509, 2010.

46. E. Rashedi and H. Nezamabadi-pour, "A stochastic gravitational approach to feature based color image segmentation," *Engineering Applications of Artificial Intelligence*, vol. 26, no. 4, pp. 1322–1332, 2013.

47. Y. Yang, P. Stafford, and Y. J. Kim, "Segmentation and intensity estimation for microarray images with saturated pixels," *BMC Bioinformatics*, vol. 12, pp. 462–471, 2011.

7 Microarray Image Restoration and Noise Filtering

Rastislav Lukac

CONTENTS

7.1 INTRODUCTION

Microarray imaging [1, 2, 3, 4, 5, 6] is one of the most important and powerful techniques used to extract and interpret genomic information. Analysis of microarray data helps in monitoring the expression levels of thousands of genes simultaneously and provides information relevant to cell activity. Unfortunately, the microarray image formation process is often largely affected by the noise (Figure 7.1) and other impairments that can significantly degrade the value of the captured visual information and complicate the gene expression and data analysis tasks. This creates the need for using various restoration methods [7] that are effective in reducing the noise

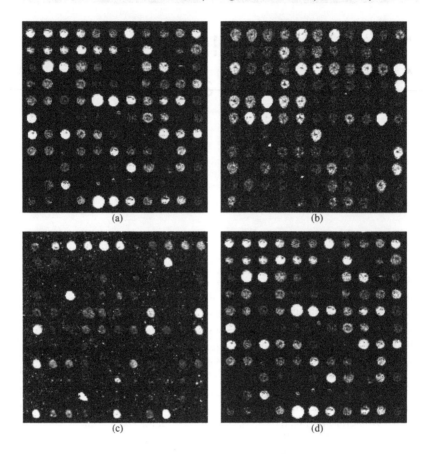

(a) (b)

(c) (d)

FIGURE 7.1 Microarray images. Spots can vary in their size and shape, and the image can be largely affected by noise.

and rectifying signal disturbances in order to produce an image that corresponds as closely as possible to the output of an ideal microarray imaging system.

This chapter surveys filtering methods suitable for restoring the desired microarray image information from the corresponding noisy measurements. To facilitate the discussion on microarray image filtering and restoration, Section 7.2 describes the problem of image formation and noise modeling, which are crucial for simulating the noise observed in microarray images and studying the effect of the filters on both the noise and the desired image features. Section 7.3 surveys various noise filtering methods that have been shown to be effective in suppressing the noise in microarray images. These methods are based on data averaging (e.g., Gaussian, bilateral, anisotropic diffusion, and nonlocal mean filters), order-statistics (e.g., median, weighted median, and combination filters), mathematical morphology (e.g., erosion, dilation, opening, and closing filters), and wavelets (e.g., hard and soft thresholding).

Since preserving the microarray spots is crucial in both noise filtering and subsequent data analysis, various popular edge detection methods (e.g., Sobel, Canny, and Laplacian) are described as well. Section 7.4 concludes this chapter.

7.2   MICROARRAY IMAGING BASICS

A single channel of the multichannel microarray image or a monochromatic microarray image x with K_1 rows and K_2 columns represents a two-dimensional matrix of samples $x_{(r,s)}$ occupying the spatial location (r,s), with $r = 1,2,...,K_1$ denoting the image row and $s = 1,2,...,K_2$ denoting the image column. In standard sixteen-bit representation typically used in the microarray image formation process, pixel values $x_{(r,s)}$ can range from 0 to 65,535, with the large values denoting brighter pixels (higher image intensities). The process of displaying an image creates a graphical representation of the image matrix where the pixel values represent particular intensities. Note that most image displays are eight-bit devices, which can display the intensities ranging from 0 to 255, thus requiring a data conversion from a native sixteen-bit format prior to rendering an image on the screen.

As can be seen in Figure 7.1, the foreground of microarray images is constituted by their spots. The intensities of the pixels within a microarray spot are used to determine a single gene expression and to identify the genes expressed in a particular cell type [8, 9]. The gaps between spots constitute the background [10].

7.2.1   MICROARRAY IMAGE FORMATION

Various degradation processes, such as aliasing, blurring, and noise, affect the image formation [11]. This reduces the value of captured visual information and complicates many image processing and analysis tasks. Complex relations among sources of image degradation are typically modelled as follows:

$$\mathbf{x} = \mathbf{H} * \mathbf{o} + \mathbf{v} \tag{7.1}$$

where $\mathbf{x}$ denotes the corrupted image, $\mathbf{H}$ approximates blurring and aliasing effects, $*$ is the convolution operator, $\mathbf{o}$ denotes the original image, and $\mathbf{v}$ denotes the noise term. The image restoration process aims at suppressing the effects of degradation and producing an image with the desired characteristics.

Among the above types of degradation, aliasing is due to sampling the visual information by the image sensor. Blurring effects are often caused by out-of-focus, diffraction, lens aberrations, spatial averaging built in the image sensor, and motion due to the long exposure time relative to the motion in the scene. The main sources of noise in microarray images are photon and electronic noise introduced during scanning, washing artifacts leading to streaks and blobs of signal on the microarray surface, and artifacts caused by laser light reflection and dust particles.

7.2.2 NOISE MODELING

Considering just the noise processes, which is the goal of this chapter, the above imaging model can be simplified at the pixel level as follows [12, 13]:

$$x_{(r,s)} = o_{(r,s)} + v_{(r,s)} \tag{7.2}$$

where (r,s) denotes the spatial position, $x_{(r,s)}$ represents the observation (noisy) sample, $o_{(r,s)}$ is the desired (noise-free) sample, and $v_{(r,s)}$ describes the noise process.

Image sensors are the usual source of noise in digital imaging [14, 15]. Noise is caused by random processes associated with quantum signal detection, signal independent fluctuations, and inhomogeneity of the responsiveness of the sensor elements. Noise increases with the temperature and sensitivity setting of the imaging device, as well as the reduced length of the exposure. It can vary within an individual image; darker regions usually suffer more from noise than brighter regions. The level of noise also depends on the physical size of photosites in the sensor; larger photosites usually have better light-gathering abilities, thus producing a stronger signal and higher signal-to-noise ratio. Noise can also appear as abrupt local changes in the image, caused, for example, by malfunctioning of the sensor elements, electronic interference, and flaws in the data transmission process.

Noise can often be approximated using the Gaussian noise model. Such approximations are useful, for example, when the noise is caused by thermal degeneration of materials in image sensors. Changes of the pixel intensities from the foreground to the background can also be attributed to the Gaussian nature of noise corrupting microarray images [16], whereas isolated discrete artifacts and outliers present in the microarray image can be seen as impairments that are impulsive in nature [17]. Such a noise corruption can be modelled through a mixture of additive Gaussian noise and impulsive noise [18, 19]:

$$v_{(r,s)} = \begin{cases} v^A_{(r,s)} + v^I_{(r,s)} & \text{with probability } p_v \\ v^A_{(r,s)} & \text{with probability } 1 - p_v \end{cases} \tag{7.3}$$

where $v^A_{(r,s)}$ denotes the Gaussian noise, $v^I_{(r,s)}$ denotes the impulsive noise, and p_v is the impulsive noise corruption probability. The impulsive noise contribution can be seen in the form of short-time duration high-energy spikes attaining large amplitudes with probability higher than predicted by a Gaussian density model [20, 21]. Impulsive noise is assumed to be independent from pixel to pixel, resulting in generally much larger or smaller amplitude of the corrupted pixels compared to that of their neighboring pixels. Figure 7.2 shows several examples of simulated noise corruption.

7.3 MICROARRAY IMAGE FILTERING

The filtering methods typically operate on the premise that an image can be subdivided into small regions, each of which can be treated as stationary [12, 22]. A processing window $\Psi_{(r,s)} = \{x_{(i,j)}; (i,j) \in \zeta\}$ is used to determine a localized area

(a)

(b)

(c)

(d)

FIGURE 7.2 Microarray image noise modeling: (a) noise-free image, (b) additive Gaussian noise, (c) impulsive noise, and (d) mixed noise comprised of additive Gaussian noise followed by impulsive noise.

of the input image. This window slides over the entire image, placing successively every pixel $x_{(r,s)}$, for $r = 1, 2, ..., K_1$ and $s = 1, 2, ..., K_2$, at the center of a local neighborhood denoted by ζ and replacing $x_{(r,s)}$ with the output $y_{(r,s)} = f(\Psi_{(r,s)})$ of a filter function $f(\cdot)$ operating over samples located inside the filter window.

The performance of a filtering method is generally influenced by the size and the shape of the processing window. Different techniques may require a different processing window, in terms of its size and shape, to achieve an optimal performance. However, the rectangular-shaped windows, such as a 3×3 window described by $\zeta = \{(r-1, s-1), (r-1, s), ..., (r+1, s+1)\}$, are most frequently used.

7.3.1 LINEAR FILTERS

The averaging-like filters are efficient for smoothing high-frequency variations and transitions. The arithmetic mean filter [23, 24] represents the simplest form of spatial averaging:

$$y_{(r,s)} = \frac{1}{|\zeta|} \sum_{(i,j) \in \zeta} x_{(i,j)} \tag{7.4}$$

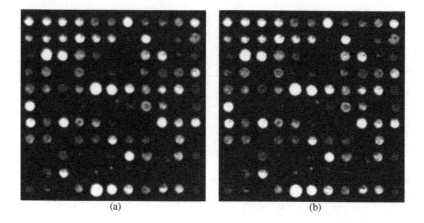

(a)　　　　　　　　　　　　　　　　　　(b)

FIGURE 7.3　Spatial averaging of the image shown in Figure 7.1d: (a) arithmetic mean filtering and (b) Gaussian filtering.

where $y_{(r,s)}$ denotes the filter output, $x_{(i,j)} \in \Psi_{(r,s)}$ is the pixel inside the filter window, and $|\zeta|$ denotes the number of samples in $\Psi_{(r,s)}$. By replacing the pixel $x_{(r,s)}$ with the average of pixels inside a localized image area centered in the pixel location (r,s) and repeating this operation of all pixels in the image, the filter suppresses pixel values that are unrepresentative of their surroundings.

This concept can be further generalized by performing the weighted averaging operations:

$$y_{(r,s)} = \sum_{(i,j) \in \zeta} w_{(i,j)} x_{(i,j)} / \sum_{(i,j) \in \zeta} w_{(i,j)} \tag{7.5}$$

where $w_{(i,j)}$ is the weight associated with the input sample $x_{(i,j)} \in \Psi_{(r,s)}$. Obviously, the arithmetic mean filter can be obtained by setting all the weights to one (or generally to the same number), that is, $w_{(i,j)} = 1$, for $\forall(i,j) \in \zeta$.

To reduce the blurring of the desired image content due to spatial averaging (Figure 7.3), the samples occupying spatial locations that are closer to the window center (sample being filtered) are often associated with the higher weights compared to the samples that reside farther from the window centers. Such design characteristics can be obtained using a Gaussian function [23, 24] as follows:

$$w_{(i,j)} = \frac{1}{2\pi\sigma} \exp\left(\frac{-\|(r,s) - (i,j)\|^2}{2\sigma^2}\right) \tag{7.6}$$

In general, linear filters are relatively easy to implement. For instance, as in the following approximation of the Gaussian filter using a 3×3 convolution mask:

$$\mathbf{w} = \begin{bmatrix} 1 & 2 & 1 \\ 2 & 4 & 2 \\ 1 & 2 & 1 \end{bmatrix}, \tag{7.7}$$

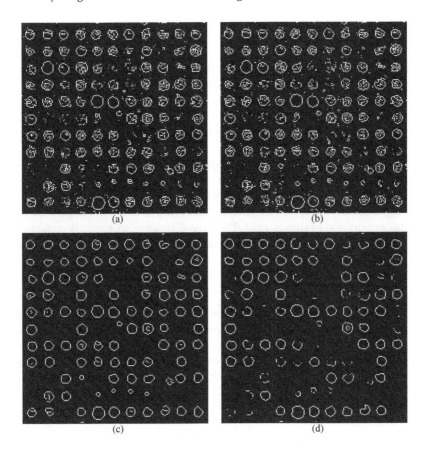

FIGURE 7.4 Edge detection performed on the microarray image shown in Figure 7.1d: (a) Prewitt, (b) Sobel, (c) Canny, and (d) Laplacian detector.

where $\mathbf{w} = \{w_{(i,j)}; (i,j) \in \zeta\}$ denotes the weight vector, both the filter weights and the sum of weights can be constrained to be a power of two in order to implement multiplications and division by bit shifting, which is very fast. A large two-dimensional filter kernel can also be approximated by sequentially applying various multiple smaller kernels. Probably the biggest advantage of various linear filters is that their two-dimensional kernels can be separated into two one-dimensional kernels; the first one applied to the rows (or columns) of the input image to generate an intermediate output and then the other one applied to the columns (or rows) of that intermediate output to generate the final filtered image.

7.3.1.1 Edge Detectors

Many filters do not directly account for the structural content of the image, such as edges and fine details. However, edges convey essential information about a scene. Edges can be defined as the boundaries of distinct image regions that differ in intensity [23, 25]. Determination of object boundaries is very important in microarray imaging in order to localize the spots [26, 27, 28, 29] for further microarray data analysis. A process of transforming an input digital image into an edge map is termed as edge detection (Figure 7.4).

Similar to various noise-reduction techniques, edge-detection techniques usually generate the edge map by processing the pixels located inside the filter window and do not use any prior information about the image structure [30]. In practice, many popular edge detectors are approximated through the use of convolution masks, with the weights constrained to sum to zero. The edge-detector output $y_{(r,s)}$ is compared with a predefined nonnegative threshold ξ to determine if a given pixel belongs to an edge or not:

$$E_{(r,s)} = \begin{cases} y_{(r,s)} & \text{if } y_{(r,s)} \geq \xi \\ 0 & \text{otherwise} \end{cases} \tag{7.8}$$

where $E_{(r,s)}$ is a resulting edge map value. Since edge operators are sensitive to noise and small variations in intensity, using the appropriate threshold helps to extract the true structural content of an image. The omitted edges are usually caused by using a very large threshold value, whereas the presence of noise-like pixels and redundant details in the resulting edge map is due to using a too-small threshold value.

The so-called gradient edge-detection methods [23, 30] rely on the first-order directional derivatives:

$$\nabla I(r,s) = \left[\frac{\partial I(r,s)}{\partial r}, \frac{\partial I(r,s)}{\partial s} \right] \tag{7.9}$$

where $\nabla I(r,s)$ denotes the gradient of the function $I(r,s)$. The gradient magnitude

$$|\nabla I(r,s)| = \sqrt{\left(\frac{\partial I(r,s)}{\partial r}\right)^2 + \left(\frac{\partial I(r,s)}{\partial s}\right)^2} \tag{7.10}$$

denotes the rate of change of the image intensity, whereas the gradient direction

$$\theta = \arctan\left(\frac{\partial I(r,s)}{\partial r} \bigg/ \frac{\partial I(r,s)}{\partial s}\right) \tag{7.11}$$

denotes the orientation of an edge.

Two convolution masks are typically used to determine the gradient magnitude in two orthogonal directions. For example, the masks for the well-known Prewitt, Sobel, and Canny operators are defined as follows:

$$\mathbf{w}_{\text{Prewitt}} = \left\{ \begin{bmatrix} -1 & 0 & 1 \\ -1 & 0 & 1 \\ -1 & 0 & 1 \end{bmatrix}, \begin{bmatrix} -1 & -1 & -1 \\ 0 & 0 & 0 \\ 1 & 1 & 1 \end{bmatrix} \right\} \tag{7.12}$$

$$\mathbf{w}_{Sobel} = \left\{ \begin{bmatrix} -1 & 0 & 1 \\ -2 & 0 & 2 \\ -1 & 0 & 1 \end{bmatrix}, \begin{bmatrix} -1 & -2 & -1 \\ 0 & 0 & 0 \\ 1 & 2 & 1 \end{bmatrix} \right\}, \quad (7.13)$$

$$\mathbf{w}_{Canny} = \left\{ \begin{bmatrix} 0 & 0 & 0 \\ -1 & 0 & 1 \\ 0 & 0 & 0 \end{bmatrix}, \begin{bmatrix} 0 & 1 & 0 \\ 0 & 0 & 0 \\ 0 & -1 & 0 \end{bmatrix} \right\} \quad (7.14)$$

Note that the Canny operator [31] does not solely rely on the intensity variations, but attempts to limit the effect of the noise using Gaussian pre-smoothing in order to improve the quality of the edge maps, and improve the appearance of the edge maps using the hysteresis-based thinning of the thresholded edge maps. Figure 7.4 shows the resulting edge maps for various edge detectors after applying similar preprocessing and postprocessing steps.

Another popular class of edge-detection methods, often termed as zero-crossing methods [30, 32], is based on the second-order directional derivatives. When the first derivative achieves a maximum, the second derivative of $I(r,s)$ is zero. An example of such detection is the Laplacian operator

$$\Delta I(r,s) = \nabla^2 I(r,s) = \frac{\partial^2 I(r,s)}{\partial r^2} + \frac{\partial^2 I(r,s)}{\partial s^2} \quad (7.15)$$

which is approximated in practice using the convolution masks

$$\mathbf{w} = \begin{bmatrix} 0 & 1 & 0 \\ 1 & -4 & 1 \\ 0 & 1 & 0 \end{bmatrix} \text{ or } \mathbf{w} = \begin{bmatrix} 1 & 1 & 1 \\ 1 & -8 & 1 \\ 1 & 1 & 1 \end{bmatrix} \quad (7.16)$$

defined for a four- and eight-neighborhood, respectively. Other popular zero-crossing based methods combine Laplacian and Gaussian operators or use the difference of Gaussians [33].

7.3.2 DATA-ADAPTIVE FUZZY FILTERS

To obtain trade-off between reducing random noise and preserving the desired image content (Figure 7.5a), the data-adaptive fuzzy filters [34, 35] perform fuzzy weighted averaging inside a localized image area:

$$y_{(r,s)} = f\left(\sum_{(i,j)\in\zeta} w_{(i,j)} x_{(i,j)} \right) \quad (7.17)$$

where $f(\cdot)$ is a nonlinear function and

$$w_{(i,j)} = \mu_{(i,j)} / \sum_{(g,h)\in\zeta} \mu_{(g,h)} \quad (7.18)$$

is the normalized filter weight associated with the input sample $x_{(i,j)} \in \Psi_{(r,s)}$. The filter is constrained via $w_{(i,j)} \geq 0$ and $\sum_{(i,j)\in\zeta} w_{(i,j)} = 1$ to ensure that its output is unbiased.

The weights $w_{(i,j)}$ are calculated using fuzzy membership function terms $\mu_{(i,j)}$, which can have a form of an exponential function [34, 35]:

$$\mu_{(i,j)} = \beta \left(1 + \exp\left\{D_{(i,j)}\right\}\right)^{-r} \tag{7.19}$$

where r is a parameter adjusting the weighting effect of the membership function, β is a normalizing constant, and

$$D_{(i,j)} = \sum_{(m,n)\in\zeta} \left|x_{(i,j)} - x_{(m,n)}\right| \text{ or } D_{(i,j)} = \sum_{(m,n)\in\zeta} \left(x_{(i,j)} - x_{(m,n)}\right)^2 \tag{7.20}$$

is the aggregated absolute or square difference between the sample $x_{(i,j)}$ under consideration and all other samples inside the filter window, that is, $x_{(m,n)}$, for $\forall(m,n) \in \zeta$. The filter operates without the requirement for fuzzy rules and can adapt to various noise characteristics by tuning the parameters of its membership function.

7.3.3 ANISOTROPIC DIFFUSION

Anisotropic diffusion [36, 37, 38] constitutes an iterative filtering process. Depending on the filter settings and the characteristics of the image being processed, a number of iterations may be needed to obtain the desired result (Figure 7.5b). Considering just a single iteration, the anisotropic diffusion filter can be implemented as follows:

$$y_{(r,s)} = x_{(r,s)} + \lambda \sum_{(i,j)\in\zeta} f(x_{(i,j)} - x_{(r,s)})(x_{(i,j)} - x_{(r,s)}) \tag{7.21}$$

where λ is the parameter used to control the rate of the diffusion process, the term $\zeta = \{(r-1,s),(r,s-1),(r,s+1),(r+1,s)\}$ denotes the neighboring pixels located above, left, right, and below with respect to the pixel location (r,s) under consideration.

The term $f(\cdot)$ denotes the nonlinear conduction function that determines the behavior of the diffusion process. This function is defined to perform an edge-sensing operation based on image gradients by producing the diffusion coefficients that can be seen as the weights associated with the neighboring vectors $\mathbf{x}_{(i,j)}$, for $(i,j) \in \zeta$. The two most common forms of the conduction function are

$$f(a) = \frac{1}{1 + (a/\tau)^2} \tag{7.22}$$

$$f(a) = \exp(-(a/\tau)^2) \tag{7.23}$$

where τ states the diffusion rate and a denotes the image gradient.

The conduction function should produce large values in homogenous regions to encourage image smoothing and small values in edge regions to preserve the structural content. Ideally, this function should satisfy four properties: $f(0) = M$ for $0 < M < \infty$, which ensures isotropic smoothing in regions of similar intensity; $f(a) = 0$ for $a \to \infty$, which ensures edge preservation; $f(a) \geq 0$ to ensure a forward diffusion process; and that $af(a)$ is a strictly decreasing function to avoid numerical instability. However, many functions successfully used in practice accomplish just the first three properties.

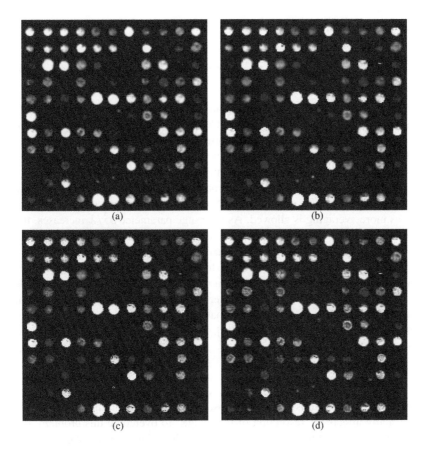

FIGURE 7.5 Adaptive spatial averaging of the image shown in Figure 7.1d: (a) data-adaptive fuzzy filtering, (b) anisotropic diffusion, (c) bilateral filtering, and (d) nonlocal mean filtering.

7.3.4 BILATERAL FILTERS

A non-iterative formulation of anisotropic image smoothing can be achieved by bilateral filtering [39, 40, 41]. This filter (Figure 7.5c) is well known for its ability to reduce image noise and simultaneously preserve the edges and image details. The filter weights indicate range and spatial closeness between the pixel located in the window center and its neighbors:

$$y_{(r,s)} = \sum_{(i,j)\in\zeta} w_{(i,j)}u_{(i,j)}x_{(i,j)} \Big/ \sum_{(i,j)\in\zeta} w_{(i,j)}u_{(i,j)} \qquad (7.24)$$

where ζ denotes pixel locations inside the filter window, $w_{(i,j)}$ are the spatial weights defined in Equation 7.6 used to reduce the contribution of pixels that reside farther from the pixel being filtered, and

$$u_{(i,j)} = \exp\left(-\frac{(x_{(r,s)} - x_{(i,j)})^2}{\sigma_r^2/\lambda}\right)$$
(7.25)

are the range weights used to evaluate the range closeness through a function of differences in the range of the signal, often referred to as a photometric space. The term λ is a scaling factor and σ_r^2 denotes the noise variance. Unlike the range weights, which are not known *a priori* since they depend on actual samples inside the filter window, the spatial weights can be precomputed.

Parameters λ, σ_r, and σ (used in the spatial weight calculations) control the amount of image smoothing. Increasing the spatial parameter σ_d smoothes larger features, as more averaging is allowed. As the range parameter σ_r^2/λ increases, the bilateral filter approximates Gaussian convolution more closely, because the range weights become more constant over the intensity interval of the image.

The bilateral filter has a solid theoretical foundation [42, 43], as it has some relation to robust filtering, anisotropic diffusion, and local mean filtering. It can be applied in a noniterative manner, but also iteratively. To enhance its computational efficiency, bilateral filtering can be approximated using separable kernels [44], local histograms [45], layered approximation [46], bilateral grid [47], and bilateral pyramid [48].

7.3.5 NONLOCAL MEAN FILTERS

Although spatial averaging is one of the most popular tools for reducing the image noise, it usually affects the resolution of the image. To overcome this problem, the so-called nonlocal mean filters [49, 50] make use of the redundancy in images arising from repeated patterns and average only pixels that belong to similar patches. In this way, improved detail-preserving noise suppression characteristics can be obtained (Figure 7.5d).

The nonlocal mean filtering process is guided by the weights that are calculated by comparing the similarity of patches or neighborhoods of pixels:

$$w_{(i,j)} = \exp\left\{-\frac{1}{h^2}\left\|\Psi_{(r,s)} - \Psi_{(i,j)}\right\|_2^2\right\}$$
(7.26)

where $\Psi_{(r,s)}$ and $\Psi_{(i,j)}$ denote the two patches centered in the spatial locations (r,s) and (i,j), respectively. The term h controls the noise reduction strength; the amount of noise reduction increases with the increased value of h. This parameter can be set, for instance, between 10σ and 15σ where σ denotes the standard deviation of the noise in the image [49].

Unlike local smoothing filters, the weight calculations in nonlocal filtering can potentially include pixels from all over the image. Since pixels that reside closer to patch boundaries are usually less important in the filtering process than the pixels that reside closer to the center of a patch, the patch distance calculations can be defined to incorporate the Gaussian weights as follows:

$$\left\|\Psi_{(r,s)} - \Psi_{(i,j)}\right\|_2^2 = \sum_{(u,v) \in \rho} G_{(u,v)} \left(x_{(r+u,s+v)} - x_{(i+u,j+v)}\right)^2 / \sum_{(u,v) \in \rho} G_{(u,v)} \qquad (7.27)$$

where ρ is a set of offsets of a pixel in an image patch from the center pixel; for instance, $\rho = \{(u,v); 0 \geq u \geq 2, 0 \geq v \geq 2$ for a 5×5 patch. Obviously, the computational cost of nonlocal filtering can be reduced by limiting the search area within certain distance T from the actual pixel location (r,s); this can be achieved by considering only the pixel locations (i,j), which satisfy $((r-i)^2 + (r-i)^2)^{1/2} < T$.

Similar to the bilateral filter, the nonlocal mean filter can be extended by applying it iteratively; such an iterative process is often associated with minimization of a cost function [50]. Recent implementation advances in nonlocal filtering include faster computing of the distance between patches [51, 52, 53] and adaptive selection of sizes of the searching windows [54].

7.3.6 ORDER-STATISTIC FILTERS

Each sample $x_{(i,j)}$ from the set of samples in $\Psi_{(r,s)} = \{x_{(i,j)}; (i,j) \in \zeta\}$ can be compared to other samples in $\Psi_{(r,s)}$ to produce the ordered sequence [55, 56]:

$$x_{(1)} \leq x_{(2)} \leq \cdots \leq x_{(\tau)} \leq \cdots \leq x_{(|\zeta|)} \qquad (7.28)$$

where $x_{(\tau)}$, for $\tau = 1, 2, ..., |\zeta|$, denotes the so-called τ-th order statistics. Samples that diverge greatly from the data population usually occupy higher and lower ranks in the ordered set, whereas the most representative samples of the local image area are associated with middle ranks. Since the ordering can be used to determine the positions of the different samples without any *a priori* information regarding the signal distributions, filters based on order statistics are considered to be robust estimators [22, 57, 58].

The median of the population $\Psi_{(r,s)}$ corresponds to the middle ranked sample $x_{((|\zeta|+1)/2)}$ in Equation 7.28. Equivalently, the median of $\Psi_{(r,s)}$ can be obtained using the minimization concept, which leads to the definition of the well-known median filter [59, 60]:

$$y_{(r,s)} = \arg \min_{x_{(g,h)} \in \Psi_{(r,s)}} \sum_{(i,j) \in \zeta} \left| x_{(g,h)} - x_{(i,j)} \right| \qquad (7.29)$$

where $y_{(r,s)}$ denotes the filter output. Due to minimizing the aggregated absolute difference between $x_{(g,h)} \in \Psi_{(r,s)}$ and all the other samples in $\Psi_{(r,s)}$, the median filter is effective (Figure 7.6a) in suppressing impulsive noise.

In order to take into account the importance of the specific samples inside the filter window and/or the structural contents of the image, the contribution of the input samples $x_{(i,j)}$ to the aggregated absolute differences can be controlled using the associated weights $w_{(i,j)}$. This concept is employed in the weighted median filter (Figure 7.6b) defined as follows [61, 62, 63, 64]:

$$y_{(r,s)} = \arg \min_{x_{(g,h)}} \sum_{(i,j) \in \zeta} w_{(i,j)} \left\| \mathbf{x}_{(g,h)} - \mathbf{x}_{(i,j)} \right\|_L \qquad (7.30)$$

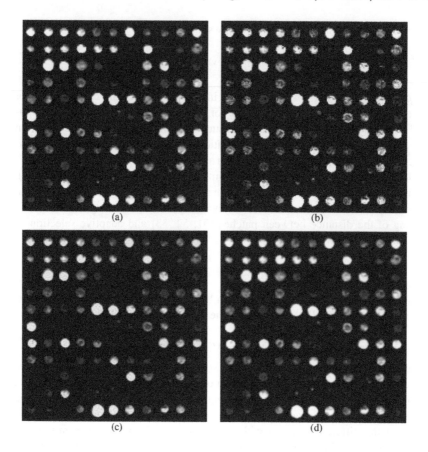

(a)   (b)

(c)   (d)

FIGURE 7.6 Order-statistic filtering of the image shown in Figure 7.1d: (a) median filtering, (b) weighted median filtering, (c) α-trimmed filtering, and (d) combination filtering.

where each setting of the weights represents a unique filter that can be used for a specific task. If all the weights are set to the same value, the filter reduces to the median filter. Various methods for optimizing the weights can be found in References [65, 66, 67, 68].

In the presence of additive Gaussian noise, the noise reduction characteristics of the median filter can be improved by combining it with the linear filter, for instance, as follows [60, 63]:

$$
y_{(r,s)} = \begin{cases} y_{(r,s)}^{AF} & \text{if } \sum_{(i,j)\in\zeta} |y_{(r,s)}^{AF} - x_{(i,j)}| \leq \sum_{(i,j)\in\zeta} |y_{(r,s)}^{MF} - x_{(i,j)}| \\ y_{(r,s)}^{MF} & \text{otherwise} \end{cases} \tag{7.31}
$$

where $y_{(r,s)}^{MF}$ denotes the median obtained in Equation 7.29, whereas $y_{(r,s)}^{AF}$ corresponds to an arithmetic mean filter defined in Equation 7.4. This hybrid filter tends to apply the median filter near a signal edge to preserve the structural information of the image and the arithmetic mean filter in the smooth areas to improve noise attenuation.

Since the averaging operation tends to smooth fine details and it is prone to outliers, improved design characteristics can be obtained by replacing it with its alpha-trimmed version (Figure 7.6c) [60, 63]:

$$y_{(r,s)} = \frac{1}{|\zeta| - 2\alpha} \sum_{\alpha}^{|\zeta| - \alpha} x_{(\tau)} \tag{7.32}$$

where α is a design parameter that can have values $\alpha = 0, 1, ..., (|\zeta| - 1)/2$. By ordering the input samples and then removing the extreme order statistics, the filter usually averages only the samples whose values are close to the median of the input set.

The linear combination of the ordered input samples is also used in the filter design (Figure 7.6d) [69, 70, 71, 72, 73]:

$$y_{(r,s)} = \sum_{\tau=1}^{|\zeta|} w_{\tau} x_{(\tau)} \tag{7.33}$$

where w_{τ} is the weight associated with the τ-th ordered sample $x_{(\tau)} \in \Psi_{(r,s)}$. Assuming the weight vector $\mathbf{w} = [w_1, w_2, ..., w_{|\zeta|}]$ and the unity vector $\mathbf{e} = [1, 1, ..., 1]$ of the dimension identical to that of $\mathbf{w}$, the filter weights can be optimally determined using the mean-square error criterion as follows:

$$\mathbf{w} = \frac{\mathbf{R}^{-1}\mathbf{e}}{\mathbf{e}^T \mathbf{R}^{-1}\mathbf{e}} \tag{7.34}$$

where $\mathbf{w}^T\mathbf{e} = 1$ is the constraint imposed on the solution and $\mathbf{R}$ is a $|\zeta| \times |\zeta|$ correlation matrix of the ordered noise variables. The optimization process can speed up by using the least mean square formulation $\mathbf{w} = \mathbf{w} + 2\mu e_{(r,s)} \Psi^r_{(r,s)}$ based on the ordered input set $\Psi^r_{(r,s)}$ instead.

7.3.7 MORPHOLOGICAL FILTERS

In various image processing and computer vision applications, mathematical morphology [74, 75] is usually used to analyze and extract object features, such as shape and size. The basic morphological operations are dilation and erosion, which can be combined in sequence to produce other operations, such as opening and closing.

The traditional binary morphological operators apply the structuring element (filter window) of a predetermined shape to the input image; the structuring element and the image can be seen as the two sets. Erosion combines these two sets using vector subtraction of set elements, whereas dilation combines the two sets using vector addition of set elements; thus, erosion and dilation are dual to each other. Iteratively applying dilation followed by erosion or erosion followed by dilation eliminates specific image details whose sizes are smaller than the structuring element, while it preserves the geometric characteristics of unsuppressed features.

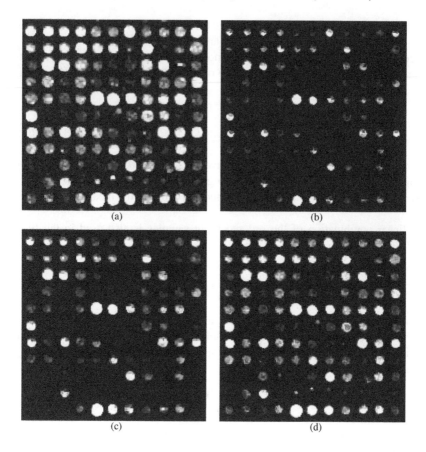

(a)   (b)

(c)   (d)

FIGURE 7.7 Morphological filtering of the image shown in Figure 7.1d: (a) dilation, (b) erosion, (c) opening, and (d) closing.

In the case of continuous-tone images, the dilation (Figure 7.7a) and erosion (Figure 7.7b) can be effectively implemented using the maximum and minimum operators, respectively, as follows [60, 75, 76]:

$$y_{(r,s)} = \max\{x_{(i,j)}; (i,j) \in \zeta\} \tag{7.35}$$

$$y_{(r,s)} = \min\{x_{(i,j)}; (i,j) \in \zeta\} \tag{7.36}$$

where the filter window, spanning over the spatial locations described by ζ, can be seen as a structuring element. Erosion followed by dilation constitutes the so-called opening operation (Figure 7.7c), which eliminates all the pixels in the regions that are too small to contain the structuring element. Dilation followed by erosion results in the so-called closing operation (Figure 7.7d), which fills in holes and concavities smaller than the structuring element. The opening and closing operations are thus dual to each other. A morphological top-hat operation can be obtained by calculating the difference between the input image and its opened version; this difference can be used to enhance the high-frequency information of the image.

Advanced morphological solutions include alternating sequential filters [77] that combine iterative morphological operators with increasing size of structuring elements. In some situations, iterative processing can be avoided by using the morphological operators in a recursive manner [78], that is, considering the previously processed pixels together with the unprocessed pixels as the input set. There are also various soft morphological filters [75, 79, 80, 81] that employ the structuring element with the soft boundary; this concept is often combined with the weighted order statistics and recursive processing to obtain an additional flexibility and improved performance characteristics. Finally, the so-called regulated morphological filters [82] can adapt their performance to the image content, whereas fuzzy morphological filters [83, 84] rely on fuzzy set theory, aiming to benefit from the fact that fuzziness is an intrinsic property of the image.

7.3.8 WAVELETS-BASED FILTERS

The wavelet transform [85, 86, 87] is known for its good energy compaction characteristics and ability to separate the noise from the signal. Applying the wavelet transform to the image produces a set of coefficients; small coefficients usually correspond to the noise whereas the large coefficients correspond to the signal. This is due to the fact that a signal has its energy concentrated in a small number of wavelet dimensions, whereas noise spreads its energy over a large number of wavelet coefficients. Thus, thresholding the wavelet coefficients or shrinking the amplitude of the coefficients can suppress the noise while preserving the desired image features (Figure 7.8).

The wavelet transform decomposes the image into subimages with different frequency components using the orthogonal functions to ensure a unique reconstructed signal. This process can be implemented via two-dimensional transform or one-dimensional transform applied, for example, first along the rows of the image to produce intermediate results and then along the columns of these intermediate results to produce the actual subimages. At each decomposition level, four subimages, usually referred to as L (approximation), H (horizontal), V (vertical), and D (diagonal) or LL (low-low), LH (low-high), HL (high-low), and HH (high-high) are produced. These subimages represent different frequency locations of the original image. To obtain the subimages at the next decomposition level, only the subimage L or LL from the previous decomposition level is further decomposed. As a result, the representation of the input image at different scales or resolution is obtained.

The choice of wavelets determines the final waveform shape; popular wavelet functions are Haar, Daubechies, Coiflets, Symlet, and Biorthogonal. Other factors that influence the performance of wavelet-based noise reduction are the wavelet decomposition depth, the thresholding function, and the actual threshold value(s). Typically, the noise reduction process first requires applying the wavelet transform to the noisy image to produce the noisy wavelet coefficients. Then, the appropriate thresholding method and threshold value need to be selected at each decom-

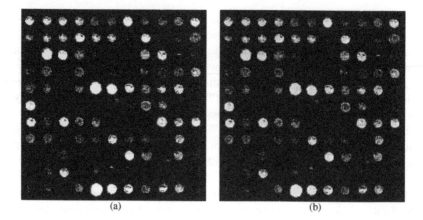

(a)   (b)

FIGURE 7.8 Wavelet-based filtering of the image shown in Figure 7.1d: (a) hard thresholding and (b) soft thresholding.

position level, or sometime even for each subimage, to suppress the noise. Finally, the process concludes with applying the inverse wavelet transform to the thresholded wavelet coefficients to restore the image. There are two basic thresholding approaches [88, 89, 90, 91]; hard thresholding

$$y_{(r,s)} = \begin{cases} x_{(r,s)} & \text{for } |x_{(r,s)}| \geq \delta \\ 0 & \text{otherwise} \end{cases} \tag{7.37}$$

sets zeros for all wavelet coefficients whose absolute value is less than the specified threshold, whereas soft thresholding

$$y_{(r,s)} = \begin{cases} \text{sgn}(x_{(r,s)})(x_{(r,s)} - \delta) & \text{for } |x_{(r,s)}| \geq \delta \\ 0 & \text{otherwise} \end{cases} \tag{7.38}$$

affects all the coefficients to some degree. Although hard thresholding can preserve the edge information better than soft thresholding, it often creates discontinuities in the processed image. Therefore, soft thresholding is generally more preferred than hard thresholding, at least when it comes to non-computer generated images, such as microarray and natural images. It is also possible to combine the two approaches; for example, detail levels can be processed via soft thresholding, whereas hard thresholding can be applied to other levels of wavelet decomposition.

The actual threshold value can be set in numerous ways. In the case of Gaussian noise, the transformed signal preserves the Gaussian nature of the noise and thus the threshold can usually be set between three and five times the standard deviation of the wavelet-transformed signal at each wavelet decomposition level. Eventually, a single threshold value

$$\delta = \sigma\sqrt{2\log(N)} \tag{7.39}$$

can be used at all wavelet decomposition levels to obtain a fast and effective solution [92]. The term N denotes the length of the input signal and σ denotes the standard deviation of the noise, usually estimated from the detail subimage at the highest resolution level. If the procedure requires estimating the standard deviation of the noise from the wavelet coefficients, another solution is to set the threshold as follows [88]:

$$\delta = MAD/0.6745 \tag{7.40}$$

where MAD denotes the median of the absolute values of the wavelet coefficients. More advanced solutions aim at minimizing certain numerical criteria and/or objective functions [93, 94, 95].

7.4 CONCLUSION

This chapter surveyed popular filtering methods that can effective restore the microarray image data from noisy measurements. Understanding the process of image formation is essential in establishing faithful noise approximations in order to simulate the noise observed in microarray images and to study the effect of the filters on both the noise and the desired image features. The noise, usually modeled as a mixture of additive Gaussian noise and impulsive noise, is a common source of degradation of microarray images; it reduces the value of captured visual information and complicates many image processing and analysis tasks.

Averaging the pixels in a localized image area constitutes an efficient approach for smoothing high-frequency variations and transitions in a digital image. The two basic methods in this category, the mean filter and the Gaussian filter, perform a fixed amount of smoothing in each pixel location. Advanced methods, such as the fuzzy filters, anisotropic diffusion-based filters, bilateral filters, and nonlocal mean filters, can adapt their performance to both the image content and varying noise levels. To effectively deal with impulsive noise or outliers present in the image, the filters methods based on order statistics or data ordering are generally preferred over the averaging-based filters. Various median filters, weighted median filters, and morphological filters are discussed as examples of the filters that benefit from ranking the input data. Finally, wavelets-based filters rely on the ability of the wavelet transform to separate the signal from the noise, which is then reduced by thresholding or shrinking the amplitude of the wavelet coefficients.

The surveyed filtering methods take advantage of robust estimation and data-adaptive image processing, which make them effective in suppressing the image noise while preserving the microarray spots. Since the edge information plays an important role in both noise filtering and spot extraction, various popular edge detection methods were also discussed. The efficient filter design methodology and the possibility to adjust the filtering performance based on some qualitative criteria make the presented filters very valuable for a variety of imaging applications, including microarray imaging.

REFERENCES

1. M.B. Eisen and P.O. Brown, "DNA arrays for analysis of gene expression," *Methods in Enzymology*, vol. 303, pp. 179–205, 1999.
2. M. Schena, *Microarray Biochip Technology*. Natick, MA, USA: Eaton Publishing Company / Biotechniques Books, 2000.
3. A.K. Whitchurch, "Gene expression microarrays," *IEEE Potentials*, vol. 21, pp. 30–34, Feb.-March 2002.
4. R.S.H. Istepanian, "Microarray image processing: Current status and future directions," *IEEE Transactions on Nanobioscience*, vol. 2, pp. 173–175, 2003.
5. P. O'Neill and G.D. Magoulas, "Improved processing of microarray data using image reconstruction techniques," *IEEE Transactions on Nanobioscience*, vol. 2, pp. 176–183, December 2003.
6. D.A. Adjeroh, Y. Zhang, and R. Parthe, "On denoising and compression of DNA microarray images," *Pattern Recognition*, vol. 39, pp. 2478–2493, 2006.
7. R. Lukac: "Color image restoration using vector filtering operators," in *Image Restoration: Fundamentals and Advances*, B.K. Gunturk and X. Li (eds.), Boca Raton, FL, USA: CRC Press / Taylor & Francis, September 2012, pp. 249–283.
8. X.Y. Zhang, F. Chen, Y.T. Zhang, et al., "Signal processing techniques in genomic engineering," *Proceedings of the IEEE*, vol. 90, pp. 1822–1833, December 2002.
9. V. Filkov, S. Skiena, and J. Zhi, "Analysis techniques for microarray time-series data," *Journal of Computational Biology*, vol. 9, pp. 317–330, April 2002.
10. J.H. Kim, D.M. Shin, and Y.S. Lee, "Effect of local background intensities in the normalization of cDNA microarray data with a skewed expression profiles," *Experimental and Molecular Medicine*, vol. 34, pp. 224–232, 2002
11. P. Milanfar, *Super-Resolution Imaging*. Boca Raton, FL, USA: CRC Press / Taylor & Francis, September 2010.
12. R. Lukac, B. Smolka, K. Martin, K.N. Plataniotis, and A.N. Venetsanopulos, "Vector filtering for color imaging," *IEEE Signal Processing Magazine*, vol. 22, no. 1, pp. 74–86, January 2005.
13. J. Astola, P. Haavisto, and Y. Neuvo, "Vector median filters," *Proceedings of the IEEE*, vol. 78, no. 4, pp. 678–689, April 1990.
14. S.T. McHugh, "Digital camera image noise." Available online, http://www.cambridgeincolour.com/tutorials/noise.htm.
15. R. Lukac, "Single-sensor digital color imaging fundamentals," in *Single-Sensor Imaging: Methods and Applications for Digital Cameras*, R. Lukac (ed.), Boca Raton, FL, USA: CRC Press / Taylor & Francis, September 2008, pp. 1–29.
16. R. Nagarajan, "Intensity-based segmentation of microarrays images," *IEEE Transactions on Medical Imaging*, vol. 22, pp. 882–889, July 2003.
17. R. Lukac, K.N. Plataniotis, B. Smolka, and A.N. Venetsanopoulos, A multichannel order-statistic technique for cDNA microarray image processing, *IEEE Transactions on Nanobioscience*, vol. 3, no. 4, pp. 272–285, December 2004.
18. K. Tang, J. Astola, and Y. Neuvo, "Nonlinear multivariate image filtering techniques," *IEEE Transactions on Image Processing*, vol. 4, no. 6, pp. 788–798, June 1995.
19. R. Lukac and K.N. Plataniotis, "A taxonomy of color image filtering and enhancement solutions," in *Advances in Imaging and Electron Physics*, P.W. Hawkes (ed.), Elsevier/Academic Press, vol. 140, June 2006, pp. 187–264.

20. V. Kayargadde and J.B. Martens, "An objective measure for perceived noise," *Signal Processing*, vol. 49, no. 3, pp. 187–206, March 1996.

21. J. Zheng, K.P. Valavanis, and J.M. Gauch, "Noise removal from color images," *Journal of Intelligent and Robotic Systems* vol. 7, no. 3, pp. 257–285, 1993.

22. I. Pitas and A.N. Venetsanopoulos, *Nonlinear Digital Filters, Principles and Applications*. Boston, MA, USA: Kluwer Academic Publishers, 1990.

23. R. Gonzalez and R.E. Woods, *Digital Image Processing*. Reading, MA, USA: Prentice Hall, 3rd edition, 2007.

24. J.C. Russ, *The Image Processing Handbook*, Boca Raton, FL, USA: CRC Press / Taylor & Francis, 5th Edition, 2007.

25. K.N. Plataniotis and A.N. Venetsanopoulos, *Color Image Processing and Applications*, New York, NY, USA: Springer Verlag, 2000.

26. R. Lukac and K.N. Plataniotis, "Vector edge operators for cDNA microarray spot localization," *Computerized Medical Imaging and Graphics*, vol. 31, pp. 510–522, 2007.

27. J.H. Kim, H.Y. Kim, and Y.S. Lee, "A novel method using edge detection for signal extraction from cDNA microarray image analysis," *Experimental and Molecular Medicine*, vol. 33, pp. 83–88, 2001.

28. M. Katzer, F. Kummert, and G. Sagerer, "Methods for automatic microarray image segmentation," *IEEE Transactions on Nanobioscience*, vol. 2, pp. 202–213, December 2003.

29. A.W.C. Liew, H. Yana, and M. Yang, "Robust adaptive spot segmentation of DNA microarray images," *Pattern Recognition*, vol. 36, pp. 1251–1254, 2003.

30. R. Lukac, K.N. Plataniotis, A.N. Venetsanopoulos, R. Bieda, and B. Smolka, *Color Edge Detection Techniques*, in "Signaltheorie und Signalverarbeitung, Akustik und Sprachakustik, Informationstechnik," W.E.B. Universität Verlag, Dresden, vol. 29, 21-47, 2003.

31. J.F. Canny, "A computational approach to edge detection," *IEEE Transactions on Pattern Analysis and Machine Intelligence*, vol. 8, pp. 679–698, 1986.

32. D. Ziou and S. Tabbone, "Edge detection techniques: An overview," *Pattern Recognition Image Analysis*, vol. 8, no. 4, pp. 537-559, 1998.

33. J. Gomes and L. Velho, *Image Processing for Computer Graphics*. New York, USA: Springer-Verlag, 1997.

34. K.N. Plataniotis, D. Androutsos, and A.N. Venetsanopoulos, "Adaptive fuzzy systems for multichannel signal processing," *Proceedings of the IEEE*, vol. 87, no. 9, pp. 1601–1622, September 1999.

35. R. Lukac, K.N. Plataniotis, B. Smolka, and A.N. Venetsanopoulos, "cDNA microarray image processing using fuzzy vector filtering framework," *Fuzzy Sets and Systems*, vol. 152, no. 1, pp. 17–35, May 2005.

36. P. Perona and J. Malik, "Scale space and edge detection using anisotropic diffusion," *IEEE Transactions on Pattern Analysis and Machine Intelligence*, vol. 12, no. 7, pp. 629–639, July 1990.

37. G. Sapiro and D.L. Ringach, "Anisotropic diffusion of multivalued images with applications to color filtering," *IEEE Transactions on Image Processing*, vol. 5, no. 11, pp. 1582–1586, November 1996.

38. B. Smolka, R. Lukac, K.N. Plataniotis, and A.N. Venetsanopoulos, "Modified anisotropic diffusion framework," *Proceedings of SPIE*, vol. 5150, pp. 1657–1666, June 2003.

39. V. Aurich and J. Weule, "Non-linear Gaussian filters performing edge preserving diffu-

sion," in *Proceedings of the DAGM Symposium*, pp. 538–545, 1995.

40. S.M. Smith and J.M. Brady, "SUSAN - A new approach to low level image processing," *International Journal of Computer Vision*, vol. 23, no. 1, pp. 45–78, May 1997.

41. C. Tomasi and R. Manduchi, "Bilateral filtering for gray and color images," in *Proceedings of the IEEE International Conference on Computer Vision*, Bombay, India, January 1998, pp. 839–846.

42. S. Paris, P. Kornprobst, J. Tumblin, and F. Durand, "Bilateral filtering: Theory and applications," *Foundations and Trends in Computer Graphics and Vision*, vol. 4, no. 1, pp. 1–73, 2008.

43. B.K. Gunturk, "Bilateral filter: Theory and applications," in *Computational Photography: Methods and Applications*, R. Lukac (ed.), Boca Raton, FL, USA: CRC Press / Taylor & Francis, October 2010, pp. 339–366.

44. T.Q. Pham and L.J. van Vliet, "Separable bilateral filtering for fast video preprocessing," in *Proceedings of the IEEE International Conference on Multimedia and Expo*, Amsterdam, The Netherlands, July 2005.

45. B. Weiss, "Fast median and bilateral filtering," *ACM Transactions on Graphics*, vol. 25, no. 3, pp. 519–526, July 2006.

46. F. Durand and J. Dorsey, "Fast bilateral filtering for the display of high dynamic-range images," *ACM Transactions on Graphics*, vol. 21, no. 3, pp. 257–266, July 2002.

47. S. Paris and F. Durand, "A fast approximation of the bilateral filter using a signal processing approach," *International Journal of Computer Vision*, vol. 81, no. 1, pp. 24–52, January 2009.

48. R. Fattal, M. Agrawala, and S. Rusinkiewicz, "Multiscale shape and detail enhancement from multi-light image collections," *ACM Transactions on Graphics*, vol. 26, no. 3, August 2007.

49. A. Buades, B. Coll, and J.M. Morel, "A non-local algorithm for image denoising," in *Proceedings of the IEEE International Conference on Computer Vision and Pattern Recognition*, June 2005.

50. P. van Beek, Y. Su, and J. Yang, "Image denoising and restoration based on nonlocal means," in *Image Restoration: Fundamentals and Advances*, B. Gunturk and X. Li (ed.), Boca Raton, FL, USA: CRC Press / Taylor & Francis, September 2012.

51. T. Brox, O. Kleinschmid, and D. Cremers, "Efficient nonlocal means for denoising of structural patterns," *IEEE Transactions on Image Processing*, vol. 17, no. 7, pp. 1083–1092, July 2008.

52. T. Mahmoudi and G. Sapiro, "Fast image and video denoising via nonlocal means of similar neighborhoods," *IEEE Signal Processing Letters*, vol. 12, no. 12, pp. 839–842, December 2005.

53. T. Tasdizen, "Principal neighborhood dictionaries for nonlocal means image denoising," *IEEE Transactions on Image Processing*, vol. 18, no. 12, pp. 2649–2660, December 2009.

54. C. Kervrann and J. Boulanger, "Optimal spatial adaptation for patch-based image denoising," *IEEE Transactions on Image Processing*, vol. 15, no. 10, pp. 2866–2878, October 2006.

55. V. Barnett, "The ordering of multivariate data," *Journal of Royal Statistical Society A*, vol. 139, pp. 318–354, 1976.

56. K. E. Barner and G. R. Arce, CRC Press, *Nonlinear Signal and Image Processing: Theory, Methods and Applications*. Boca Raton, FL, 2004.

57. J. Astola and P. Kuosmanen, *Fundamentals of Nonlinear Digital Filtering*. Boca Raton,

FL, USA: CRC Press, 1997.

58. S. Mitra and J. Sicuranza, *Nonlinear Image Processing*. San Diego, CA, USA: Academic Press, 2001.

59. J.W. Tukey, "Nonlinear (nonsuperposable) methods for smoothing data," *The Collected Works of J.W. Tukey, Volume II: Time Series, 1965 - 1984*, pp. 837-856. Monterey, CA: Wadsworth Advanced Books & Software.

60. I. Pitas and A.N. Venetsanopoulos, "Order statistics in digital image processing," *Proceedings of the IEEE*, vol. 80, no. 12, pp. 1892–1919, December 1992.

61. R.C. Hardie and C.G. Boncelet, "LUM filters: A class of rank-order-based filters for smoothing and sharpening," *IEEE Transactions on Signal Processing*, vol. 41, no. 3, pp. 1061–1076, March 1993.

62. G.R. Arce, "A general weighted median filter structure admitting negative weights," *IEEE Transactions on Signal Processing*, vol. 46, no. 12, pp. 3195–3205, December 1998.

63. T. Viero, K. Oistamo, and Y. Neuvo, "Three-dimensional median related filters for color image sequence filtering," *IEEE Transactions on Circuits, Systems and Video Technology*, vol. 4, no. 2, pp. 129-142, April 1994.

64. R. Lukac and K.N. Plataniotis, "cDNA microarray image segmentation using root signals," *International Journal of Imaging Systems and Technology*, vol. 16, no. 2, pp. 51–64, April 2006.

65. L. Yin, R. Yang, M. Gabbouj, and Y. Neuvo, "Weighted median filters: A tutorial," *IEEE Transactions on Circuits and Systems -II*, vol. 43, no. 3, pp. 157–192, March 1996.

66. R. Lukac, K.N. Plataniotis, B. Smolka, and A.N. Venetsanopoulos, "Generalized selection weighted vector filters," *EURASIP Journal on Applied Signal Processing*, vol. 2004, no. 12, pp. 1870–1885, September 2004.

67. R. Lukac, B. Smolka, K.N. Plataniotis, and A.N. Venetsanopoulos, "Selection weighted vector directional filters," *Computer Vision and Image Understanding*, vol. 94, no. 1-3, pp. 140–167, April-June 2004.

68. Y. Shen and K.E. Barner, "Fast adaptive optimization of weighted vector median filters," *IEEE Transactions on Signal Processing*, vol. 54, no. 7, pp. 2497–2510, July 2006.

69. C. Kotropoulos and I. Pitas, *Nonlinear Model-Based Image/Video Processing and Analysis*, New York, NY, USA: J. Wiley, 2001.

70. N. Nikolaidis and I. Pitas, "Multichannel L filters based on reduced ordering," *IEEE Transactions on Circuits and Systems for Video Technology*, vol. 6, no. 5, pp. 470–482, October 1996.

71. E.J. Coyle, "Rank order operators and the mean absolute error criterion," *IEEE Transactions on Acoustics, Speech, and Signal Processing*, vol. 36, no. 1, pp. 63–76, January 1988.

72. A. Bovik, T. Huang, D. Munson Jr., "A generalization of median filtering using linear combinations of order statistics," *IEEE Transactions on Acoustics, Speech, and Signal Processing*, vol. 31, no. 6, pp. 1342–1350, December 1983.

73. I. Pitas and A.N. Venetsanopoulos, "Adaptive filters based on order statistics," *IEEE Transactions on Signal Processing*, vol. 39, no. 2, pp. 518–522, December 1991.

74. J. Serra, *Image Analysis and Mathematical Morphology*. New York, USA: Academic Press, 1982.

75. F.Y. Shih, *Image Processing and Mathematical Morphology: Fundamentals and Applications*. Boca Raton, FL, USA: CRC Press / Taylor & Francis, March 2009.

76. R. Hirata Jr., J. Barrera, R.F. Hashimoto, D.O. Dantas, and G.H. Esteves, "Segmenta-

tion of microarray images by mathematical morphology," *Real-Time Imaging*, vol. 8, pp. 491–505, 2002.

77. S.R. Sternberg, "Grayscale morphology," *Computer Vision, Graphics, and Image Processing*, vol. 35, pp. 333–355, September 1996.

78. F.Y. Shih, C.T. King, and C.C. Pu, "Pipeline architectures for recursive morphological operations," *IEEE Transactions on Image Processing*, vol. 4, no. 1, pp. 11–18, January 1995.

79. P. Kuosmanen and J. Astola, "Soft morphological filtering," *Journal of Mathematical Imaging and Vision*, vol. 5, no. 3, pp. 231–262, September 1995.

80. P. Maragos and R. Schafer, "Morphological filters – Part II: Their relations to median, order-statistic, and stack filters," *IEEE Transactions on Acoustics, Speech, and Signal Processing*, vol. 35, no. 8, pp. 1170–1184, August 1987.

81. S. Pei, C. Lai, and F.Y. Shih, "Recursive order-statistic soft morphological filters," *IEEE Proceedings – Vision, Image and Signal Processing*, vol. 145, no. 5, pp. 333–342, October 1998.

82. G. Agam and I. Dinstein, "Regulated morphological operations," *Pattern Recognition*, vol. 32, no. 6, pp. 947–971, June 1999.

83. D. Sinha and E.R. Dougherty, "Fuzzy mathematical morphology," *Journal of Visual Communication and Image Representation*, vol. 3, no. 3, pp. 286–302, September 1992.

84. D. Sinha and E.R. Dougherty, "A general axiomatic theory of intrinsically fuzzy mathematical morphologies," *IEEE Transactions on Fuzzy Systems*, vol. 3, no. 4, pp. 389–403, November 1995.

85. I. Daubechies, Ten Lectures on Wavelets. Philadelphia SIAM, 1992.

86. M. Vattereli and J. Kovacevic, *Wavelets and Subband Coding*. Englewood Cliffs, NJ, USA: Prentice Hall, 1995.

87. S. Mallat, *A Wavelet Tour of Signal Processing*, San Diego, CA, USA: Academic Press, 2nd Edition, 1999.

88. D.L. Donoho and I.M. Johnstone, "Ideal spacial adaption by wavelet shrinkage," *Biometrika*, vol. 81, pp. 425–455, 1994.

89. D.L. Donoho, "De-noising by soft-thresholding," *IEEE Transactions on Information Theory*, vol. 41, no. 3, pp. 613–627, 1995.

90. M. Janse, *Noise Reduction by Wavelet Thresholding*. Volume 161, New York, NY, USA: Springer Verlag, 2001.

91. X.H. Wang, R.S.H. Istepian, and Y.H. Song, "Microarray image enhancement using stationary wavelet transform," *IEEE Transactions on Nanobioscience*, vol. 2, pp. 184-189, December 2003.

92. D.Q. Dai and H. Yan, "Wavelets and face recognition," in *Face Recognition*, K. Delac and M. Grgic (eds.), Vienna, Austria: I-Tech Education and Publishing, pp. 59-92, June 2007.

93. J. Portilla, V. Strela, M. Wainwright, and E.P. Simoncelli, "Image denoising using scale mixtures of Gaussians in the wavelet domain," *IEEE Transactions on Image Processing*, vol. 12, no. 11, pp. 1338–1351, November 2003.

94. A. Chambolle, R.A. DeVore, N.Y. Lee, and B.J. Lucier, "Nonlinear wavelet image processing: Variational problems, compression and noise removal through wavelet shrinkage," *IEEE Transactions on Image Processing*, vol. 7, no. 3, pp. 319–335, 1998.

95. M.A.T. Figueiredo and R.D. Nowak, "Wavelet-based image restoration," *IEEE Transactions on Image Processing*, vol. 12, no. 8, pp. 906–916, August 2003.

8 Compression of DNA Microarray Images

Miguel Hernández-Cabronero, Michael W. Marcellin, and Joan Serra-Sagristà

CONTENTS

8.1 INTRODUCTION

Different parts of the microarray image analysis process are still under active research, as described in Chapter 1. As new approaches for the extraction of genetic expression data are developed, it will be beneficial to reanalyze the images in order to obtain more accurate data. In this scenario, repeating the whole microarray experiment will most probably not be a practical option due to the high economic costs and also to the possible unavailability of the needed biological samples. Thus, it is highly desirable to keep the microarray images for future use.

The high spatial resolution of microarray image scanners results in relatively large image dimensions, easily exceeding 4000×13000 pixels in size in current commercial hardware. Due to the wide range of possible gene expression intensities, 16 bits are required to store each image pixel. In consequence, over 230 Megabytes can be required to store in raw or uncompressed tagged image file format (TIFF) the output of a single microarray experiment. The space requirements are multiplied when several experiments are performed as a part of a microarray study, as is usually the case. Therefore, the large amounts of data that are produced in laboratories around the world need to be efficiently stored and transmitted among computers in the same or in different centers. Data compression is the natural approach to this problem.

This chapter discusses the state of the art in compression of microarray images, beginning, in Section 8.2, with the properties of the images used for benchmarking in most of these schemes. Both lossless and lossy compression schemes have been proposed in the literature. Lossless compression guarantees perfect image fidelity after decompression, at the cost of larger compressed image file sizes. The most successful lossless approaches are described in Section 8.3, paying special attention to image compression standards. Lossy compression can generate several-times-smaller compressed sizes, but some information is lost in the process. In Section 8.4, we describe the most important lossy compression approaches and address the problem of measuring the amount of relevant information lost when a microarray image is distorted. Finally, some conclusions are drawn in Section 8.5.

8.2   IMAGE CORPORA

Before discussing the different compression approaches, some image properties important from the point of view of the coding process are discussed. This information will situate the reader in a better position to understand the problematic nature of microarray image compression, the techniques proposed in the literature, and the obtained performance results.

Different image sets have been employed for assessing the performance of the different published compression schemes. None of these sets has been used consistently across all publications, even though three sets are more common: the MicroZip [3], the ApoA1 [4], and the ISREC [5] image sets. In recent publications, at least three other sets have also been used in order to provide a more variate and updated selection of images: the Yeast [6], the Stanford [7], and the Arizona [8] image sets. In particular, the last set contains images that were registered at least 7 years later than any of the other ones and thus are more representative of the output of modern commercial DNA microarray scanners.

The most important properties of the image sets are provided in Table 8.1. The number of images of each set, considering each Cy3/Cy5 channel independently, is indicated in the *Images* column. It should be noted that Cy3/Cy5 channel image pairs are not available in the MicroZip and the Stanford sets. Instead, they consist of single grayscale images obtained in different experiments. All grayscale images are unsigned and require 16 bits per pixel so that the wide genetic expression ranges can be accurately registered. Thus, it is possible to differentiate between 2^{16} different gray levels. The percentage of those values that is actually used has been computed independently for each image, and the average value of these percentages for each set is reported in the *Average intensities* column. It can be observed that all sets except

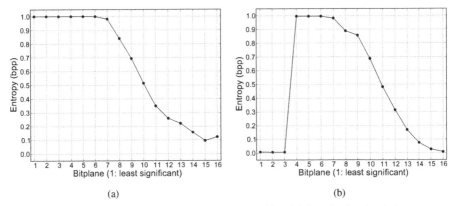

FIGURE 8.1 Average per-bitplane binary entropy profiles. (a) Results for the Arizona set; (b) Results for the Yeast set.

for the Arizona set employ less than 40% of all possible intensities. In microarray images, high intensities are infrequent because most pixels belong to the background and to low intensity spots. Further discussion on the pixel intensity distribution can be found in Section 8.3.6. Images not from the Arizona set have relatively small sizes and low numbers of spots, so that the total number of high intensity pixels is small and not all possible intensity values appear.

The average entropy of the images in each set is provided in the *Average entropy* column. Even though 16 bits are required to store each pixel in raw format, average entropies only rarely exceed 11 bits per pixel (bpp) because high intensity values are comparatively scarce. The bitplane binary entropy of most microarray images is similar to the one shown in Figure 8.1(a), corresponding to the average of all images in the Arizona set. The least significant bitplanes (between 6 and 10, depending on the set) have entropies close to the maximum possible value of 1 bpp, while the most significant bitplanes are almost constant, so that their entropy is close to 0 bpp.

It has been claimed that the least significant byte of microarray images has an almost random nature [9]. When the statistical hypothesis test χ^2 is used to assess the goodness of fit of the bitplanes to the uniform distribution, it is revealed that only three to five of the bitplanes truly exhibit this random behavior. The results of this test are reported in Table 8.2. In practice, compressing the least significant byte in a lossless manner is still a challenging task. It can be observed that images from the Yeast set exhibit an abnormally low fraction of used intensities and entropy values. In Figure 8.1(b), an explanation for this can be found: the binary entropy of the three least significant bitplanes is close to zero, which means that they are almost constant. This phenomenon does not occur in any of the other sets.

8.3 LOSSLESS COMPRESSION OF MICROARRAY IMAGES

With lossless compression schemes, it is possible to recover an exact copy of the original image from the compressed file. This property is valuable when working

TABLE 8.1
Image sets used in this work.

(a)

Property	MicroZip [3]	ApoA1 [4]	ISREC [5]
Year	2004	2001	2001
Images	3	32	14
Size	$> 1800 \times 1900$	1044×1041	1000×1000
Spot layout	square	square	square
Spot count	$\sim 9 \cdot 10^3$	$\sim 6 \cdot 10^3$	$\sim 2 \cdot 10^2$
Average intensities	37.71%	39.51%	33.34%
Average entropy	9.831	11.033	10.435

(b)

Property	Yeast [6]	Stanford [7]	Arizona [8]
Year	1998	2001	2011
Images	109	20	6
Size	1024×1024	$> 2000 \times 2000$	4400×13800
Spot layout	square	square	hexagonal
Spot count	$\sim 9 \cdot 10^3$	$\sim 4 \cdot 10^3$	$\sim 2 \cdot 10^5$
Average intensities	5.39%	28.83%	82.82%
Average entropy	6.628	8.293	9.306

Note: *All original images are grayscale 16 bpp. (a) Most frequently used sets in the literature. (b) Additional image sets used in this work.*

with microarray images, since all original information will be available for current and new analysis techniques.

8.3.1   MICROARRAY-SPECIFIC TECHNIQUES

Most published works in the area of microarray image compression address the topic of lossless coding and propose specific techniques designed to exploit the properties of microarray images. We first provide an overview of the most recent microarray-specific techniques and then discuss their compression performance.

In 2004, Lonardi and Luo presented their MicroZip compression algorithm [3]. In their work, the image is first gridded and segmented into foreground and background. To do so, spot regions are discovered by studying the period and minima of the sum of intensities by rows and by columns. The spot centers are then estimated based on the region centroid, and pixels are tagged as belonging to a spot following a spiral path. Once the image is segmented into 16-bit spot and background subimages, each is divided into two 8-bit substreams corresponding to the most and least

TABLE 8.2

Average significance levels for the χ^2 test for bitplane randomness using 64×64 **regions, as described in [1].**

Set	LSBP	BP2	BP3	BP4	BP5	BP6	...	MSBP
ISREC	**0.929**	**0.930**	**0.928**	**0.619**	**0.219**	0.000	...	0.000
ApoA1	**1.000**	**1.000**	**1.000**	**0.512**	0.001	0.000	...	0.000
Yeast	0.000	0.000	0.000	0.000	0.000	0.000	...	0.000
MicroZip	**1.000**	**0.995**	**0.333**	**0.333**	0.003	0.000	...	0.000
Stanford	**0.850**	**0.150**	**0.100**	0.000	0.000	0.000	...	0.000
Arizona	**1.000**	**0.838**	**0.167**	0.000	0.000	0.000	...	0.000

Note: *Higher significance indicates a more uniform distribution in a bitplane (BP). Values that indicate that the randomness hypothesis for a given bitplane cannot be rejected with a confidence of 5% are highlighted in bold.*

significant bytes (MSB and LSB) of each sample. The Burrows-Wheeler Transform is then used separately to compress each substream. In 2005, Zhang et al. [10] proposed a context-based lossless approach also based on segmentation. The spot and background subimages are also divided into the MSB and LSB substreams, and compressed independently. A simple prediction scheme that considers only two neighbor pixels is applied to the MSB of each subimage. The residuals of this prediction are coded using a method called prediction by partial approximate matching (PPAM) [11], a lossless codec of the prediction by partial matching (PPM) family with variable contexts. The LSB are coded directly with PPAM without using the prediction scheme. In 2006, Neves and Pinho [12] published a technique that encodes the bitplanes by order of significance (most significant first) until the first bitplane that cannot be compressed. In their approach, 3D context models are used to drive an arithmetic coder. The contexts applied when coding one bit consist of bits from the same and previous bitplanes, and hence the name of 3D contexts. The way in which pixels are used in the context (the context template) is static and calculated a priori by a trial and error procedure so that the average compression performance of their method is maximized. In 2009 [2], they extended their method so that the context template is built specifically for each image, at the expense of higher computational costs. In 2007, Neekabadi et al. [13] proposed a technique that first calculates an optimal threshold for segmenting the images. These are divided into three subsets (background, edge and spot pixels) by applying morphological operations after the threshold-based segmentation. Each set is scanned first independently in order to obtain a robust linear predictor that minimizes the energy of the produced errors for that set. The resulting predictor is then applied to the corresponding segment and the errors are compressed using Huffman coding. In 2009, Battiato and Rundo [14] presented an approach that employs cellular neural networks (CNNs) to segment the

TABLE 8.3

Average lossless compression results for the most recent microarray-specific approaches.

Algorithm	MicroZip (9.831)	Yeast (6.628)	ApoA1 (11.033)	ISREC (10.435)
MicroZip [3], 2004	9.843	—	—	—
PPAM [10], 2005	9.587	6.601	—	—
Neves [12], 2006	8.840	—	10.280	10.199
Neekabadi [13], 2007	8.856	—	10.250	10.202
Neves [2], 2009	8.667	5.521	10.223	10.199
Battiato [14], 2009	8.369	—	9.520	9.490

Note: *All values are expressed in bits per pixel as reported by the original authors, except the data for [2], which has been calculated using the authors' original codec (ftp://ftp.ieeta.pt/~ap/codecs). All images are 16 bpp. The average first-order entropy in bits per pixel for each image set is provided under its name.*

images into foreground and background. The foreground subimage is compressed directly using portable network graphics (PNG), a generic lossless codec. The 16-bit background subimage is transformed into two 8-bit indexed images using a custom algorithm for palette size reduction. The palette of these indexed images is then reordered using an algorithm that minimizes the entropy of the local differences, and the resulting index matrices are also coded using PNG. More on palette reindexing is discussed in Section 8.3.4.

Compression performance results for the algorithms described above are provided in Table 8.3. To the best of our knowledge, these algorithms yield the best performance results for this type of image. The results are expressed in bits per pixel (bpp), and hence lower values are better. The dashes mean that the results were not provided by the authors for a particular image set. In most cases, there is not any available implementation of these methods. Therefore, results cannot be generated for the newer, more representative datasets. Attending to the available data, Battiato's technique is the one with the best compression performance, claiming 8.619 bpp, 9.52 bpp, and 9.49 bpp for the MicroZip, ApoA1, and ISREC sets, respectively. These results are between 0.3 bpp and 0.7 bpp better than the second best performance, by Neves. The improvement obtained by designing image-specific context templates enhances their original proposal by less than 0.2 bpp on average. These compression results, which only rarely achieve 2:1 compression ratios, are poor compared to those obtained in other types of images.

8.3.2   IMAGE COMPRESSION STANDARDS

The main goal of the compression of microarray images is the long-term storage for future reanalysis. The approaches described in the previous section produce the best lossless compression performance results. Nevertheless, the software implementations are not likely to be maintained or further developed. As a result, the availability of working decoders in future platforms might be an issue. In addition, as microarray images are adopted for clinical purposes, it will be necessary to adhere to the digital imaging and communication in medicine standard (DICOM) [15] so that microarray images can be used in health centers. For this reason, we have focused on the best performing image compression standards, with special attention to those accepted by the DICOM standard with implementations that support 16 bpp images (JPEG-LS and JPEG2000).

In our experiments, we have tested the lossless compression performance of different generic and standard image coders: Bzip2[1], JBIG[2], JPEG-LS[3], and JPEG2000.[4] The results are reported in Table 8.4. The data for JPEG2000, JBIG, and JPEG-LS are almost identical to those reported in [16], one of the latest publications regarding the usage of image compression standards for microarray images. It can be observed that the microarray-specific compressors produce results that are only between 0.4 bpp and 1.4 bpp better than the best image compression standards, depending on the image set. In this direct approach, JPEG-LS is the best standard for three out of six sets, even though its results are comparable to the other DICOM-compatible standards. In particular JPEG2000, the latest image compression standard [17], offers a wide range of features that can be employed for the compression of microarray images. Thus, in the following we focus on enhancing the lossless compression performance of JPEG2000 by considering several approaches.

The results for JPEG2000 provided in Table 8.4 are for the best choice of decomposition levels for each set. The performance of lossless JPEG2000 for DNA microarray images when different 5/3 discrete wavelet transform (DWT) spatial decomposition levels are selected is also analyzed. The results are reported in Table 8.5. It can be observed that for four out of six image sets, increasing the number of decomposition levels from 0 to 5 yields bitrates approximately 0.5 bpp better. Using more decomposition levels produces only minor performance improvements, which are not included in the table. For the other two sets—the Yeast and the ISREC sets— increasing from 0 to 1 DWT decomposition levels causes a significant performance reduction, while increasing from one to five levels yields small performance gains of 0.05 and 0.16 bpp. When the DWT is applied to the Yeast set, the three least

---

[1] Version v1.0.6 using the `--best` parameter.

[2] Version v2.0 of JBIG-KIT, `http://www.cl.cam.ac.uk/~mgk25/jbigkit/`, using the `-q` parameter.

[3] Version v1.00X of the Hewlett-Packard implementation, `http://www.hpl.hp.com/loco/locodown.htm`, with default parameters.

[4] Version v7.2 of Kakadu, `http://www.kakadusoftware.com/`, with default parameters for lossless compression.

significant bitplanes are no longer almost constant, which explains the observed performance degradation. The generally poor performance of wavelet-based methods is explained by two facts. First, the abrupt intensity changes caused by the presence of bright spots on a dark background reduces the energy compaction of the wavelet transform, and thus the efficiency of coders like JPEG2000. Second, the large number of possible intensities that can be described with the 16 bpp used for microarray images limits the entropy reduction for this type of transform. The transformed images exhibit the expected Laplacian-like distributions, but the obtained tails are heavy in comparison to the original distribution of the images, in which large intensities are scarce.

TABLE 8.4

Average lossless compression results for generic compressors and standard image coders.

Algorithm	MicroZip (9.831)	Yeast (6.628)	ApoA1 (11.033)	ISREC (10.435)	Stanford (8.293)	Arizona (9.306)
Bzip2	**9.394**	**6.075**	11.067	10.921	7.867	8.944
JBIG	9.747	6.888	10.852	10.925	7.776	8.858
JPEG-LS	9.441	8.580	**10.608**	11.145	**7.571**	**8.646**
JPEG2000	9.467	6.829	10.999	**10.888**	7.969	9.064
Neves [2]	8.667	5.521	10.223	10.199	7.335	8.275
Battiato [14]	8.369	—	9.520	9.490	—	—

Note: *All values are expressed in bits per pixel and all images are 16 bpp. The results for JPEG2000 have been obtained with the best number of wavelet decomposition levels for each set. The best results for generic and standard compressors are highlighted in bold font. The results for the best-performing microarray-specific techniques are included at the bottom for ease of reference.*

TABLE 8.5

Compression results in bpp for lossless JPEG2000 and different DWT decomposition levels.

DWT levels	MicroZip	Yeast	ApoA1	ISREC	Stanford	Arizona
0	10.027	**6.829**	11.525	**10.888**	8.567	9.548
1	9.542	9.089	11.088	11.476	8.146	9.221
3	9.472	9.042	**10.999**	11.312	7.985	9.068
5	**9.467**	9.038	**10.999**	11.314	**7.969**	**9.064**

Note: *The best choice on average for each image set is highlighted in bold font.*

8.3.3 MULTICOMPONENT COMPRESSION

The green and red channel images of a microarray experiment are obtained by scanning the same microarray chip. Hence, spots appear in almost coincident positions and many of them have similar intensities. In addition, the different experiments that are carried out as a part of a study usually employ the same microarray chip model and the output images are usually stored together. Therefore, it is reasonable to try to take advantage of the image similarities. Only brief mentions to this possibility are found in the microarray literature [18]. Multicomponent compression can help to exploit the redundancy present in two or more images by applying spectral transforms that decorrelate their information and then coding them in a single file.

As a first step to maximize the multicomponent compression performance, the correlation in pairs of DNA microarray images has been analyzed. Since size homogeneity is a requirement for multicomponent compression, only the four image sets whose images have all the same dimensions—Yeast, ApoA1, ISREC, and Arizona—have been considered. To quantify the correlation among image pairs, the images have been considered to be ordered sequences of pixel values, and the correlation coefficient has been calculated as

$$r = \frac{\sum_{i=1}^{n}(X_i - \bar{X})(Y_i - \bar{Y})}{\sqrt{\sum_{i=1}^{n}(X_i - \bar{X})^2}\sqrt{\sum_{i=1}^{n}(Y_i - \bar{Y})^2}},$$

where X and Y are the pixel sequences, n the length of these sequences, and $\bar{X}$ and $\bar{Y}$ their mean values. The correlation coefficient is always in $[-1, 1]$, where larger absolute values mean a larger linear correlation between X and Y.

The correlation coefficient has been calculated for every possible pair of different images from the same set. In Table 8.6(a), statistical information about these correlations is displayed for each set. It can be observed that, in general, the correlation among pairs is not very high: most values are under 0.2 except for the Arizona set, which contains pairs with larger r values. Even though average values are low or very low, the maximum correlation for each set is larger than 0.89 for all of the image sets. Hence, some pairs are much more correlated than others. The larger correlation values for the Arizona set when all pairs are considered is due to the larger proportion of Cy3/Cy5 pairs in the set. Since Cy3/Cy5 channel pairs are created using the same DNA microarray chip, they are more likely to exhibit larger r coefficients. In Table 8.6(b), some representative statistics of the correlations obtained considering only Cy3/Cy5 channel image pairs are provided. As is obvious from these data, correlation is now considerably larger. Tenfold increases in the average values can be observed, and they are larger than 0.75 for all sets. In addition, all pairs have positive r values. Also, maximum values remain constant, which indicates that Cy3/Cy5 pairs have the largest r values inside each set. From these data we can conclude that it is reasonable to use multicomponent compression only on Cy3/Cy5 channel image pairs since they have larger correlation, whereas all possible pairs in each set are not so correlated.

The lossless compression performance of multicomponent JPEG2000 is now investigated. The multicomponent compression capabilities of JPEG2000 are included

TABLE 8.6

Statistical properties of the correlation coefficients obtained by comparing (a) all image pairs in each set and (b) only Cy3/Cy5 pairs from the same experiment.

(a)

Set	Max	Min	Average	Variance
Yeast	0.8961	-0.0329	0.0390	0.0070
ApoA1	0.9678	0.0105	0.2043	0.0468
ISREC	0.8901	-0.0186	0.0735	0.0501
Arizona	0.9332	0.2845	0.6153	0.0383

(b)

Set	Max	Min	Average	Variance
Yeast	0.8961	0.6084	0.7821	0.0048
ApoA1	0.9678	0.7969	0.9229	0.0029
ISREC	0.8901	0.3250	0.7822	0.0359
Arizona	0.9332	0.5937	0.7508	0.0195

in Part 2 of the standard [17]. Each Cy3/Cy5 channel image pair is compressed independently as a two-component image. Four different spectral transforms—the 5/3 DWT, the pairwise-orthogonal transform (POT) [19], the differential pulse-code modulation (DPCM) transform, and the reversible Haar transform (R-Haar)—have been tested and compared to the case in which no multicomponent transform is used. Other transforms with longer filters have not been employed because images only have two components, and thus they are not expected to yield better results. Compression results are reported in Table 8.7. The main observation that can be made is that, in general, the tested multicomponent transforms do not produce significantly better results, as compared to not using any multicomponent transform. Compression performance gains can be noticed in approximately half of the cases. No single transform is clearly better than the others, even though the DPCM transform achieves some compression performance gain for more sets than any other. It can also be noted that the 5/3 DWT and the R-Haar transform produce very similar compression results, which is explained by the mirror effect that appears when the multicomponent DWT is used with only two components. The largest performance improvement is 0.568 bpp, obtained by the POT for the ApoA1 set, while other gains are smaller. For the ISREC set, no multicomponent transform is able to improve upon not using any transform. In all cases, applying any number of spatial 5/3 DWT decomposition levels yields poorer results. Furthermore, compressing together the Cy3/Cy5 channel image pairs has proven to be the most effective way of grouping images for multicomponent compression.

The bitrate differences obtained when using lossless JPEG2000 and the afore-

TABLE 8.7

Average lossless multicomponent compression results in bpp for Cy3/Cy5 channel image pairs and different multicomponent transforms.

Set	No transform	5/3 DWT	POT	DPCM	R-Haar
Yeast	6.829	**6.786**	9.279	**6.439**	**6.790**
ApoA1	11.524	**11.217**	10.956	11.289	11.218
ISREC	10.887	11.451	11.468	11.203	11.452
Arizona	9.548	9.649	**9.439**	**9.386**	9.649

Note: *The bold font has been used when applying a transform improves the compression performance. One decomposition level was employed for each multicomponent transform. No spatial DWT decomposition levels were used.*

TABLE 8.8

Correlation between compression improvements in bpp of multicomponent JPEG2000 and r values for the Cy3/Cy5 pairs of each set.

Set	Pairs	DWT	POT	DPCM	R-Haar
Yeast	54	0.296	0.245	0.222	0.294
ApoA1	16	-0.066	0.042	-0.200	-0.065
ISREC	7	-0.051	-0.156	-0.331	-0.053
Arizona	3	-0.774	-0.835	-0.994	-0.776
All	80	0.333	0.461	-0.075	0.335

Note: *Results considering every Cy3/Cy5 pair in all sets are included at the bottom.*

mentioned transforms, as compared to not using any spatial or spectral transform, has been calculated. The relationship of these differences and the correlation coefficient of each pair have been analyzed. The bitrate difference for each transform and image set are plotted in Figure 8.2 as a function of r for each channel pair. The correlation between these differences and the r values—hereafter referred to as bitrate-r correlation—has also been computed. The slope of the gray line in the figure equals the bitrate-r correlation for each multicomponent transform considering the images from all sets. Further bitrate-r correlation results are provided in Table 8.8.

All transforms except for the DPCM exhibit positive bitrate-r correlation values over 0.3. This suggests that the DPCM transform is not affected as much as the other transforms by the amount of correlation present in the pairs. With this transform, pairs with lower correlation are compressed with relative performance similar to that of the pairs with higher correlation, which explains why the DPCM is able to provide improvements over the no-transform case for more image sets than any of the other transforms. The 5/3 DWT and the R-Haar transform exhibit very similar bitrate-r correlations because they yield almost identical compression results, as discussed in

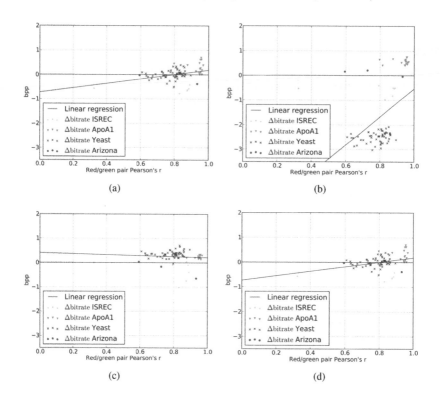

FIGURE 8.2 Bitrate differences using lossless JPEG2000 on the Cy3/Cy5 pairs without any spatial or spectral transform (baseline) and with different spectral transforms, as a function of the correlation of each pair. (a) Results for the 5/3 DWT; (b) Results for the POT; (c) Results for the DPCM transform; (d) Results for the reversible Haar transform. The slope of the blue line indicates the correlation between the bitrate differences and Pearson's r for each pair.

Section 8.3.3. The POT is the transform with the largest global bitrate-r correlation. This can be explained by the fact that the POT is designed after the Karhunen-Loeve transform (KLT), which is similar to the principal component analysis. It is noteworthy that the POT produces considerably poorer results for the Yeast set, as compared to other sets and to the other transforms. This can be clarified by attending to the three almost constant least significant bits of the Yeast set and the definition of the other transforms. The DPCM is defined as

$$\text{DPCM}(y_0, y_1) = (y_0, y_1 - y_0).$$

Thus, when it is applied to images from the Yeast set, these three least significant bitplanes remain almost constant. The R-Haar transform, defined as

$$\text{RHaar}(y_0, y_1) = (y_0 + \lfloor y_1/2 \rfloor, y_1 - y_0),$$

maintains these bitplanes essentially constant except for the third least significant bitplane of the low-pass components, which is mixed with the fourth least significant bitplane. Very similar results are yielded by the DWT due to the mirroring effect caused by the images having only two components. On the contrary, the POT does not use powers of two as divisors in the lifting steps, hence no bitplane is kept almost constant.

In conclusion, multicomponent transforms can be used to improve the compression performance of JPEG2000 on DNA microarray images to some extent, although the obtained bitrate gains are narrow. Improvements are always smaller than 0.6 bpp, and no transform is able to improve average compression results for all sets. In consequence, better approaches are needed to make JPEG2000 a competitive alternative to the best-performing microarray-specific and image compression standards.

8.3.4   IMAGE SEGMENTATION

The most successful microarray-specific compressors rely on a segmentation stage so that spot and background pixels are coded separately, in order to obtain better compression results. We will now analyze the suitability of the segmentation of microarray images as a way to improve the compression performance for microarray images.

As discussed in Chapter 1, griding and segmentation are one of the most active research topics in the analysis of microarray images. Spot and background pixels exhibit different statistical properties, and hence coding them separately can produce better performance results. Several segmentation approaches have been employed in the microarray image compression literature. Many of these approaches are largely inspired by different analysis-oriented methods described in Chapters 5 and 6. Some compression-oriented segmentation techniques exploit the spot layout in grids and subgrids using methods based on the computer vision field [9, 20, 21]. Others choose specific intensity thresholds to characterize spot pixels [13, 22], while the best-performing microarray-specific technique employs cellular neural networks (CNN) [14].

When designing new compression techniques for DNA microarray images, it is a key issue to determine how effective it can be to use segmentation before coding the images. In the experiments that we have run to analyze this question, only threshold-based segmentation, i.e., tagging pixels above a certain intensity threshold as spot pixels, has been considered. Analysis-oriented segmentation approaches can produce more accurate results, although this accuracy may not necessarily yield better compression results. Figure 8.3 shows crops of example threshold-based segmentations of microarray images. Different threshold values are needed depending on the image. Pixels identified as background are displayed in grayscale, while pixels tagged as belonging to a spot are shown in white. It can be observed that most spots are properly detected, while background pixels contain almost no spot-like zones. Furthermore, background and spot regions obtained with threshold-based techniques possess desirable statistical properties like a smaller dynamic range. For this reason, it can be justified to consider only threshold-based segmentation in our tests.

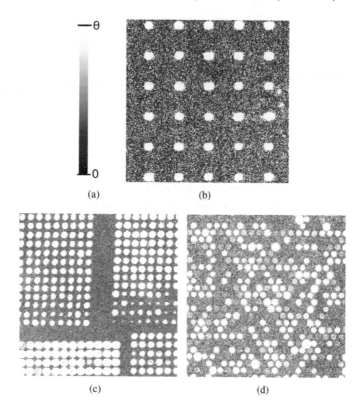

FIGURE 8.3 Threshold-based segmentations of DNA microarray images. (a) Grayscale employed to represent the background pixels. Values larger than the threshold θ are shown in white; (b) Segmentation of a crop of *Def667Cy3* from the ISREC set with $\theta = 512$; (c) Segmentation of a crop of *y744n89-ch1* from the Yeast set with $\theta = 512$; (d) Segmentation of a crop of *slide-2-red* from the Arizona set with $\theta = 400$.

In order to assess the utility of segmentation from the image compression perspective, a number of experiments have been run. The images have been segmented into background and spots using different threshold values $\theta \in \{2^0, 2^1, \ldots, 2^{16}\}$. As a result, two pixel sequences X and Y of length $|X|$ and $|Y|$, respectively are produced for each image I and threshold value. The entropy $H(X)$ and $H(Y)$ of these sequences has been calculated, and then combined into a single value $H(I)$ based on

$$H(I) = \frac{|X|}{|X| + |Y|} H(X) + \frac{|Y|}{|X| + |Y|} H(Y),$$

which is a weighted average of $H(X)$ and $H(Y)$ considering the fraction of pixels of the original image that belong to X and Y, respectively. The conditional entropy for a pixel p in a sequence W has been computed using the average neighborhood pixel value as a conditioning event, or context. Specifically, the average neighborhood pixel value is given by

$$C(p) = \left\lfloor \log_2 \frac{1}{|N(p)|} \sum_{n \in N(p)} n \right\rfloor,$$

where $N(p)$ is the set of eight nearest neighbors of x that belong to W, and $|N(p)|$ the number of such neighbors. The binary logarithm and the floor function are applied as a simple quantization after computing the mean to obtain values in $\{0,\dots,15\}$ to avoid the context dilution problem. With this context definition, the conditional entropy in a sequence W is given by

$$H(W|C) = \sum_{c=0}^{15} P(C = c)H(W|C = c),$$

where $P(C = c)$ is the probability of the context C being equal to c, and $H(W|C = c)$ the entropy considering only the pixels whose context is equal to c. Note that W can be any of the two sequences produced by segmenting the image. The conditional entropy employing a causal context—using only the pixels that appear earlier in a raster scan of the image to calculate the context—has also been measured. The results for all tested θ values for two image sets are plotted in Figure 8.4, while the values obtained for the best selection of θ for all sets are provided in Table 8.9. The conditional entropy is lower than the first-order entropy when all pixels are tagged either as background or spot. It can be observed that when different threshold values are used, the conditional entropy presents a behavior analogous to that of the first-order entropy. It is also noteworthy that the conditional entropy results are very similar using causal or non-causal contexts. Altogether, these data provide a theoretical explanation of why most compression techniques found in the literature benefit from a segmentation process: the combined entropy is lower when considering spot and background pixels separately. On the other hand, the obtained results indicate that the margin for improvement relying solely on segmentation is not very wide since entropy values are only reduced by at most 0.6 bpp. In addition, the difference between the first-order entropy when no segmentation is used and the conditional entropy when using an optimal threshold value—the worst and best entropy results—is approximately 1 bpp for all image sets. This improvement is not large enough to obtain combined entropies below the bitrates produced by the best microarray-specific technique except for the ISREC set. Furthermore, even though a practical coder could use a causal context without any mayor drawback, it would need to either encode the produced segmentation mask or to compress the background and spot subimages independently. In consequence, part of the compression performance increase would be canceled. Considering the obtained entropy differences, it cannot be expected to obtain much larger performance improvements through the use of segmentation techniques.

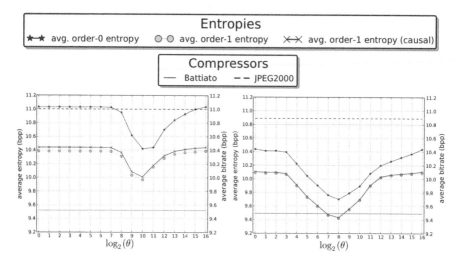

FIGURE 8.4 Combined first-order and conditional entropy $H(I)$ of images segmented using different threshold values (θ). (left) Average results for the ApoA1 set; (right) Average results for the ISREC set.

TABLE 8.9

Combined entropy first-order, conditional, and conditional using causal context H(I) results using the best choice of θ.

	MicroZip	Yeast	ApoA1	ISREC	Stanford	Arizona
$\log_2(\theta)$, best θ	8	9	10	8	6	8
first-order H(I)	9.28	5.86	10.42	9.70	7.70	8.58
conditional H(I)	8.85	5.28	9.97	9.42	7.11	8.11
causal H(I)	8.91	5.34	10.01	9.43	7.15	8.15

8.3.5 PALETTE REINDEXING

Color images can be represented using an index matrix together with a color palette that associates the matrix indices with the original colors. Grayscale images like the ones employed in microarray experiments can be described in a similar fashion using a palette that maps pixel intensities instead of colors. In both cases, the colors or intensities can be permuted in the palette so that the index matrix can acquire desirable properties and be coded more efficiently. This process is known as palette reindexing or palette reordering, and it is the basis of the microarray-specific proposal that claims the best compression results [14]. We analyze now the utility of reindexing applied to microarray images from the point of view of standard compressors. In particular, we focus on JPEG2000, which admits the compression of palettized images in Part 1 of the standard.

For a palette $\mathscr{P}$ of size n, there exist $n!$ possible orderings, and hence exhaustively searching for an optimal permutation from the point of view of image compression is unfeasible even for modest palette sizes. Since each grayscale microarray image needs 16 bpp, palette sizes are large even when only a fraction of all possible intensity values is used, thus a heuristic algorithm must be employed to search for good palette permutations. The best compression results for microarray images are claimed for a technique that applies a heuristic that consists in minimizing the first-order entropy of the differences of consecutive indexes. By doing so, the PNG algorithm can code the produced index matrix more efficiently. However, this approach is not necessarily useful for bitplane-based coders like JPEG2000. For an unsigned image of B bpp, the first step performed by a JPEG2000 Part 1 encoder is to subtract 2^{B-1} from each pixel value [17] in the so-called level offset stage. This usually results in pixel intensities nominally distributed symmetrically about the origin, for which JPEG2000 has been designed to be more efficient. With natural images, this distribution is typically obtained by applying some type of DWT to the image data. However, as previously discussed in Section 8.3.2, the DWT is not very effective for DNA microarray images. Notwithstanding, palette reindexing techniques can be used to adapt the microarray image histograms to JPEG2000.

A reindexing scheme with the goal of producing highly symmetrical index distributions has been tested. First, an auxiliary list of the intensities employed in the original image is created and sorted by frequency. After that, another list is created by inserting the elements of the auxiliary list alternatively in the left and right hand of the new list. Let S_i be the i-th element of the auxiliary list of size $n+1$, then the palette is defined as

$$\mathscr{P} = \{S_n, S_{n-2}, \ldots, S_2, S_0, S_1, S_3, \ldots, S_{n-3}, S_{n-1}\}.$$

When this reindexing algorithm is applied to DNA microarray images, the produced indexes follow a Laplacian-like distribution similar to the one expected by JPEG2000 before the level offset stage. Hereinafter, we will refer to this method as Laplacian reindexing. Table 8.10 reports the average compression results for lossless JPEG2000 after applying either our Laplacian reindexing scheme or the best number of 5/3 DWT decomposition levels. Results for the Laplacian technique do not include any overhead due to palette storage, even though it is needed to recover the original image from the compressed index matrix. No DWT decomposition levels were applied to obtain the data provided in the table, since more levels decrease the compression performance. It can be observed that JPEG2000 can compress the index matrices produced by the Laplacian reindexing algorithm better than the images transformed with the best number of DWT decomposition levels for all the image sets: the bitrate differences range from 0.186 bpp for the ApoA1 set to 1.040 bpp for the Yeast set. In addition to these results, the Laplacian reindexing scheme is very simple and does not call for many computational resources or a previous segmentation scheme. In contrast, the best-performing microarray-specific compressor uses neural networks—with as many neurons as there are pixels in the image—both for the reindexing scheme and the required segmentation stage. On the other hand, the

TABLE 8.10

Average compression results in bpp of lossless JPEG2000 after applying a Laplacian reindexing scheme or the best number of 5/3 DWT decomposition levels.

Method	MicroZip	Yeast	ApoA1	ISREC	Stanford	Arizona
Laplacian	9.198	5.789	10.813	10.658	7.754	8.844
Best 5/3 DWT	9.467	6.829	10.999	10.888	7.969	9.064

Note: *Results for the Laplacian reindexing assume zero wavelet decomposition levels and do not include any overhead due to palette storage.*

best microarray-specific compression techniques outperform the Laplacian reindexing scheme for the image sets for which results exist.

8.3.6 THE HISTOGRAM SWAP TRANSFORM

In Sections 8.3.3 to 8.3.5, we have discussed three different possibilities for the compression of microarray images, with palette reindexing being the most promising option. In this section we propose a very fast invertible transform based on Laplacian reindexing with almost identical compression performance and fewer drawbacks.

DNA microarray images present very asymmetric intensity distributions. Figure 8.5 shows the average pixel value distributions for two of the image sets. The majority of the pixels of all sets have small intensities, while higher values are several orders of magnitude less frequent. In general, 95% of the pixels are below 2^{12}, that is, 6.25% of the maximum intensity. This fact can be exploited to obtain histograms as symmetrical as possible. We now define a novel point transform, which we call the histogram swap transform or HST [23]. If the left and right parts of the histogram are swapped, the resulting image exhibits a more symmetrical distribution, centered about 2^{15}, as shown in Figure 8.6. After the level offset stage, a Laplacian-like distribution centered about the origin is obtained, for which JPEG2000 is best designed. The histogram swap transform of a pixel x in an image of bitdepth B can be defined as

$$HST(x) = \begin{cases} x + 2^{B-1} & \text{if } x < 2^{B-1} \\ x - 2^{B-1} & \text{if } x \geq 2^{B-1} \end{cases},$$

which is equivalent to flipping the most significant bit of the binary representation of the pixel, or (for 16 bpp images like microarray images) adding or subtracting the hexadecimal value 0x8000. As a result, the left and right parts are translated to the right and left, respectively, preserving the shape of each part. In the original image a majority of the pixels have very low intensities, and hence most transformed intensities are close to the middle of the histogram. The histogram swap transform can be regarded as a special type of reindexing algorithm, with the benefit that its sim-

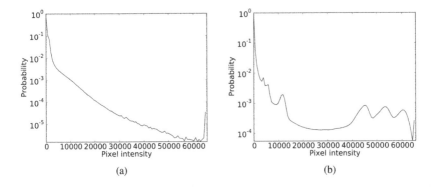

FIGURE 8.5 Average intensity distributions for (a) the MicroZip set; (b) the Arizona set.

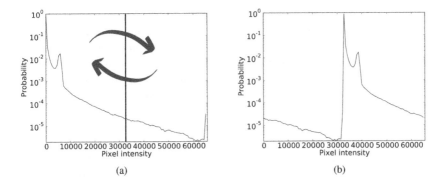

FIGURE 8.6 Pixel value distribution for DNA microarray images. (a) Effect of the HST on *slide-2-red* from the Arizona set. (b) Distribution for *slide-2-red* after applying the HST.

plicity avoids the need for palette storage. By definition, the HST is its own inverse transform. Therefore, it can be used in lossless compression schemes.

The lossless compression performance of JPEG2000 after applying the HST has been tested. Table 8.11 provides the compression performance for different numbers of DWT decomposition levels, as well as the bitrate differences as compared to not using the HST. The results for the best choice of decomposition levels with and without applying the HST are also compared. It can be observed that the best results for each set improve upon JPEG2000 without the HST from 0.213 bpp to 0.918 bpp, i.e., from 1.97% to 15.53%, and are obtained by using zero wavelet decomposition levels. When the HST is applied, the performance of JPEG2000 decreases with the number of DWT decomposition levels for all sets, while it is generally improved when the HST is not employed, as discussed in Section 8.3.2. The following observation can help explain this behavior. When the HST is applied on an unsigned image with bit-depth B, pixel values slightly smaller than 2^{B-1} become close to 2^B, while

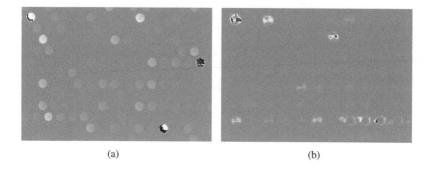

(a)   (b)

FIGURE 8.7 Sudden intensity changes after applying the HST. (a) Detail of the *slide-1-red* image from the Arizona set. (b) Detail of the *array3* image from the MicroZip set.

values slightly greater than 2^{B-1} become close to zero. In other words, mid-gray values before the HST cause abrupt intensity differences between near black and near white after the HST. Some examples of this behavior are shown in Figure 8.7. As a result, increasing the number of DWT decomposition levels results in a performance reduction, since it is not very effective for images with high frequencies [3].

After applying the histogram swap transform, JPEG2000 becomes arguably the overall best compression standard for microarray images, which produces the best results for three out of six image sets. In particular, this is true for the Arizona set—the most realistic of the benchmarking sets—even though JPEG2000 without the HST performs worse than the other standards. Nevertheless, JPEG2000 after the HST still produces results from 0.788 to 1.266 bpp worse than the best microarray-specific technique, proposed by Battiato.

The HST has an advantage in terms of implementation, as compared to other point transforms that lead to similar histograms, like the Laplacian reindexing transform. The HST can be implemented without any explicit changes in the original image data or JPEG2000. If the unmodified data are interpreted as two's-complement signed values, they have the same values that are obtained after the HST and the JPEG2000 level offset process. When interpreted as twos-complement values, pixels between 0x0000 and 0x7FFF produce decimal values between 0 and 32767, while pixels between 0x8000 and 0xFFFFF yield values between -32768 and -1. When data are interpreted as unsigned and the HST is applied, values between 0x0000 and 0x7FFF are transformed to values from 0x8000 to 0xFFFF, or 32768 to 65535 in decimal. After the JPEG2000 level offset, values from 0 to 32767 are obtained. Likewise, values from 0x8000 to 0xFFFF become 0x0000 to 0x7FFF (or 0 to 32767) after the HST and -32768 to -1 after the level offset. Hence, applying the HST and then compressing with JPEG2000 can be performed by simply applying JPEG2000 to the unsigned data as if it were signed. Encoding, decoding, and the resulting codestreams are all JPEG2000 Part 1 compliant, and no computational overhead is introduced. In comparison, the best-performing microarray-specific technique performs expensive

TABLE 8.11

Lossless compression results in bpp for JPEG2000 after applying the HST and then using different numbers of DWT decomposition levels.

levels	MicroZip	Yeast	ApoA1	ISREC	Stanford	Arizona
0	9.157	5.911	10.786	10.624	7.685	8.795
	(-0.87)	(-0.92)	(-0.74)	(-0.26)	(-0.88)	(-0.75)
1	9.297	8.862	10.917	11.238	7.851	8.967
	(-0.25)	(-0.23)	(-0.17)	(-0.24)	(-0.30)	(-0.25)
3	9.455	9.026	11.003	11.300	7.950	9.058
	(-0.02)	(-0.02)	(+0.00)	(-0.01)	(-0.04)	(-0.01)
5	9.466	9.035	11.012	11.313	7.958	9.070
	(-0.00)	(-0.00)	(+0.01)	(-0.00)	(-0.01)	(+0.01)
Best	9.157	5.911	10.786	10.624	7.685	8.795
	(-0.31)	(-0.92)	(-0.21)	(-0.26)	(-0.28)	(-0.27)

Note: *The bitrate differences between compressing the transformed and original images are reported in brackets for each number of decomposition levels. The bitrate differences, considering the best number of decomposition levels for transformed and untransformed images, are included at the bottom. Negative values imply better results using the HST transformed images.*

routines based on neural networks, which might not scale well as image sizes are increased due to continuing improvements in microarray scanner resolution.

The histogram swap transform can be combined with other techniques in order to try to improve the overall compression performance. We have performed some experiments to calculate the compression performance obtained when the HST is applied to the output of the multicomponent transform stage. The Cy3/Cy5 channel pairs have been processed with the same spectral transforms discussed in Section 8.3.3. Then the HST has been applied using two different setups and finally the pairs have been compressed as two-component images using lossless JPEG2000 with zero DWT decomposition levels. In the first setup, which we call *HST-unsigned*, the HST is applied only on components that remain unsigned after the spectral transform. The intensity distribution of the unsigned components is more similar to that of the original images, and hence the compression performance is more likely to be improved after the HST is applied. In the second setup, *HST-full*, the HST is also applied to the signed components for comparison reasons. Table 8.12 reports the bitrates yielded when using the different transforms and HST setups, and also the results obtained without applying the HST for ease of reference. It can be observed that combining the HST and the multicomponent compression capabilities of JPEG2000 does not generally produce any performance gain, as compared to applying the HST on the original images. On the other hand, when the HST is applied on

TABLE 8.12

Average lossless JPEG2000 multicomponent compression results in bpp with and without applying the HST for Cy3/Cy5 channel image pairs.

Set	Spectral transform	HST setup		
		no HST	HST-unsigned	HST-full
Yeast	None	6.829	5.911	5.911
Yeast	DWT 5/3	6.786	6.391	7.991
Yeast	POT	9.279	9.279	12.474
Yeast	DPCM	6.439	6.000	6.697
Yeast	R-Haar	6.790	6.396	7.121
ApoA1	None	11.524	10.785	10.785
ApoA1	DWT 5/3	11.217	10.789	11.668
ApoA1	POT	10.956	10.956	14.064
ApoA1	DPCM	11.289	10.892	11.244
ApoA1	R-Haar	11.218	10.773	11.137
ISREC	None	10.887	10.623	10.623
ISREC	DWT 5/3	11.451	11.295	12.477
ISREC	POT	11.468	11.468	14.638
ISREC	DPCM	11.203	11.109	11.478
ISREC	R-Haar	11.452	11.281	11.624
Arizona	None	9.548	8.795	8.795
Arizona	DWT 5/3	9.649	9.173	10.498
Arizona	POT	9.439	9.439	12.622
Arizona	DPCM	9.386	9.000	9.420
Arizona	R-Haar	9.649	9.195	9.545

Note: *Zero spatial DWT decomposition levels have been employed in all cases, as more levels decrease the performance.*

the unsigned transformed components, the performance is improved as compared to applying the spectral transforms only. When the HST is also applied to the signed transformed components, the compression results become worse than when only the spectral transforms are applied.

The segmentation and the reindexing approaches discussed in Sections 8.3.4 and 8.3.5 could also be considered as candidates to be combined with the HST. When a segmentation scheme is applied on a microarray image, spot and background pixels are separated. Even if the segmentation technique is not threshold based, most bright pixels are tagged as spot and most dark pixels are tagged as background. For this reason, it is not expected that applying the HST on the background or the spot subimages separately will produce better compression results than applying it on the original image. As previously discussed, the Laplacian reindexing and the HST have an analogous effect on the histogram of DNA microarray images, and hence combining them is not a very promising idea either. It is nonetheless interesting to

TABLE 8.13

Compression results in bpp for lossless JPEG2000 after applying the HST and Laplacian reindexing, using zero DWT decomposition levels.

Transform	MicroZip	Yeast	ApoA1	ISREC	Stanford	Arizona
HST	9.157	5.911	10.786	10.624	7.685	8.795
Laplacian	9.198	5.789	10.813	10.658	7.754	8.844

Note: *Results for the reindexing scheme do not include any overhead due to palette storage.*

compare their results, which are reported in Table 8.13. The histograms produced by the HST are not necessarily monotonic at each side of the most frequent intensity. In contrast, histograms obtained with the Laplacian reindexing are monotonic by definition. However, both techniques yield very similar results for all sets, with differences below 0.122 bpp. It is noteworthy that the compression performance after the Laplacian reindexing scheme is worse than after applying the HST except for one of the sets, even if no overhead due to the storage of the palette is considered. This suggests that the large number of value discontinuities introduced by the Laplacian reindexing transform affect the performance of the contexts used to drive the arithmetic coder of JPEG2000.

8.4   LOSSY COMPRESSION OF MICROARRAY IMAGES

The results discussed in the previous section demonstrate that lossless compression of microarray images is a challenging task and only moderate performance results are obtained with either image compression standards or microarray-specific schemes. In this section, we analyze the lossy compression of microarray images.

8.4.1   ADVANTAGES AND DISADVANTAGES OF LOSSY COMPRESSION METHODS

When a lossy compressor is used to code any type of images, better compression performance is obtained, and thus smaller file sizes and transmission times as compared to lossless methods. However, when lossy compression is used, it is not possible to recover an exact copy of the original image from the compressed file, but instead only a distorted version. Modern lossy schemes allow almost arbitrary compression ratios, at the cost of introducing larger distortions in the recovered image as the selected bitrates are decreased. In the microarray image compression literature, compression ratios of 45:1 have been reported in [24]. As previously discussed, lossless compression methods do not usually produce ratios better than 2:1.

The main drawback of applying lossy compression methods on microarray images is the fact that the introduced distortions are not always acceptable. If the decompressed versions differ too much from the original images, relevant information needed by present and future microarray image analysis techniques could be lost. For this reason, it is necessary to consider whether lossy compression processes are suitable for microarray images. Fortunately, at least two authors have addressed this question. Jornsten in 2003 [9] and Xu in 2009 [24] stated that some types of lossy compression do not significantly affect the extracted differential gene expression levels or subsequent classification processes. In particular, Jornsten claimed that the distortion introduced by their method is smaller than the variability between repetitions of the same experiment and between different extraction algorithms. For this reason, it is justified to study the lossy compression of microarray images.

8.4.2 STATE OF THE ART ON LOSSY COMPRESSION OF MICROARRAY IMAGES

Several authors have explored the lossy compression of microarray images. In 2003, Jornsten et al. proposed a scheme called segmented LOCO (SLOCO) [9]. After gridding and segmenting the images, each spot and background region is coded separately using a modified version of the LOCO-I algorithm, the basis of JPEG-LS. Spatial segmentation is applied, and the produced errors are refined in order to obtain a progressive lossy-to-lossless scheme. In 2004, Hua et al. published their BASICA software [21]. The images are also gridded and segmented, and a modified version of the embedded block coding with optimal truncation (EBCOT) algorithm, the basis of the entropy coding in the JPEG2000 standard, is applied to the spot and background pixels of each grid subblock. As a result, a lossy-to-lossless scheme is obtained. Also in 2004, Lonardi et al. presented their MicroZip algorithm [3]. After segmenting the whole image into spot and background pixels, the most and least significant bytes (MSB and LSB, respectively) of each subimage are separated and coded independently. The least significant byte of the background pixels is coded using the lossy set partitioning in the hierarchical trees (SPIHT) algorithm, while the remaining information is coded using a lossless algorithm based on the Burrows-Wheeler transform. In 2007, Peters et al. proposed a version of the singular value decomposition (SVD) applied to microarray images [25]. The image matrix I is decomposed into three matrices, $I = USV$, where S is a diagonal matrix containing the singular values. Only some of these values are kept and a lossy method is obtained. In 2011, Avanaki et al. applied fractal and wavelet-fractal techniques to exploit the self-similarity of microarray images and define two lossy methods [26].

In some of the aforementioned papers, the impact of the distortion introduced by the proposed techniques is discussed. To do so, they rely on the results for specific microarray image analysis methods and study the differences in either the extracted differential gene expression levels, or the disagreement rates for some classification, and clustering algorithms. Although this approach may be valid in some cases, these loss assessment techniques are largely influenced by the particularities of the employed image analysis, classification, and clustering algorithms. This creates difficulties in objectively comparing the results provided in the literature and, more importantly, limits the validity of the results as new microarray image and data analysis methods are developed.

The disadvantages of analysis-dependent information assessment techniques suggest that it is desirable to design distortion metrics based on image properties rather than particular analysis algorithms. Existing distortion metrics could be considered for this task. Metrics based on the pixel-wise absolute error like MSE or PSNR are not a particularly attractive option since they consider all pixels to be equally important, while spot pixels are clearly more important than background pixels. Alternatives like the weighted PSNR are not suitable either because they do not take into account the relative distortion of the pixels, which directly affects the extracted expression ratios. On the other hand, it is not justified to use metrics inspired in the human visual system like the structural similarity index (SSIM) and the multi-scale SSIM (MS-SSIM) [27] because microarray images are analyzed by computer programs, not humans.

8.4.3   THE MICROARRAY DISTORTION METRIC

For the reasons discussed above, new, microarray-specific distortion metrics should be designed. We now propose such a metric, the microarray distortion metric (MDM) [28]. It is intended to be a proof of concept and to contain the basic ideas that may be used in future, more sophisticated metrics. The MDM is a full-reference metric, that is, its calculation requires both the original and the distorted versions of the image.

Any successful distortion metric for microarray images must consider the key image properties that affect the extracted expression intensities of present or future analysis processes. If any of those properties are altered in a distorted microarray image, the extracted genetic data will vary and subsequent data analysis processes will be affected. It is impossible to predict how exactly future analysis methods will extract the genetic information, but the most basic image features that are intrinsic to almost all current analysis algorithms are also likely to be part of future methods. In particular, it is interesting to study the features that determine the obtained spot ratios and the way in which they are corrected before any data analysis process.

It is possible to identify three such features: *1)* The average intensity of the pixels inside each spot μ_{spot}, *2)* the average intensity of the pixels tagged as local background $\mu_{localBG}$, and *3)* the average intensity of the image μ_{Img}. Only average quantities are used for μ_{spot} since the genetic expression intensities are extracted in this way in most analysis methods. The same can be said about $\mu_{localBG}$ and the background-correction process, and also about μ_{Img} and the removal of systematic variations to dye biases or laser calibration inaccuracies. To calculate these image features, a prior segmentation process is needed. It is out of the scope of this section to design ad-hoc segmentation techniques or to choose an existing one described in the literature. For this reason, the existence of a sound image segmentation of the image is assumed in the rest of the section.

To compute the MDM, the three aforementioned features must be calculated for both the original and the distorted images. Absolute variations in μ_{spot}, $\mu_{localBG}$ and μ_{img} are not important, since it is the ratio of these quantities in the Cy3 and Cy5 channels that affects the final corrected spot ratios. Hence, we consider only the relative differences in the original and distorted versions of the image. For a spot i, we calculate the relative distortion of the spot ratio and local background, respectively, as

$$d_{\text{spot}}^i = \max\left(\mu_{\text{spot}}^{\text{Cy3}}/\mu_{\text{spot}}^{\text{Cy5}}, \mu_{\text{spot}}^{\text{Cy5}}/\mu_{\text{spot}}^{\text{Cy3}}\right)$$

$$d_{\text{localBG}}^i = \max\left(\mu_{\text{localBG}}^{\text{Cy3}}/\mu_{\text{localBG}}^{\text{Cy5}}, \mu_{\text{localBG}}^{\text{Cy5}}/\mu_{\text{localBG}}^{\text{Cy3}}\right),$$

so that values equal to or greater than 1 are obtained independently of whether the average intensity is reduced or increased in the distorted version of the image. There exists no *a priori* knowledge of the relative importance of the genes associated with a spot. Therefore, it should be assumed that all spots are the most important in order to define a metric that can guarantee that no important information has been lost. Thus, the spot and local background distortion of the image and the relative global intensity distortion can be defined as

$$d_{\text{spot}} = \max_i d_{\text{spot}}^i$$

$$d_{\text{localBG}} = \max_i d_{\text{localBG}}^i,$$

$$d_{\text{img}} = \max\left(\mu_{\text{img}}^{\text{Cy3}}/\mu_{\text{img}}^{\text{Cy5}}, \mu_{\text{img}}^{\text{Cy5}}/\mu_{\text{img}}^{\text{Cy3}}\right).$$

The proposed MDM has been defined so that a compact and easy-to-interpret output is produced:

$$MDM = 10\log_{10}\frac{\text{maxintensity}^2}{\text{ME}}.$$

To calculate the "signal-to-noise ratio" given by this metric, the maximum intensity in the image is used as a measure of the signal, while the noise is estimated as the microarray error (ME). Ideally, the MDM should produce high signal-to-noise ratios only if the image has not been distorted in a way that affects the genetic data extraction. As any of the key image features is being distorted, the ME should grow fast so that the produced output approaches 0 dB. To achieve this, we define the ME as

$$\text{ME} = \text{maxintensity}^p - \text{maxintensity} + \min(\text{maxintensity}, \text{MSE}_{\text{image}}),$$

where

$$p = 2/(1 + exp(-\alpha(d_{\text{spot}} + d_{\text{localBG}} + d_{\text{img}} - 3))).$$

To calculate the power p, the relative distortions of the key image features are combined, scaled with a parameter $\alpha > 0$ and used as the argument of a function of the logistic family. Since this argument is always non-negative, the power p is contained in $[1,2)$. The parameter α controls how fast the power p approaches 2 as the relative distortions are increased. In our experiments, we have found $\alpha = 3$ to be a balanced decision that is not too sensitive to slight variations in the relative distortions but causes

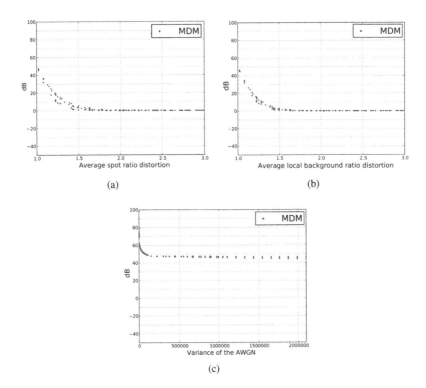

(a)

(b)

(c)

FIGURE 8.8 Distortion results for our proposed MDM, when applied to three sample images from the ApoA1 set. (a) Results after modifying pixels inside spots. (b) Results after modifying pixels inside the spot local backgrounds. (c) Results after applying additive white Gaussian noise to pixels outside spots and local backgrounds.

the MDM to drop to values close to 0 dB fast enough as these variations become important. The remaining terms, maxintensity and $\min(\text{maxintensity}, \text{MSE}_{\text{image}})$, are included for normalization purposes. When the two images compared by the metric are exactly the same, $-$maxintensity guarantees that the signal-to-noise ratio is ∞. The second normalization term guarantees that the MDM can output values as close to 0 as necessary, and uses the MSE of the whole image to identify any change in the distorted image.

In order to assess the suitability of our metric for the task of measuring the loss of relevant information, some experiments have been run. Images from all benchmarking sets have been distorted in controlled ways and the MDM results have been analyzed.

In a first experiment, the pixels inside all spots have been multiplied by different coefficients in $(1/3, 3)$. For each coefficient, one distorted image has been produced. The results for this experiment are shown in Figure 8.8(a). This type of distortion would have a dramatic impact on the extracted genetic data. It can be observed that the MDM is able to detect it and approaches 0 dB as larger coefficients are applied on the spot pixels. In a second experiment, the same multiplying coefficients have been applied on the pixels of the local background of all spots. This would affect the background correction step, and therefore the final corrected spot ratios. As shown in Figure 8.8(b), the MDM is also able to detect this type of distortion. In the last experiment, additive white Gaussian noise (AWGN) of zero mean and different variances has been applied on background pixels that do not belong to spots or their local backgrounds. This type of distortion should not affect the extracted genetic information because the extracted average intensities remain unmodified, the background correction process would be carried out just as in the original image, and the zero mean of the AWGN results in the same average image intensity. Results for this experiment are shown in Figure 8.8(c), where the x-axis represents the used AWGN variances. In this case, the MDM detects that the image has been modified when small variances are used. After that, it remains almost constant at high values, reflecting that the distortion is not important to the data extraction process, even though large amounts of noise have been introduced.

8.5 CONCLUSIONS AND FUTURE DIRECTIONS

Data extraction methods for microarray images are under active development and new techniques are constantly being proposed. It is desirable to store the microarray images so that they can be analyzed by the new techniques because repeating the whole microarray experiment is not practical due to economic and logistic reasons. Large amounts of microarray image data are produced each year, increasing the needs for efficient storage and transmission. Image compression emerges as a natural approach.

Lossless compression offers exact image fidelity after decompression, at the expense of larger compressed files. The best-performing microarray-specific techniques only rarely obtain compression ratios of 2:1, even after expensive computation processes. Image compression standards are desirable for their long-term support and their compliance to standards like DICOM, but the obtained compression results are not as good as those obtained by microarray-specific proposals. Several approaches to improve the performance of JPEG2000, the most promising image compression standard, have been discussed. The multicomponent compression of microarray images is most effective when Cy3/Cy5 pairs are compressed together. However, none of the studied spectral transforms is able to significantly improve JPEG2000's performance. Independently coding the spots and the background of microarray images has been exploited by the most successful microarray-specific techniques, even though the discussed theoretical results suggest that the total gain of this method cannot exceed 1 bpp. Palette reindexing has proven to be an effective method for enhancing JPEG2000's performance for all benchmarking sets, although the palette must be stored as side information. The histogram swap transform or HST is able to produce very similar performance improvements on JPEG2000, with the advantage of not needing any side-information and a very fast implementation using any unmodified JPEG2000-compliant coder. These performance enhancements range from 0.213 bpp to 0.918 bpp, i.e., from 1.97% to 15.53%, and situates JPEG2000 as the overall best image compression standard for microarray images, with bitrates only 0.788 to 1.266 bpp larger than the best results claimed in the literature.

Lossy compression does not guarantee exact image fidelity, but can produce almost arbitrary compressed file sizes. Several lossy methods for microarray images have been proposed, but none of them provides an objective way of measuring the relevancy of the compression loss whose results can be extrapolated to new analysis techniques. Traditional image distortion metrics like PSNR or SSIM are not suitable for this task. Thus, new metrics specific for microarray images must be devised. The microarray distortion metric (MDM) has been proposed. Its design is based on the image features that are used in any image analysis algorithm. Experimental results suggest that this metric is suitable for the task of measuring the loss of relevant information, and is able to differentiate important and unimportant distortions in microarray images.

In the foreseeable future, it is not likely that great advances can be achieved for the lossless compression of DNA microarray images. Small improvements may be obtained by combining and refining existing approaches like pixel prediction and per-bitplane context-based compression. However, all data gathered in Section 8.3 suggest that lossless compression ratios of 3:1 are not going to be obtained anytime soon. On the other hand, there is a large margin for improvement for the lossy compression of microarray images. New distortion metrics like the MDM defined in Section 8.4 are currently being researched and validated. Once a consensus is reached and some of these metrics are generally accepted by biologists and physicians, the attention of the microarray image coding community will likely be drawn to lossy compression. Therefore, great advances in the microarray image compression performance will probably be observed.

8.6 ACKNOWLEDGMENTS

The MicroZip image set was kindly provided by Neves and Pinho from the University of Aveiro. The Arizona image set was provided by David Galbraith and Megan Sweeney from the University of Arizona. The research leading to this chapter has been partially funded by the European Union, by the Spanish Ministry of Economy and Competitiveness (MINECO), and the Catalan Government under projects FP7-PEOPLE-2009-IIF FP7-250420, FPU AP2010-0172, TIN2009-14426-C02-01, TIN2012-38102-C03-03 (LIFE-VISION), and 2009-SGR-1224.

REFERENCES

1. K. Chuang and H. Huang, "Assessment of noise in a digital image using the join-count statistic and the Moran test," *Physics in Medicine and Biology*, vol. 37, no. 2, p. 357, 1992.

2. A. J. R. Neves and A. J. Pinho, "Lossless compression of microarray images using image-dependent finite-context models," *IEEE Transactions on Medical Imaging*, vol. 28, pp. 194–201, February 2009.

3. S. Lonardi and Y. Luo, "Gridding and compression of microarray images," in *Proceedings of the Computational Systems Bioinformatics Conference*, pp. 122–130, 2004.

4. Speed Berkeley Research Group, "ApoA1 microarray image set."
http://www.stat.berkeley.edu/users/terry/zarray/Html/apodata.html. Last access: July 2013.

5. SIB Computational Genomic Group, "ISREC microarray image set."
Originally available at
http://www.isrec.isb-sib.ch/DEA/module8/P5_chip_image/images. Last access: June 2013.

6. Yeast Cell Cycle Analysis Project, "Yeast microarray image set." Available at genome-www.stanford.edu/cellcycle/data/rawdata/individual.html. Last access: June 2013.

7. Stanford Microarray Database, "Stanford microarray image set." Available at ftp://smd-ftp.stanford.edu/pub/smd/transfers/Jenny. Last access: June 2013.

8. David Galbraith Laboratory, "Arizona microarray image set."
http://deic.uab.es/~mhernandez/materials. Last access: July 2013.

9. R. Jornsten, W. Wang, B. Yu, and K. Ramchandran, "Microarray image compression: SLOCO and the effect of information loss," *Signal Processing*, vol. 83, pp. 859–869, April 2003.

10. Y. Zhang, R. Parthe, and D. A. Adjeroh, "Lossless compression of DNA microarray images," in *Proceedings of the IEEE Computational Systems Bioinformatics Conference*, pp. 128 – 132, August 2005.

11. Y. Zhang and D. A. Adjeroh, "Prediction by partial approximate matching for lossless image compression," *IEEE Transactions on Image Processing*, vol. 17, pp. 924 –935, June 2008.

12. A. J. R. Neves and A. J. Pinho, "Lossless compression of microarray images," in *Proceedings of the International Conference on Image Processing, ICIP*, pp. 2505–2508, 2006.

13. A. Neekabadi, S. Samavi, S. A. Razavi, N. Karimi, and S. Shirani, "Lossless microarray image compression using region based predictors," in *Proceedings of the International Conference on Image Processing*, pp. 349–352, 2007.

14. S. Battiato and F. Rundo, "A bio-inspired CNN with re-indexing engine for lossless DNA microarray compression and segmentation," in *Proceedings of the 16th International Conference on Image Processing*, vol. 1-6, pp. 1717–1720, 2009.

15. National Electrical Manufacturers Association, "Digital Image and Communication in Medicine standard, DICOM." Available at http://medical.nema.org/. Last access: July 2013.

16. A. J. Pinho, A. R. C. Paiva, and A. J. R. Neves, "On the use of standards for microarray lossless image compression," *IEEE Transactions on Biomedical Engineering*, vol. 53, pp. 563–566, March 2006.

17. D. S. Taubman and M. W. Marcellin, *JPEG2000: Image compression fundamentals, standards and practice*. Boston: Kluwer Academic Publishers, 2002.

18. D. A. Adjeroh, Y. Zhang, and R. Parthe, "On denoising and compression of DNA microarray images," *Pattern Recognition*, vol. 39, pp. 2478–2493, December 2006.

19. I. Blanes and J. Serra-Sagrista, "Pairwise orthogonal transform for spectral image coding," *IEEE Transactions on Geoscience and Remote Sensing*, vol. 49, pp. 961–972, March 2011.

20. N. Faramarzpour, S. Shirani, and J. Bondy, "Lossless DNA microarray image compression," in *Proceedings of the 37th Asilomar Conference on Signals, Systems and Computers*, vol. 2, pp. 1501–1504, November 2003.

21. J. Hua, Z. Liu, Z. Xiong, Q. Wu, and K. Castleman, "Microarray BASICA: Background adjustment, segmentation, image compression and analysis of microarray images," *EURASIP Journal on Applied Signal Processing*, vol. 2004, pp. 92–107, January 2004.

22. R. Bierman, N. Maniyar, C. Parsons, and R. Singh, "MACE: lossless compression and analysis of microarray images," in *Proceedings of the ACM Symposium on Applied Computing*, pp. 167–172, 2006.

23. M. Hernandez-Cabronero, J. Muñoz-Gómez, I. Blanes, M. W. Marcellin, and J. Serra-Sagrista, "DNA microarray image coding," in *Proceedings of the IEEE International Data Compression Conference, DCC*, pp. 32–41, 2012.

24. Q. Xu, J. Hua, Z. Xiong, M. L. Bittner, and E. R. Dougherty, "The effect of microarray image compression on expression-based classification," *Signal Image and Video Processing*, vol. 3, pp. 53–61, February 2009.

25. T. Peters, R. Smolikova-Wachowiak, and M. Wachowiak, "Microarray image compression using a variation of singular value decomposition," in *Proceedings of the Annual International Conference of the IEEE Engineering in Medicine and Biology Society*, vol. 1-16, pp. 1176–1179, 2007.

26. M. Avanaki, A. Aber, and R. Ebrahimpour, "Compression of cDNA microarray images based on pure-fractal and wavelet-fractal techniques," *ICGST International Journal on Graphics, Vision and Image Processing, GVIP*, vol. 11, pp. 43–52, March 2011.

27. Z. Wang, E. P. Simoncelli, and A. C. Bovik, "Multiscale structural similarity for image quality assessment," in *Conference Record of the Thirty-Seventh Asilomar Conference on Signals, Systems and Computers*, vol. 2, pp. 1398–1402, 2003.

28. M. Hernandez-Cabronero, V. Sanchez, M. W. Marcellin, and J. Serra-Sagrista, "A distortion metric for the lossy compression of DNA microarray images," in *Proceedings of the IEEE International Data Compression Conference, DCC*, 2013.

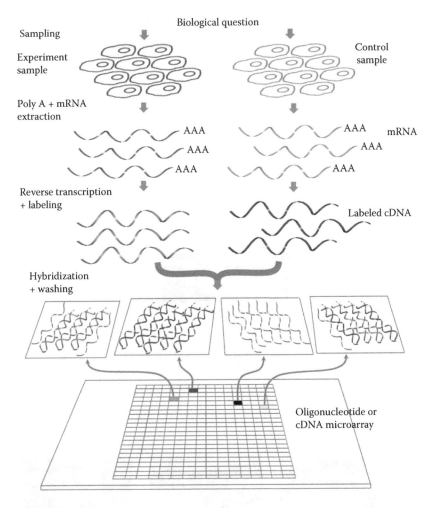

Sampling

Experiment
sample

Biological question

Control
sample

Poly A + mRNA
extraction

AAA   AAA
AAA   AAA   mRNA
AAA   AAA

Reverse transcription
+ labeling

Labeled cDNA

Hybridization
+ washing

Oligonucleotide or
cDNA microarray

FIGURE 1.3 Schematic view of microarray hybridization assay and analysis
for comparison of two samples (e.g., experiment versus control).

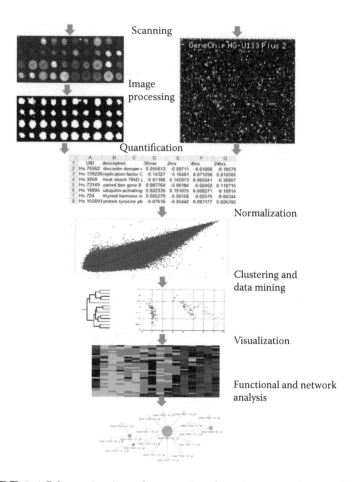

FIGURE 1.4 Schematic view of processing the microarray data and its analysis.

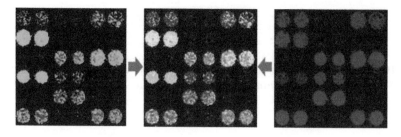

FIGURE 1.6 Portions of two DNA microarray images (green and red channels), along with the corresponding composite image.

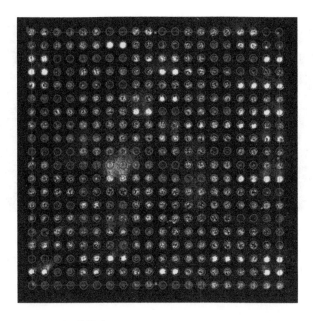

FIGURE 5.3 Part of a DNA microarray image segmented using the circular-shape-based approach.

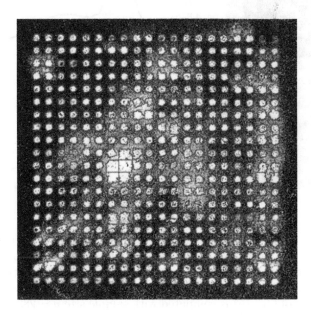

FIGURE 5.5 Part of a DNA microarray image segmented using the region growing approach.

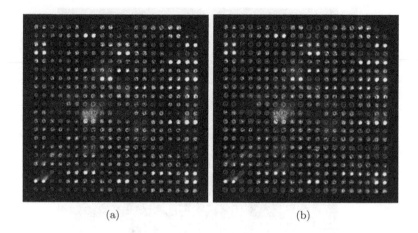

(a)   (b)

FIGURE 5.8 (a) A DNA microarray image. (b) Segmentation results using the graph-based method.

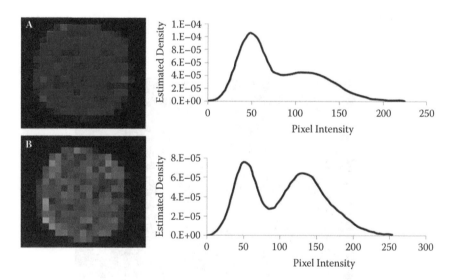

FIGURE 6.3 Two estimated density curves for spots of Cy5 (a) and Cy3 (b) dyes. Both Cy5 and Cy3 images have two intensity distributions $\hat{f}(y_j)$, which consist of background and foreground pixels. The intensity distribution of Cy5 (a) is computed with sample size $n = 289$ and bandwidth $h = 35$, and the distribution of Cy3 (b) is computed with sample size $n = 289$ and bandwidth $h = 30$. The local minimum is used as the cut-off point for segmenting the spot.

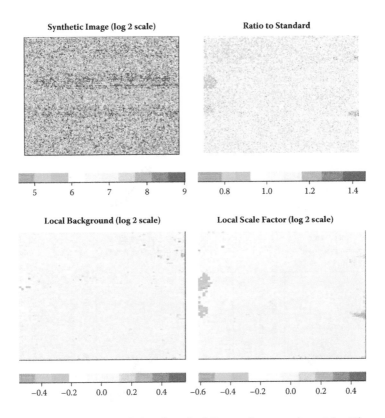

FIGURE 9.2 Example of the SmudgeMiner plots used to identify spatial flaws in a HG-U133A array. Four heatmaps are presented for each array to show different aspects of bias. For each plot a color scale is visible below the heatmap.

FIGURE 9.3 caCORRECT plot to identify spatial flaws in a HGU133A array. Black colored probes represent artifacts, and blue colored probes represent

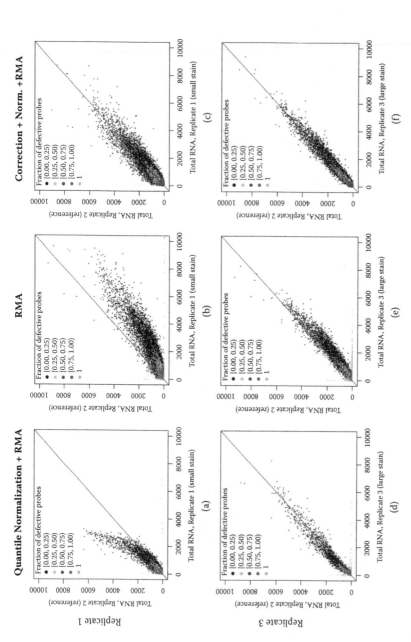

FIGURE 10.2 Replicate scatter plots comparing total RNA for replicates 1 (a–c) and 3 (d–f) against the artifact-free replicate 2 for the exon array measurement in DG75-10/12 cells. Figures are shown separately for results using both *RMA* and quantile normalization (a, d), using only *RMA* without quantile normalization (b, e) and after probe correction (c, f). Probesets are colored based on the percentage of their probes that were flagged as corrupted according to the ε-*criterion* (see Section 10.3.5) based on noise scores calculated using newly transcribed and pre-existing RNA as control. For replicate 1 there is a bias even for the uncorrupted probesets (a) that can be reduced by omitting quantile normalization (b). If artifact correction is applied prior to normalization and summarization (c, f), this bias is removed.

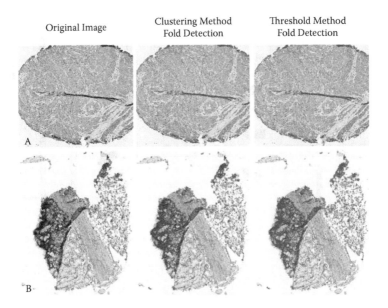

FIGURE 11.3 Tissue fold detection in TMA spots. (A-B) Original TMA spots, (center) tissue folds detecting using clustering method, and (right) tissue folds detecting using the threshold method. The threshold method removes slightly more tissue regions than the clustering method.

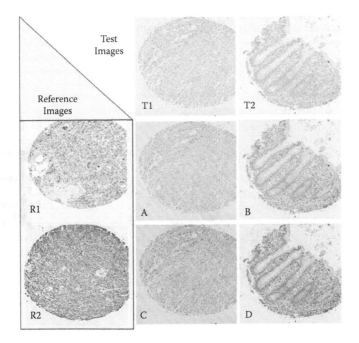

FIGURE 11.4 Color map normalization of TMA spots. Test TMA spot images (T1 and T2) are normalized to two reference images (R1 and R2)

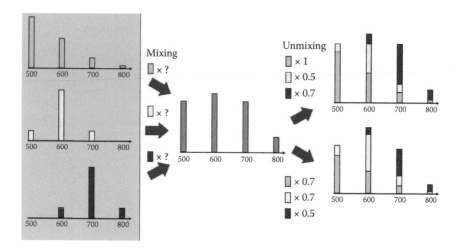

FIGURE 11.5 Simple Example of Unmixing Uncertainty. Unmixing determines how to allocate a mixture of given source signals to arrive at the closest estimation of the observed signal. There are many possible solutions to any unmixing problem, but the solutions will vary in how much they differ from the observed signal. In this example, the top solution has less error than the bottom solution. However, the solution with lowest error may not be the most accurate if: 1) there is uncertainty in the exact shape of the source signals, 2) the observed signal contains saturated channels, or 3) the solution violates constraints such as signal positivity. Note that in this example, the three source signals are measured using four channels, indicating that the system is not under-specified.

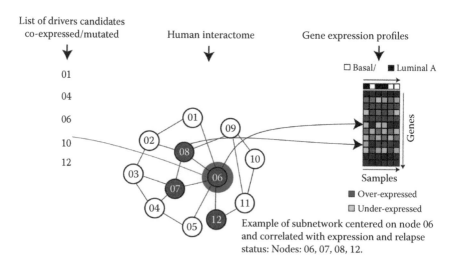

FIGURE 12.3 ITI algorithm and data organization. The 471 genes selected as seeds by the previous step are tested by ITI. Subnetworks are aggregated recursively around these seeds if their expression is correlated with the sub-

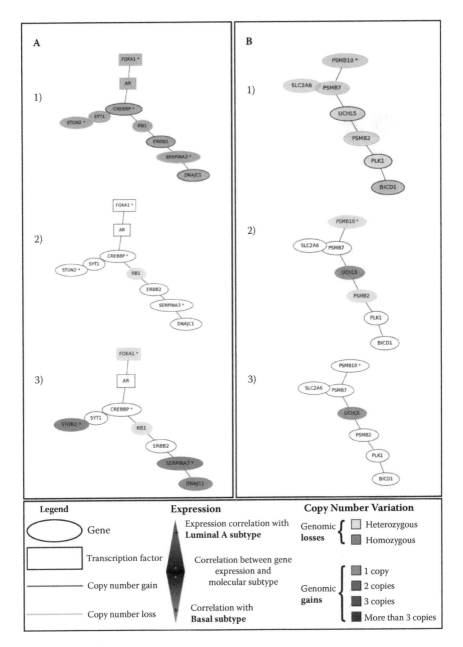

FIGURE 12.4 Examples of subnetworks detected with ITI, expressed in the luminal A (A) and basal (B) subtypes. The two subnetworks are represented overlaid with three types of information related to gene expression and copy number variation. 1) Subnetwork with nodes colored as the gene expression correlated score with the phenotype (blue = correlated with luminal A subtype, red = correlated with basal subtype). 2) Subnetwork with nodes colored with CNA information (genomic loss or gain) with respect to luminal A subtype. 3) Subnetwork with nodes colored with CNA information (genomic loss or gain) with respect to basal subtype.

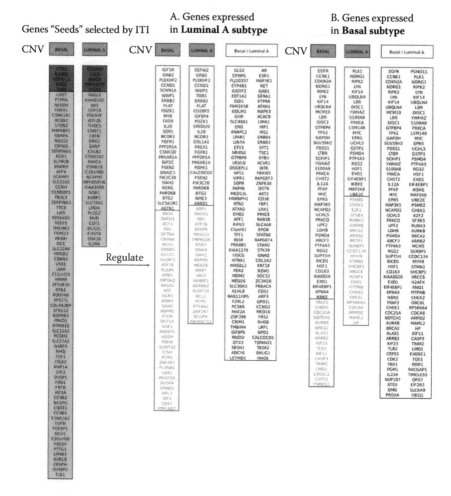

FIGURE 12.6 Detected driver genes with the CNV-ITI pipeline. On the left are represented the list of genes specifically expressed in luminal A and basal subtypes that have been retained as seeds by ITI (red = amplified, green = deleted). These are considered as putative driver genes, either tumor supressors or oncogenes. On the right, the list of genes regulated by these genes. In A, the list of genes found in subnetworks expressed in luminal A tumors. In B, the list of genes found in subnetworks expressed in basal tumors. For the A and B lists, we detailed the list of genes mutated in basal and luminal A tumors, or neither. In each of these sublists, genes marked as red are amplified and genes marked as green are deleted.

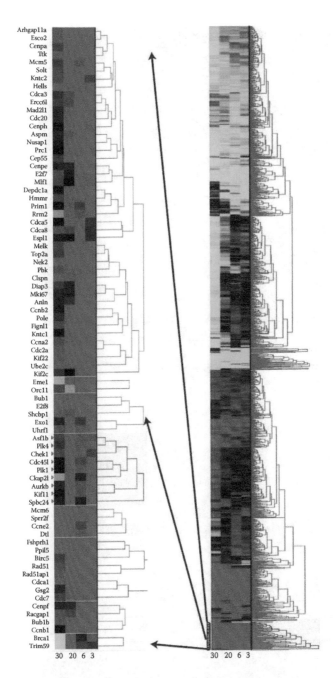

FIGURE 16.2 Heatmap and dendrogram of the hierarchical clustering of the four time points (3, 6, 20, 30 weeks). Genes that are upregulated, downregulated or unchanged, respectively, in the UPII-SV40 mice when compared with the WT littermates are shown. The right image contains approximately 1,900 genes, and its last cluster of genes (more strongly expressed in WT bladders) has been enlarged and rescaled (genes highly up-regulated at all four time points [1]). Dendrograms on the right of the heatmaps show the hierarchical clustering of the genes.

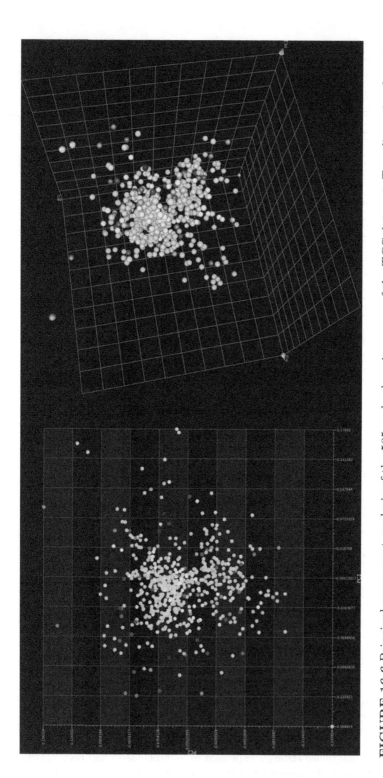

FIGURE 16.6 Principal component analysis of the 585 early changing genes of the TCC data set. Two-dimensional scatter plot of PC_1 vs. PC_2 (left) and three-dimensional scatter plot of PC_1 vs. PC_2 vs. PC_3 (right) show relationships among the records in a high-dimensional dataset, revealing relationships that usually remain hidden when visualized with classic visualization techniques.

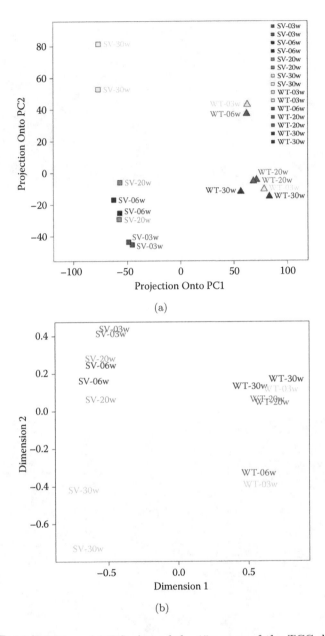

FIGURE 16.7 PCA and MDS plots of the 15 arrays of the TCC data set using the complete set of genes on the array generated using the R project [4]. (a) Triangles represent the WT arrays while squares represent the SV40 arrays, with a clear separation of the two sets. The majority of the replicates are positioned in close proximity to each other, indicating good replication of the data. (b) Similar observations can be made about the MDS plot of the arrays based on the top 5000 genes, which shows the between-object distances of the arrays. We can conclude that the WT are most similar to each other, and similar relationships between the individual time points can be observed, just as in the PCA plot.

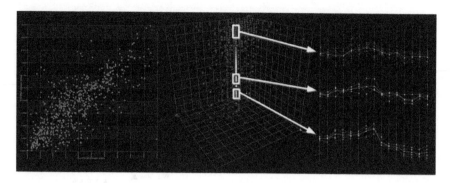

FIGURE 16.10 A comparison of a traditional scatter plot (left) and a self-organizing map enhanced SmartJitter Stacks visualizations with records related to the 585 early changing genes in the TCC data set [3] being pulled into the third dimension (middle). This technique alleviates the issue of over-plotting in a classic scatter plot visualization. Brushing three randomly chosen clusters within an arbitrary stack (middle) reveals the properties of co-located records—they show close association of co-located records colored by the z-value (right). White lines denote related cluster centroids (right). Dimensional values mapped to x and y-coordinates were selected arbitrarily (left and middle image) to demonstrate the power of this technique. The z-values in the middle image were calculated using a SmartJitter algorithm. Separation of otherwise overlapping records is obvious.

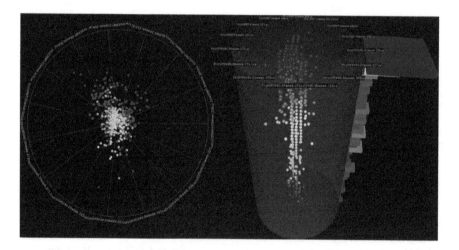

FIGURE 16.13 Two (left) and three-dimensional (right) Radviz visualization of dimensional values of the 585 early changing genes in the TCC data set [3]. Three-dimensional version of Radviz utilizes a similar approach as exhibited in Radviz 2D. Primary mapping locates the winning output node on a two-dimensional surface (left). The record is then pulled into the third dimension based on all or a subset of dimensional values in a dataset. This technique alleviates heavy overplotting noticeable in a two-dimensional Radviz (left).

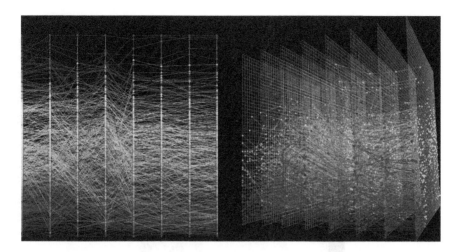

FIGURE 16.15 Two-dimensional parallel coordinates representation of gene expression data (left) and the corresponding three-dimensional self-organized parallel coordinates of dimensional values of the 585 early changing genes in the TCC data set [3] (right). Records are pulled into the third dimension based on dimensional values resulting with records exhibiting similar properties being plotted in proximity to each other.

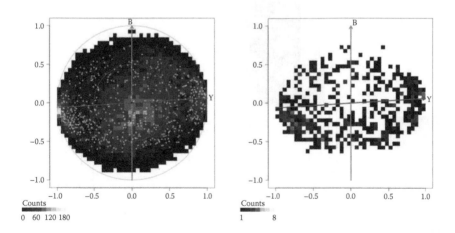

FIGURE 16.16 Biasograms of the 45,101 expression values with the 585 early changing genes (indicated by the circles and shown separately in the right image). We used the genotype as the outcome vector (Y) and the fraction of "present" calls as the source of technical bias (B).

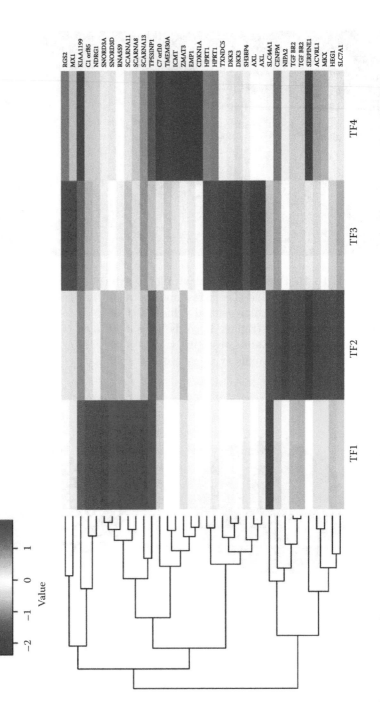

FIGURE 17.4 Clustering of the top ten differentially expressed genes from each main contrast. Red and blue correspond to up-and down-regulation, respectively.

9 Image Processing of Affymetrix Microarrays

Jose Manuel Arteaga-Salas

CONTENTS

One of the purposes of several statistical techniques is to reduce random noise and improve the quality of the data. In microarray experiments random noise can introduce errors that affect gene expression measurements and obscure the interesting biological variation. In recent years, several authors have described common techniques to minimize the effects of systematic variation incorporated into microarray data at different stages of the manufacturing process (see for example [3] and [4]). There is though a particular step in the microarray data analysis workflow that is often overlooked and that should be performed before other calibration or normalization steps: the image processing of raw data.

In this chapter, we present an overview of image processing methods for Affymetrix microarrays. All GeneChips are affected to some extent by spatially coherent defects, and image processing has a number of potential impacts on the downstream analysis of GeneChip data [5]. A number of algorithms to identify such defects have been proposed in recent years, and their use should be incorporated in the analysis of every microarray experiment.

9.1 OVERVIEW

The Affymetrix GeneChip comprises a set of multiple independent probes, which are utilized to measure the expression levels of the investigated genes or of their interactions. Each probe is a 25-nucleotide long oligomer, designed in perfect match (PM) and corresponding mismatch (MM) pairs. The manufacturing process of a microarray involves hybridizing DNA samples on a *surface* (i.e., the glass slide or chip), and then capturing a scanned image of that surface with a high resolution instrument. The output (stored as a set of pixel glow intensities in an image data file denoted *DAT file*) shows for each probe a brightness proportional to the amount of genetic material hybridized (e.g., RNA or DNA) that represents whether gene activity levels are absent or present. These intensities are subsequently converted into electronic signals and transcribed into numeric values (stored in a file denoted as *CEL file*), representing the expression intensity levels for different genes.

The correct processing of images of microarray experiments has the potential of minimizing inaccuracies introduced in the data during the manufacturing process. In order to obtain pristine scanned measurements it is necessary that the hybridization image is properly aligned and that the array surface is free of impurities. Understanding microarray data preprocessing steps is critical for performing optimal microarray data analysis [6]. According to the authors of [7], microarray data are susceptible to a wide range of artifacts introduced from the ambient environment, such as dust particles, fingerprints, scratches, and fluorescent residues. The same authors also suggested that a different type of artifact could be caused by the manufacturing process itself (during storage, washing, hybridization, or scanning), since physical variables (such as temperature or reagent concentration) rarely remain constant across the microarray surface. Such artifacts can create blemishes that affect regions of the array, which might affect subsequent analysis. We refer to these regional blemishes as *spatial flaws* or *spatial biases*.

The design of Affymetrix oligonucleotide arrays, which include probesets containing between 11 and 20 probe pairs that are placed separately at scattered locations in the array, was conceived in such a way as to minimize the effect of spatial biases in the array. However, several authors (see for example [8] and [9]) have observed the existence of a variety of spatial blemishes present in many arrays. These defects are described in [10] as "large, spatially contiguous clusters of signal from high or low intensity distributions, presumably resulting from extrinsic sources independent of transcription levels that could render useless the information in the affected area." In [11], it is suggested as a potential source of spatial flaws in which the bubbles being trapped inside the Affymetrix cassette do not travel uniformly over the chip during hybridization. The most common spatial artifacts are described in [12] as scratches, edge effects, and bubble effects that manifest as visible localized variations in the array. If the spatial flaws affect a subset of the probes within a probeset, then this will impact upon the gene expression measurements assigned to that probeset.

There are a number of computational packages (e.g., MAS 5.0 by Affymetrix) that provide tools and criteria to assess overall chip quality. However, the authors of [13] note that researchers should carefully inspect chip images visually since the criteria and tools provided by quality control packages have little ability to detect blemishes concentrated in regions of the array. Unfortunately, spatial flaws are usually difficult to see directly by eye on a GeneChip (only the starkest defects might be directly visible). Therefore, to unmask spatial blemishes in microarrays, it is necessary to inspect potential physical blemishes in microarrays with appropriate methods and criteria.

According to the authors of [5] there are three types of algorithms for the image processing of microarrays: by examining the transformation from the original hybridization DAT image to the representative CEL file; by utilizing pattern recognition routines to identify spatial blemishes in replicate arrays; or by exploiting the information available in public repositories of GeneChips. We now describe these three types of algorithms in detail.

9.2 FROM DAT TO CEL FILE

The expression levels of the genetic material hybridized in the microarray are acquired with the scanner and software supplied by Affymetrix, and subsequently converted into an electronic signal and numerical values. Finally, these values are stored as pixel intensities in an image file with the suffix *DAT*. This file consists of a 512 byte header that contains information detailing the size of the image, the scanning equipment, and other technical information, followed by the array of scanned pixel intensities stored as 2-byte unsigned little-endian integers that can take values in the range [0,65535] (see [5]).

After the hybridized array is scanned and the DAT file is obtained, it is not known *a priori* which pixels will correspond to which probe cells. The process of attributing pixel elements to probe cells is an image segmentation problem. Due to the rectangular shape of the array, the segmentation process is usually referred to as *grid alignment*. In Section 9.2.1 we discuss the process of grid alignment for Affymetrix microarrays in detail. For more details on methods for microarray image gridding and segmentation utilized in DNA microarrays see Chapters 3–6 of this book.

9.2.1 GRID ALIGNMENT

Microarray grid alignment is one of the basic pre-processing steps that affect the quality of gene expression information and could impact the conclusions of the biological experiment. According to the author of [6], the "ideal" image content of a microarray is characterized by constant grid geometry, known background intensity with zero uncertainty, infinite spatial resolution, predefined spot shape and constant spot intensity that has zero uncertainty for all spots. The process of grid alignment attempts to compensate for deviations from this "ideal" microarray image model.

After the fluorescent genetic material is hybridized to probes on a microarray, the hybridization data is extracted with a scanner and recorded as a large two-dimensional array of pixel data. Each probe cell occupies an area comprised by approximately 8×8 pixels of the scanned image. In theory, most of these pixels can be attributed to a single probe cell. In practice, however, the process of attributing pixels and pixel intensities to probe cells must be an automated pre-processing step on the extracted array of pixels. This process is usually referred to as grid alignment, although it is also known as addressing or spot finding [14] or gridding [15].

The gridding algorithm implemented in the Affymetrix GeneChip starts by locating the four checkerboard patterns of alternating bright and dim probes at each of the four corners of the array surface (Figure 9.1). Using the checkerboard patterns as anchors, Affymetrix's image processing software first locates the four checkerboard sequences and then defines a preliminary grid through bi-linear interpolation between the four corners. Several chip designs incorporate a lattice of 2×2 checkerboard patterns on the hybridization surface to aid or validate the interpolation [5]. After the preliminary interpolation, the grid alignment is then locally refined (Affymetrix documentation does not describe the refinement process in detail, though).

After the local refinement of the grid alignment is complete, the feature extraction procedure is used to assign a set of pixels to each probe cell. In early versions of Affymetrix chips the pixels comprised a rectangular array; for current versions all pixels corresponding to probe cells are square. The perimeter pixels mapped to each probe cell are deemed the least reliable and are thus discarded. One reason for doing this is that, even if the alignment was perfect, pixels in the perimeter may contain signal for more than one feature. In case the grid is incorrectly aligned (i.e., *grid*

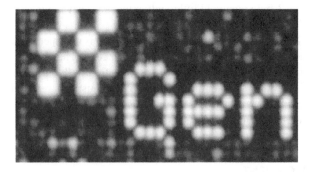

FIGURE 9.1   Top-left corner of the DAT file of a GeneChip Human Exon 1.0 ST Array showing a checkerboard pattern used for alignment.

misalignment; see Section 9.2.2 for more details), those pixels around the perimeter of probe cells will be most affected by misalignment. A second reason for discarding the pixels is that the signal of a probe cell tends to be the weakest around its edges. As a result, if there were a 5×5 array of pixels attributed to a probe cell, then this would be reduced to a 3×3 array of central pixels. Using the reduced set of pixels after discardment, the 75th percentile is reported as the estimate of the probe cell intensity. The estimated intensity, the corresponding standard deviation, and the number of pixels used to compute the 75th percentile are later stored in the cell intensity file with the suffix *CEL*. A description of other relevant files for GeneChip analysis can be found in the Affymetrix Developers Network at http://www.affymetrix.com.

From the information contained in the CEL file, the 75th percentile of pixel intensities is typically the only part of the summary of gene expression used in downstream analysis methods. The pixels near the center of a probe cell with hybridization levels above the background noise tend to have much higher signals than those around the edges. After the pixels in the perimeter of the probe are discarded, the remaining pixels represent a heterogeneous population and thus the standard deviation reported in the CEL file is of little use in downstream analysis of gene expression levels. However, according to the authors of [16], the standard deviation can be used in aiding the detection of outliers, grid misalignment, and defects in the hybridization surface. Since the photon count for each pixel is amplified, distributional assumptions such as Poisson counts for probe cell intensities are not close to being valid and cannot be usefully applied to downstream estimates of gene expression [5]. Therefore, the 75th percentiles of pixel intensities attributed to a probe cell are effectively the hybridization summary of gene expression.

Since the CEL file provides a summary of the information in the DAT file, any inaccuracies in this summary can only be recovered by reprocessing the DAT file. Such errors could be important but are not always evident. Thus, the correct allocation of pixel intensities to probe cells has a number of potential beneficial impacts on the downstream analysis of GeneChip data. To examine alternative methods and algorithms that could assist in this process it is first necessary to understand the common sources of grid misalignment in microarrays. We now discuss these sources in detail.

9.2.2 GRID MISALIGNMENT

As discussed previously, the most important source of errors in grid alignment is the inadequate segmentation of the hybridization stored in the DAT file, and some authors have proposed alternative grid alignment algorithms to those supplied by Affymetrix (see Section 9.2.3 for more details). The author of [17] also notes that grid alignment algorithms should be automated and "parameter free" so that the grid alignment obtained for all the arrays in the experiment are free of bias.

Another important source of error is *image blur*, which is originated when the image is subject to some sources of degradation during the acquisition process. This occurs when each pixel records signals not only from the region of hybridization surface that it is covering but also from its surrounding neighborhood. The pixels covering high intensity regions will tend to lose more signal to their surrounding pixels than they gain, and pixels covering low intensity regions will tend to accumulate extra signal from their neighbors. According to the authors of [5], the proportion of unreliable information in the hybridization image of successive versions of the Affymetrix chips has increased since the physical size of probe cells has decreased from approximately 24×24 microns to 5×5 microns (hence increasing the ratio of the perimeter to area of a probe cell). Manufacturers should attempt to directly reduce blur instead of increasing the number of pixels per probe cell. Nevertheless, some algorithms have also been proposed to reduce blur. For example, the author of [18] evaluates how to eliminate blur in GeneChips by application of the Lucy-Richardson algorithm discussed in [19] and [20] by using a point spread function based on the size of the probe and how the undegraded spots should look like. When the grid alignment is accurate, image blur is not an important factor and the image is free of spatial flaws and hybridization defects, and the CEL file should adequately summarize the contents of the DAT file for estimating gene expression.

The author of [17] suggests that other variation sources that could cause inadequate segmentation of the hybridization image include the regularity of the dipping prints of spotting machines and the rotational offset of the image with respect to the image edge. These variations can be caused by mechanical strains, strong background signals, or strong signal interference of neighboring spots. Furthermore, the same author also notes that another potential challenge in grid alignment is produced if non-circular pixel shapes are used in the hybridization process. An automatic grid alignment algorithm that attempts to reduce these effects is discussed in [17].

As mentioned previously, the recorded pixel values are truncated in the range [0,65535] for 16-bit unsigned integers. Truncation of the signal at the high end is called *fluorescent saturation*. Saturation can be controlled by amplification of the signal in the scanner photomultiplier tube and circuit board [5]. However, if the amplification is reduced, the resolution of the signal at the low ends is also reduced. Fluorescent saturation should not be confused with chemical saturation [21], which is a function of inhibition of binding caused by pre-existing hybridization duplexes. According to the authors of [22], since probe affinity is highly probe sequence dependent, chemical saturation can be restricted to some extent by selecting against probes with extremely high affinity in the design of a chip.

9.2.3   ALTERNATIVE GRID ALIGNMENT METHODS

In the presence of errors generated during the grid alignment process, any inaccurate information stored in the CEL file can only be recovered by reprocessing the DAT file. For this purpose, alternative solutions to those proposed by Affymetrix supplied grid alignment method are required.

9.2.3.1   Grid Alignment Transformation

The authors of [23] proposed a grid alignment transformation that improves the allocation of pixel intensities to probe cells, resulting in improved attribution of probe cell summaries. The method starts by defining the 3×3 rectangular grid of evenly spaced pixel locations that comprise the set of pixel intensities for each probe in the array. Define these nine locations as a_{uv} with $u, v \in \{-1, 0, 1\}$, $c = a_{00}$ as the centroid of the rectangular grid, and $\delta \in (0, 1]$ as the distance separating adjacent locations. At each a_{uv}, the method obtains the penalty for locating the probe cell out of alignment, t_{uv}, as

$$t_{uv} = \sqrt{u^2 + v^2} \tag{9.1}$$

and the variance s_{uv}^2 using the pixel intensities within a 6×6 grid centered at each a_{uv}. The decision of which a_{uv} to choose as the revised estimate of c is based on minimizing a weighted average of the penalties t_{uv} and s_{uv}^2. After evaluating the algorithm with a series of trial runs, the authors found it best to first log transform the pixel intensities and utilize the effective weighted penalty t_{uv}^* given by

$$t_{uv}^* = \delta t_{uv} + 5 \frac{s_{uv}^2}{S^2} \tag{9.2}$$

where S^2 is the mean of s_{00}^2 over all probe cells.

The alignment algorithm weights the penalties so that in early iterations probe cells with substantial information about their boundaries influence the deformation of the array of estimated probe locations. Decrementing the value of δ reinforces the smoothness of the deformation in later interactions [23]. Thus, prior to the first iteration of the alignment algorithm the authors set δ to 0.5. After each iteration, δ was decremented by 0.05 and the iterations ceased when δ was no longer greater than 0.

According to the authors of [5], by reprocessing a DAT file to a CEL file with the grid alignment transformation described above it is possible to improve the sensitivity of summarization algorithms (e.g., RMA [24]) to detect genes having a fold change of 2. This suggests that image processing can be used to clean up data, which will improve the accuracy of some gene expression measurements and the implications biologists draw from them.

9.2.3.2 Adaptive Pixel Selection

The authors of [16] introduced an "adaptive pixel selection" algorithm (APS) to compensate for failures to extract the central part of true probe features due to grid misalignment caused by the effects of image blur.

The first step of the algorithm is to remove pixels with extreme intensities (pixels more than three standard deviations from the mean pixel value within a feature). The edge of the grid defining each probe feature whose removal results in the greatest reduction of the coefficient of variation (CV) of the remaining pixels is discarded. This is then repeated until the feature size has been reduced to a predefined minimum (by default 4×4 pixels). According to the authors of the method, this procedure tends to select the most homogeneous group of pixels whose mean value is used as an estimate of the given feature.

The authors of [16] compare the APS method to the standard grid alignment algorithm provided by Affymetrix. The authors concluded that the APS method led to a reduction in the standard deviations of feature intensities, which in turn led to a reduction in the intensity standard deviation across replicate arrays. Therefore, the method improves the computation of feature intensities and reduces the measurement error. This results in an increase of sensitivity and specificity of the experiment when the effects of grid misalignment are removed from the data.

9.3   BLEMISH DETECTION THROUGH REPLICATE ARRAYS

Several authors have clearly observed the existence of a variety of defects affecting GeneChip images (e.g., [8, 25]). These are described in [13] as "dirt," "dark and bright spots," "dark clouds," and "shadowy circles," while [2] refer to "blob," "lines," "rectangular enhancement," and "coffee rings." These unwanted spatial patterns are not easy to see by eye directly on a single Affymetrix GeneChip because of the large dynamic range of probe intensities. However, by identifying probes with outlier values and observing how these cluster spatially in the array, it is possible to detect spatial blemishes.

In order to decide whether or not a probe in the array is an outlier, it is first necessary to have some notion of what value should be expected in that probe. We refer to such a set of expected values as a *reference set*. By comparing the raw microarray intensities with a reference set of values it is possible to identify effects that would occur only rarely by chance, but would occur more frequently in the presence of spatial blemishes. One choice for a reference set in experiments with replicate arrays is the median of all replicates. However, other alternatives are also possible.

In this section we discuss in detail several methodologies to visualize spatial flaws in experiments with replicate arrays. All the methods assume that the available replicates have been presented with the same genetic material under the same manufacturing conditions. If that is the case then, at any location of the array, the majority of variation between the replicates is produced by random errors. The majority of these methodologies require at least three replicates to produce plots where spatial flaws are noticeable. To illustrate the plots these methodologies produce we utilize data publicly available at the Gene Expression Omnibus (GEO) [26, 27] and Affymetrix (www.affymetrix.com).

9.3.1   SMUDGEMINER

The SmudgeMiner tool [11] is a piece of software for the visualization and quantification of spatial artifacts and regional biases. SmudgeMiner is implemented in R utilizing several routines included in the *affy* package [28] of Bioconductor [29]. To assist with the visualization of spatial flaws, SmudgeMiner provides four different plots for each array in the replicate set. Each of these plots is presented as a *heatmap* (a graphical representation of a data matrix where the values are presented as colors) to show different aspects of bias. Figure 9.2 shows an example of these plots produced using a Human Genome U133A array. Interestingly, the selected array is part of the Affymetrix Latin Square Experiment and illustrates that spatial flaws can appear even in experiments with high quality control protocols.

The top-left plot in Figure 9.2 shows the $\log_2(I_{ijr})$ values, where I_{ijr} is the raw intensity value of the probe at location (i, j) in array r. Spatial biases are difficult to discern by solely observing this plot. Neighboring probes often show a wide range of intensities next to one another, and hence it is difficult to discern clusters of probes

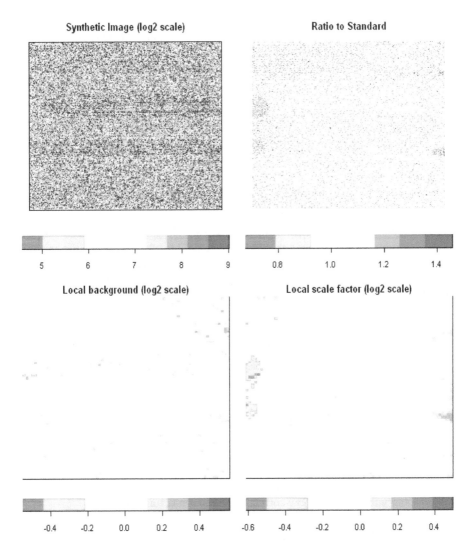

FIGURE 9.2 (SEE COLOR INSERT.) Example of the SmudgeMiner plots used to identify spatial flaws in a HG-U133A array. Four heatmaps are presented for each array to show different aspects of bias. For each plot a color scale is visible below the heatmap.

with the same color. However, in this plot we can observe some horizontal patterns resembling straight lines. These patterns are caused by alternating light and dark spots, produced with probes of similar sequences (i.e., probes with high GC content).

The upper right plot in Figure 9.2 shows the differences h_{ijr} given by

$$h_{ijr} = \log_2(I_{ijr}) - T_{ij} \tag{9.3}$$

where T_{ij} is the reference value at location (i, j), and is obtained as

$$T_{ij} = \text{trim}(\log_2(I_{ij1}), \log_2(I_{ij2}), ..., \log_2(I_{ijR})) \tag{9.4}$$

where trim denotes the trimmed mean operator (by default, SmudgeMiner uses a 20% trim), and R is the number of replicates available in the set. The h_{ijr} values reflect the discrepancies between replicate r and the set of reference values T.

In the plot described above it is possible to identify clusters of probes with the same color, suggesting the presence of a spatial bias. However, the authors of [11] note that this plot is only effective to visualize discrepancies from a reference value, and does not investigate in detail how the h_{ijr} values vary between probes with different intensity levels. To address this issue SmudgeMiner examines the background factor b and the scale factor SF, obtained with

$$b_{ijr} = \text{trim}(\log_2(I_{ijr}) - T_{ij}|T_{ij} < q_{0.2}) \tag{9.5}$$

$$SF_{ijr} = \text{trim}(\log_2(I_{ijr}) - T_{ij}|T_{ij} > q_{0.8}) \tag{9.6}$$

where $\text{trim}(\chi|\Lambda)$ is the 20%-trimmed mean of variable χ restricted to the set Λ, and $q_{0.2}$ and $q_{0.8}$ represent the 20th and 80th percentiles of probe intensities in the reference set T. The bottom-left plot in Figure 9.2 shows the b_{ijr} values, and the bottom-right plot shows the SF_{ijr} values. The authors of [11] recommend placing these plots side by side to identify regions in which the background is raised but the scale factor is unaffected, and vice versa.

In addition to the four plots to assist with the visualization of spatial biases, SmudgeMiner also provides a measure of the distortions introduced to the data by those biases. This measure is obtained with the correlation ρ, given by

$$\rho = \text{corr}\left(h_{ij}, \frac{h_{(i-1)j} + h_{(i+1)j}}{2}\right) \tag{9.7}$$

where $\text{corr}(\chi, \kappa)$ refers to the Pearson correlation between variables χ and κ. Equation (9.7) calculates the correlation using only the h-values in row-neighboring locations because neighbors within columns include probes with very similar sequence (i.e., PM and MM probes), which are (theoretically) highly correlated. In the absence of spatial biases the authors of [11] suggest that ρ should be close to 0 and that a "good" chip will typically show correlations between 0.1 and 0.2.

9.3.2 caCORRECT

The caCORRECT tool [12] is a Web-based application for the detection, analysis, and correction of spatial artifacts in microarrays. In a similar fashion to SmudgeMiner, caCORRECT displays the spatial flaws detected in an array as a heatmap.

caCORRECT first pre-processes the data with quantile normalization [30] and then implements an iterative procedure to identify spatial flaws in the array. The first step of the procedure is to obtain the "artifact-weighted independent mean" μ_{ijr} for location (i, j) of replicate r as

$$\mu_{ijr} = \frac{\sum_{t \neq r} I_{ijt}\, \alpha_{ijt}}{\sum_{t \neq r} \alpha_{ijt}} \tag{9.8}$$

where I represents the probe intensity, and α is a weighting factor defined as

$$\alpha_{ijr} = \begin{cases} 1, & \text{no artifact at location } (i,j,r) \\ 0 \leq \lambda < 1, & \text{an artifact at location } (i,j,r) \end{cases} \tag{9.9}$$

where λ is a constant. In the first iteration no artifacts have been identified, and thus Equation (9.8) reduces to the sample mean.

Using the values μ_{ijr} and the weights α_{ijr}, caCORRECT obtains the "artifact independent deviation score" σ_{ijr} with

$$\sigma_{ijr} = \sqrt{\frac{\sum_{t \neq r} (I_{ijt} - \mu_{ijr})^2\, \alpha_{ijt}}{\left(\sum_{t \neq r} \alpha_{ijt}\right) - 1}} \tag{9.10}$$

which is used to obtain the variance statistic S_{ijr}, given by

$$S_{ijr} = \frac{|I_{ijr} - \mu_{ijr}|}{\sigma_{ijr}}. \tag{9.11}$$

According to the authors of [12], S-values with high magnitudes indicate statistical improbability, and thus artifactual tendency. caCORRECT then plots the S-values as a heatmap. S-values exceeding a threshold (by default, the 80th percentile of the S-values) are colored in black to represent spatial blemishes, and low values are colored in blue to represent clean data. Figure 9.3 shows an example of the caCORRECT heatmap, produced with the same array of Figure 9.2. In this heatmap, we observe spatial flaws similar to those in the "Ratio to Standard" heatmap of SmudgeMiner, corresponding to regions with large S-values.

In addition to the heatmap for spatial flaw visualization, caCORRECT also provides two scores to measure the distortions introduced by spatial flaws in each available array: the *Uniformity Score* (UP) and the *Artifact Coverage Percent* (ACP). According to the authors of [12], the UP is obtained with "a pairwise normalized cross-correlation algorithm." Chips free of spatial biases will have a UP score close to 1, and chips highly contaminated with spatial biases will have a UP score close to 0. The ACP is obtained with the ratio between blemished locations and total locations,

FIGURE 9.3   (SEE COLOR INSERT.) caCORRECT plot to identify spatial flaws in a HG-U133A array. Black colored probes represent artifacts, and blue colored probes represent clean data.

and can be utilized as a criterion to remove highly suspect chips from the dataset (the authors recommend discarding a chip when ACP> 20%).

caCORRECT does not obtain a unique set of reference values to compare the replicate data and identify probes with unusual values. Instead, the method obtains artifact-weighted independent means μ_{ijr} for each available replicate, which are used to identify outlier probes. If no artifacts are detected in a location after the iterative procedure, then the artifact-weighted mean reduces to the sample mean of probe intensities.

A recently developed improved version of this method (caCORRECT2 [31]) implements a new bad-data-replacement algorithm and better integration tools with gene expression calculation protocols. Additionally, the authors of [31] also implement an "artifact-aware quantile normalization" that identifies artifacts before quantile normalization and sets them aside temporarily, so that the intensity distributions are not incorrectly warped in the presence of a faulty array. The authors also state that the more similar arrays are available for blemish identification, the more powerful caCORRECT is (preferably, biological or technical replicates). More details on the artifact correction and detection methods utilized in caCORRECT2 can be found in Chapter 10 of this book.

9.3.3 HARSHLIGHT

The Harshlight package [13] is a tool for the automatic detection and correction of suspicious spatial patterns in microarrays using diagnostic tests based on image processing methods. Harshlight is implemented in R and is freely available as a package in Bioconductor. In contrast to SmudgeMiner, Harshlight does not plot the spatial flaws detected in an array as heatmaps.

For the visualization of spatial flaws, Harshlight first calculates the h_{ijr} values defined in Equation (9.3), albeit with the reference set T_{ij} given by

$$T_{ij} = M\{\log_2(I_{ij1}), \log_2(I_{ij2}), ..., \log_2(I_{ijR})\} \qquad (9.12)$$

where $M\{\cdot\}$ is the median operator. The Harshlight package analyzes any suspicious patterns among the h-values to identify three types of spatial defects: extended, compact, and diffuse. We now describe these three types of defects in more detail. A further description of Harshlight and of the methodology it utilizes to correct spatial defects can be found in Chapter 10 of this book.

9.3.3.1 Extended Defects

Extended defects usually affect large areas of the chip, and can cause substantial variation in the overall intensity levels from one region of the array to another. To quantify extended defects Harshlight decomposes the error differences h as

$$h = F + \varepsilon \qquad (9.13)$$

where F represents the features in the h-values, and ε represents the random noise in the h-values. Assuming that the random noise and the features are uncorrelated, the variance of the h-values, σ_h^2, can be decomposed as

$$\sigma_h^2 = \sigma_F^2 + \sigma_\varepsilon^2 \qquad (9.14)$$

where σ_F^2 is the variance of the features, and σ_ε^2 is the variance of the random noise. According to the authors, σ_F^2 is a constant (since the features F originate in differentially expressed genes, and are expected to be spatially randomly distributed in the array with mean zero and variance σ_F^2), and σ_h^2 can be obtained directly from the data. To estimate σ_ε^2, Harshlight creates a median filter through a sliding median kernel by using a discretized circular window centered at location (i, j) with a user defined radius (by default, radius = 10 pixels) [32] . For the probes near the border of the array that lie outside of the window, Harshlight fills the window by mirroring the image at the borders.

According to the authors of [13], in an unblemished array the variance σ_h^2 is mainly due to σ_F^2, and σ_ε^2 should suffer small variations across the chip (i.e., $\sigma_F^2 > \sigma_\varepsilon^2$). Thus, the variation in the h-values explained by the random noise, given by $\sigma_\varepsilon^2/\sigma_h^2$, quantifies the extended defects in a replicate array. The authors recommend discarding a chip from the analysis if the extended defects exceed a threshold defined by the user (appropriate threshold levels are expected to be between 17% and 33%).

9.3.3.2   Compact Defects

If the array is not discarded after quantifying the extended defects, Harshlight continues the analysis by detecting compact defects. Compact defects occur when all the probes in a small or medium size region are blemished, whether an extended defect is previously identified in the same region or not. The authors of [13] suggest that these defects are caused by mechanical and optical imperfections, and are visible as "pieces of dirt" on the face of the chip. To identify compact defects Harshlight first analyzes the distribution of the h-values, and declares as outliers all the locations for which h is smaller than the quantile q_p or bigger than $(1 - q_p)$ (by default, $p = 2.5\%$). Binary images of outlier values are then created (where 1 represents locations declared as outliers), so that clusters of connected outliers can be detected with the FloodFill algorithm [33]. By default, Harshlight considers two outliers to be connected if they share an edge or a corner (i.e., any of the 8 surrounding locations). Cluster sizes are then compared to the minimum acceptable cluster size (by default, minimum cluster size = 15 pixels) to eliminate clusters made only of a few pixels. The clusters remaining are labeled as compact defects, and are excluded from the array before identifying diffuse defects.

9.3.3.3   Diffuse Defects

Diffuse defects are defined as regions with a high density of blemished probes, although the probes are not necessarily connected. The authors of [13] found evidence that diffuse defects are probe-sequence dependent (with typically more C's and T's and fewer A's in the sequences of contaminated probes), and suggest that they are caused by defects in the hybridization stage. To identify diffuse defects Harshlight analyzes two binary images of outlier values, one for unusually high values (bright outliers) and the other for unusually low values (dark outliers). For each location (i, j) of array r, the outlier values O_{ijr} are obtained with

$$O_{ijr} = \begin{cases} 1, & h_{ijr} \leq -\log_2(1 + x) \\ 0, & \text{otherwise} \end{cases} \tag{9.15}$$

where x is a percentage of decrease in intensities defined by the user (by default, $x = 40$ for bright outliers and $x = 35$ for dark outliers). The outlier images are scanned separately with a discretized circular sliding window of user-defined radius (by default, radius = 10 pixels) to compute the proportion of outliers d_{ijr} at each location (similarly as described for the extended defects, Harshlight mirrors the image for probes near the borders of the array). A binomial test is then used to obtain the diffuse-defected locations D_{ijr} as

$$D_{ijr} = \begin{cases} 1, & d_{ijr} > b_{1-\alpha}(d_0, n_{ijr}) \\ 0, & \text{otherwise} \end{cases} \tag{9.16}$$

where $b_{1-\alpha}(d_0, n_{ijr})$ is the $(1 - \alpha)$ percentile of a binomial distribution (by default, $\alpha = 0.001$), d_0 is the overall proportion of outliers in the image, and n_{ijr} is the number of probes in the discretized circular window centered at location (i, j). As in the analysis of compact defects, the FloodFill algorithm is then used to identify clusters of connected pixels in the image D to subsequently discard clusters of small size. The size limit of the clusters can be set by the user, but the authors recommend using a size "three times the area of the sliding window" (i.e., 942 pixels). As a final step, the image undergoes a "closing procedure" [32] to close up breaks in the features of an image and to better outline the area of the diffuse blemishes. The process is repeated separately for the bright and dark outlier images.

9.3.3.4 The Harshlight Report

After the identification of the three types of defects Harshlight produces a PostScript report for each available replicate. Figure 9.4 shows an example of the Harshlight report, produced with the same array of Figure 9.2. The bright spots in the plots represent flaws of unusually high values, and the dark spots represent flaws of unusually low values. The report shows four plots: the original image (i.e., the $\log_2(I_{ijr})$ values), the error image (i.e., the h_{ijr} values), the compact defects, and the extended defects. The first two plots are presented as two-colored heatmaps (which are obtained as for the "Synthetic Image" and "Ratio to Standard" heatmaps of SmudgeMiner), and the last two plots show only outlier locations (that exceed a defined threshold). The report also includes the number of defects and the percentage of the array affected by these defects. The spatial patterns visible in the diffuse defects plot and in the error image are similar to the patterns observed in the SmudgeMiner and caCORRECT heatmaps.

We observe similarities between Harshlight and other methods discussed previously (the original and error images are similar to the heatmaps produced by SmudgeMiner), but also some differences (outlier plots to identify compact and diffuse defects, and a quality measurement of the flaws detailed by defect type). Although this tool utilizes several parameters for flaw visualization, the implementation is user-friendly and the PostScript report provides more information about detected blemishes than other tools discussed previously. A disadvantage of Harshlight noted in [31] is that the method tends to be aggressive in flagging regions of chips as artifactual, which causes "ghosting effects" (i.e., the false appearance of artifacts on clean chips after being compared to true artifacts on other chips of the batch).

9.3.4 OTHER METHODS

Upton and Lloyd [2] introduced a simple method to identify deficiencies in the data by plotting outlier values in each available replicate (as for the compact and diffuse defects plots in Harshlight). The method identifies which of the R available arrays provide the largest value at each location and examines regional biases by dividing the microarray into non-overlapping 3×3 sub-arrays (the same criterion applies for the minimum values). Spatial flaws can then be visualized by plotting values that exceed a certain threshold, using two reference sets (the maxima and minima intensities per location) to identify unusual values. Figure 9.5 (a) shows an example of the plot produced by the method using the array of Figure 9.2.

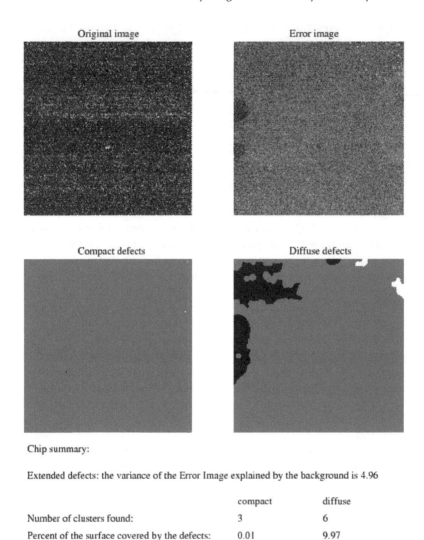

Chip summary:

Extended defects: the variance of the Error Image explained by the background is 4.96

	compact	diffuse
Number of clusters found:	3	6
Percent of the surface covered by the defects:	0.01	9.97

FIGURE 9.4   Example of the Harshlight report to identify spatial flaws.

Another simple method to visualize spatial flaws was introduced by Arteaga-Salas et al. [34]. The method is similar to that of [2] but utilizes the median value of all available replicates as reference value to identify outliers at each location of the array. The method then plots those cells that differ by more than 25% of the median array value in that location. Not all cells so identified will correspond to flaws, since there will be many cells whose unusual magnitude is a result of a biological signal. However, these genuinely interesting cells are unlikely to be spatially clustered. Figure 9.5 (b) shows an example of the plot produced by the method using the array of

Figure 9.2.

It is clear that the methods introduced in both [2] and [34] produce similar plots to efficiently visualize spatial flaws. The two circular patterns at the left-margin of the array (or "finger prints" as the authors denote them) identified in these plots are also visible in the plots produced by the other methods described previously. However, a disadvantage of both methods is that they require at least three replicates per batch to produce a spatial flaw identification plot, and that they work better when technical replicates are available. After blemish identification, the authors of [34] also suggested two methodologies for artifact correction: the local probe effect (LPE) adjustment, and the complementary probe pair (CPP) adjustment. Details on these methods can be found in Chapter 10 of this book.

Another method published recently in [35] introduces two approaches for the identification of corrupted probes. The authors first obtain *probe noise scores* that quantify the deviation from a control (e.g., replicates that are artifact-free in the relevant region) or from a linear model (e.g., residuals from the RMA model). The probe noise scores are then obtained either for each probe alone (the $\varepsilon - criterion$) or as a

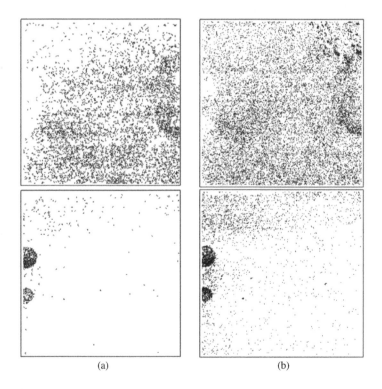

(a)   (b)

FIGURE 9.5 Spatial flaws in an HG-U133A array visualized with: (a) the Upton and Lloyd method, (b) the Arteaga-Salas et al. method. The upper row shows locations of unusually high values, and the lower row the locations of unusually low values.

distance-weighted mean of the noise scores within a window around the probe. The method then visualizes blemishes by plotting the probe noise scores. The authors show the efficiency of their method when applied to the recent Gene and Exon ST arrays. More details on this method can be found in Chapter 10 of this book.

Various other methods that are only applicable to specific platforms have also been proposed by some authors (e.g., the method developed by [36], which is specific to CGH arrays, and the method proposed in [37], which is specific to SNP arrays).

Several gene expression analysis tools (such as PLIER [38], dCHIP [16] and RMA [24]) provide visualization features that highlight outlier probes in the array. These features can lead to the identification of scratches, contaminated or misaligned arrays, and concentrations of unusual values. However, these procedures are dependent on the appropriateness of the models and can only be implemented after estimating the model parameters (no spatial information is included in their outlier detection schemes).

9.3.5 COMPARISON OF SPATIAL FLAW DETECTION METHODS

Each of the methods for spatial flaw detection employ different statistical and image processing techniques to unmask blemishes in the replicate arrays, and thus some differences in the results obtained with each method are expected. We now discuss these differences briefly, as well as the advantages and disadvantages of each method. The discussion that we now present is based on a comparison of the methods when utilized with default parameters (modifying the default values could result in different blemish visualization).

Table 9.1 shows a comparison of the main features of each methodology. We observe four aspects in this comparison: the pre-processing of the data, the choice of reference set, the graphical procedure for spatial flaw visualization, and the quality measure to estimate the effects introduced to the data by spatial flaws.

With R replicates available in the experiment, the SmudgeMiner method obtains a reference set using a 20% trimmed mean (which is the standard mean operator when $R < 5$). The method does not require pre-processing the data since the spatial flaws are plotted as heatmaps. SmudgeMiner is implemented in a user-friendly environment, and it provides four heatmaps to assist in the analysis of the spatial biases. However, if one of the available replicates is contaminated with a "serious" flaw (i.e., a flaw that causes large differences between the observed intensities and the reference set), then the SmudgeMiner heatmaps will show this flaw for all the available replicates. In such cases it is not clear if the spatial blemish affects one replicate only or all of them equally.

The caCORRECT tool recommends pre-processing the data with quantile normalization before calculating the reference set with an "artifact-weighted independent mean" [12]. The spatial blemishes detected by this tool in Figure 9.2 are very similar to the flaws detected by Smudgeminer. Nonetheless, a disadvantage of caCORRECT is that it does not differentiate between unusually low and unusually high blemishes in its heatmaps (both types of flaws are plotted with the same color).

TABLE 9.1

Summary of the main features of six spatial flaw visualization methods.

Method	Features			
	Pre-processing	Reference Set	Visualization	Quality Measurement
SmudgeMiner	not required	20% trimmed mean	heatmap	correlations ρ
caCORRECT	quantile	artifact weighted mean	heatmap	UP and ACP
Harshlight	none	median	outliers only	% of three types of defects
Upton and Lloyd	constant	maximum or minimum	outliers only	none
Arteaga-Salas et al.	constant	median	outliers only	none
$\varepsilon - criterion$	quantile	probe noise scores	heatmaps	precision recall curves

The Harshlight package calculates the median of all replicates as the reference set, and identifies three types of defects if the h_{ijr} values in a local neighborhood exceed a defined threshold. The Harshlight report provides useful information about the spatial biases by plotting three types of defects and by calculating the percentage of the array affected by the flaws. However, the method also has disadvantages. The final step in the detection of diffuse defects involves a "closing procedure" to better outline the area covered by blemishes. This has the intention of improving the visualization of flaws, but it often outlines blemishes in the borders of the array that are caused by small concentrations of points (probably caused by control samples), and it could also increase the size of certain blemishes. In addition to this, the authors of [13] do not recommend any pre-processing routine, which is necessary if the replicate arrays have differences in their overall intensities.

The Upton and Lloyd and Arteaga-Salas et al. methods result in very similar spatial flaw plots, and provide the simplest approach for blemish visualization among the methods considered in this section. Both methods recommend pre-processing the data with constant normalization, and both methods obtain a reliable reference set. The difference, though, is that the default threshold in the second method could be "less demanding" in identifying outlier values, which could result in a greater number of points in the plot. This could lead to unmasking blemishes that are not easily detected by other methods. However, it could also lead to a large number of points scattered in the plot. In the latter case, the user must remember that not all the points in the plots correspond to spatial blemishes, but only to outlier values.

The $\varepsilon - criterion$ first obtains probe noise scores to examine the global residual level of an array, to subsequently indicate which arrays are likely to contain artifacts.

The method utilizes quantile normalization before blemish detection and visualizes the detected spatial biases by plotting the probe noise scores by location with color codes (i.e., as a heatmap). The evaluation of the artifact detection is performed with precision/recall curves (see Chapter 10 of this book for more details).

9.4 BLEMISH DETECTION WITHOUT REPLICATE ARRAYS

The methods discussed in Section 9.3 obtain a set of reference values that indicates how the data should be in the absence of spatial flaws. By comparing the replicate values with the reference set the methods identify unusual probe values. If these unusual values are concentrated in compact regions, then spatial blemishes can be detected. Typically, the set of reference values is obtained using all the available replicates in the experiment. However, in the absence of replicate arrays the reference set cannot be obtained, and hence an alternative is required.

9.4.1 USE OF NO REPLICATES

With the advent of large microarray data studies, repositories of data generated from several experiments, or laboratories, consisting of arrays that have been analyzed by previous researchers are becoming publicly available. Using all available arrays in one of such repositories (the GEO) , the authors of [8] proposed an alternative to obtain a reliable set of reference values in the absence of replicate arrays. The authors utilized the geometric mean and the variance of the logarithm of the raw intensity values for each probe in each available chip (after removing the upper and lower 0.5% of the values to avoid the potential presence of outliers) to obtain an "average GeneChip" and a "variance GeneChip," separately by species and chip type.

Using the reference set given by the "average chip" A and the "variance chip" V, it is possible to identify spatial biases in an observed array of logarithm values following the steps described in [1]:

1. For each location (i, j) in the array calculate the values h_{ij}, given by:

$$h_{ij} = \frac{\log(I_{ij}) - A_{ij}}{\sqrt{V_{ij}}}. \qquad (9.17)$$

   where I denotes the observed intensity values.

2. Sort the h_{ij} values by column j. For each sorted value assign a rank, and store them in array K. The values are sorted by column j and not by row i because all the probes in column j are of the same type (i.e., PM or MM probes). By sorting the h-values by column the visualization of "horizontal line patterns" across the array is prevented (as in the top-left heatmap in Figure 9.2).

3. To examine the spatial information in K, divide the microarray into non-overlapping sub-arrays centered at (i, j). Each sub-array needs to include enough spatial information in a neighborhood. The authors of [1] suggest that a sub-array size 11×11 is efficient for this purpose, although a different

sub-array size can also be used. The sub-arrays are reduced for probes near the borders of the array (minimum sub-array size: 3×3).

4. The sub-array centered at K_{ij} contains K-values for PM and MM probes. Typically, the PM and MM intensities are highly correlated, and hence their ranks will be correlated too. To avoid using correlated values, consider only PM-locations in the sub-array if K_{ij} corresponds to a PM-probe (or only MM-locations if K_{ij} corresponds to an MM-probe), thus selecting 55 locations from the total 121 in the 11×11 sub-array. Using the K-values for PMs (or MMs) only obtain the scores Z for each location, given by:

$$Z_{ij} = \frac{\sum_{l=1}^{55} K_l - 55\mu}{\sqrt{55\sigma^2}} \qquad (9.18)$$

where μ is the mean and σ^2 is the variance of the discrete uniform distribution $(1, J)$, J is the number of columns in the array, and l denotes the selected PM-locations (or MM-locations) in the sub-array. The scores Z have a normal distribution with mean 0 and variance S^2 (in the absence of spatial biases $S^2 = 1$).

5. Plot the center cell of all sub-arrays where $|Z_{ij}| \geq 2S$. These locations are centered in neighborhoods with unusually high (or unusually low) values. If these locations are concentrated in the same region of the array then a spatial bias is present.

Figure 9.6 shows an example of the plot produced following the five steps described above using the array of Figure 9.2. The methods discussed in Section 9.3 identified "fingerprint"-shaped blobs on the left-margin of the array, which are also visible in this figure (as expected). This confirms the reliability of the reference set proposed in [8].

9.4.2 METHODS FOR REPLICATE ARRAYS APPLIED TO INDEPENDENT ARRAYS

If three or more arrays of the same type are available, it is possible to utilize any of the methods described in Section 9.3 to identify spatial biases in individual arrays. However, the results obtained with this approach are variable and depend on the appropriate choice of arrays.

Figure 9.7 shows a blemish visualization example using three non-replicate Human Genome U133 Plus 2.0 arrays from GEO, with accession numbers GSM49942, GSM117700, and GSM96262. For this example the method by [2] to visualize spatial flaws is utilized since it is the only routine that does not calculate the reference set using either the mean or the median (it employs the maximum or minimum of all replicates), thus reducing the possibility of obtaining biased estimate values in the reference set.

It is evident that no immediately clear spatial flaws are visible in Figure 9.7 for these arrays. The dark arrays shown in this figure are not caused by differences in

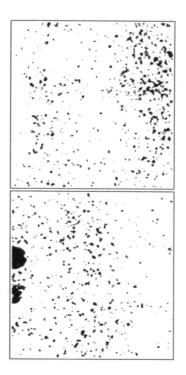

FIGURE 9.6 Spatial flaws in an HG-U133A array visualized with the method of [1] without using replicates; interpretation as for Figure 9.5.

overall intensity levels between the arrays, since the method proposed in [2] pre-processes the data with constant normalization. Figure 9.8 shows another spatial flaw visualization with the same method, albeit now choosing different arrays of the same type. Exchanging array GSM117700 for GSM96264 (a biological replicate of GSM96262) and repeating the analysis, the results are now different.

A small blob of unusually high values is now detected near the center-right of the array for GSM49942 (displayed as a "cut out circle" in the lower panel at the same zone of the array) not clearly visible in Figure 9.7. Additionally, an arc of unusually high values near the top-left edge of the array is now visible in GSM96262. These blemishes are similar to the flaws detected with the method of [1] in GSM49942 (Figure 9.9 (a)) and by the method of [2] when analyzing GSM96262 along with its biological replicates (Figure 9.9 (b)), and were not observed in Figure 9.7.

The example above illustrates that when the methods discussed in Section 9.3 are utilized with non-replicates to identify spatial flaws the results can be ambiguous. An explanation for this discrepancy is the biological variation between arrays that are non-replicates. An underlying assumption of all the methods discussed in Section 9.3 is that replicate arrays are presented with the same genetic material, and thus the variability between their intensity values is expected to be small. With non-replicates

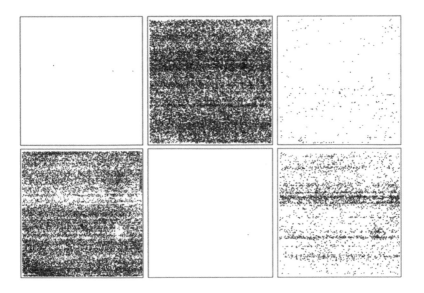

FIGURE 9.7 Spatial flaws detected in arrays GSM49942 (left), GSM117700 (center), and GSM96262 (right) with the method of [2]; interpretation as for Figure 9.5.

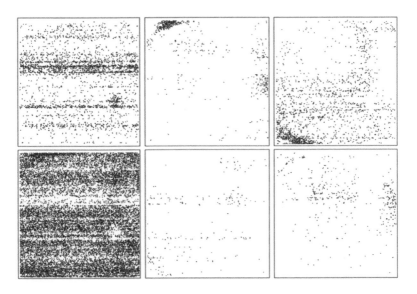

FIGURE 9.8 Spatial flaws detected in arrays GSM49942 (left), GSM96262 (center), and GSM96264 (right) with the method of [2]; interpretation as for Figure 9.5.

this assumption does not hold. Therefore, due to the small sample size and the high variability between non-replicate intensity values, the reference set calculated with any of the methodologies discussed in Section 9.3 will estimate biased values at some locations. The amount of this bias will depend on the arrays chosen in the sample, and hence different spatial biases could be identified.

9.5   SUMMARY AND CONCLUSIONS

In spite of their potential and popularity, the use of microarrays as a tool for gene expression studies also has inconveniences since their manufacturing process is complex and expensive. Undesired random variation incorporated during the various steps of the manufacturing process could affect gene expression measurements and compromise the biological quality of the data. In this chapter, we have discussed several methods for the image processing of Affymetrix microarrays that attempt to minimize biases incorporated to the data during the grid alignment process and when the data is affected by spatial blemishes. Image processing methods have a number of potential impacts on the analysis and interpretation of the data, and their importance

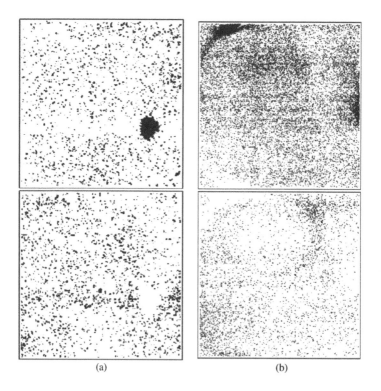

(a)   (b)

FIGURE 9.9 Spatial flaws visualized with: (a) the [1] method in GSM49942 (left) and (b) the [2] method in GSM96262 (right); interpretation as for Figure 9.5.

should not be overlooked by data analysts.

Microarray grid alignment is one of the basic processing steps of microarray images. Understanding this step is critical for performing an optimal microarray data analysis and obtaining conclusions from every experiment. There are several sources of error during the grid alignment process that could cause the information stored in the CEL file from the DAT file to be erroneous, and thus affect the data analysis workflow. Therefore, the grid alignment process should not be overlooked in the analysis of Affymetrix arrays.

Every oligonucleotide array contains spatial flaws to some degree [34], and efforts should be directed to evaluate and assess the reproducibility of microarray data (e.g., [39]). The most common flaws appear to be the consequence of trapped bubbles and are manifested as rings or arcs [13], irregular shaped blobs, or occasional "scratches" [2]. These are usually seen towards a side of the array [8].

As noted in [31], several computational tools (e.g., RMA) whose purpose is to assess overall chip quality and improve the accuracy of microarray gene expression fail to incorporate adequate spatial information into their outlier detection methods or normalization routines. The image processing algorithms described in this chapter seek to address these deficiencies and complement existing methods to improve the accuracy of gene expression measurements and the implications biologists draw from them. According to the authors of [35], it is usually not clear how to proceed once spatial artifacts have been detected, and the researcher must decide if the contaminated chip is to be included or excluded from further analysis. However, the decision of excluding information from an experiment could be cost-intensive.

The identification of spatial biases using image processing techniques is a topic often neglected in microarray data analysis. A possible explanation for this is that several scientists regard this problem as part of the manufacturing stage rather than as part of the data analysis stage. In other words, scientists often ignore the problem because they believe that the manufacturers have taken appropriate measures to prevent the existence of blemishes. Since this is not always the case, we suggest both encouraging microarray manufacturers to improve their quality protocols to prevent the incorporation of these biases, and incorporating image processing techniques and spatial normalization algorithms in the microarray data analysis work flow so that these algorithms are applied routinely.

REFERENCES

1. J. M. Arteaga-Salas, G. J. G. Upton, and A. P. Harrison, "Identification of spatial biases in Affymetrix oligonucleotide arrays," in *Proceedings of the 8th International Conference for the Critical Assessment of Microarray Data Analysis (CAMDA 2008)*, Vienna, Austria, pp. 25–29, December 2008.
2. G. J. G. Upton and J. C. Lloyd, "Oligonucleotide arrays: information from replication and spatial structure," *Bioinformatics*, vol. 21, pp. 4162–4168, November 2005.
3. R. A. Irizarry, B. M. Bolstad, F. Collin, et al., "Summaries of Affymetrix Genechip probe level data," *Nucleic Acids Research*, vol. 31, p. e15, February 2003.

4. Z. J. Wu, R. A. Irizarry, R. Gentleman, et al., "A model-based background adjustment for oligonucleotide expression arrays," *Journal of the American Statistical Association*, vol. 99, pp. 909–917, December 2004.

5. J. M. Arteaga-Salas, H. Zuzan, W. B. Langdon, et al., "An overview of image-processing methods for Affymetrix Genechips," *Briefings in Bioinformatics*, vol. 9, pp. 25–33, January 2008.

6. P. Bajcsy, "An overview of DNA microarray grid alignment and foreground separation approaches," *EURASIP Journal on Applied Signal Processing*, no. 80163, pp. 1–13, 2006.

7. D. S. Yuan and R. A. Irrizarry, "High-resolution spatial normalization for microarrays containing embedded technical replicates," *Bioinformatics*, vol. 22, pp. 3054–3060, December 2006.

8. W. B. Langdon, G. J. G. Upton, R. D. Camargo, et al., "A survey of spatial defects in Homo Sapiens Affymetrix Genechips," *IEEE/ACM Transactions on Computational Biology and Bioinformatics*, vol. 7, pp. 647–653, October-December 2010.

9. M. Suárez-Fariñas, A. Haider, and K. Wittkowski, "'Harshlighting' small blemishes on microarrays," *BMC Bioinformatics*, vol. 6, p. 65, March 2005.

10. J. S. Song, K. Maghsoudi, W. Li, et al., "Microarray blob-defect removal improves array analysis," *Bioinformatics*, vol. 23, pp. 966–971, April 2007.

11. M. Reimers and J. M. Weinstein, "Quality assessment of microarrays: visualization of spatial artefacts and quantitation of regional biases," *BMC Bioinformatics*, vol. 6, p. 166, July 2005.

12. T. H. Stokes, R. A. Moffitt, J. H. Phan, et al., "Chip artifact *CORRECT*ion (caCOR-RECT): A bioinformatics system for quality assurance of genomics and proteomics array data," *Annals of Biomedical Engineering*, vol. 35, pp. 1068–1080, June 2007.

13. M. Suárez-Fariñas, M. Pellegrino, K. Wittkowski, et al., "Harshlight: a "corrective make-up" program for microarray chips," *BMC Bioinformatics*, vol. 6, p. 294, December 2005.

14. N. Brandle, H. Bischof, and H. Lapp, "Robust DNA microarray image analysis," *Machine Vision and Applications*, vol. 15, pp. 11–28, October 2003.

15. C. W. Whitfield, A. M. Cziko, and G. E. Robinson, "Gene expression profiles in the brain predict behavior in individual honey bees," *Science*, vol. 302, pp. 296–299, October 2003.

16. E. E. Schadt, C. Li, B. Ellis, et al., "Feature extraction and normalization algorithms for high-density oligonucleotide gene expression array data," *Journal of Cellular Biochemistry Supplement*, vol. 37, no. 294, pp. 120–125, 2001.

17. P. Bajcsy, "GridLine: automatic grid alignment DNA microarray scans.," in *IEEE Transactions on Image Processing*, vol. 13, pp. 15–25, 2004.

18. L. Perez-Torres, "Deblurring Affymetrix Genechip images by applying the Lucy-Richardson algorithm," Master's Thesis. Unpublished. September 2008.

19. W. H. Richardson, "Bayesian-based iterative method of image restoration," *Journal of the Optical Society of America*, vol. 62, pp. 55–59, September 1970.

20. L. B. Lucy, "An iterative technique for the rectification of observed distributions," *The Astronomical Journal*, vol. 79, pp. 745–754, June 1974.

21. D. Hekstra, A. R. Taussig, M. Magnasco, et al., "Absolute mrna concentrations from sequence-specific calibration of oligonucleotide arrays," *Nucleic Acids Research*, vol. 31, pp. 1962–1968, April 2003.

22. F. Naef and M. O. Magnasco, "Solving the riddle of the bright mismatches: Labelling

and effective binding in oligonucleotide arrays," *Physical Review E*, vol. 68, p. 011906, July 2003.

23. H. Zuzan, C. Blanchette, H. Dressman, et al., "Estimation of probe cell locations in high-density synthetic-oligonucleotide DNA microarrays." Working paper, October 2001.

24. R. A. Irizarry, B. Hobbs, F. Collin, et al., "Exploration, normalization, and summaries of high density oligonucleotide array probe level data," *Biostatistics*, vol. 4, pp. 249–264, April 2003.

25. M. Hulsman, A. Mentink, E. P. van Someren, et al., "Delineation of amplification, hybridization and location effects in microarray data yields better-quality normalization," *BMC Bioinformatics*, vol. 11, p. 156, March 2010.

26. T. Barrett, D. B. Troup, S. E. Wilhite, et al., "NCBI GEO: archive for functional genomics data sets—10 years on," *Nucleic Acids Research*, vol. 39, pp. D1005–D1010, January 2011.

27. T. Barrett, S. E. Wilhite, P. Ledoux, et al., "NCBI GEO: archive for functional data sets—update," *Nucleic Acids Research*, vol. 41, pp. D991–D995, January 2013.

28. L. Gautier, L. Cope, B. M. Bolstad, et al., "affy—analysis of Affymetrix Genechip data at the probe level," *Bioinformatics*, vol. 20, pp. 307–315, February 2004.

29. R. C. Gentleman, V. J. Carey, D. M. Bates, et al., "Bioconductor: Open software development for computational biology and bioinformatics," *Genome Biology*, vol. 5, p. R80, September 2004.

30. B. M. Bolstad, R. A. Irizarry, M. Astrand, et al., "A comparison of normalization methods for high density oligonucleotide array data based on variance and bias," *Bioinformatics*, vol. 19, pp. 185–193, January 2003.

31. R. A. Moffitt, Q. Yin-Goen, T. H. Stokes, et al., "caCORRECT2: Improving the accuracy and reliability of microarray data in the presence of artifacts," *BMC Bioinformatics*, vol. 12, p. 383, September 2011.

32. J. C. Russ, *The Image Processing Handbook*. Boca Raton: CRC Press, 2002.

33. T. H. Cormen, C. E. Leiserson, R. L. Rivest, et al., *Introduction to algorithms*. Boston: The MIT Press, 2001.

34. J. M. Arteaga-Salas, A. P. Harrison, and G. J. G. Upton, "Reducing spatial flaws in oligonucleotide arrays by using neighborhood information," *Statistical applications in Genetics and Molecular Biology*, vol. 7, p. 29, October 2008.

35. T. Petri, E. Berchtold, R. Zimmer, et al., "Detection and correction of probe-level artefacts on microarrays," *BMC Bioinformatics*, vol. 13, p. 114, May 2012.

36. P. Neuvial, P. Hupé, I. Brito, et al., "Spatial normalization of array-CGH data," *BMC Bioinformatics*, vol. 7, p. 264, May 2006.

37. H. S. Chai, T. M. Therneau, K. R. Bailey, et al., "Spatial normalization improves the quality of genotype calling for Affymetrix SNP 6.0 arrays," *BMC Bioinformatics*, vol. 11, p. 356, June 2010.

38. A. Inc., "Guide to probe logarithmic intensity error (PLIER) estimation.," tech. rep., Affymetrix, Santa Clara, USA, 2005.

39. L. Shi, G. Campbell, W. D. Jones, et al., "The microarray quality control (MAQC)-ii study of common practices for the development and validation of microarray-based predictive models," *Nature Biotechnology*, vol. 28, p. 827, October 2010.

10 Treatment of Noise and Artifacts in Affymetrix Arrays

Caroline C. Friedel

CONTENTS

10.1 INTRODUCTION

Hybridization-based DNA microarrays are a key technology for high-throughput quantification of expression levels for thousands of genes [1, 2]. State-of-the-art microarrays now allow the genome-wide analysis of transcript abundance not only for entire genes but also individual exons, alternatively spliced transcripts, and even a large fraction of non-coding genomic regions [3, 4]. Thus, despite the increasing prevalence of alternative methods such as RNA-seq [5], RNA microarrays remain important for the analysis of many biological processes such as miRNA-based regulation [6], alternative splicing patterns across human tissues [7], or the role of alternative splicing in stem cell differentiation [8] and cancer [9].

Recently, Langdon et al. [10] reported that *all* human Affymetrix microarrays available in the Gene Expression Omnibus (GEO) [11] contain spatial defects to some degree. Thus, quality control for microarrays remains a major issue. Although many methods and software tools have been developed for quality assessment of microarrays [12, 13, 14, 15], detection of spatial artifacts is not yet routinely applied. Furthermore, it is usually not clear how to proceed once such artifacts have been detected. The two alternatives are (1) to either completely exclude or (2) to include the corresponding arrays for any subsequent analysis. In the first case, the corresponding measurements are not available for gene expression profiling and may even have to be repeated if they are crucial to the analysis. This can be cost-intensive, in particular, if corresponding samples have been used up. In the second case, one has to assume that normalization and summarization methods can correct for the measurement errors.

The latter assumption is based on the construction of microarrays where probes of the same probeset are not contiguous on the array. Thus, smaller artifacts due to uneven hybridization or other experimental problems may only affect a subset of probes for a probeset. It is usually assumed that summarization methods such as *RMA* [16]—which combine the values for individual probes to a probeset value—can estimate the probeset value correctly despite measurement errors for some probes.

In a recent study [17], we showed that this assumption is often invalid as even small artifacts on the array can have a significant effect on the overall expression values of many probesets, not only the ones affected by the artifact. Although this was illustrated only for Affymetrix exon arrays, this problem likely affects a wider range of array types. To address this problem, we introduced two simple but effective approaches for the identification of corrupted probes: (1) a threshold-based approach and (2) an extension of this approach that takes into account the neighborhood of a probe, i.e., spatial information of the array. The use of spatial information significantly improved the identification of defective probes as well as reproducibility of probeset intensities after summarization. Finally, two strategies were proposed to either correct probe values using probeset information or filter corrupted probes, both of which improved summarization accuracy as well as reproducibility between replicates. In this way, we could recover even arrays with large artifacts for downstream analysis that otherwise would have had to be discarded.

The focus of this chapter is to provide an overview on the problems probe-level artifacts pose for gene expression profiling using (Affymetrix) microarrays as well as on methods developed to address these problems. It is based on our recently published study on artifacts in Affymetrix microarrays [17]. In Section 10.2, a case study from this article illustrates the detrimental effect of experimental artifacts on gene expression profiling results. Here, state-of-the-art summarization methods could not appropriately correct for these artifacts. Section 10.3 provides an overview of current methods for detecting and correcting such artifacts, including our own recently developed methods. Finally, results on the performance of these methods are shown in Section 10.4 and discussed in Section 10.5.

10.2 INFLUENCE OF ARTIFACTS ON GENE EXPRESSION PROFILING

10.2.1 AFFYMETRIX MICROARRAY MEASUREMENTS WITH ARTIFACTS

For our recent study on experimental artifacts on Affymetrix microarrays, we used RNA measurements for two cell lines with both Affymetrix GeneChip Human Gene 1.0 ST and Exon 1.0 ST arrays: 1) the B-cell line DG75 transduced to express 10 out of 12 miRNAs encoded by the Kaposi's sarcoma-associated herpesvirus (KSHV) (DG75-10/12), and 2) DG75 transduced to express eGFP (DG75-eGFP) as control [18]. For each cell line and array type, total RNA was quantified. In addition, RNA synthesis and decay was measured using a recently developed method for labeling newly transcribed RNA using 4-thiouridine (4sU) [19, 20]. This allows separating total cellular RNA (T) into the labeled newly transcribed RNA (N) and the unlabeled pre-existing RNA (P) as well as quantification of *de novo* transcription and decay in a single experimental setting. As newly transcribed and pre-existing RNA should sum up to total RNA, these experiments provide a true biological control for the assessment of quality problems and their correction.

For each cell line and each RNA fraction 3 replicates were measured resulting in a total of 18 arrays for each microarray platform. The Gene 1.0 ST measurements were recently published [21]. Exon 1.0 ST measurements were performed in the same way. However, in this case considerable experimental artifacts were observed for several of the 18 arrays resulting in distinctive stains visible in the array images (see Figure 10.1 for examples). These artifacts were probably a consequence of a drying out of the central part of the array during the hybridization step resulting in artificially high values for the corresponding probes.

In our study, we focused mostly on the measurements of the DG75-10/12 cells. In this case, two out of three total RNA measurements showed substantial spatial

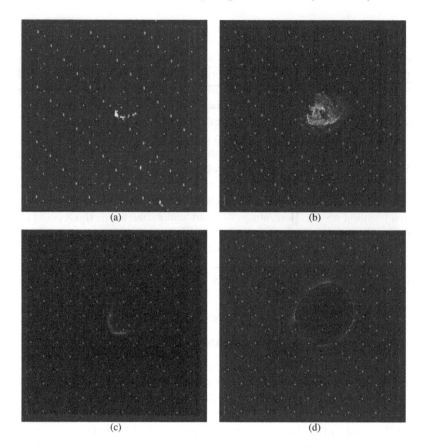

FIGURE 10.1 Measurement artifacts observed on different arrays of our data set: total RNA for replicates 1 (a) and 3 (b) in DG75-10/12 cells; total RNA for replicate 2 (c) in DG75-eGFP cells; newly transcribed RNA for replicate 1 (d) in DG75-eGFP cells.

artifacts in the images of the arrays but the corresponding measurements of newly transcribed and pre-existing RNA were free of defects or showed only very small artifacts, allowing us to use these as biological control. The largest artifact affecting a sizable amount of probes was observed in replicate 3 and a smaller one in replicate 1 (see Figure 10.1). Replicate 2 was artifact-free in total RNA, although slight defects were observed in pre-existing and newly transcribed RNA.

10.2.2 INSUFFICIENT CORRECTION BY SUMMARIZATION

For any microarray experiment, two steps are usually performed first in its analysis: normalization and summarization. Normalization is applied to allow the comparison of results from different replicates and conditions. Summarization estimates overall expression values for each probeset from the individual probe measurements. As summarization uses information from several probes to determine the probeset value, it is generally assumed that it can correct appropriately for individual defect probe measurements, in particular as probes for the same probeset are not located close to each other but distributed across the whole array. In the following, results are shown for the exon array measurements described above that illustrate that this is not necessarily the case and that even small artifacts can have substantial effects.

For our study, we used the following methods for normalization and summarization. For normalization, we used quantile normalization, which is commonly used in combination with *RMA* summarization. If newly transcribed (N) as well as pre-existing (P) RNA have been quantified in addition to total cellular RNA (T), an additional normalization step was applied to account for the different amounts of RNA between the fractions [20]. Since $T = N + P$ has to hold approximately for all probes, the linear model $T = \lambda_1 N + \lambda_2 P$ minimizes the sum of residuals for $\lambda_i \in R^+, i \in \{1,2\}$. The corresponding λ_i can be found by linear regression [20], which can be applied both on the summarized probeset values as well as the individual probe values themselves. When analyzing fold-changes, LOESS normalization is additionally applied before fold-change calculation.

For summarization, we used *RMA* [16], one of the most widely used summarization methods, which estimates both an overall expression value for each probeset and the probe-specific measurement error by fitting a linear model to the probe values. Thus, this method implicitly estimates the noise level for each probe and effectively subtracts the estimated noise from the probe when calculating the overall probeset values. This explains why it is commonly assumed that this method can correct for measurement artifacts [22]. For our purposes, the Affymetrix Power Tools were used for summarization (http://www.affymetrix.com). An alternative implementation is provided by the *affyPLM* library [12], which also provides access to the estimated residuals. However, due to its considerable memory usage when working on exon arrays, the *affyPLM* library could not be applied to all exon array measurements together.

To evaluate whether the final probeset values can be estimated correctly by summarization, we used *replicate scatter plots* that compare probeset levels between the affected array and a control (see Section 10.3.5). If summarization were to appropriately correct for artifacts, summarized probeset values should be highly reproducible between the replicates. However, this was clearly not the case as for both examples with artifacts considerable deviations were observed (see Figure 10.2 a,d).

Not surprisingly, the deviation to the control was substantial for probesets with all probes affected as no reasonable estimation was possible. Here, the fraction of probes per probeset affected by the artifact were determined using the ε-*criterion* approach described in Section 10.3.5. In contrast, if only 75% or less (0-3 probes for most probesets) probes were affected, we did not see a correlation between the number of defective probes and the deviation from the diagonal. Instead, all probeset levels were affected to some degree. Strikingly, the deviation for replicate 1 with the small stain was stronger than for replicate 3 with the largest stain. Furthermore, this deviation was more pronounced for probesets with high expression values, which did

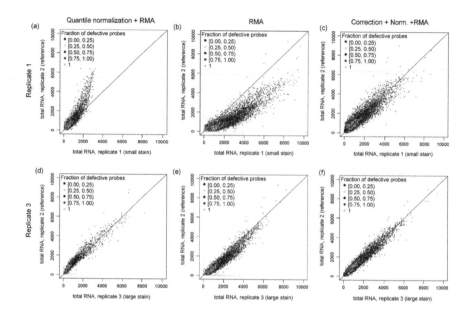

FIGURE 10.2   (SEE COLOR INSERT.) Replicate scatter plots comparing total RNA for replicates 1 (a–c) and 3 (d–f) against the artifact-free replicate 2 for the exon array measurement in DG75-10/12 cells. Figures are shown separately for results using both *RMA* and quantile normalization (a, d), using only *RMA* without quantile normalization (b, e) and after probe correction (c, f). Probesets are colored based on the percentage of their probes that were flagged as corrupted according to the ε-*criterion* (see Section 10.3.5) based on noise scores calculated using newly transcribed and pre-existing RNA as control. For replicate 1 there is a bias even for the uncorrupted probesets (a) that can be reduced by omitting quantile normalization (b). If artifact correction is applied prior to normalization and summarization (c, f), this bias is removed.

not even contain probes affected by the stain.

One possible explanation is that this is an effect of the normalization—in this case quantile normalization—that has to be performed before summarization. It might compensate for the extremely high values for some of the probes by reducing the levels of the remaining probes. Indeed, when omitting quantile normalization, the strong deviation for highly expressed genes in replicate 1 was reduced (Figure 10.2 b,e). Nevertheless, even without normalization, the nonlinear behavior for both replicates in comparison to replicate 2 was still observed.

Exon arrays also offer the possibility to summarize probe values to meta-probesets that correspond to genes. As there are more probes per meta-probeset the effect of the artifacts should be smaller. Nevertheless, we still observed a systematic shift from the diagonal in the corresponding replicate scatter plot, although for replicate 3 the deviation was much smaller (shown in the original publication [17]). In contrast to probeset-level summarization, however, omitting quantile normalization did not reduce this deviation.

10.2.3 SENSITIVITY OF SUMMARIZATION TO NOISE

To systematically analyze the influence of measurement artifacts on summarization, we performed the following experiment using three replicates of total RNA measured with exon arrays for the DG75-eGFP cell lines. These measurements were basically artifact-free with only a very small stain in one replicate, which could be easily corrected using our ε-*criterion* (see Section 10.3.5 for a description). Here, only 6380 probes (out of >5.5 million features on the array) were identified as corrupted and 6335 probesets had 1 corrupted probe, 21 had 2 and only one had 3. This is a much smaller number than observed for the substantial artifacts on the DG75-1012 arrays.

We then introduced artificial measurement artifacts into the corrected DG75-eGFP arrays by *spiking* probes in the following way. First, a noise level δ was selected, i.e., a probability that a random probe was spiked. For each probe, it was then decided randomly with probability δ whether it was spiked. For spiked probes, the raw measured values were replaced by an artificial level drawn from a log-normal distribution with mean μ and standard deviation σ. In our case, $\mu = \log_2(850)$ and $\sigma = 1$ were inferred from intensities within the real artifacts identified by the ε-*criterion* on the DG75-10/12 total RNA measurements. This was done to provide a realistic level of noise. Furthermore, only probes corresponding to core probesets defined by Affymetrix were spiked. After spiking the raw values on the array, we then performed summarization and normalization. Simulations for each selected value of δ were repeated 100 times. To assess the effect on the resulting probeset levels, we evaluated the average $\log_2$ fold-changes in probeset levels between each pair of spiked array and unspiked control for noise levels in the range of 0.01 to 0.1.

Comparing the $\log_2$ fold-change against the number of spiked probes for each probeset (Figure 10.3), we found a very clear trend: if only one probe was spiked, the median fold-change was slightly higher than for probesets not affected by spiking. However, if more than one probe was spiked, fold-changes increased substantially. Thus, variance of the probeset levels was considerably higher even if only a few probes were affected. This larger variance can lead to low or no statistical significance for differentially expressed genes and as a consequence can reduce the sensitivity of gene expression profiling.

10.3 METHODS FOR ARTIFACT DETECTION AND CORRECTION

Although not commonly included in standard microarray analysis pipelines, a number of methods have been previously proposed for visualization of microarray artifacts as well as identification and/or correction of corrupted probe measurements (see also [23] for an overview of methods published before 2008). In the following, the most important methods are introduced in more detail: (1) Harshlight by Suárez-Fariñas et al. [24], (2) Microarray blob remover by Song et al. [25], (3) Local probe effect adjustment by Arteaga-Salas et al. [26], (4) caCORRECT2 by Moffit

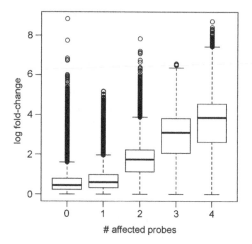

FIGURE 10.3 Boxplot of the $\log_2$ fold changes for probesets with 0-4 spiked probes for the 100 simulations with $\delta = 0.05$. Here, probesets with the same number of spiked probes were pooled across all simulation results. For the case of 0 spiked probes, probesets were selected randomly from the pooled set of all simulations as there were too many probesets for loading into R. In this case, each probeset was selected with a probability of 0.01. We observed a very strong correlation between the number of affected probes and fold-change biases on probeset level, which may seriously harm downstream analyses.

el al. [27], and (5) a window-based method that we presented recently [17]. Please note that some of the artifact detection methods are also presented in Chapter 9, in particular Harshlight and local probe effect adjustment as well as the caCORRECT method on which caCORRECT2 is based. These artifact detection methods are described here again, as it is necessary for the understanding of the artifact correction methods.

10.3.1 HARSHLIGHT

One of the first methods developed for identifying artifacts is Harshlight [24], which identifies and masks local artifacts ("blemishes") based on statistical and image processing methods. Basically, three types of artifacts are identified one after the other: extended defects that are very large and may invalidate the whole array, compact defects affecting only a few probes close together, and diffuse defects affecting larger areas but clustered less tightly. Probes within defects can either be excluded from the analysis or be replaced by the median intensity across replicates. In the following, the procedures for identifying extended, compact and diffuse defects are briefly described. In this case, all steps are based on the so-called Error Image (E) for each array [28], which is defined as

$$E^{(i)} = L^{(i)} - median_j L^{(j)}.$$ (10.1)

Here, $L^{(i)}$ is the logarithm of the intensity matrix for array i. Thus, $E^{(i)}$ describes the deviation of the i-th array to the median of all arrays in the experiment. Here, the median is defined such that for each probe the median over all arrays is used.

10.3.1.1 Extended Defects

For the identification of extended defects, two values are determined for each array: σ_E^2, which is defined as the variance of the deviations from the median array, and $\sigma_{B_E}^2$, which quantifies the variance in the background of the error image E. The latter is estimated by first determining the so-called median filtered image $\tilde{E}$ and then calculating its variance. Here, $\tilde{E}$ is determined by sliding a window, e.g., a circular one, across the error image and calculating the value for each probe in the center of the window as the median of all probe values within the window. The ratio $\sigma_{B_E}^2 / \sigma_E^2$ then quantifies the proportion of the variance of E that can be explained by the background variation. Ideally, this ratio should be very small and large ratios indicate substantial artifacts, suggesting that the corresponding arrays should be discarded.

10.3.1.2 Compact Defects

In the first step of Harshlight only the presence of extended defects is tested but not the individual probes affected by the defects. This is done to exclude arrays that show such substantial defects that there is no hope of correcting them and that consequently have to be discarded. In the second step, smaller artifacts and the corresponding probes are identified. These smaller artifacts are denoted as compact defects and are defined as small connected clusters of outliers in the error image E. Here, the following steps are pursued to identify these clusters.

First, outlier probes are identified whose values in E are either smaller than the α-percentile or larger than the $(1 - \alpha)$-percentile of E (e.g., $\alpha = 2.5\%$). Second, clusters of connected outliers are identified using the FloodFill algorithm [29], an algorithm commonly used in computer graphics to identify areas of connected pixels of the same color. Finally, significance of the identified clusters is evaluated as small clusters might also be identified in case of randomly distributed outliers. For this purpose, the empirical distribution of cluster size under the null hypothesis that outliers are randomly distributed was determined based on simulations for arrays of the same dimension and with the same proportion of outliers. Based on this empirical distribution, significance of cluster size is then evaluated and only clusters with significant size are considered as compact defects. The corresponding probes are then excluded for the third analysis step.

10.3.1.3 Diffuse Defects

Although diffuse defects are also defined as areas with densely distributed outliers, these outliers do not have to be connected. As a consequence, the major part in the identification of diffuse defects is to connect outlier probes, such that the corresponding area can be identified by the FloodFill algorithm. For this purpose, bright (increased intensities) and dark outliers (decreased intensities) are treated separately from each other.

For each type of outliers a so-called *outlier image* is identified first, in which the respective outliers have the value 1 and all other probes the value 0. These outlier images are then scanned with a circular window similar to the identification of the median filtered image. If the proportion of outliers in this window is significantly larger than the overall proportion of outliers on the array, the central probe of this window is assigned the value 1 in a new image D. In this image, connectivity of outliers is increased as non-outlier probes in regions of high outlier density have now also been flagged as outliers. The FloodFill algorithm is then applied to D to identify clusters of probes with value 1. Just as for compact defects, small clusters are discarded.

A final step is added to better outline the area of the diffuse defect using a closing procedure, a standard image processing technique. The approach used by Harshlight consists of two parts: dilation, which fills up empty spaces within the defect leading to a growth of the defect borders, and erosion, which limits the extension of the defect to the convex hull of the original cluster. Both for dilation and erosion, a circular window is centered at each probe. In the dilation step, a probe is flagged if any of the probes in the window are flagged. In the subsequent erosion step, a probe is flagged only if all the probes in the window were flagged in the dilation step.

10.3.1.4 Contiguity Test and Correction of Defects

As the definitions of compact and diffuse defects are somewhat overlapping, diffuse defects may contain small regions with higher outlier density that are flagged as compact defects. As the corresponding probes are removed before identification of diffuse defects, this may lead to problems in recognizing the larger diffuse defects. As a consequence, the authors included a contiguity test between identification of compact and diffuse defects, which again consists of applying a closing procedure to the identified compact defects. If the area of the defect is increased substantially by closing, this indicates that the compact defect is actually part of a larger diffuse defect. In this case, the corresponding probes are not eliminated before the next step.

In contrast to the defect identification procedure, the approach to correcting these defects implemented in Harshlight is quite simple and similar to the approach used for our method. The standard approach is to declare probes as missing if they are contained within a compact or diffuse defect. If down-stream analysis tools cannot deal with missing data, the corresponding probe values can be substituted by the median intensity of the corresponding probes on the other arrays.

10.3.2 MICROARRAY BLOB REMOVER

Microarray blob remover (MBR) is a method for identifying artifacts from raw intensity values [25]. This is in contrast to all other approaches we are aware of, which use other arrays for comparison. Here, MBR operates in two steps. First, broad areas are determined in which more than half of the probes are above the kth percentile of probe intensities, where k may be in the range of 60 to 100. For this purpose, they also use a sliding window, in this case a 100×100 square, which is moved in steps of 50 probes in both dimensions. If more than half of the probes in this square are above the threshold, a second refinement step is applied to the square.

For this refinement step, a circle of radius 20 is used to scan the square, this time taking steps of 2 probes. For this step the $(k-5)$th percentile is used as threshold and probes within the circle are flagged as corrupted if at least $p\%$ of these probes are above the threshold. Here, p is set to 90 by default, but can be adjusted by the user to values between 80 and 100. As a consequence of the shape of the window used for the refinement step, MBR is limited to the detection of circular or oval defects as may be caused by bubbles during array production. Finally, probes flagged as artifacts are added to the "outlier entries" section in the original CEL files, leaving it to downstream algorithms how to address these defects.

10.3.3 LOCAL PROBE EFFECT ADJUSTMENT

Local probe effect (LPE) adjustment was developed by Arteaga-Salas et al. [26] and is based on a visualization of spatial flaws by plotting a matrix D_r for array r defined as

$$d_{klr} = L_{klr} - M_{kl}. \tag{10.2}$$

Here, L_{klr} is the logarithm of the probe intensity at position (k,l) on array r and M_{kl} the logarithm of a reference intensity for position (k,l). For the latter, different alternatives can be used, such as the trimmed mean of reference arrays as proposed by Reimers and Weinstein [30]. Arteaga-Salas et al. recommend using the median value of the reference arrays, as this measure is more robust. If this is done, D_r is the same as the error image E used by Harshlight.

Before the development of Affymetrix Exon and Gene ST arrays, Affymetrix arrays contained both perfect match (PM) and mismatch (MM) probes. The MM probes differed only at the central position from the corresponding PM probe to identify unspecific hybridization. As a consequence, overall intensities of MM and PM probes could differ significantly. This problem was addressed by Arteaga-Salas et al. by defining standardized d-values d_{klr}^* as

$$d_{klr}^* = \frac{d_{klr}}{S_{kl}} \tag{10.3}$$

with S_{kl} the standard deviation of the L_{klr} values across the arrays.

10.3.3.1 Artifact Detection

To identify spatial artifacts, first a matrix E is calculated as

$$E_{kl} = |d_{klr}| \times \text{sign}(d_{klr}) \tag{10.4}$$

where $r = \text{argmax} |d_{klr}|$. Thus, for each position (k,l), E_{kl} indicates for which array the absolute value of d_{klr} is the largest as well as whether this maximal value is positive or negative. Accordingly, E_{kl} is limited to values in $\{-R, -R + 1, \ldots, -2, -1, 1, 2, \ldots, R - 1, R\}$ if R is the number of arrays. To identify artifacts, LPE then evaluates the matrix E in 5×5 windows centered on each probe (excluding quality control probes and positions outside of the array). If the same value in E (e.g., r or $-r$) is observed for a majority of probes in such a window, the central probe is marked for adjustment on array r. This approach uses a strategy first presented by Upton and Lloyd [31].

10.3.3.2 Artifact Correction

In contrast to the other artifact correction methods presented in this chapter, LPE does not replace values of corrupted probes completely or remove them from the analysis but uses an adjustment procedure based on the d^*-values. If a probe at position (k, l) on array r is marked for adjustment, all d^*-values in the 5×5 window centered on the probe are calculated for array r. The average of these values $\bar{d}^*$ is then used to adjust the value on position (k, l, r) provided d^*_{klr} and $\bar{d}^*$ have the same sign. Otherwise, no adjustment is performed. The adjusted value L^a_{klr} is calculated as

$$L^a_{klr} = L_{klr} - S_{kl}\bar{d}^*. \tag{10.5}$$

This adjustment is based on the assumption that the artifact results in some sort of multiplicative error (additive in logarithmic space) and, thus, can be subtracted out. If this is not the case, this adjustment procedure will likely fail. Furthermore, no adjustment is performed if the signs of d^*_{klr} and $\bar{d}^*$ differ. Here, a more reliable approach would be to completely eliminate the corresponding probe from subsequent analyses.

10.3.3.3 Complementary Probe Pair Adjustment

Arteaga-Salas et al. also proposed a second adjustment procedure that can be used separately from LPE or in combination. This procedure, called complementary probe pair (CPP) adjustment, evaluates pairs of PM and MM probes for potential artifacts, as the MM probes are located next to PM probes and, thus, both of them should be affected by an artifact. As the Mismatch probe design has been abandoned for the new generation of Affymetrix arrays, this approach is no longer applicable and not presented in more detail here.

10.3.4 CACORRECT2

10.3.4.1 Artifact Detection

The most recently published approach, *caCORRECT2* [27], uses a so-called h-score to estimate whether a probe value on a given array is consistent with the observed distribution for all other arrays. For this purpose, an iterative process is applied in which artifact identification and quantile normalization are alternated ("artifact-aware quantile normalization"). This means that probes predicted to be corrupted in one iteration are set aside and not included in quantile normalization to avoid biasing normalization by the corrupted probes. In the current implementation, caCORRECT2 uses four iterations. In each iteration, the h-score for a probe j is calculated from a z-value-like statistic defined as

$$z_j = \frac{x_j - \mu_j}{\sigma_j}. \tag{10.6}$$

Here, x_j is the value of probe j on the array considered and μ_j and σ_j are the mean and standard deviation for the probe on all other arrays, excluding those arrays for which the probe has been flagged as corrupted. A nonlinear scaling procedure is applied to calculate h from z_j, which takes into account the number of arrays used and results in h-values between 0 an 1. Artifacts are then identified from the calculated h-scores using again a sliding window approach that flags probes as corrupted if they both have high h-scores themselves and are contained in regions of high h-scores.

10.3.4.2 Artifact Correction

After identifying the final set of corrupted probes on each array, data imputation is used to replace the missing values for the corrupted probes appropriately. For this purpose, probe intensities are modeled as

$$x_{b,p,j} = \theta_{p,j} a_{b,p} + \varepsilon_{b,p,j}. \tag{10.7}$$

Here, $x_{b,p,j}$ is the intensity of the bth probe in the pth probeset on the jth array, $\theta_{p,j}$ denotes the gene expression, $a_{b,p}$ the probe affinity, and $\varepsilon_{b,p,j}$ an additive error term.

For each probeset p including B_p probes, the set of equations for N arrays are then described in the following matrix form

$$X_p = \theta_P a_p + \varepsilon_p, \tag{10.8}$$

where X_p and ε_p are $N \times B_p$-matrices, θ_P is an $N \times 1$-matrix, and a_p an $1 \times B_p$-matrix. Using singular value decomposition (SVD), the authors then find the solution to this equation system that minimizes the Frobenius norm of ε_p, which is defined as

$$\|\varepsilon_p\|_F = \sqrt{\sum_{j=1}^{N} \sum_{i=1}^{B_p} (\varepsilon_{b,p,j})^2}. \tag{10.9}$$

The values θ_P and a_p learned from the SVD can then be used to calculate the expected probe intensities according to the model as $\theta_P a_p$. Artifact correction in caCORRECT2 is then performed using an expectation-maximization approach, in which alternately θ_P and a_p are estimated using SVD and known artifact values in X_p are replaced with the corresponding values in the matrix $\theta_P a_p$.

10.3.5   ARTIFACT DETECTION AND CORRECTION USING PROBE NOISE SCORES

10.3.5.1   Probe Noise Scores

The method we presented recently for artifact detection and correction is based on an assessment of the noise level for individual probes [17]. For this purpose, different criteria can be applied. If measurement errors are explicitly modeled as in the *RMA* approach, the estimated residuals can be used to assess the effect of measurement artifacts [12]. The higher the absolute values of the residuals for a probe, the stronger the effect of measurement errors on this probe. The global residual level (e.g., calculated by the Affymetrix Power Tools subroutine *qcc*, [22]) can be used to indicate which arrays are suited as control and which are likely to contain artifacts.

A general probe-level noise score for probe j can be calculated as the fold-change compared to a control:

$$s_j = \left| \log_2 \frac{v_j + c}{v'_j + c} \right| \tag{10.10}$$

Here, v_j is the intensity for the probe on the corrupted array and v'_j is their value on the control. The pseudocount c corresponds to the estimated detection limit (in our case $c = 16$). Both v_j and v'_j can be measured directly or can be derived values, e.g., using measurements of newly transcribed and pre-existing RNA as described in Section 10.2.2. In the latter case, the normalized sum of N and P serves as a control for the measurement of T, i.e., $v_j = T$ and $v'_j = \lambda_1 N + \lambda_2 P$. Alternatively, technical or biological replicates may serve as a control. If it is not possible to determine a suitable control, the fold-change against the median probe intensities of all replicates can be used, resulting in the absolute value of the error image used by Harshlight [24]. Accordingly, our approach differs from the error image used by Harshlight and the D_r matrix used by LPE only in the type of control used to calculate the probe noise scores as well as the use of the absolute value of the scores. The latter means that we do not distinguish between artifacts leading to either abnormally increased or decreased probe intensities.

10.3.5.2   Probe Noise Plots

As each probe has a defined location on the array, the noise level of individual probes can be visualized by plotting the noise score of the probe against this location. For a more intuitive visualization, the noise score is color-coded and the location represented by the *x*- and *y*-axis. If residuals from the *RMA* model are plotted, this corresponds to the *residual plot* proposed by Bolstad et al. [12]. The *residual plot* was introduced as a measure to identify the presence of potential artifacts but not to identify the corresponding corrupted probes. To calculate residuals for a *residual plot*, the *affyPLM* implementation of *RMA* can be used, which provides access to these residuals.

10.3.5.3   Replicate Scatter Plots

Probe noise plots using different types of noise scores indicate the presence of artifacts. However, these artifacts would not necessarily have an effect on the final probeset values if normalization and summarization procedures were capable of correcting for them. To evaluate whether this is the case or not, we use the replicate scatter plots introduced in Section 10.2.2. For this purpose, summarized probeset values for a microarray experiment with artifact are plotted against corresponding values for a control measurement. Such a control measurement without artifacts can be easily identified based on probe noise plots. Here, the same type of control measurements can be used as for calculating probe noise scores. Alternatively, an average microarray measurement can be used as a control, which can be calculated from sets of published microarray experiments downloaded from the GEO. This approach was used in the study of Langdon et al. [10].

10.3.5.4   Artifact Detection

Based on probe noise scores, we developed two alternative approaches to identify corrupted probe measurements [17]. The first method is based on a simple threshold criterion (*ε-criterion*), the second approach extends this method by including the neighborhood information on the array (*Window Criterion*).

10.3.5.4.1   ε-criterion

The *ε-criterion* is based on the noise score defined in Equation 10.10 and simply applies a threshold t on this score. Thus,

$$\varepsilon(s_j) = \begin{cases} \text{true} & \text{if } s_j > t \\ \text{false} & \text{otherwise} \end{cases} \tag{10.11}$$

If $\varepsilon(s_j)$ is true, probe j is flagged as corrupted. Thresholds can be adjusted manually by analyzing both probe noise and replicate scatter plots.

10.3.5.4.2   Window Criterion

As measurement artifacts usually affect a specific region on the array and, accordingly, a set of probes closely located to each other, the second method takes into account the neighborhood information of a probe. This is based on the observation that measurement artifacts, e.g., due to uneven hybridization, usually affect several closely located probes and not only individual probes. Thus, for estimating the reliability of a specific probe the values of the probes in a window around this probe are evaluated as well. For this purpose, a 2D window of dimension $(2k+1) \times (2k+1)$ is used with the probe considered in the center of the window. For the presented results, we used $k = 25$, but this can be adjusted by the user. For the central probe in this window a new probe noise score is then calculated as the weighted average of the probe noise scores in this window:

$$sw_j = \frac{\sum_{p \in P} s_p \cdot w(p, j)}{\sum_{p \in P} w(p, j)}. \tag{10.12}$$

Here, P is the set of probes in the window, s_p is the noise score of the probe p, and $w(p, j)$ is the weight of probe p in the window for j. The weight is calculated as $1/d(p, j)$ if $p \neq j$ where $d(p, j)$ is the distance between probes p and j. In our study, we used the Euclidean distance on the probe coordinates but alternative distances can be used. If $p = j$, the weight is set to 2. If residuals from *RMA*-like methods are used as noise scores, s_p is set to the absolute value of the residuals. For probes close to the borders of the array, the window will be cut off at the respective sides. Subsequently, the *ε-criterion* is applied to the window-based noise scores.

10.3.5.5   Artifact Correction

For correction of corrupted probe values, we developed two alternative approaches. In the first case, the intensities of the corrupted probes are replaced by the mean intensity of the remaining probes of the corresponding probeset in the CEL file. This correction only takes into account probe values measured with the same array, thus, differences in intensity distributions between arrays do not have to be considered. If all probes of a probeset are corrupted it is not possible to infer a meaningful probeset intensity. Thus, we set all probe intensities to 0, resulting in a probeset intensity of 0. These probesets should be excluded from further analysis.

The alternative method consists of removing corrupted probes from the probeset definition by modifying the PGF annotation file provided by Affymetrix. However, as downstream analysis tools including the Affymetrix Power Tools may request the missing probes and cannot handle missing values appropriately and even the *de facto* standard of present and absent flags is often ignored, probe value correction is far more robust than removal. It should be noted that it is also possible to completely exclude affected probes from the summarization procedure using the "--kill-list" option of the Affymetrix Power Tools, but it is still experimental and does not always work.

10.4   PERFORMANCE OF ARTIFACT DETECTION AND CORRECTION METHODS

10.4.1   IDENTIFICATION OF ARTIFACTS USING PROBE NOISE PLOTS

In the examples presented in Section 10.2, the array images already provided a first indication of experimental artifacts. However, this was only due to the high intensity values of the affected probes and not all defects can be identified so easily. An alternative to simple intensity plots are the *probe noise plots* introduced in Section 10.3. These are used to visualize the level of noise for individual probes quantified in different ways, e.g., in terms of the deviation from a control or a linear model as used by *RMA*.

Probe noise plots for the exon array examples are shown in Figures 10.4 and 10.5 and compared to the original intensity plots. In this case, we used as controls for total RNA either the normalized sum of newly transcribed and pre-existing RNA of the corresponding sample or an artifact-free replicate. Comparison of the different probe noise measures evaluated show that all of them were capable of picking up the artifacts in the two exon array measurements shown. Here, the comparison to the artifact-free replicate gave the strongest signal, whereas the *RMA* residuals indicated only few probes with large levels of noise. This may also explain why *RMA* summarization was not sufficient to correct for these artifacts as measurement errors were likely underestimated for most probes. Remarkably, the *RMA* residual plots for the artifact-free replicate showed an additional stain in the center of the array, which was neither observed in the original image nor the other two types of probe noise plots (data shown in [17]). Most likely, the *RMA* model, which is based on several replicates (in this case three measurements of total RNA), was biased by the stains on the two corrupted arrays at this location, thus leading to large residuals for the artifact-free replicate. This may also explain the poor performance of artifact correction using summarization.

10.4.2 IDENTIFICATION AND CORRECTION OF CORRUPTED PROBES

As described in Section 10.3, we recently proposed a two-step approach to correct for artifacts by first identifying corrupted probes and then correcting for these corrupted probes by either replacing their values using the remaining unaffected probes for the given probeset or removing the probe from the analysis, for instance by re-defining probeset definitions. To identify the corrupted probes a simple threshold criterion is used based on probe noise scores calculated either from fold-changes to a control or *RMA*-derived residuals. Here, we developed two alternative approaches that calculate the probe noise score either for each probe alone (*ε-criterion*) or as a distance-weighted mean of the noise scores within a 2D-window around the probe (*window-criterion*) (see Section 10.3 for details).

To illustrate the performance of this simple approach, the two DGF75-10/12 measurements with artifacts were used. As we did not know which probes have been truly affected by the artifacts, evaluation was performed by comparing it against a control measurement. In Section 10.2, the deviation was shown to be considerable if no attempts at artifact correction apart from summarization were made. We compared this against the deviations observed after correcting for the artifacts with our approach. In order to avoid overfitting, we pursued the following procedure to correct the DGF75-10/12 measurements and to evaluate the performance of correction appropriately. If we compared the corrected and summarized probeset values between replicates (Figure 10.2 c,f), detection of corrupted probes was based on the ratio of total RNA and the normalized sum of newly transcribed and pre-existing RNA. If we used the ratio of total RNA and the normalized sum of newly transcribed and pre-existing RNA to evaluate correction, replicates were used to determine the corrupted probes. Results for the latter case are shown in the supplement to the original publication [17], but are very similar to the results for the first case shown here.

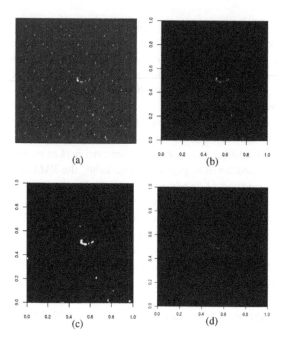

FIGURE 10.4 Probe noise scores for replicates 1 of total RNA in DG75-10/12 cells measured with exon arrays. (a) Original intensities, (b) noise scores based on the fold-changes of total RNA to the sum of newly transcribed and pre-existing RNA, (c) noise scores based on the fold-change between replicates (replicate 2 was used as control), and (d) *affyPLM* residuals. The shades of gray in Subfigures (b–d) indicate the fold-changes s_j to the control (b–c) or the *affyPLM* residuals (d), with black indicating low deviation and white high deviation.

In both cases, even the simple ε-*criterion* already showed a significant improvement after probe value correction. Both with and without quantile normalization, the distinctive deviation for large expression values seen before in replicate 1 was no longer observed. Instead, for both defective replicates 1 and 3, the variance was symmetrical on both sides of the diagonal. This was true both for the correction using mean values of unaffected probes of the same probeset as well as for the filtering approach in which the affected probes were removed from the probeset definition (not shown). Here, the mean absolute deviation from the diagonal decreased from 12.2 in the original data to 7.34 and 4.7 for the first and second correction methods, respectively. Thus, even a simple thresholding approach could successfully identify the defective probes and probeset values could be corrected appropriately, with slightly better results obtained by removing affected probes from the probeset definition instead of re-using values from unaffected probes for the affected ones.

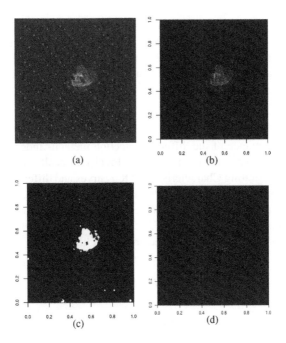

FIGURE 10.5   Probe noise scores for replicates 3 of total RNA in DG75-10/12 cells measured with exon arrays. The type of probe noise scores used for each subfigure are the same as in Figure 10.4 and explained there.

10.4.3   ACCURACY OF ARTIFACT IDENTIFICATION APPROACHES

To perform a systematic analysis of the performance in detecting measurement artifacts, we used Gene ST array measurements of the same samples that were measured with the exon arrays. The Gene ST measurements were free of artifacts and have been published recently [21]. Artificial stains were spiked into these artifact-free Gene ST measurements by projecting the artifact observed in total RNA of replicate 3 for the DG75-10/12 cells from the exon arrays to one sample of total RNA measured with gene arrays as described in the following. We used the pattern of the stain on a real-life example instead of random selection or some other spatial pattern to perform a realistic simulation of noise and fair comparison of the approaches.

10.4.3.1   Introduction of Artificial Noise on Gene ST Arrays

Real-life stains were projected onto the Gene ST arrays to create realistically shaped artifact patterns in the following way. First, our artifact detection approach described in Section 10.3.5.4 was applied to an exon array measurement with artifacts to detect the location of the corrupted probes. The exon array artifacts were then scaled down to the dimensions of the gene arrays and transferred to the artifact-free gene arrays. For this purpose, 2×2 rectangles of probes on the exon arrays were mapped to one probe on the gene arrays and the maximum value of any of the probes in this rectangle was used for the spiked probe. To account for the overall larger intensities on the gene arrays, the resulting value was multiplied by the ratio of the 75th percentile of the intensity distribution on the gene arrays relative to the corresponding 75th percentile for the exon arrays.

10.4.3.2   Performance Measures

To evaluate the performance of the different detection methods independent of any particular threshold applied, we used so-called Precision-recall curves. These are similar to Receiver Operating Characteristic (ROC) curves and differ only in the performance measures plotted against each other for decreasing values of the threshold. This requires calculating *true positives* (spiked probes that are filtered, TP), *false positives* (probes not spiked but filtered, FP), *true negatives* (probes neither spiked nor filtered, TN), and *false negatives* (spiked probes not filtered, FN) at each possible threshold. Precision-Recall curves were then created by plotting

$$precision = TP/(TP+FP) \tag{10.13}$$

on the *y*-axis against

$$recall = TP/(TP+FN) \tag{10.14}$$

on the *x*-axis for all possible thresholds.

10.4.3.3   Compared Methods

We compared the *ε-criterion* and *window-criterion* using probe noise scores based on

1. fold-changes between replicates, calculated from all three replicates of total RNA for the DG75-eGFP cells including the spiked replicate.
2. fold-changes between total RNA and normalized sum of newly transcribed and pre-existing RNA corresponding to the spiked replicate.
3. *RMA* residuals calculated based on all 18 replicates using the *affyPLM* library.

These approaches were compared against Harshlight [24] and *MBR* [25] applied to the six array measurements of total RNA. Harshlight does not provide noise scores per se but relies on downstream algorithms to decide on affected probes. To compute precision and recall values, probes were ranked by their fold-change to the corresponding median probe value across all arrays. This score is used by Harshlight in its initial step. For our purposes, it was additionally incremented by a constant value for probes flagged as defects by Harshlight such that all flagged probes ranked higher than any other probe. As *MBR* is only available as a GUI, we investigated only a small number of values for the parameter k (60-80 in increments of 5; values larger than 80 were found to have only very small recall).

Additionally, we planned to evaluate performance of *caCORRECT2* [27] as well as the method by Reimers and Weinstein [30], which are both available as Web

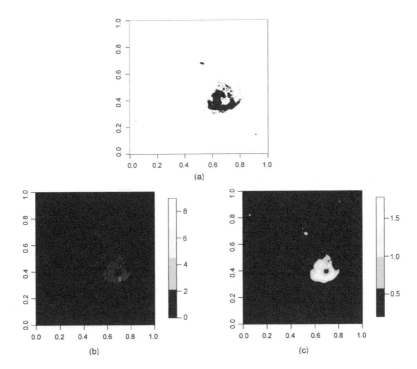

FIGURE 10.6 Illustration of the results on the spiked Gene ST arrays. Both shape of the artifact and intensities of the spiked probes were transferred from exon arrays containing artifacts. (a) shows the spiked probes in black and (b) and (c) the probe scores based on fold changes between replicates using only the probe information itself (b) or also its neighborhood (c). Here, the shade of gray indicates the value of the fold-changes s_j (b) or sw_j (c). For both (b) and (c) the overall shape of the spiked stain can easily be identified, but only when using the *window-criterion* (c) are all probes within this area identified. Furthermore, in (b) there are more probes with high noise scores that were not spiked (false positives).

servers. However, as none of the two programs had yielded a result 24 h after uploading the data to the Web servers, we aborted the evaluation. Thus, it appears that these methods did not scale well to the size of the Gene ST arrays used in this study, which are substantially larger than older Affymetrix arrays but still much smaller than the exon arrays. Alternatively, in particular for the Reimers and Weinstein method, the Web servers might no longer be maintained. Finally, the LPE method by Arteaga-Salas et al. [26] could not be evaluated as no software implementing this method was available.

10.4.3.4 Evaluation Results

Figure 10.6 illustrates the spiked artifact as well as the probe noise scores calculated using either only the probe information alone or also including the probe neighbor-

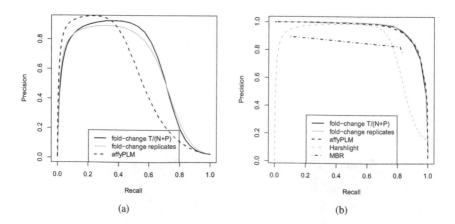

FIGURE 10.7 Precision-recall curves for spiked Gene ST measurements. Here, artifacts were projected from the exon array measurements onto the gene arrays to produce realistic noise patterns. Three different scoring approaches were compared for both the simple threshold approach, the *ε-criterion* (a), and its cumulative variant, the *window-criterion* (b), which takes into account the probe neighborhood information. The scoring approaches compared are: (i) absolute log fold-change between total RNA and normalized sum of newly transcribed and pre-existing RNA (*fold-change (T/N + P)*, see methods); (ii) absolute log fold-change between replicates (*fold-change replicates*); (iii) residuals determined with the *RMA* summarization approach using the *affyPLM* model (*affyPLM*). These results show that the window-based approach improved the performance of all used methods, resulting in almost identical performance for all of them, which was also superior to the performance of both Harshlight and *MBR*.

hood using the window-based approach. Here, the probe scores were calculated from fold-changes between replicates. The window approach resulted in a much smoother change of scores and high noise scores within the complete spiked area. If scores were calculated on each probe alone, we observed large variations in the spiked area with not all probes having high scores. Similar results were observed for the other types of probe noise scores, indicating that the window approach resulted in higher sensitivity in identifying defective probes.

Precision-recall curves comparing the different detection approaches are shown in Figure 10.7. Here, several interesting observations can be made. First, the noise scores based on fold-changes to either replicate or newly transcribed plus pre-existing RNA samples performed almost identically using the *ε-criterion*. In contrast, the scores based on the *RMA* residuals showed a higher precision for low recall values but this precision deteriorated more rapidly for increasing recall values.

Second, performance of all probe scores improved considerably when the *window-criterion* was applied. By taking the local information of a probe's neighborhood into account, recall could be increased significantly while the number of probes mistakenly flagged as corrupted was reduced. Furthermore, when using the *window-criterion* the differences between the scoring approaches disappeared and all scoring methods showed a very similar performance. Here, the reason for the poor performance of the ε-*criterion* at low recall were a few isolated probes with high noise scores on the arrays that were not spiked and, thus, were counted as false positives. While these outliers might also be interesting, they do not indicate a systematic artifact. Accordingly, smoothing over the scores in the neighborhood of these probes reduced their noise level, making it easier to find an appropriate threshold between spiked and unspiked probes independent of the scoring method used.

Finally, the performance of the different *window-criterion* variants was compared to Harshlight and *MBR*. While Harshlight performed similarly well for intermediate recall values, precision was very low when trying to reach full recall. At a recall of 85% of spiked probes, the fraction of probes flagged correctly was only less than 50%, whereas for the *window-criterion* more than 90% of the flagged probes had been spiked. Thus, it appears that Harshlight uses too strict requirements on probe quality and, accordingly, tends to flag too many probes as defective. Additionally, modern platforms like gene and exon arrays appear to cause problems to Harshlight due to either calibration or technical issues. Using default settings large diffuse defects were detected even for artifact-free arrays and spike-in probes used for calibration were detected as compact defects.

MBR also performed worse than all *window-criterion* variants at all recall values but outperformed Harshlight in a small range. It should be noted that the parameter k used by *MBR* allowed only very little tuning of performance. For $k = 80$ (the largest value investigated), recall was as low as 0.1, then increased dramatically to 0.81 for $k = 75$ and then only increased moderately up to 0.83 for the smallest allowed value of $k = 60$. At the same time, precision varied only between 0.90 for $k = 80$ and 0.77 for $k = 60$.

10.5 DISCUSSION

A recent study [10] showed that an alarming number of arrays deposited in the Gene Expression Omnibus (GEO) contain spatial defects. As deposition in the GEO generally occurs only prior to publication, this does not even include the array measurements that were discarded before analysis due to larger artifacts. If artifacts are not considered substantial enough to discard the array, normalization and summarization procedures are routinely relied upon to correct for these smaller defects.

Lately, we showed that this reliance is risky as even small artifacts can have serious impacts on the results of normalization and summarization procedures [17]. This was illustrated using a set of exon array measurements that contained several artifacts of various sizes in the form of stains in the center of the arrays. In this case, replacement of the corresponding arrays by measurements of the same samples was not an option as for some samples all RNA material had been used up. Thus, the only alternatives were to either repeat the whole experiment and all array measurements, which would be highly cost-intensive, or try to use at least the measurements of the probes not affected by the artifacts.

However, the comparison of replicates containing artifacts with corresponding artifact-free measurements clearly showed that even a robust summarization procedure such as *RMA* could not appropriately correct for the defective probes as considerable non-linear deviations were observed between replicates. Interestingly, the degree of the deviations between replicates did not necessarily depend on the number of probes affected in total on the array. One of the arrays showed only a minor stain, which would normally be considered acceptable. Yet, deviations compared to the control were substantial, in particular for high expression values. Although this effect could be traced partially to quantile normalization, it did not disappear completely when omitting normalization. Although we analyzed the effect of artifacts on the summarized probeset values only for *RMA* summarization, we expect similar problems for other summarization methods, as *RMA* is one of the most robust summarization methods available to date.

While simulations showed a clear correlation between the number of probes affected within a probeset and the deviation between replicates of the estimated probeset values, this was not quite reflected in the real-life measurements. Here, probesets were similarly affected no matter whether 1, 2 or 3 probes were corrupted. Only probesets for which all probes were defective were clear outliers. Furthermore, even probeset measurements with no corrupted probes were biased. To some degree this is due to the normalization step required before summarization, which is based on the global intensity distribution on the arrays. As even small artifacts on the arrays can have a substantial influence on the distribution of intensities, all normalized probe values are affected to some degree. Although it is not clear why this effect is also seen without normalization, it can clearly be attributed to the artifacts as it disappears after correction of corrupted probes.

As outlined in this chapter, several methods have previously been proposed for identifying such probe-level artifacts. However, they are usually highly complex using, e.g., image analysis as in the case of Harshlight or singular value decomposition in case of *caCORRECT2*. Furthermore, they were often developed for previous generations of Affymetrix arrays and do not scale well to the new Gene and Exon ST array designs. In contrast, the approach we proposed recently is much simpler and relies only on the availability of an artifact-free control measurement, which is usually available in the form of technical or biological replicates. Instead of replicates, a biological control can alternatively be used as in the case of simultaneous measurements of total, newly transcribed and pre-existing RNA.

We showed that even a simple comparison of fold-changes between corrupted and control measurements was sufficient to identify the corrupted probes on our exon array measurements and correct them by either replacing their values or removing them from the analysis. While noise scores calculated only from the probe values themselves were already successful for this purpose, an extension of this approach that also uses the neighborhood of the probe improved the performance even more. In this case, the probe noise score is calculated as a distance-weighted average of all noise scores in a window around the probe considered. In this way, the layout of the array and the spatial nature of most artifacts can be taken into account. This approach was also superior to *MBR* as well as Harshlight, which in general appeared to be too stringent on probe quality as it also flagged many probes on arrays without apparent defects.

It should be noted here that the results of this evaluation are independent of the particular summarization method used later to calculate probeset values, as methods were applied to the individual probe values before summarization. Furthermore, performance of the approaches was evaluated in terms of the number of spiked probes identified correctly and *not* the fold-changes of the summarized probeset values. However, probe values were normalized before applying artifact detection using quantile normalization. Thus, we cannot completely exclude the possibility that the relative performance of the methods might be different for a different normalization method. Nevertheless, as quantile normalization is one of the most commonly used normalization methods, this evaluation is likely representative for most standard applications.

In summary, these results illustrate the importance of properly addressing artifacts in microarray analysis. Furthermore, they show that quite simple methods already perform very well for this purpose and can outperform more complicated approaches. The main challenge is now not to develop even more sophisticated methods but to establish artifact detection and correction methods as an integral part of microarray expression analysis. Furthermore, as microarray analysis is more and more replaced by RNA sequencing (RNA-seq) due to rapidly decreasing costs of sequencing, it will be important to establish similar quality control procedures for RNA-seq. Although the artifacts will likely be different in this case, the importance of addressing such artifacts will remain the same.

REFERENCES

1. D. J. Lockhart, H. Dong, M. C. Byrne, et al., "Expression monitoring by hybridization to high-density oligonucleotide arrays," *Nature Biotechnology*, vol. 14, pp. 1675–1680, Dec 1996.
2. D. Shalon, S. J. Smith, and P. O. Brown, "A DNA microarray system for analyzing complex DNA samples using two-color fluorescent probe hybridization," *Genome Research*, vol. 6, pp. 639–645, Jul 1996.
3. P. Bertone, V. Stolc, T. E. Royce, et al., "Global identification of human transcribed sequences with genome tiling arrays," *Science*, vol. 306, pp. 2242–2246, Dec 2004.
4. P. J. Gardina, T. A. Clark, B. Shimada, et al., "Alternative splicing and differential gene expression in colon cancer detected by a whole genome exon array," *BMC Genomics*, vol. 7, p. 325, 2006.
5. Z. Wang, M. Gerstein, and M. Snyder, "RNA-Seq: a revolutionary tool for transcriptomics," *Nature Reviews Genetics*, vol. 10, pp. 57–63, Jan 2009.
6. Y. Fang, C. Shi, E. Manduchi, M. Civelek, and P. Davies, "MicroRNA-10a regulation of proinflammatory phenotype in athero-susceptible endothelium in vivo and in vitro," *Proceedings of the National Academy of Sciences*, vol. 107, no. 30, p. 13450, 2010.

7. P. de la Grange, L. Gratadou, M. Delord, M. Dutertre, and D. Auboeuf, "Splicing factor and exon profiling across human tissues," *Nucleic Acids Research*, vol. 38, pp. 2825–2838, May 2010.

8. N. Salomonis, C. R. Schlieve, L. Pereira, et al., "Alternative splicing regulates mouse embryonic stem cell pluripotency and differentiation," *Proceedings of the National Academy of Sciences, USA*, vol. 107, pp. 10514–10519, Jun 2010.

9. A. Lapuk, H. Marr, L. Jakkula, et al., "Exon-level microarray analyses identify alternative splicing programs in breast cancer," *Molecular Cancer Research*, vol. 8, pp. 961–974, Jul 2010.

10. W. B. Langdon, G. J. G. Upton, R. da Silva Camargo, and A. P. Harrison, "A survey of spatial defects in Homo Sapiens Affymetrix GeneChips," *IEEE/ACM Transactions on Computational Biology and Bioinformatics*, vol. 7, no. 4, pp. 647–653, 2010.

11. T. Barrett, D. B. Troup, S. E. Wilhite, et al., "NCBI GEO: archive for functional genomics data sets—10 years on," *Nucleic Acids Research*, vol. 39, pp. D1005–D1010, Jan 2011.

12. B. Bolstad, F. Collin, J. Brettschneider, et al., "Quality Assessment of Affymetrix GeneChip Data," in *Bioinformatics and Computational Biology Solutions Using R and Bioconductor* (M. Gail, K. Krickeberg, J. Samet, A. Tsiatis, W. Wong, R. Gentleman, V. J. Carey, W. Huber, R. A. Irizarry, and S. Dudoit, eds.), Statistics for Biology and Health, pp. 33–47, Springer New York, 2005.

13. C. L. Wilson and C. J. Miller, "Simpleaffy: a BioConductor package for Affymetrix Quality Control and data analysis," *Bioinformatics*, vol. 21, pp. 3683–3685, Sep 2005.

14. G. V. C. Freue, Z. Hollander, E. Shen, et al., "MDQC: a new quality assessment method for microarrays based on quality control reports," *Bioinformatics*, vol. 23, pp. 3162–3169, Dec 2007.

15. A. Kauffmann, R. Gentleman, and W. Huber, "arrayQualityMetrics–a bioconductor package for quality assessment of microarray data," *Bioinformatics*, vol. 25, pp. 415–416, Feb 2009.

16. R. A. Irizarry, B. M. Bolstad, F. Collin, L. M. Cope, B. Hobbs, and T. P. Speed, "Summaries of Affymetrix GeneChip probe level data," *Nucleic Acids Research*, vol. 31, p. e15, Feb 2003.

17. T. Petri, E. Berchtold, R. Zimmer, and C. C. Friedel, "Detection and correction of probe-level artefacts on microarrays," *BMC Bioinformatics*, vol. 13, p. 114, 2012.

18. S. E. Wang, F. Y. Wu, H. Chen, M. Shamay, Q. Zheng, and G. S. Hayward, "Early activation of the Kaposi's sarcoma-associated herpesvirus RTA, RAP, and MTA promoters by the tetradecanoyl phorbol acetate-induced AP1 pathway," *Journal of Virology*, vol. 78, pp. 4248–4267, Apr 2004.

19. M. Kenzelmann, S. Maertens, M. Hergenhahn, et al., "Microarray analysis of newly synthesized RNA in cells and animals," *Proceedings of the National Academy of Sciences, USA*, vol. 104, pp. 6164–6169, Apr 2007.

20. L. Dölken, Z. Ruzsics, B. Rädle, et al., "High-resolution gene expression profiling for simultaneous kinetic parameter analysis of RNA synthesis and decay," *RNA*, vol. 14, pp. 1959–1972, Sep 2008.

21. L. Dölken, G. Malterer, F. Erhard, et al., "Systematic analysis of viral and cellular microRNA targets in cells latently infected with human gamma-herpesviruses by RISC immunoprecipitation assay.," *Cell Host Microbe*, vol. 7, pp. 324–334, Apr 2010.

22. H. E. Lockstone, "Exon array data analysis using Affymetrix power tools and R statistical software," *Briefings in Bioinformatics*, vol. 12, pp. 634–644, Nov 2011.

23. J. M. Arteaga-Salas, H. Zuzan, W. B. Langdon, G. J. G. Upton, and A. P. Harrison, "An overview of image-processing methods for Affymetrix GeneChips," *Briefings in Bioinformatics*, vol. 9, pp. 25–33, Jan 2008.

24. M. Suárez-Fariñas, M. Pellegrino, K. M. Wittkowski, and M. O. Magnasco, "Harshlight: a "corrective make-up" program for microarray chips," *BMC Bioinformatics*, vol. 6, p. 294, 2005.

25. J. S. Song, K. Maghsoudi, W. Li, E. Fox, J. Quackenbush, and X. S. Liu, "Microarray blob-defect removal improves array analysis," *Bioinformatics*, vol. 23, pp. 966–971, Apr 2007.

26. J. M. Arteaga-Salas, A. P. Harrison, and G. J. G. Upton, "Reducing spatial flaws in oligonucleotide arrays by using neighborhood information," *Statistical Applications in Genetics and Molecular Biology*, vol. 7, no. 1, p. Article29, 2008.

27. R. A. Moffitt, Q. Yin-Goen, T. H. Stokes, et al., "caCORRECT2: Improving the accuracy and reliability of microarray data in the presence of artifacts," *BMC Bioinformatics*, vol. 12, p. 383, 2011.

28. M. Suárez-Fariñas, A. Haider, and K. M. Wittkowski, ""Harshlighting" small blemishes on microarrays," *BMC Bioinformatics*, vol. 6, p. 65, 2005.

29. T. H. Cormen, C. Stein, R. L. Rivest, and C. E. Leiserson, *Introduction to Algorithms*. New York: McGraw-Hill Higher Education, 2nd ed., 2001.

30. M. Reimers and J. N. Weinstein, "Quality assessment of microarrays: visualization of spatial artifacts and quantitation of regional biases," *BMC Bioinformatics*, vol. 6, p. 166, 2005.

31. G. J. G. Upton and J. C. Lloyd, "Oligonucleotide arrays: information from replication and spatial structure," *Bioinformatics*, vol. 21, pp. 4162–4168, Nov 2005.

11 Quality Control and Analysis Algorithms for Tissue Microarrays as Biomarker Validation Tools

Todd H. Stokes, Sonal Kothari, Chih-wen Cheng, and May D. Wang

CONTENTS

11.1 INTRODUCTION

11.1.1 ABSTRACT

Tissue Microarrays (TMAs) are a common tool for evaluating protein expression patterns *in situ*. Genomics researchers use them to verify that subtle changes in phenotype survive the translational and post-translational modification processes in the cell. Proper analysis of TMA experiments is critical to support explanations of complex protein interactions. Variations of TMAs include cores from tissue biopsies, heterogeneous protein lysates, whole cells, and homogeneous pure protein mixtures. Quality control and analysis algorithms for TMAs share much in common with the analysis of DNA microarray spots. Quality problems such as non-specific binding, auto-fluorescence, fluorescent photo bleaching, quenching, and missing samples due to laboratory processing are points of commonality that require different analytical approaches. This chapter covers algorithms most suited to address these variations, including spot-finding algorithms, artifact detection, multi-spectral analysis, controls to compare across experiments, and annotation and metadata sharing.

11.1.2 INTRODUCTION TO TISSUE MICROARRAYS

A typical tissue microarray is comprised of a microscope slide covered with 6 to 1200 small discs of tissue samples, usually laid out systematically by a robot. The discs are prepared in the same way that blocks of whole tissue are prepared for histological analysis and the laboratory processing of TMAs for staining should match previously developed protocols. TMAs have gained wide adoption because, though they contain far less tissue for a given sample, it is generally agreed that they contain enough tissue to evaluate protein expression and tissue morphology [2, 3].

The ability to perform one staining protocol and evaluate the results across more than one hundred tissue samples without the variability due to the laboratory environment makes protein expression analysis more statistically robust. This is commonly referred to as high throughput analysis, though in many cases, tissue microarray spots must be manually characterized by trained experts. The size of the discs (more commonly called spots) ranges from 0.5 to 2.5 mm in diameter. The number of cells in that area will vary based on the tissue type, disease state, and tissue heterogeneity, but the numbers range from approximately 1000-75000 cells. Thus, TMAs are a viable tool for studying sub-cellular characteristics of protein expression.

TMAs have helped scientists and clinicians that collect and bank tissue samples to share their valuable resources. Tissue samples can be subdivided to enable more studies, which should lead to more collaboration and access to scarce tissue. Private companies (e.g., U.S. Biomax, Rockville, MD, USA and Cybrdi, Rockville, MD, USA) sell TMAs for healthy and diseased human tissue (or a combination of both) at prices ranging from $50 to $1000, depending on the spot density and the amount of associated clinical data that is needed. They also provide incomplete arrays with missing or broken spots at roughly 1/3 of the price of complete arrays for testing new protocols.

The cross-disciplinary nature of tissue microarray analysis has led to multiple field-specific sets of terminology. In this chapter, we use terminology that will be most easily understood by experts in DNA microarray (DMA) analysis. Table 11.1 gives a cross-reference to help the reader in translating terminology that they will encounter in papers cited in this chapter. It is important to note that not all related terms are exact synonyms, and the choice of a term depends on the importance of defining the field-specific characteristics of the term.

11.1.2.1 TMA Analysis Software

The popularity of TMAs has led to the development of software tools specifically designed to handle imaging data captured from TMAs. This software contains features specific to TMA characteristics that are different from whole tissue section analysis, including: spot identification and segmentation, comparative analysis of signal between spots, multi-resolution analysis within one spot, and quality control to handle incomplete, missing, or folded spots.

TMA analysis software tools are becoming increasingly automated [4]. Of special note are the commercial systems Automated Cellular Imaging System (ACIS) by Clarient (now part of GE Healthcare, Aliso Viejo, CA, USA) [5] and Automated Quantitative Analysis (AQUA) by HistoRx (now marketed by Genoptix, a division of Novartis, Carlsbad, CA, USA) [6]. Another important TMA analysis system is the ScanScope (Aperio Technologies, Inc., Vista, CA, USA). This software is packaged only with Aperio scanning microscopes.

Current challenges for TMA analysis software include reproducible classification of staining patterns into sub-cellular compartments [7] and managing access to huge libraries of high-resolution images. This analysis requires multispectral fluorescence imaging to be fully accurate as the sub-cellular compartments of interest may never be reliably identifiable by traditional brightfield stains.

TABLE 11.1

Common terminology for tissue microarrays.

Term (as used in this paper)	Definition	Common synonymns and related terms found in literature
Spot	A circular slice taken from a core, usually 0.5–2 mm in diameter and 3–10 microns in thickness	
Stain	A laboratory reagent used to label a tissue component for imaging	Contrast agent, chromagen, fluorophore, counterstain, organic dye
Targeting molecule	A molecule with functions similar to antibodies, including: binding to specific proteins in a specific species, a common region for secondary labeling and/or binding sites for chemically conjugating stains	Antibody, DNA aptamer, primary antibody, antibody fragment
Target protein	The molecule in the tissue that will be characterized and compared across samples	Antigen, biomarker
Protein expression	The characterization of a protein in a tissue sample, including: sub-cellular location, quantity, percentage of cells containing the protein, and possible interactions with other proteins	Molecular profiling, antigenicity
Antigen retrieval	Raising the temperature and modifying the pH environment around a tissue microarray to break sulfatide bonds caused by fixing tissue and expose targeting antibody binding sites on the tissue surface	Unmasking, decloaking, heat-induced antigen retrieval

11.1.2.2 Less Common Arrays Related to TMAs

Many arrays are being developed that are similar to TMAs and that take advantage of TMA analysis pipelines. One example is an array of spots composed of cultured cells suspended in a hydrogel and then embedded in paraffin [8]. These can be used not only to measure protein expression in the cell line but also to verify the location of the proteins in either the nucleus, membrane, or cytoplasm. The advantage is that the problem of tissue heterogeneity can be excluded from the interpretation of the data. However, many experts agree that cell lines can have significantly different protein expression from native human tissue.

Another example of a useful array is that of spots created from cell lysates suspended in a hydrogel. While data about the location of proteins in the cells are lost, proteins that are difficult to target in their native environment may show a stronger signal after the cell membranes and nuclear envelopes are lysed. Cell lysates should produce a uniform signal across a spot, so they are useful for detecting protocol quality problems. If your staining protocol is robust and your target proteins are available, the total signal from cell lysates should correlate with the total signal across intact cells across various concentrations.

Finally, a third specialty array is that of pure proteins spotted on the slide. These are called protein arrays, and they might one day reach the density of DNA microarrays, but currently they are closer to the scale of TMAs due to the difficulty of synthesizing and handling proteins. Protein arrays are used to test the specificity of new targeting antibodies and to evaluate whether a protein under study interacts with a panel of other proteins.

11.1.2.3   Complementary and Substitute Technologies for TMAs

DNA microarrays are a complementary technology for TMAs. They generate the hypotheses about protein biomarkers that drive the need for validation. Researchers that specialize in DNA microarray analysis can improve the clinical impact of their papers by including verification at the protein level. Similarly, experts in protein expression (such as pathologists) tend to specialize in one particular tissue type, and they do not have the resources to try every possible antibody on their tissue in a high-throughput fashion. Thus, the hypotheses about new biomarkers that are generated by DNA expression experiments are useful to focus their experiments.

One exciting substitute technology for TMAs is tissue imaging mass spectrometry (TIMS) [9]. The TIMS technique is to perform a raster scan across a tissue sample, sending microscopic blasts of ionizing particles at a small region, and then performing a full mass spectrometry analysis pipeline on each region to look for known proteins. TIMS is exciting because it overcomes the need for antibody targeting and can be used to discover thousands of proteins in a sample with one pass. However, drawbacks to TIMS include low resolution of the scans, destruction of the sample, high background noise levels, the expense of maintaining the sensitive equipment, and the need to calibrate a series of complex detectors based on the size and type of proteins expected. In other words, while it is possible to detect thousands of proteins at once, that group may not contain all of the proteins needed for a particular disease. For now, TIMS is prohibitively expensive for a researcher that needs to quantify expression for a small number of proteins across a large number of samples. However, if the pixel resolution of the TIMS scan can approach that of a microscope at 10x magnification (approximately $0.4 \ \mu m^2$ features), it will become a very interesting technology.

11.1.3   COMPLEMENTARY AND SUBSTITUTE TECHNOLOGIES FOR TISSUE MICROARRAYS

The primary difference between TMAs and DNA microarrays is that the placement of the sample and the querying probes are reversed. For TMAs, the sample is fixed on the slide and the querying probes are suspended in solution above the slide during hybridization. For DNA microarrays, the sample is suspended in solution and the querying probes are fixed on the slide. The reason for this reversal is that querying

TABLE 11.2

Physical similarities and differences between tissue microarrays and DNA microarrays.

Category	DMA Property	TMA Property	Similarity/Difference
Physical Size	Early chips were 20 mm x 60 mm microscope slides, modern chips are 12.8 mm x 12.8 mm	20 mm x 60 mm microscope slides	Both are high-throughput technologies, but there is not much incentive to shrink TMAs further
Spot Size	11–240 μm, depending on manufacturer (8 x 8 – 64 x 64 pixels)	0.5–1.5 mm are most common (6400 x 6400 – 20000 x 20000 pixels at 40x)	TMAs require spatial detail to interpret results, so smaller spot sizes have negative tradeoffs
Scale/Data Size	Microarrays are only limited by size of the genome, and contain many "control spots" to improve quantification accuracy	Much less dense than DNA microarrays, but typically scanned at higher resolution, generating more data	Both are high-throughput technologies (as compared to previous options)

probes for proteins (usually antibodies) are expensive to produce and are not stable at a variety of temperatures or moisture levels. The experimental constraints imposed by antibodies might be overcome in the future if libraries of small DNA molecules (called aptamers) are developed that can specifically bind to proteins [10, 11]. However, when the sample has been lysed and is suspended in a solution, it is impossible to study the morphology of the cells; therefore, TMAs show their greatest potential for studies that depend on cell morphology. Table 11.2 compares the physical characteristics of DMAs and TMAs.

Another important difference between TMAs and DNA microarrays is that TMA spots represent a heterogeneous mixture of tissue-specific cells, interstitial space (e.g., fat or secretions), necrotic regions, immune cells, and red blood cells, while DNA microarray spots represent a homogeneous layer of DNA probes. This suggests that there is a limit on how dense TMAs can become. It is conceivable that devices could be developed to build TMAs composed of spots with diameters in the micron range, but the ratio of spots that did not contain staining patterns of interest would rise. This would be especially problematic for proteins of low abundance. Additionally, protein expression experts derive useful information from features of the tissue surrounding the staining pattern; therefore, they prefer spot sizes of at least 0.5 mm [12]. Table 11.3 compares analytical methods for DMAs and TMAs.

TABLE 11.3

Similarities and differences in analytical methods for tissue microarrays and DNA microarrays.

Category	DMA Property	TMA Property	Similarity/Difference
Previous Technology	Polymerase Chain Reaction (PCR) amplification on a 96-well plate	One tissue sample per slide, viewed by experts	High-throughput technologies provide for better internal controls and discovery-based science
Approximate Data Increase over Previous Technology	200x	100x	DMAs contain many redundant features, so the increase in features is not as dramatic as increase in measurements
Companion Technology	Automated scanners evolved simultaneously and are always matched to the chip platform for intra-platform standardization	Automated scanners are available, but are not as widespread due to pre-existing slide capture infrastructure (manual microscopes)	Use of automated scanners for TMAs should slowly catch up to that of DMAs, but may never account for 100% of applications
Spot Uniformity	Spots contain very small and very accurate concentrations of DNA probes (picomoles), normally of identical length	No two spots are alike on one slide or across the same catalog number	The pattern of labeling within a spot on a TMA is important to the researcher, but for a DMA it can be modeled and abstracted out in scanning hardware
Labeling and Detection	Spots are labeled by either a single fluorescent dye, which is bound to complementary DNA/RNA of the sample, or by two spectrally distinct dyes reprensenting two samples for comparison	Targeting molecules are typically 1–3 antibodies derived from the immune systems of research animals. Their specificity is tested using liquid chromotography	Targeting molecules for TMAs are much less specific, and often require an expensive optimization process
Analysis Pipeline	Spots are converted to features and then to gene expression without human intervention, accounting for spatial variability	Each spot may be analyzed on its own, often requiring human experts to convert stain strength to a qualitative expression value	TMA analysis pipelines are much less standardized than DMAs. Algorithms accounting for spatial variability of TMAs are not as mature

11.1.4 QUALITY ISSUES WITH TMAS

The laboratory processing protocols for TMAs have many diverse and complex components that have not been standardized as completely as the protocols for DNA microarrays. There is also more variability between protocols based on the type of tissue and the target protein of interest. Quality issues with DNA microarrays are now reported frequently and are often explained by lack of standard protocols [13]. However, DNA microarrays that are processed by robots (and thus should be highly reproducible) can contain quality problems that may be caused by manufacturing or sample contamination.

11.1.4.1 Sample Preparation

Some TMA quality problems derive from the same sources as DNA microarrays. However, as sample preparation tasks are divided among researchers and technicians separated geographically, they are more difficult to address. Sample preparation starts with the surgeon that either sacrifices and dissects an animal, or performs a biopsy on a human subject. A clinician consulting with the surgeon will label the sample with a disease state and other metadata to relate it back to a record of the treatment and outcomes. The "fresh" tissue is passed to a tissue processor, which may subdivide large samples into 1 cm x 1 cm cubes. The tissue is fixed by submersion in formaldehyde (or a similar fixative), dehydrated with alcohols, and then left in a hot bath of paraffin wax for preservation. All water must be removed from the sample to avoid natural decomposition. At the end, the paraffin is allowed to cool and the sample can be stored for long periods of time without serious decomposition.

The choice of reagents, timings, and temperatures can all affect whether unstable proteins are preserved and how close they will be to their native form. A protocol with the wrong parameters can lead to tissue that looks normal under the microscope, but contains damaged proteins that will not be recognized by antibodies or other targeting molecules. Less commonly, decomposition or contamination may occur if the tissue is frozen before being fixed, or if too much time passes after removal from its natural environment. Sample decomposition is usually uniform throughout the sample. It is easily recognized by a trained pathologist and will not be found in high quality commercial TMAs because the source tissue has been inspected before spotting.

11.1.4.2 Coring and Spotting

At the end of the paraffin embedding step, a source tissue block cannot be inspected without first slicing away the surface layer of paraffin with 5–10 μm slices on a microtome. This process inevitably leads to some destruction of the sample This process is called "facing the block." This allows for staining and quality inspection of one surface of the tissue, but nothing can be determined about the tissue under the surface. Regions of the tissue may be selected by experts as "representative" based on inspection of the surface, or the tissue block may be cored to produce a maximum number of cores. The coring process is similar to coring a tree or an ice sheet to produce an axial cylinder, but a simple wide-bore syringe is typically used for tissue. If the tissue is a perfect 1 cm^3 cube, it could theoretically produce 40,000 cores of 0.5 mm diameters. Due to tissue heterogeneity, the number of cores taken per sample is closer to 15–20. If one of those cores (of length 1 cm) was sliced efficiently into 5 um slices (an almost universally accepted thickness except for especially fragile tissue types), it would produce 200,000 spotted array members (i.e., one sample in up to 200,000 experiments). However, due to the general fragility of paraffin-embedded samples, these numbers are not practical with current spotting robots, and slicing is still performed manually. In most cases, 10–20% of the arrays produced from a block will be consumed by routine quality control processes, and the results of quality control may exclude a large percentage of intermediate slices. Figure 11.1 shows a TMA block of paraffin that is ready for slicing on the microtome.

The quality issues caused by spotting are primarily those of tissue heterogeneity. If a tissue is not naturally dense with cells of interest, larger spots will be needed to capture enough cells for analysis. Thicker slices are not a viable solution because targeting molecules do not penetrate through thick tissue. Biological replicates are a useful technique to compensate for situations where a core contains a section of all necrotic or interstitial space [12]. Verification that cells of interest exist in the spots on a commercial TMA is normally left to the researcher.

One obvious spotting problem is missing spots on the array. This might be caused by a gap in that section of the core or by that region of the core being extremely fragile and curling during slicing. In borderline cases, this spotting problem may be missed by the TMA manufacturer because the paraffin will show a visible spot outline that appears to be tissue. Because only tissue adheres to the microscope slide and the paraffin is dissolved during the staining protocol, the missing spot will become obvious during the deparaffinization step (see below). This is useful because if that spot is necessary for analysis, that slide can be replaced before treating the slide with expensive targeting molecules.

A related spotting problem is incomplete spots on the array. These spots naturally occur in many tissue types, or they may be the result of folding or tearing of the tissue during slicing or staining. Spotting errors can be detected after imaging by calculating the circularity of each spot after segmentation of the tissue from the background.

11.1.4.3 Mounting the TMA Slices

After slicing the TMA block, the thin ribbons of paraffin that include the tissue spots are mounted onto glass microscope slides that have a special coating for tissue adhesion. Some slides have a hydrophilic positive charged coating and some have a polylysine coating. The same slides are used for TMAs that are used for larger sections of tissue prepared for clinical histology analysis. However, very small and very dense TMAs may suffer from peeling or folding during the immunohistochemical staining protocol , especially during the heat-induced antigen retrieval step when the slide is typically heated to 90°C. This problem might be resolved with higher quality adhesion coatings on the microscope slides [14].

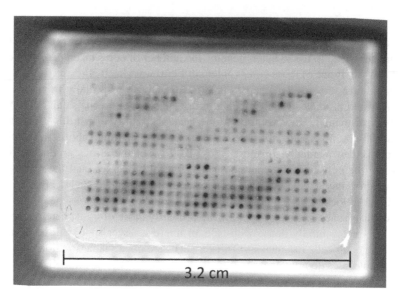

FIGURE 11.1 A photograph of a paraffin block composed of 378 TMA cores, each with a 0.6 mm diameter. This block is ready for slicing and mounting onto microscope slides. It contains 27 columns and 14 rows of tissue spots. This picture is from http://en.wikipedia.org/wiki/File:Rob7_melanoma.jpg and is reproduced with permission.

11.1.4.4 Immunohistochemical Staining Protocol

The immunohistochemical (IHC) staining protocol is named after the antibodies that bind to specific proteins native to a particular species. The antibodies are produced by injecting whole protein or large peptides into a laboratory animal (most commonly a mouse or a rabbit). The plasma of the animal is then run through a separation column to extract antibodies against the target protein. This process is very expensive and difficult to scale up for production efficiencies. Thus, these targeting antibodies are rarely directly conjugated with a specific stain because of the large variety of stains that might be chosen for a specific purpose. Thus, a secondary antibody is used to detect the targeting antibody. The secondary antibody is normally produced in a larger animal and will detect the part of the targeting antibody that is shared by all animals in its species. For example, a goat secondary antibody will bind to all mouse targeting antibodies, but will not bind to most human proteins. Figure 11.2 shows a typical IHC staining protocol.

The most common quality problem with IHC protocols is false negative results. It is difficult to interpret negative results because of the multitude of reasons why the signal might be impaired. First, the antigenicity may have been destroyed in the sample. This can happen during the staining protocol when using harsh antigen retrieval conditions (e.g., very high or low pH). Second, either the primary or secondary antibodies might have lost their binding power. Finally, the concentrations of the primary

FIGURE 11.2 Diagram of a typical IHC staining protocol on paraffin embedded tissue microarrays. Primary antibody incubation is at 4°C for 16 hours. Secondary antibody incubation is at 21°C for 2 hours. Washing can be 3-5 minutes per wash with 3-5 water changes. This protocol is derived from the one presented by Tran et al. [1].

and secondary antibodies might be too low or are mismatched.

The Canadian Immunohistochemistry Quality Control consortium reported on best practices and common errors that it found in the process of performing proficiency testing with 62 IHC laboratories in Canada [15]. The most common problems they found were false negative results. Their recommendations include proper selection of positive controls, which require the use of specialized tissue microarrays. This is useful when using well-understood and well-validated target proteins to make clinical decisions.

For exploratory research on new target proteins, multiplexing can be used to simultaneously query well-characterized proteins in the same experiment and rule out some factors that produce false negatives. However, many samples and protocol variations will need to be tested before a new target protein can be considered reliable. For example, antigen retrieval parameters, such as the temperature and pH, are considered to be specific to each target protein [16] and claims that a universal buffer can efficiently retrieve most protein targets should be viewed skeptically [17].

11.1.4.5 Age of Samples

Sample age is an issue because target proteins may have limited durability even after paraffin embedding. Stable target proteins may prove to be extremely durable when stored in block form. Camp et al. showed in 2000 that breast cancer tissue stored in their archive since 1942 had similar positive identification rates as tissue from all subsequent decades for three common breast cancer biomarkers [12]. However, not all proteins with clinical predictive value will be so durable. Many of today's most established biomarkers represent the "low-hanging fruit," i.e., the most durable and most easily reproducible target proteins.

There is a significant difference in the aging effects between storing tissue in its original paraffin block form and as a slice from a TMA block mounted on a microscope slide. This difference may be due to the increased exposed surface area for oxidation or decomposition. Fergenbaum et al. found lower stain intensity and higher false negative rates in the same target proteins after only 6 months of storing TMAs at room temperature [18]. Follow-up studies compared different techniques for storing tissue microarrays to best preserve unstable target proteins, including: paraffin coating, nitrogen dessication, or both [19]. The observed 10–35% reductions in protein expression even only a few months after slicing and mounting indicate that when purchasing TMAs from commercial vendors, it is a good idea to request a guarantee of their "freshness."

11.1.5 BIOMARKER VALIDATION TECHNIQUES

The primary research application of TMAs today is to validate biomarker findings from gene expression analysis [20]. Protein expression and gene expression do not always agree because of epigenetic effects, complex regulatory mechanisms that prevent translation of the mRNA transcripts. It is commonly believed that while clinical diagnostic tests based on gene expression will prove useful for broad-based screening of disease risk, protein expression tests are a more direct indicator toward a treatment plan (e.g., using a drug that also targets the protein) [21]. Thus, TMAs provide a faster way to develop new diagnostic tests for clinical use [22]. TMAs speed this process in two key areas: targeting molecule optimization and multiplexing.

11.1.6 TARGETING MOLECULE OPTIMIZATION

TMAs provide a means to quickly evaluate the relative merits of a pool of targeting molecules [23]. These molecules may bind to the same epitope of the target protein, but display different binding affinities. Molecules with lower binding affinities will require higher concentrations to acheive strong signal intensity. Higher concentrations increase the cost of the test, as well as the risk of negative interactions between targeting molecules and non-specific binding. We refer the reader to the collaborative Web portal, Antibodypedia, to explore this issue further [24, 25, 26].

A different way to apply TMAs to optimization is to evaluate whether molecules that target different proteins have similar diagnostic power [27]. The goal of this process is to find the minimal number of proteins that must be characterized to definitively classify a sample into a treatment group. As the number of target proteins in a staining protocol goes up, the chance of developing a reliable diagnostic panel for clinical use goes down. This is because of the difficulty of developing a tissue preparation protocol that is compatible with a wide variety of target proteins. This issue is discussed further in the next section on multiplexing.

11.1.6.1 Multiplexing

TMAs achieve their full potential when tissues can be probed for many target proteins simultaneously, called multiplexing. Multiplexing improves quality control methods by raising confidence in the identification of false negatives. It can also improve the sensitivity of a diagnostic test by introducing new dimensions that further separate difficult cases. Multiplexing with traditional brightfield IHC is possible by sending separate samples from a single tissue through a different staining protocol for each protein target, then collecting the protein expression values at the end.

A second generation of multiplexing uses fluorescent stains that can label one piece of tissue in multiple ways that are then separated at the end by filters on the microscope (similar to two-channel microarrays). This second generation has a maximum panel size of 3 or 4 protein targets because of the large degree of overlap between fluorscent stains in the light spectrum.

A third generation of multiplexing uses quantum dot nanoparticles, which can be designed for very precise emission characteristics. This IHC staining technology may push the maximum panel size to 10 or more protein targets. However, as the number of fluorescent stains on a single sample grows, the problem of quality control arises once again as filters may not be able to perfectly separate the signals. A process called spectral unmixing is used to assign the correct signals to the correct fluorophores, but this process may also introduce error.

11.2   ANALYSIS ALGORITHMS FOR IMAGE QUALITY IMPROVEMENT

For an accurate diagnosis using a TMA, it is essential to maintain standards and high image quality. However, TMA images acquired using high-throughput imaging systems often suffer from quality issues. The most reported quality issues in literature are as follows: (1) insufficient tissue, (2) tissue folds, (3) insufficient stain intensity, and (4) blurred regions. Among these issues, researchers have suggested methods for correcting stain intensity, methods for eliminating spots with insufficient tissue, and methods for eliminating regions of a spot with tissue folds or blurriness.

11.2.1   SPOT DETECTION AND SEGMENTATION

Many spot detection and indexing algorithms have been proposed for low-density DNA microarrays [28, 29]. Most assume that the image is a fluorescent signal with a black background. These algorithms could be easily converted to process brightfield images with minor modifications to the color segmentation process or by inverting the TMA image colors before processing. More details on spot detection and segmentation algorithms can be found in Chapters 3–6 of this text.

We have investigated a spotting quality score that incorporates the total spot area and the circularity. Circularity can be found by segmenting the tissue from the background pixels, formulating a mathematical description of the perimeter, and calculating the circularity [30]. There are many options for how to accomplish these steps, and the correct parameters to use are largely dependent on image resolution (i.e., scale). The results of this investigation can be found on the Web site http://tissuewiki.bme.gatech.edu.

In general, segmentation of brightfield TMA images is more difficult than fluorescent TMAs or DNA microarrays because the tissue color can be very faint and difficult to distinguish from dust on the slide. We use a freely available algorithm called statistical region merging (SRM) for segmentation [31, 32]. SRM provides a single parameter, Q, that governs the coarseness of the segmentation. Using an algorithm that gives a coarse boundary is useful because with simple circularity calculations, the perimeter can be exaggerated by many small variations in the border. Similar to comparing the border of a country on a globe versus a wall-sized map, the more detailed borders produce very low circularity scores. This problem can be overcome by down-sampling the images before calculating circularity, by fitting a circle or ellipse to the pixel boundary, or using dilation and erosion morphological operations on a binary image to remove rough boundaries.

11.2.2 TISSUE FOLD ELIMINATION

A tissue fold is an image artifact caused by folding a thin slice of tissue when placing it on a glass slide. Tissue folds may also arise from heating of the slide during the deparaffinization and antigen retrieval steps in an IHC protocol. Tissue folds alter morphological properties, including stain intensity, nuclear shape, and texture in the local area of the fold. Therefore, it is essential to detect and eliminate tissue folds before calculating stain intensity for a TMA spot. Tissue folds are common in whole-slide images, and researchers avoid tissue folds by manually selecting images or regions-of-interest (ROIs) [33, 34]. Although manual selection ensures the quality of selected tissue regions, it limits the speed and objectivity of computer-aided analysis by introducing a user-interactive step and by adding user subjectivity. Moreover, manual selection is a tedious process for large TMA datasets.

Because tissue fold regions have multiple layers of stained tissue, these regions have high saturation and low intensity. Palokangas et al. proposed an unsupervised method for tissue-fold detection using k-means clustering [35]. This method is composed of three steps: preprocessing, segmentation, and the discarding of extra objects. In the first step, smoothed and contrast-enhanced saturation and intensity images are subtracted to calculate a difference image. In the second step, the pixels of the difference image are clustered using k-means. Then, the cluster containing pixels with center at the maximum difference value is assigned to tissue folds. Finally, small extra objects are discarded using an area threshold. For k-means clustering, the number of clusters, k, can be optimized based on the change in the average sum of the difference (variance) over all clusters. This method detects most of the prominent folds if a variety of folds are present on the slide. However, the method fails if no folds are present (i.e., the method assumes that folds are present, resulting in false positives).

Bautista and Yagi proposed thresholding a difference image with a fixed threshold [36, 37]. Unlike unsupervised clustering, this method does not fail for images without folds. However, a fixed threshold is not effective for all images, especially if there are data batch effects (e.g., images acquired with different microscopes). We found tissue folds in three TMA spots by applying clustering and a fixed threshold (threshold $= -0.3$) methods on the difference value images. Figure 11.3 illustrates the results.

Original Image   Clustering Method Fold   Threshold Method Fold
                             Detection   Detection

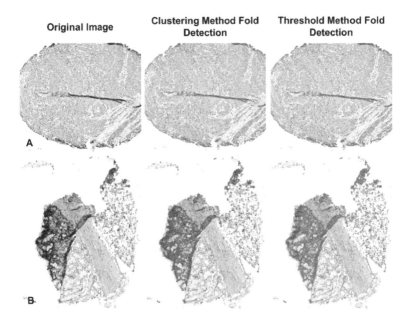

FIGURE 11.3   (SEE COLOR INSERT.) Tissue fold detection in TMA spots. (A-B) Original TMA spots, (center) tissue folds detecting using clustering method, and (right) tissue folds detecting using the threshold method. The threshold method removes slightly more tissue regions than the clustering method.

11.2.3 STAIN INTENSITY NORMALIZATION

Non-biological factors may lead to variation in stain color and intensity of digital TMA images. These include variations in slide preparation, age of the sample, the microscope and camera model, and post-processing. When these variations result in detectable patterns that can easily distinguish one dataset from another, that is called batch effect . Batch effects often overwhelm machine learning algorithms and could lead to misdiagnosis if left uncorrected. Batch effects reveal themselves most often in collaborative research projects because of variations in laboratory equipment and protocols. Researchers have suggested two possible solutions for addressing color batch effects: (1) color normalization to a standard reference, or (2) color conversion to a batch-effect invariant color space.

Color normalization is a process that shifts the distribution of pixel colors in an image to match the colors in a standard reference. Ideally, similar tissue objects such as nuclei in the test and reference images will have the same color after normalization. However, this goal introduces a circular dependency on the non-trivial task of tissue object segmentation, which also depends on the color properties. Researchers have proposed pixel-level color normalization methods, where pixel colors are normalized irrespective of the objects using quantile normalization [38]. This can distort the colors because each image contains a different number of objects. For example,

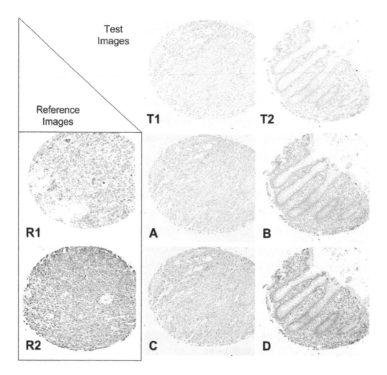

FIGURE 11.4   (SEE COLOR INSERT.) Color map normalization of TMA spots. Test TMA spot images (T1 and T2) are normalized to two reference images (R1 and R2) using colormap pixel-level normalization to produce images A–D.

if a reference image contains cells with enlarged nuclei and the test image does not, then normalization may cause cytoplasmic pixels in the test image to have nuclear colors. Kothari et al. illustrate that considering unique colors rather than all pixels during normalization can remove bias due to object frequencies [38]. Figure 11.4 shows two TMA spot images normalized to two reference images using colormap normalization.

Researchers have also proposed normalization of individual stains rather than RGB colors [39]. Magee et al. proposed following four-step normalization strategy: (1) Separate stains using color de-convolution, (2) cluster pixels in each stains image as foreground and background using k-means or Bayesian methods, (3) normalize colors for both clusters for every stain using Reinhard method, and (4) combine stains to give normalized image [40].

Because the primary objective of TMA analysis is to quantify stain, researchers may consider using the Luv color space [41, 39] or the Lab color space [42, 43], which are robust to color batch effects.

11.2.4 BLURRED REGION DETECTION

During high-throughput scanning of the slides, the focus of the microscope may be set manually at certain points on the slide, e.g., the four corners of the slide or the center. However, if the platform is slightly tilted or the tissue on the slide is uneven, the microscope may lose focus inside TMA spots. Most modern high throughput microscopes apply corrections at the hardware and software levels to prevent blurred regions, but some spots may still be blurred in older TMA datasets. Researchers can eliminate blurred spots using methods similar to autofocus detection methods based on image sharpness or texture properties [44].

11.3 ANALYSIS ALGORITHMS FOR QUANTIFICATION

11.3.1 BRIGHTFIELD STAINS

An unstained slice of formalin fixed paraffin embedded (FFPE) tissue will not contain much visible information under a normal light microscope. The only cells that are clearly visible without staining are red blood cells, which will appear bright red to copper in color. Other observable features are calcium or other mineral deposits in certain tissue, but these provide minimal information compared to what is available through staining.

Brightfield stains work by blocking certain wavelengths of white light from a broad-spectrum light source opposite the objective of the microscope. The remaining light is detected by the camera as a distinct color. Mixing of brightfield stains tends to lead to darker colors, converging on black, which is very difficult to separate into component parts. The mixing type is called subtractive mixing. This effect means there is an upper limit on the number of stains that can be combined for brightfield imaging. That limit is in the range of 5–7, depending on how much the stains are expected to overlap. A well-known multiplexed brightfield stain is the Papanicolaou stain for detecting malignant cancer cells in smear preparations of bodily fluids. This stain is composed of five dyes and produces images of cells stained the entire spectrum of visible color (due to mixing of the dyes). It was formulated to be interpreted directly by human observers looking through the microscope, but automated computer analysis algorithms of pap smear images have been developed [45].

11.3.1.1 H&E Stain

Hemotoxylin and eosin (H&E) are the two most common stains in histology or pathology labs. Hemotoxylin stains nuclei blue and eosin stains cytoplasm pink. H&E provides nearly all of the information needed for morphological analysis of the tissue, including cell counts, nuclear sizes and shapes, cell size, cell density, and nuclear texture. However, H&E provides no information about protein expression, except by inference.

11.3.1.2 DAB Stain

Diaminobenzidine (DAB) is a common stain used in combination with H&E to target specific proteins. It is very inexpensive and can be easily bound to primary antibodies. The hydrogen peroxidase enzyme reacts with this dye to produce a brown color. As long as this stain is not too dark, it will not be confused with the dark purple produced when H&E stained regions overlap [46]. The TMAs provided for public analysis by the Human Protein Atlas are all stained with hemotoxylin and DAB.

11.3.2 FLUORESCENT STAINS

Fluorescent stains work by absorbing photons of one wavelength and emitting that energy as photons of a different wavelength. White light is passed through a filter to isolate the excitation wavelengths and then through the objective of the microscope to focus on one point in the sample. Because the emission light returning from the sample will be mixed with the reflection of the excitation light, bandwidth filters are used to isolate the emission signal. When multiple fluorescent stains are present in the sample, the increase in photons of various types increases the overall signal detected. This mixing type is called additive mixing. In practice, fluorescent molecules are matched to their filters so precisely that they can be isolated from all other stains for imaging. However, the greatest problem with signal detection is that many proteins have a natural fluorescence behavior, called auto-fluorescence . This signal is mostly in the green region of the spectrum but also has a long tail that spans the entire spectrum. Removing auto-fluorescent signal from tissue images is a major open problem for protein expression quantification.

11.3.2.1 DAPI

DAPI is the short form identifier for 4,6-diamidino-2-phenylindole. This molecule binds very well to adenosine and thymine DNA base pairs. It is primarily used for identification and characterization of nuclei in applications where hemotoxylin is not suitable. We have observed that hemotoxylin interferes with the emission of DAPI signal. This interference is not dependent on in which order the DAPI and hemotoxylin treatments are applied, so competition for binding sites is not the main cause. Thus, hemotoxylin might interefere with the emission of other fluorescent dyes. Studies to determine whether DAPI gives the same morphological information about nuclei as hemotoxylin were not found in the literature.

11.3.2.2 Organic Dyes

Fluorescent organic dyes have been conjugated with antibodies since 1942 [47]. Common fluorsecent dyes include cyanine, fluorescein, rhodamine, Alexa Fluors and Dylight fluors. Fluorescent dyes suffer from two phenomena that make quantification of staining difficult: photobleaching and quenching. Quenching results when photons emitted from one dye are absorbed by another dye before they leave the sample for detection by the microscope. It is one of the factors that limits the multiplexing capacity of organic dye stains. Photobleaching results from the limited capacity for excitation of fluorescent biomolecules. Excitation energy inevitably leads to the destruction of these fragile molecules and reduction in signal. This means that, over time, the signal will degrade and confound quantitative analysis.

The impact of these problems may not be significant enough to impact experimental design if the goal of the experiment is only to evaluate presence or absence of a signal. However, if the goal of the experiment is to determine relative quantities of two source signals in a single image or to evaluate the relationship between a source signal and a standard that is separate from the image, these problems can lead to faulty results.

11.3.2.3   Quantum Dots

Quantum dots (QDs) present many favorable characteristics when compared to other contrast agents for multi-spectral microscopy. First, quantum dots have narrow emission bandwidth, so it is possible to select filters that each capture signal from only one QD even when many QDs are present. Second, quantum dots are resistant to photo-bleaching. This means slides can be excited for long periods during multi-channel or time-lapse experiments without significant signal reduction. This feature also allows experimenters to use the photo-bleaching of auto-fluorescence signal to further separate dim signals from background noise.

11.3.3   PHASE CONTRAST MICROSCOPY

Phase contrast microscopy (PCM) is part of an exciting trend toward analyzing tissue and cell samples without the need for expensive or unreliable staining protocols. PCM makes use of subtle changes in the polarity of light striking structures in the sample and being scattered. A series of filters isolate scattered light from direct light and further separate the light based on polarity. The result is a grayscale image that shows cellular structures that might be useful in further classifying the location of protein expression. Additionally, PCM might be useful in experiments where other common dyes for identifying structures interact with the dyes that generate the primary signal. Many fluorescent microscopes already contain the necessary insertion points to add lenses and filters to enable PCM. Thus, PCM is a relatively inexpensive option for "stain-free" analysis.

11.3.4   MULTI-SPECTRAL UNMIXING

Multi-spectral imaging is a technique with applications spanning astronomy [48], geology [49], medicine [50], and forensic research [51]. Though traditional three-color (RGB) cameras are already multi-spectral, the term is frequently used to indicate imaging with four or more channels often including light beyond the visible spectrum (e.g., ultra-violet or infra-red light). Less commonly, the term may refer to non-light-based spectroscopy, such as mass spectroscopy [52]. The key motivations for introducing multi-spectral imaging are: 1) to better separate the desired signals from background noise, 2) to discover new elemental signal sources that were previously overlooked, and 3) to expand the number of signals that can be compared in a single experiment for better quantification and classification accuracy.

Multi-spectral imaging increases the complexity of an experiment in many ways. First, the volume of data increases and often the time to acquire also increases. This introduces problems for experimental design, especially in the case of time-lapse experiments. Second, new standards must be adopted to specify the source of each channel in the data for sharing experimental results. Specifically, a data analysis pipeline designed for between one and three channels can make use of common data storage and visualization formats (e.g., JPG and PNG) that do not allow for four or more channels. This is especially important because there is such a diversity of multi-spectral acquisition instruments available. Each instrument will have different total quantum efficiencies (the percent of photons detected at each wavelength), which can be calculated by calibration using a very well-characterized sample (e.g., using a spectrometer). Third, the spatial alignment of objects in the image may be altered due to filter design or chromatic aberration in the camera lens or microscope objective. In most cases, the alterations will not be simple linear translations, but instead will be non-linear, effecting pixels near the borders in different ways from pixels in the center.

However, the complexities of maintaining accurate focus and uniform light intensity at high magnification and across large regions may be increased in a multi-spectral experiment. The best case scenario is when all of the desired signals are present in a single "common channel" that can be used as a single reference for focusing and registering. However, when signals vary throughout the tissue depth and are spectrally distinct, it may be necessary to acquire multiple reference channels in the acquisition process. Ultimately, the task of obtaining focused signals across many wavelengths may require acquiring each channel at a different focal length. This process requires a high-precision microscope stage controller.

Processing of multi-spectral images varies in complexity based on the assumptions that can be made about signal sources and observations. If source signal wavelengths are narrow-band and perfectly matched to bandpass filters, then each channel can be directly used for analysis after subtraction of background noise. In this case, it is possible to have a one-to-one correspondence between source signals and captured channels. However, it is much more common for source signals to appear at various levels in multiple channels, resulting in a multi-point spectral profile for each signal source that overlaps with other signal sources. In this case, an unmixing (or source separation) algorithm must be employed to obtain accurate signal for analysis. This process requires that the system not be under-specified (i.e., one additional channel is acquired beyond the number of source signals) and that each channel acquired provides useful information about at least one source signal.

Three important assumptions for straightforward unmixing are additive signal mixing, known profiles of pure signal sources, and spatial uniformity of signal source profiles. When these assumptions are valid, accurate separation of the source signals

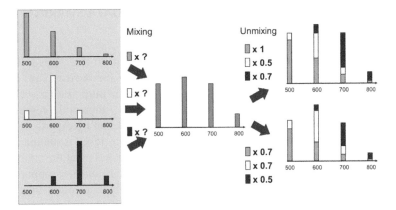

FIGURE 11.5 (SEE COLOR INSERT.) Simple Example of Unmixing Uncertainty. Unmixing determines how to allocate a mixture of given source signals to arrive at the closest estimation of the observed signal. There are many possible solutions to any unmixing problem, but the solutions will vary in how much they differ from the observed signal. In this example, the top solution has less error than the bottom solution. However, the solution with lowest error may not be the most accurate if: 1) there is uncertainty in the exact shape of the source signals, 2) the observed signal contains saturated channels, or 3) the solution violates constraints such as signal positivity. Note that in this example, the three source signals are measured using four channels, indicating that the system is not under-specified.

is a simple linear least-squares regression procedure. In addition to these assumptions, it is important that the detector is never saturated during acquisition of any channel, as this will produce artificially low peaks that do not accurately represent the mixture. This consideration should be easy to accommodate with modern high-precision cameras. Figure illustrates a simple additive mixing model for source separation.

If the assumption of spatial uniformity of signal sources is valid, multi-spectral images can be processed on a pixel-by-pixel basis, greatly reducing the complexity of the unmixing task. This assumption is more difficult to validate, especially when one of the source signals is composed of unknown background noise. In Figure 11.5, the "gray" source signal (top of Figure 11.5) has a distribution similar to that of tissue auto-fluorescence. Auto-fluorescence is a signal that may include more than 10 commonly occurring biological molecules (e.g., collagen, NADPH, and various flavins). Mammalian tissue is heterogeneously composed of these molecules, so the auto-fluorescence signal will change from one region of a sample to another. The impact of this variation will depend on how much overlap the auto-fluorescence spectral profile has with your other signals. Minimizing this interference involves selecting sources that emit light in wavelengths far from the green auto-fluorescence source.

11.4 BEST PRACTICES FOR ANNOTATION AND DATA SHARING

11.4.1 DATA SHARING

The largest public collection of TMA images is the Human Protein Atlas (HPA), hosted by Atlas Antibodies in Sweden [53, 54] at http://proteinatlas.org. HPA has developed a series of high-throughput assays for the purpose of assessing the quality of antibodies, including: immunohistochemistry, immunofluorescence, Western blot, and protein arrays. The raw data and interpretation by experts can be searched and downloaded in a number of ways [55]. Each assay result is summarized by a three-degree scale of whether the data support the hypothesis that the antibody is specific to its target. The three levels are supportive, uncertain, and non-supportive. Reliability of an antibody is assessed on a four-degree scale: high, medium, low, and very low. This score is based on how similar the staining patterns are to expectations and to other antibodies that target those same proteins. It is possible to evaluate reliability for such a large range of antibodies because of TMAs [56]. IHC experiments in HPA include two biological replicates and simultaneously test the antibody against 66 human tissues and 47 human cell lines. If the staining pattern of two antibodies is similar across this diversity of samples, it is unlikely that the similarity is due to chance.

HPA is a data resource of impressive scale. A new version of the database is released roughly every year, each time with a significant increase in antibodies. In our random sampling of antibodies from version 6 (released in 2010), we estimate that each antibody has 560 accompanying TMA images and 40 cell line images. The most recent release of HPA at the time of writing is version 11 and includes more than 18,000 antibodies. That means that more than 10 million TMA images of human tissue are available for analysis. While the images are not high magnification, they are detailed enough to analyze staining patterns. Dammeyer and Arner give an example of how subsets of this data can be created to ask interesting questions about related proteins [57].

There are other efforts to create distributed repositories that leverage TMA data collected by smaller laboratories [58]. The success of these efforts depends on the ability for smaller laboratories to agree on a common data format and an ontology for annotating images [59].

Another important effort in medical image data sharing is the open microscopy environment (OME) data model [60]. The OME data model was defined to build an open source image management system for research labs, called OME Remote Objects (OMERO). The number of features and image formats supported by the free OMERO system has driven widespread adoption of the system.

Effective and accurate image retrieval and analysis via querying annotations is important in voluminous image databases. TAMEE, a Web-based system for TMA samples, provides a data query system based on image metadata [61]. TMABoost [62] and several virtual microscope systems [63] also use metadata-based retrieval. Marchevsky et al. used metadata including patient demographics and SNOMED (Systematized Nomenclature of Medicine) terms for image retrieval [64]. The primary difference between these databases is the software platform and their data access methods. Among others, prominent data access methods include MySQL code, extensible markup language (XML) code, and user-interactive interfaces. TmaDB is a MySQL database that stores TMA images and related metadata about samples. The user can query data from the database using specific SQL identifiers or CDEs (common data elements). Identifiers here are similar to metadata terms in DICOM images [65]. The BrowseTMA system uses XML code for generating a block list of samples [66].

11.4.2 ANNOTATION OF TISSUE REGIONS

Image annotation is a process of associating context of an image or images. In the medical community, such as in radiology and pathology, image annotations can provide clinical explanatory and descriptive information that can be considered in clinical decision-making. Traditionally, medical images are annotated by human observation and captured as free text, posing challenges of direct translation between annotations and non-imaging biomedical data.

Medical image annotations can be classified into two levels: image- and pixel-level. Image-level is generally the metadata of the image, including where, when, and how the image was acquired. Pixel-level annotations contain information about a spatial location of a single point or a region of interest (ROI). Pixel-level annotations may be associated with clinical markers (e.g., tumors, bone fractures, blood clots) presented within the image. Annotations can be created observationally or computationally. Medical professionals create observational annotations by attaching text and/or depicting markups for textual information on the ROIs. Machines can create computational annotations by processing observational annotations. For example, a computer-aided system can generate a computational annotation to measure the size change of two manually drawn observational tumor ROIs.

Traditional annotation processes make it difficult for humans and computers to index and query based on free text descriptors or direct hand-drawn markups. To improve the accessibility of medical images, new technologies are required to capture annotations in a structured manner and create semantic content that users can easily search. For instance, a pathologist can query all images annotated with at least one abnormal nuclei from all patients diagnosed with kidney cancer. Therefore, following semantic annotation standards can increase reusability of the images.

Considering this, Digital Imaging and Communications in Medicine (DICOM) and Health Level Seven (HL7) are the most commonly used standards for storage and exchange of medical images [67, 68]. The DICOM Working Group Number 26, titled Digital Pathology, has proposed some extensions to DICOM (DICOM supplements 122 and 145) for pathological images [69, 70]. A group of researchers in Germany have proposed a similar DICOM extension for virtual pathology using whole slide images [63]. The international integrating health enterprise (IHE) initiative provides a template, Anatomic Pathology Structures Report (APSR), based on HL7 standards to implement synoptic reports [71]. The Tissue Microarray Data Exchange Specification) (TMA-DES) is a well-formed XML document with four required sections:

Header, Block (details about paraffin embedding), Slide (describing slide development from core), and Core (information about individual samples) [72]. XML tags consist of eight CDEs (common data elements), conforming to the ISO-11179 specifications. However, DICOM and HL7 include only image-level metadata, making them difficult for acquiring and storing pixel-level annotations.

Several standards have been developed to semantically acquire both image-level and pixel-level annotations and make them easier to share, store, and analyze. The Human Physiome Project is one effort that defines ontologies and objective associations to created annotations and modeling standards of biological data representation [73]. The Human Physiome Project framework covers biological data representation in a variety of scales from molecular to whole body data. The Annotation and Image Markup (AIM) project is another proposed annotation standard [74]. It established a standard format that is semantically interpretable with the infrastructure of the caBIG^{TM} to expedite data sharing and collaboration networks for cancer researchers. In the AIM framework, annotations are associated with a semantic term taken from a predefined ontology. The AIM framework also provides coverage of image-level (e.g., patient, observer, equipment), pixel-level (e.g., two- and three-dimensional coordinates of defined geometric shapes, markup, text annotation, and comments), and arbitrary computational annotations.

Researchers involved with the AIM project have also developed a publicly available application programming interface (API), called the "AIM API" [75]. It was developed in Java to create objective-oriented models for the AIM model. Developers can use AIM API to create structural annotations that are stored as an XML format called AIM XML, and can be transformed into DICOM-SR, HL7-CDA XML, and AIM ontology instances. The AIM project also provides a semantic tool that acquires annotations directly from researchers and physicians as they view the images on medical imaging workstations [76].

ImageMiner, a software platform for TMA analysis, allows image retrieval using texton image features, clinical data, and results of previous analysis [77]. The "pathology analytical imaging standards" (PAIS) is a database model supporting both TMA and whole slide pathology images. It allows data queries based on metadata, analysis results, segmented objects, features, and annotated image objects [78].

11.4.2.1 TissueWikiMobile

The touch-sensitive screen of smartphone and tablet furnishes a richer human computer interface that is often quicker and more convenient in image annotation. TissueWikiMobile is one touch-enabled histological image browsing and annotation tool, aiming to furnish researchers with a ubiquitous way to collaborate and share their expert opinions not only on the performance of various antibody stains but also on histology image annotations [79]. Truthmarker is another system that promises structured and efficient annotation data gathering in diverse types of images. Truthmarker produces annotations in a standardized format, allowing them to be efficiently parsed, analyzed, and used in research [80].

11.4.3 VISUALIZATION

Multi-spectral imaging presents new problems for visualization, similar to other high-dimensional data analysis. Image visualization is grounded in three-color mixing techniques, where two-color combinations (e.g., yellow and purple) represent mixtures of two signals and white represents a mixture of three signals). There is no familiar way to represent mixtures of more than three signals. Many publications accept pseudo-colored images that use cyan and white to represent signals after red, green, and blue are assigned. These images are only interpretable when signals have minimal spatial overlap, as cyan and white will appear naturally as mixtures of other signals. For this reason, authors choose images with the best spatial separation for publication, adding some bias to the interpretation of the experiment. With the increased use of whole-slide imaging and virtual microscopy, these short-cuts do not serve as well for someone using software to interpret images.

The alternative to pseudo-coloring is to present every signal as a grayscale image and allow the user to activate and deactivate channels and assign them to red, green, or blue through the user interface. A great example of this functionality is the image viewing software provided by OMERO [81]. This can be computationally expensive and even prohibitive via online software. The alternative to this technique, avoiding the requirement of specialized software, is to pre-compute the most relevant combinations of channels and make those accessible through an interactive viewer.

Liu et al. used the heatmap with dendrograms visualization (TreeView software) to discover clusters among TMA samples based on antibody stain intensities. In their case study, they observed that lymphoma TMA samples with similar diagnosis cluster together based on 27 different antibodies [82].

11.5 CONCLUSION

Throughout this chapter, we have reviewed the many applications and analysis pipeline options for tissue microarrays (TMAs). Along the way, we have noted a number of important issues for future research. The two most important efforts in this field are towards standardization of the equipment used to acquire TMA images, and towards automation of stain characterization, including quantification and subcellular localization. These issues are much more complex than they are with DNA microarrays because of the heterogeneity of the sample spots. Future solutions to these research problems are needed as use of TMAs becomes more widespread for the development of clinically diagnostic protein expression assays.

11.6 ACKNOWLEDGMENTS

This research has been supported by Microsoft Research, as well as grants from National Institutes of Health (Bioengineering Research Partnership R01CA108468, Molecular Imaging Exploratory Center P20GM072069, Emory-Georgia Tech NCI Center for Cancer Nanotechnology Excellence U54CA119338). M.D.W. acknowledges the Georgia Cancer Coalition for a Distinguished Cancer Scholar Award; T.H.S. acknowledges the NSF IGERT program for fellowship support.

REFERENCES

1. J. K. Tran, E. N. Hubbard, T. H. Stokes, R. A. Moffitt, and M. D. Wang, "Feasibility of multiplex quantum dot stain using primary antibodies from four distinct host animals," in *Proceedings of the Annual International Conference of the IEEE Engineering in Medicine and Biology Society (EMBC 2012)*, San Diego, CA, USA, pp. 6576–6579, 2012.

2. J. Kononen, L. Bubendorf, A. Kallioniemi, et al., "Tissue microarrays for high-throughput molecular profiling of tumor specimens," *Nature Medicine*, vol. 4, no. 7, pp. 844–847, 1998.

3. S. L. Sallinen, P. K. Sallinen, H. K. Haapasalo, et al., "Identification of differentially expressed genes in human gliomas by DNA microarray and tissue chip techniques," *Cancer Research*, vol. 60, no. 23, pp. 6617–6622, 2000.

4. J. M. Giltnane and D. L. Rimm, "Technology insight: identification of biomarkers with tissue microarray technology," *Nature Clinical Practice Oncology*, vol. 1, no. 2, pp. 104–111, 2004.

5. K. D. Bauer, J. de la Torre-Bueno, I. J. Diel, et al., "Reliable and sensitive analysis of occult bone marrow metastases using automated cellular imaging," *Clinical Cancer Research*, vol. 6, no. 9, pp. 3552–9, 2000.

6. A. McCabe, M. Dolled-Filhart, R. L. Camp, and D. L. Rimm, "Automated quantitative analysis (AQUA) of in situ protein expression, antibody concentration, and prognosis," *Journal of the National Cancer Institute*, vol. 97, no. 24, pp. 1808–1815, 2005.

7. R. L. Camp, G. G. Chung, and D. L. Rimm, "Automated subcellular localization and quantification of protein expression in tissue microarrays," *Nature Medicine*, vol. 8, no. 11, pp. 1323–1327, 2002.

8. A. C. Andersson, S. Stromberg, H. Backvall, et al., "Analysis of protein expression in cell microarrays: a tool for antibody-based proteomics," *Journal of Histochemistry and Cytochemistry*, vol. 54, no. 12, p. 1413, 2006.

9. M. Stoeckli, P. Chaurand, D. E. Hallahan, and R. M. Caprioli, "Imaging mass spectrometry: A new technology for the analysis of protein expression in mammalian tissues," *Nature Medicine*, vol. 7, no. 4, pp. 493–496, 2001.

10. S. D. Jayasena, "Aptamers: an emerging class of molecules that rival antibodies in diagnostics," *Clinical Chemistry*, vol. 45, no. 9, pp. 1628–50, 1999.

11. E. N. Brody, M. C. Willis, J. D. Smith, S. Jayasena, D. Zichi, and L. Gold, "The use of aptamers in large arrays for molecular diagnostics," *Journal of Molecular Diagnostics*, vol. 4, no. 4, pp. 381–8, 1999.

12. R. L. Camp, L. A. Charette, and D. L. Rimm, "Validation of tissue microarray technology in breast carcinoma," *Laboratory Investigation*, vol. 80, no. 12, pp. 1943–9, 2000.

13. T. Stokes, J. T. Torrance, H. Li, and M. Wang, "ArrayWiki: an enabling technology for sharing public microarray data repositories and meta-analyses," *BMC Bioinformatics*, vol. 9, no. Suppl 6, p. S18, 2008.

14. E. Metwalli, D. Haines, O. Becker, S. Conzone, and C. G. Pantano, "Surface characterizations of mono-, di-, and tri-aminosilane treated glass substrates," *Journal of Colloid and Interface Science*, vol. 298, no. 2, pp. 825–31, 2006.

15. M. Copete, J. Garratt, B. Gilks, et al., "Inappropriate calibration and optimisation of pan-keratin (pan-CK) and low molecular weight keratin (LMWCK) immunohistochemistry tests: Canadian immunohistochemistry quality control (CIQC) experience," *Journal of Clinical Pathology*, vol. 64, no. 3, pp. 220–225, 2011.

16. H. Kajiya, S. Takekoshi, M. Takei, et al., "Selection of buffer pH by the isoelectric point

of the antigen for the efficient heat-induced epitope retrieval: re-appraisal for nuclear protein pathobiology," *Histochemistry and Cell Biology*, vol. 132, no. 6, pp. 659–667, 2009.

17. S. Namimatsu, M. Ghazizadeh, and Y. Sugisaki, "Reversing the effects of formalin fixation with citraconic anhydride and heat: A universal antigen retrieval method," *Journal of Histochemistry & Cytochemistry*, vol. 53, no. 1, pp. 3–11, 2005.

18. J. H. Fergenbaum, M. Garcia-Closas, S. M. Hewitt, et al., "Loss of antigenicity in stored sections of breast cancer tissue microarrays," *Cancer Epidemiology, Biomarkers & Prevention*, vol. 13, no. 4, pp. 667-72, 2004.

19. K. A. DiVito, L. A. Charette, D. L. Rimm, and R. L. Camp, "Long-term preservation of antigenicity on tissue microarrays," *Laboratory Investigation*, vol. 84, no. 8, pp. 1071–8, 2004.

20. S. C. Kao, K. Lee, N. J. Armstrong, et al., "Validation of tissue microarray technology in malignant pleural mesothelioma," *Pathology*, vol. 43, no. 2, pp. 128–32, 2011.

21. D. J. Brennan and W. M. Gallagher, "Prognostic ability of a panel of immunohistochemistry markers—retailoring of an 'old solution'," *Breast Cancer Research*, vol. 10, no. 1, p. 102, 2008.

22. W. Chen and D. J. Foran, "Advances in cancer tissue microarray technology: Towards improved understanding and diagnostics," *Analytica Chimica Acta*, vol. 564, no. 1, pp. 74–81, 2006.

23. E. D. Hsi, "A practical approach for evaluating new antibodies in the clinical immunohistochemistry laboratory," *Archives of Pathology and Laboratory Medicine*, vol. 125, no. 2, pp. 289–294, 2001.

24. V. Kiermer, "Antibodypedia," *Nature Methods*, vol. 5, no. 10, pp. 860–860, 2008.

25. K. Cottingham, "Antibodypedia seeks to answer the question: "how good is that antibody?"," *Journal of Proteome Research*, vol. 7, no. 10, p. 4213, 2008.

26. E. Björling and M. Uhlen, "Antibodypedia, a portal for sharing antibody and antigen validation data," *Molecular Cell Proteomics*, vol. 7, no. 10, pp. 2028–37, 2008.

27. S. C. Kao, K. Griggs, K. Lee, et al., "Validation of a minimal panel of antibodies for the diagnosis of malignant pleural mesothelioma," *Pathology*, vol. 43, no. 4, pp. 313–7, 2011.

28. H. Y. Jung and H. G. Cho, "An automatic block and spot indexing with k-nearest neighbors graph for microarray image analysis," *Bioinformatics*, vol. 18 Suppl 2, pp. S141–51, 2002.

29. A. N. Jain, T. A. Tokuyasu, A. M. Snijders, R. Segraves, D. G. Albertson, and D. Pinkel, "Fully automatic quantification of microarray image data," *Genome Research*, vol. 12, no. 2, pp. 325–32, 2002.

30. C. Di Ruberto and A. Dempster, "Circularity measures based on mathematical morphology," *Electronics Letters*, vol. 36, no. 20, pp. 1691–1693, 2000.

31. R. Nock, "Fast and reliable color region merging inspired by decision tree pruning," in *Proceedings of the IEEE Computer Society Conference on Computer Vision and Pattern Recognition*, Kauai, HI, USA, pp. 271–276, 2001.

32. R. Nock and F. Nielsen, "Statistical region merging," *IEEE Transactions on Pattern Analysis and Machine Intelligence*, vol. 26, no. 11, pp. 1452–1458, 2004.

33. H. Chang, G. V. Fontenay, J. Han, et al., "Morphometic analysis of TCGA glioblastoma multiforme," *BMC Bioinformatics*, vol. 12, no. 1, p. 484, 2011.

34. L. A. Cooper, J. Kong, D. A. Gutman, et al., "Integrated morphologic analysis for the identification and characterization of disease subtypes," *Journal of the American Medi-*

cal *Informatics Association*, vol. 19, no. 2, pp. 317–323, 2012.

35. S. Palokangas, J. Selinummi, and O. Yli-Harja, "Segmentation of folds in tissue section images," in *Proceedings of the IEEE Engineering in Medicine and Biology Society*, Lyon, France, pp. 5642–5645, 2007.

36. P. A. Bautista and Y. Yagi, "Detection of tissue folds in whole slide images," in *Proceedings of the IEEE Engineering in Medicine and Biology Society*, Minneapolis, USA, pp. 3669–3672. 2009.

37. P. A. Bautista and Y. Yagi, "Improving the visualization and detection of tissue folds in whole slide images through color enhancement," *Journal of Pathology Informatics*, vol. 1, p. 25, 2010.

38. S. Kothari, J. H. Phan, R. A. Moffitt, et al., "Automatic batch-invariant color segmentation of histological cancer images," in *Proceedings of the IEEE Internatonal Symposium on Biomedical Imaging*, Chicago, IL, USA, pp. 657–660, 2011.

39. M. Macenko, M. Niethammer, J. S. Marron, et al., "A method for normalizing histology slides for quantitative analysis," in *Sixth IEEE International Symposium on Biomedical Imaging (ISBI 2009)*, Boston, MA, USA, pp. 1107–1110, 2009.

40. D. Magee, D. Treanor, D. Crellin, et al., "Colour normalisation in digital histopathology images," in *Proceedings of the 12th International Conference on Medical Image Computing and Computer Assisted Intervention (MICCAI 2009)*, London, England, pp. 100–111, 2009.

41. L. Yang, O. Tuzel, W. Chen, et al., "PathMiner: a web-based tool for computer-assisted diagnostics in pathology," *IEEE Transactions on Information Technology in Biomedicine*, vol. 13, no. 3, pp. 291–299, 2009.

42. V. Kovalev, A. Dmitruk, I. Safonau, M. Frydman, and S. Shelkovich, "A method for identification and visualization of histological image structures relevant to the cancer patient conditions," *Lecture Notes in Computer Science*, vol. 6854, pp. 460–468, 2011.

43. J. Kong, O. Sertel, H. Shimada, K. L. Boyer, J. H. Saltz, and M. N. Gurcan, "Computer-aided evaluation of neuroblastoma on whole-slide histology images: Classifying grade of neuroblastic differentiation," *Pattern Recognition*, vol. 42, no. 6, pp. 1080–1092, 2009.

44. D. Gao, D. Padfield, J. Rittscher, and R. McKay, "Automated training data generation for microscopy focus classification," *Medical Image Computing and Computer-Assisted Intervention*, vol. 13, no. Pt 2, pp. 446–453, 2010.

45. M. E. Plissiti, C. Nikou, and A. Charchanti, "Automated detection of cell nuclei in pap smear images using morphological reconstruction and clustering," *IEEE Transactions on Information Technology in Biomedicine*, vol. 15, no. 2, pp. 233–41, 2011.

46. E. M. Brey, Z. Lalani, C. Johnston, et al., "Automated selection of DAB-labeled tissue for immunohistochemical quantification," *Journal of Histochemistry & Cytochemistry*, vol. 51, no. 5, pp. 575–84, 2003.

47. A. H. Coons, H. J. Creech, R. N. Jones, and E. Berliner, "The demonstration of pneumococcal antigen in tissues by the use of fluorescent antibody," *Journal of Immunology*, vol. 45, no. 3, pp. 159–170, 1942.

48. W. H. Farrand, J. F. Bell, J. R. Johnson, et al., "Visible and near-infrared multispectral analysis of rocks at Meridiani Planum, Mars, by the Mars Exploration Rover Opportunity," *Journal of Geophysical Research-Planets*, vol. 112, no. E6, 2007.

49. F. D. van der Meer, H. M. A. van der Werff, F. J. A. van Ruitenbeek, et al., "Multi- and hyperspectral geologic remote sensing: A review," *International Journal of Applied Earth Observation and Geoinformation*, vol. 14, no. 1, pp. 112–128, 2012.

50. M. E. Dickinson, G. Bearman, S. Tille, R. Lansford, and S. E. Fraser, "Multi-spectral imaging and linear unmixing add a whole new dimension to laser scanning fluorescence microscopy," *Biotechniques*, vol. 31, no. 6, pp. 1272–6, 2001.

51. P. Krishnasamy, S. Belongie, and D. Kriegman, "Wet fingerprint recognition: Challenges and opportunities," in *Proceedings of the International Joint Conference on Biometrics (IJCB 2011)*, Washington, DC, USA, pp. 1–7.

52. A. L. Lane, L. Nyadong, A. S. Galhena, et al., "Desorption electrospray ionization mass spectrometry reveals surface-mediated antifungal chemical defense of a tropical seaweed," *Proceedings of the National Academy of Sciences*, vol. 106, no. 18, pp. 7314–7319, 2009.

53. M. Uhlen, E. Bjorling, C. Agaton, et al., "A human protein atlas for normal and cancer tissues based on antibody proteomics," *Molecular & Cellular Proteomics*, vol. 4, no. 12, pp. 1920–1932, 2005.

54. M. Uhlen, P. Oksvold, L. Fagerberg, et al., "Towards a knowledge-based human protein atlas," *Nature Biotechnology*, vol. 28, no. 12, pp. 1248–50, 2010.

55. L. Berglund, E. Bjorling, P. Oksvold, et al., "A genecentric Human Protein Atlas for expression profiles based on antibodies," *Molecular Cell Proteomics*, vol. 7, no. 10, pp. 2019–27, 2008.

56. S. Hober and M. Uhlen, "Human protein atlas and the use of microarray technologies," *Current Opinion in Biotechnology*, vol. 19, no. 1, pp. 30–35, 2008.

57. P. Dammeyer and E. S. Arner, "Human protein atlas of redox systems—what can be learnt?," *Biochimica et Biophysica Acta*, vol. 1810, no. 1, pp. 111–38, 2011.

58. W. Chen, C. Schmidt, M. Parashar, M. Reiss, and D. J. Foran, "Decentralized data sharing of tissue microarrays for investigative research in oncology," *Cancer Informatics*, vol. 2, p. 373, 2007.

59. Y. Yagi and J. R. Gilbertson, "Digital imaging in pathology: the case for standardization," *Journal of Telemedicine and Telecare*, vol. 11, no. 3, pp. 109–16, 2005.

60. I. G. Goldberg, C. Allan, J. M. Burel, et al., "The open microscopy environment (OME) data model and XML file: open tools for informatics and quantitative analysis in biological imaging," *Genome Biology*, vol. 6, no. 5, p. R47, 2005.

61. G. G. Thallinger, K. Baumgartner, M. Pirklbauer, et al., "TAMEE: data management and analysis for tissue microarrays," *BMC Bioinformatics*, vol. 8, no. 357, p. 81, 2007.

62. F. Demichelis, A. Sboner, M. Barbareschi, and R. Dell'Anna, "TMABoost: an integrated system for comprehensive management of tissue microarray data," *IEEE Transactions on Information Technology in Biomedicine*, vol. 10, no. 1, pp. 19–27, 2006.

63. R. Zwonitzer, T. Kalinski, H. Hofmann, A. Roessner, and J. Bernarding, "Digital pathology: DICOM-conform draft, testbed, and first results," *Computer Methods and Programs in Biomedicine*, vol. 87, no. 3, pp. 181–8, 2007.

64. A. M. Marchevsky, R. Dulbandzhyan, K. Seely, S. Carey, and R. G. Duncan, "Storage and distribution of pathology digital images using integrated web-based viewing systems," *Archives of Pathology & Laboratory Medicine*, vol. 126, no. 5, pp. 533–539, 2002.

65. A. Sharma-Oates, P. Quirke, and D. R. Westhead, "TmaDB: a repository for tissue microarray data," *BMC Bioinformatics*, vol. 6, no. 354, p. 218, 2005.

66. D. G. Nohle, B. A. Hackman, and L. W. Ayers, "The tissue micro-array data exchange specification: a web based experience browsing imported data," *BMC Medical Informatics and Decision Making*, vol. 5, p. 25, 2005.

67. D. A. Clunie, "DICOM structured reporting and cancer clinical trials results," *Cancer*

Informatics, vol. 4, p. 33, 2007.

68. R. H. Dolin, L. Alschuler, S. Boyer, et al., "HL7 clinical document architecture, release 2," *Journal of the American Medical Informatics Association*, vol. 13, no. 1, pp. 30–39, 2006.

69. C. Le Bozec, D. Henin, B. Fabiani, T. Schrader, M. Garcia-Rojo, and B. Beckwith, "Refining DICOM for pathology–progress from the IHE and DICOM pathology working groups," *Studies in Health Technology and Informatics*, vol. 129, no. Pt 1, pp. 434–438, 2007.

70. R. Singh, L. Chubb, L. Pantanowitz, and A. Parwani, "Standardization in digital pathology: Supplement 145 of the DICOM standards," *Journal of Pathology Informatics*, vol. 2, no. 206, p. 23, 2011.

71. C. Daniel, F. Macary, M. G. Rojo, et al., "Recent advances in standards for collaborative digital anatomic pathology," *Diagnostic Pathology*, vol. 6 Suppl 1, p. S17, 2011.

72. J. J. Berman, M. E. Edgerton, and B. A. Friedman, "The tissue microarray data exchange specification: a community-based, open source tool for sharing tissue microarray data," *BMC Medical Informatics and Decision Making*, vol. 3, p. 5, 2003.

73. P. Hunter, P. Robbins, and D. Noble, "The IUPS human physiome project," *Pflugers Archiv – European Journal of Physiology*, vol. 445, no. 1, pp. 1–9, 2002.

74. D. S. Channin, P. Mongkolwat, V. Kleper, K. Sepukar, and D. L. Rubin, "The caBIG annotation and image markup project," *Journal of Digital Imaging*, vol. 23, no. 2, pp. 217–225, 2010.

75. H. Bulu and D. L. Rubin, "Java application programming interface (API) for annotation imaging markup (AIM).", http://sourceforge.net/projects/aimapi/, 2013.

76. D. L. Rubin, P. Mongkolwat, V. Kleper, K. Supekar, and D. S. Channin, "Medical imaging on the semantic web: Annotation and image markup," in *Proceedings of the AAAI Spring Symposium: Semantic Scientific Knowledge Integration (AAAI 2008)*, Stanford University, CA, USA, pp. 93-98, 2008.

77. D. J. Foran, L. Yang, W. Chen, et al., "ImageMiner: a software system for comparative analysis of tissue microarrays using content-based image retrieval, high-performance computing, and grid technology," *Journal of the American Medical Informatics Association*, vol. 18, no. 4, pp. 403–415, 2011.

78. F. Wang, J. Kong, L. Cooper, et al., "A data model and database for high-resolution pathology analytical image informatics," *Journal of Pathology Informatics*, vol. 2, no. 349, p. 32, 2011.

79. C. Cheng, T. Stokes, S. Hang, and M. D. Wang, "TissueWiki mobile: An integrative protein expression image browser for pathological knowledge sharing and annotation on a mobile device," in *Proceedings of the IEEE International Conference on Bioinformatics and Biomedicine Workshops (BIBMW 2010)*, Hong Kong, pp. 473–480, 2010.

80. M. A. Christopher, "Truthmarker: a tablet-based approach for rapid image annotation," Master's Thesis, University of Iowa, USA, 2011.

81. J. Moore, C. Allan, J. M. Burel, et al., "Open tools for storage and management of quantitative image data," *Methods in Cell Biology*, vol. 85, pp. 555–70, 2008.

82. C. L. Liu, W. Prapong, Y. Natkunam, et al., "Software tools for high-throughput analysis and archiving of immunohistochemistry staining data obtained with tissue microarrays," *American Journal of Pathology*, vol. 161, no. 5, p. 1557, 2002.

12 CNV-Interactome-Transcriptome Integration to Detect Driver Genes in Cancerology

Maxime Garcia, Raphaële Millat-Carus,
François Bertucci, Pascal Finetti, Arnaud Guille,
José Adélaïde, Ismahane Bekhouche,
Renaud Sabatier, Max Chaffanet,
Daniel Birnbaum, and Ghislain Bidaut

CONTENTS

12.1 ABSTRACT

The development of high-throughput gene-expression profiling technologies allows the definition of genomic signatures that help predict clinical condition or cancer patient outcome. However, such signatures show dependency on the training set and thus, suffer from a lack of generalization and from instability. This is the consequence of the microarray data topology and the fact that cancer is provoked by a small number of *driver* genes provoking changes to *passenger* genes. Driver genes are the genomics elements whose deregulation provokes the disease. Passenger genes have their expression affected because of misregulations and expression changes on the drivers but these changes have no impact on the disease. Separating drivers and passengers is of primary importance for the understanding of the disease and deciphering of molecular subtypes that exists for most cancers. Detecting these genes is a difficult process since cancer tumors are highly heterogeneous. In this chapter, we describe an interactome-based approach, copy-number-variation-interactome-transcriptome integration (CNV-ITI) that is used to detect driver genes that are specific to molecular subtypes in breast cancer (BC) by superimposition of a large scale protein-protein interactions (PPI) dataset (human interactome) over several gene expression datasets and array comparative genomic hybridization (aCGH) datasets. The algorithm extracts interactome regions, so-called subnetworks, that allow for predicting relapse-free survival in cancer and detection of driver genes. As an illustrative example, we specifically applied it to basal and luminal A BC subtypes. Two other methods of CGH-gene expression profiles integration for detecting driver genes in cancerology are described and compared to CNV-ITI.

12.2 INTRODUCTION

12.2.1 BREAST CANCER MOLECULAR SUBTYPES

Breast cancer (BC) is a heterogeneous disease. This explains why standard treatment does not work equally on all patients. This heterogeneity is difficult to decipher with the histoclinical criteria [2] that are used to predict prognosis. Consequently, many patients undergo overtreatment [3, 4]. The current challenge is to build better classifiers to i) separate breast cancer subtypes and ii) finely predict the prognosis associated with each subtype. The post-genomic era has seen the appearance of several tools that help gain a deeper knowledge of the molecular nature of the disease. Among these tools, microarray technology has contributed to understanding the molecular biology of cancer at the mRNA level and at the DNA level (array-comparative genomic hybridization—aCGH). These technologies allow for the discovery of markers that would refine disease classification. Using gene expression microarrays, four main BC subtypes (luminal, basal, ERBB2-like and normal-like) were identified [5], which were confirmed in an independent study [6]) that refined this classification by splitting the luminal group into two subtypes (luminal A and luminal B) [7]. This classification has evolved with the discovery of other pertinent subgroups [8, 9], including Claudin-low subtype [10]. Correlation studies between clinical outcome and gene expression profiles were done, and specific prognoses were assigned to the identified subtypes [11, 12]. However, these correlations lack power due to a low number of studied samples. Nevertheless, the classification in five major molecular subtypes proved very robust and Hu et al. [13] validated a list of 306 genes that are now the reference for establishing these main subtypes.

Luminal A and basal subtypes are two of the major subtypes that are characterized by opposite features both clinically and at the genomic level [4]. Luminal A BCs are the most frequent (45% total occurrence). They are low-grade, differentiated tumors that express hormonal receptors and the *ESR1* and *GATA3* luminal differentiation genes. They are usually associated with a relatively favorable prognosis due to their response to hormonal therapy. The basal subtype represents 15% of all BCs. Basal BCs are high-grade, proliferative tumors that are hormone receptor negative and associated with poor prognosis [4]. While basal tumors are relatively chemosensitive, the effectiveness of chemotherapy remains limited.

12.2.2 INTEGRATION OF GENOMIC DATA FOR DISCOVERING DRIVER GENES IN LUMINAL A AND BASAL BREAST CANCER MOLECULAR SUBTYPES

A fundamental issue for systematic characterization of disease in general and BC in particular is the discovery of driver genes or markers. While it has been established that BC is characterized by five major subtypes (the four main subtypes and the subdivision of luminal subtype into A and B), genomic profiles show that tumors are extremely diverse and most of them present unique characteristics. Therefore, it appears that most of the genetic lesions and gene misregulations that are found in tumors are not necessarily all at the origin of the disease at hand and that the challenge is to distinguish the driver (that provoke proliferation, treatment resistance, and metastasis at a primary level) from the passenger genes (genes whose changes are a by-product of drivers, deregulations and that do not have a direct impact on the disease). These drivers have to be established for each subtype [14]. In addition, subtyping, although already quite extensive [9], may still need to be further defined.

Breast cancer arises as a result of expansion driven by cells that acquire immortality and a survival advantage through specific mutations and expression changes.

The drivers are the genes that specifically provide for this selective advantage that enhances cancer hallmarks, including their involvement in pathways that favor proliferation and chemotherapy resistance. On the other hand, passengers, while also deregulated, are neutral to this selection process and are only involved in pathways of secondary importance in regards to cancer [15]. The main challenge is therefore to find a metric to separate drivers from passengers on the likelihood that they are hence driving the cancer.

The development of high resolution array CGH technology allowed the identification of genomic alterations on the whole genome [16]. Copy number aberrations (CNA) were associated with molecular subtypes and clinical outcome in BC [17]. However, the abundance and heterogeneity of genomic regions with significant CNA make the search for viable biological markers or therapeutic targets difficult [14]. To detect recurrent markers among tumors, most approaches are based on the frequency of alterations. If an alteration occurs more often within a gene in a set of tumors, it is likely that the gene represents a cancer key factor. For example, the Genomic Identification of Significant Targets in Cancer (GISTIC) algorithm identifies genomic regions that are aberrant more often than would be expected by chance. GISTIC was applied to several types of cancer [18, 19].

However, the use of CNA alone is not sufficient for detecting drivers. Amplified or deleted regions detected by array-CGH are usually large and cover multiple genes. Many of them are just passengers, but are indistinguishable from driver genes. Also, these approaches cannot determine the physiological or functional importance of the detected regions. These limitations highlight the need for more advanced, integrated approaches that take into account multiple types of biological information to identify drivers. In this regard, gene expression can give fundamental insight. However, the integration methodology is not obvious and must be carefully defined.

12.2.3 SCIENTIFIC GOAL

The basic postulate is that gene expression is fundamentally correlated with driver gene mutations at one time or the other of tumor history. Together, expression and gene mutation form a *genomic footprint* [14] that we wish to understand and establish for all BC subtypes. The goal is dual: (i) understand the disease biology and identify drug targets (druggable footprint), and (ii) predict patient outcome to adapt treatment to the disease and to the patient (actionable footprint). Several approaches have been developed to superimpose genomic information and gene expression deregulation to help detect driver genes. Among these, Akavia et al. [14] developed an algorithm (CONEXIC) to identify driver genes located in regions with recurring genomic changes. In their approach, each driver is associated with a gene module that is deregulated by the driver. This method was validated on a melanoma dataset of 62 tumors with paired measurement of gene expression and CNA, initially published by Lin et al. [20]. It confirmed many previously known gene drivers of melanoma and connected them with many of their targets, to identify their biological functions. In addition, new targets were predicted and experimentally confirmed. This integrative method is further detailed in the chapter. Another integrative approach was proposed by Beroukhim et al. [19]. This method, based on the detection of driver genes by analysis of their gene expression and CNA profiles, is also detailed.

However, even if the set of markers that can be detected is geared toward driver genes by an integrated analysis, microarray measurement noise and the tumoral heterogeneity are still major hurdles to obtain reliable markers. It has been established that DNA microarray signatures are inherently unstable in regard to their application to gene signature prediction [21]. For instance, two datasets of reference for breast cancer metastasis prediction, respectively studying 198 tumors [22] and 98 tumors [12], validated later on 298 tumors [23] produced signatures (76 genes for Wang, 70 genes for van de Vijver) that are different (only 3 genes were found to be common between the two signatures, and hence they had about 2% stability, as described by Chuang et al. [21]) and did not classify independent data (i.e., predict patient outcome) reliably [24].

The reasons behind instability of expression-based signatures are two-fold. The first is purely mathematical and finds its origin in the topology of genome-based measurement, which is highly prone to the curse of dimensionality. The number of measured variables is vastly superior to the number of tumors, which prevents direct use of tools from classical statistics. Second, the variability is due to the very nature of the biological information to be measured. Cancer is provoked by driver genes that are subtly deregulated or mutated and these prompt secondary changes in a vast array of genes. Since all genes are placed at the same level by microarray analysis, causality information is removed. In fact, a microarray-based analysis detects genes with the most favorable statistics, i.e., the ones that are the most differentially expressed. These are not necessarily the ones we wish to detect in priority but are rather a byproduct of driver deregulation.

To properly retrieve driver genes, it is first necessary to reduce the curse of dimensionality by increasing the number of studied samples. This could be done by filtering the genome-based information (CNAs) by the gene expression data. Second, one has to include biological and causal information to differentiate driver genes from their passengers. This could be done by including a large protein-protein interactions (PPI) map over gene expression data and, instead of measuring independently all genes by individual statistics, driver genes could be detected at a global interactome level by identification of interacting regions that are deregulated together for a particular BC clinical condition or subtype. Several methods have already been proposed for network analysis of gene expression [24, 25]. However, the multi-level nature of biology was not taken into account in the proposed methods since no integration with CNA was done.

In the present chapter, we describe the application of a variant of the interactome-transcriptome integration (ITI) method previously applied to metastasis prediction in BC [24, 25]. ITI showed significant improvement on previously published classification methods by detecting estrogen receptor (ER)-based signatures using an integrated analysis of five gene expression datasets covering more than 900 tumors and an independent validation dataset. ITI achieved greater metastasis prediction accuracy (74% on ER-positive tumors, 54% on ER-negative tumors) and a higher stability (11% on ER-negative tumors and 32% on ER-positive tumors, obtained by permutation between training and testing datasets) compared to previously published statistics. The basic principle of ITI is to superimpose the known human interactome and detect a list of candidate seeds based on differential expression analysis. The neighborhoods of these seeds are recursively explored and interacting genes are aggregated if they are co-expressed with it. ITI yields interactome regions that are differentially expressed, called subnetworks. In a second pass, subnetworks are statistically validated by confrontation to null distributions of scores. Subnetworks are then interrogated for biological function by Gene Ontology (GO) term enrichment [26] using a hypergeometric test [27] and annotations from the National Center for Biotechnological Information (NCBI) Entrez Gene database [28]. GO enrichment is defined as follows 12.1:

$$en(GO,S) = \frac{g_{GO,t} \in S/|S|}{g_{GO,t} \in Genome/|Genome|}, \tag{12.1}$$

with $en(GO,S)$ being the enrichment (or depletion) of gene ontology term GO,t in subnetwork S, $|S|$ being the number of genes in the studied subnetwork, $g_{GO,t}$ being the number of genes annotated with the GO term GO,t, $Genome$ the set of genes in the studied organism complete genome and $|Genome|$ the number of genes in this set. See [29] for the statistical test definition associated with this metric.

The original method described in [24] was heavily modified to include CNA measurement to focus the analysis on detection of driver genes. This was done by constructing a pipeline established by sequentially chaining the algorithm previously described [1] to select a first list of candidate drivers and the ITI algorithm that was used to build functional modules around these and visualize results by creating a dedicated bioinformatics resource. All results from this analysis are available on the ITI Web site.[1] This pipeline is called copy number variations-interactome-transcriptome integration (CNV-ITI) in the following sections.

Here, we applied the CNV-ITI method for the search of specific biomarkers for the basal and luminal A BC subtypes. This search yielded 123 subnetworks that included known markers for the biology of these subtypes as well as the interaction with functional modules deregulated in luminal A and basal tumors. The superimposition of CNA information helped us to distinguish passengers of the potential drivers. We analyzed the list of discovered driver genes with those known in biology of cancer, obtained from [4].

12.3 MATERIAL AND METHODS

12.3.1 GLOBAL CNV-ITI WORKFLOW

This section describes the global workflow used for this analysis and details the integrated datasets. This analysis includes data from genomic CNV profiles (DNA) and gene expression (mRNA) profiles, as well as PPI.

---

[1]http://iti.sourceforge.net/citi

The CNV-ITI workflow integrates these data to detect driver genes behind a phenotype, namely understanding the genomics difference between basal and luminal A BC subtypes. First, we identified the amplified genes being characterized both by high resolution array CGH at significant copy number and by expression arrays by a significant differential expression. These genes are considered to be putative driver genes behind the luminal A/basal subtype differentiation, and are further analyzed in a network context with ITI [24, 25]. The network analysis performs a simultaneous analysis of gene expression and PPI maps to replace the candidate genes in a biological context and hence determine the link between amplification and gene expression. These steps are shown in Figure 12.1.

The following sections detail all datasets (PPI and microarray datasets for measurement of gene expression and genomics alteration) used in the analysis.

12.3.2   PROTEIN-PROTEIN INTERACTION DATABASES

Before using ITI, a reference set of PPI must be defined. This set of interaction data must have certain properties, including being at a scale compatible with the expression data at hand. If the interaction set is too restricted, the ITI algorithm will not be able to detect pertinent sets of modules. To build this set of interaction data, it is possible to use locally-developed technologies and establish a PPI map. Among these technologies, the 2-hybrid screening in yeast [30] allows the discovery of PPI at a scalability that is compatible with other types of genomic assays, such as gene expression microarrays. Its principle is based on the fact that transcription factor (TF) binding can work with the activation and binding domains only at proximity without direct binding. The two proteins we wish to interrogate (usually called prey and bait) are introduced in a specifically designed mutant yeast that lack the biosynthesis of certain nutrients and thus will not survive on a medium without these. Two types of plasmids are engineered to produce on one side a protein product (that is typically a known protein the investigator is using to identify new binding partners, referred to as bait) in which the binding domain (BD) has been fused in, and on the other side, a protein product (that is either a single known protein or a library of known or unknown proteins, referred to as a prey) in which the activation domain (AD) has been fused in. The two plasmids are then introduced in the yeast. If the two proteins interact, the transcription of a reporter gene may occur. If the two proteins do not interact, the reporter gene is not activated and the yeast fails to survive. Among other approaches, coaffinity purification is followed by mass spectrometry [31], which consists of targeting a given protein with an antibody to simultaneously pull entire complexes out of solutions (technology referred to as "pull-down"). These technologies are now well-established and can be carried out in a large number of laboratories at a reasonable cost. However, these still have a number of disadvantages. There are large numbers of false positives and false negatives, due to the very nature of the screen. In the 2-hybrid screen, the error rate is as high as 70% [32]. The reasons for such a high number of false positives may be (among others) the fact that the 2-hybrid screen takes place in the nucleus and the measured proteins will not interact, if they do not usually interact there, due to the lack of proper localization

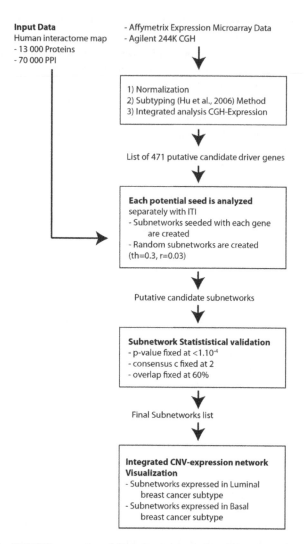

FIGURE 12.1 CNV-ITI general workflow. Gene expression data and CGH profiles help define a list of putative drivers. These are then superimposed to the human interactome for further analysis with ITI. Subnetworks regulated by the candidates are then detected and statistically validated. A visualization module allows the identification of subnetworks and their associated expression and CNA values.

signals. Also, some proteins may interact if simultaneously expressed in the yeast although they will never be present in the cell simultaneously.

For this report, we did not generate the PPI data in-house, but integrated publicly available datasets in various databases using bioinformatics methods. Publicly available databases are largely built using the previously cited technologies (yeast

TABLE 12.1
Summary of protein-protein interactions databases.

Database	#Proteins	#Interactions	Technology
HPRD [33]	9,386	36,577	Large scale assays (Y2H) associated with interactions established and curated by experts from scientific literature
Cocite [37]	6,349	15,705	In silico (Cocite algorithm)
DIP [36]	918	810	In vitro/manually curated
MINT [35]	5,559	12,143	Established and curated by experts from scientific literature
INTAct [34]	7,471	25,516	Large scale assays (Y2H, CoIP, pull-down)

2-hybrid or other high-throughput technologies) and therefore carry the same type of limitations that were just described. Building a pipeline that is based on this type of assay implies taking specific precautions in handling the data. However, many databases are built on a strong core of interactions that have been validated *in vivo* and are considered as reliable. Additionally, these sets are completed by a large number of *in silico* predicted interactions. These predictions allow the extension of the interaction data to a larger genomic footprint. For deciphering specific luminal A and basal subnetworks, we used the interaction databases reported in [24]. Our final interaction set is built from five interaction databases, including the Human Protein Reference Database [33], INTAct [34], the Molecular Interactions Database [35], the Database of Interacting Proteins [36], and an *in-silico* interaction set generated by the Cocite algorithm [37]. These total 70,530 binary interactions among 13,202 proteins (see Table 12.1).

All data were downloaded as flat files from all databases, respective Web sites, followed by similar normalization and transformation steps. These include the removal of unknown proteins (marked as such in the original file), the removal of self interactions and non-human proteins, as well as the replacement of various identifiers by standard National Center for Biotechnology Information (NCBI) Entrez Gene[2] identifiers to allow the mapping between different PPI datasets on the one hand, and between interaction and expression data on the other hand.

12.3.3   MICROARRAY GENE EXPRESSION PROFILES

To understand the molecular differences at the gene expression level between luminal A and basal BC subtypes, microarray gene expression data were analyzed and superimposed to the interaction data. We made use of publicly available expression and genomic data generated from a large pool of tumors maintained at the Institute Paoli-Calmettes on pangenomic Affymetrix microarray HG-U133 Plus 2.0. The

---

[2]http://www.ncbi.nlm.nih.gov/gene

TABLE 12.2

Summary of microarray datasets (expression and CGH) used in the analysis.

Dataset	GEO Accession	Platform Type	Basal/ luminal A
Sabatier [40]	GSE21653	Affymetrix HG-U133 Plus2.0	80/68
Bellkhouche [1]	GSE21653	Agilent Technologies Hu 244A	80/68

main advantage of using this technology is to obtain expression profiles on a very large gene set that covers most of the human genome in a single experiment. Expression data were normalized with the Robust Multi-array Average (RMA) standard method available in Bioconductor[3]. Data parsing and RMA normalization were done with the Bioconductor *Affy* package. The same tumors were used to generate CGH profiles as described in section 12.3.4. Correspondence tables for probes, IDs-NCBI Entrez Gene[4] accession numbers were generated by specific Affymetrix annotations files available from Resourcerer [38] and probes were combined using the method described by Reyal et al. [39]. For each set of Affymetrix probes corresponding to the same gene, probes carrying the "x_at" extension were filtered out and probes having the highest median signal were retained. In case of having only "x_at" marked probes, we only applied the median-based rule. On the basis of the BC subtype classifier by Hu et al. [13], all tumors were assigned to one of the five major subtypes commonly used. Among all tumors, 68 and 80 were labeled as basal and luminal A, respectively.

12.3.4 COMPREHENSIVE GENE HYBRIDIZATION PROFILES

aCGH and gene expression profiles were established for the same tumors. However, this is not a specific requirement of the presented approach, as data were separately processed and integrated under the form of statistics. After hybridization, scanning and data acquisition, standard bioinformatics analysis was applied, including initial filtering and LOESS normalization with the Feature Extraction package (Agilent Technologies, Santa Clara, CA, USA). Data were visualized and extracted under the form of log ratios using CGH Analytics 3.4 (Agilent Technology, Santa Clara, CA, USA). Then, copy number calculation was done after a binary circular segmentation [41]. Then, the detection of significantly altered genes in basal and luminal A groups were done by GISTIC [18] (for specific implementation, see [1]). GISTIC takes into account both the amplitude and frequency of alteration in the tumor dataset to attribute a *p*-value to each gene.

---

[3]http://www.bioconductor.org

[4]http://ncbi.nlm.nih.gov/gene

12.3.5   INITIAL CANDIDATE GENE DETECTION BY PRIMARY INTEGRATION OF GENE GENOMIC ALTERATION AND EXPRESSION

To understand the impact of driver genes on the deregulation of gene expression in basal and luminal A, the information associating gene expression deregulation and genomic alterations was integrated. ITI was then used to understand the impact of driver genes on the human interactome by detecting interactome regions that are deregulated by them. The list of candidate genes was first detected by adaptation of the approach described in [1]. First, we selected a subset of candidate genes with a significantly different GISTIC score among the two populations. Using the protocol described in Bekhouche et al. [1], we initially selected $n = 471$ genes significantly deregulated between luminal A and basal tumors with a False Discovery Rate $FDR = 1.10^{-3}$ (Benjamini-Hotchberg method). Other criteria have been added to select the initial gene candidates (Figure 12.2). (i) The respective frequency of genomic alteration between the two groups must be different (Fisher exact test with $p \leq 0.05$. (ii) Their expression and CNA must be correlated (student t-test with $p \leq 0.05$). (iii) These genes have to be differentially expressed between the basal and luminal A subtypes in addition to their genomic alteration differences (student t-test with $p \leq 0.05$). The final candidate gene list was submitted to ITI as initial seeds.

12.3.6   OTHER ANNOTATION DATABASES AND PIPELINE ELEMENTS

In addition to the NCBI Entrez gene and Resourcerer databases for identifier-probe conversion for microarray analysis, several other databases were used throughout the pipeline for subnetwork annotations. In particular, Gene Ontology information [26] was used to assign biological functions to subnetworks through enrichment measurement. This was done with the ErmineJ package [42]. ErmineJ specifically measures GO enrichment with a calculation of statistical significance with hypergeometric distribution and multiple testing correction. In addition, the list of human transcription factors were generated with the freely accessible for academics version of Transfac [43] (current version at the time of analysis: 7.0). To display generated subnetworks, the open-source GraphViz[5] network and graph visualization package (AT &T Research, Florham Park, NJ, USA) was used. Subnetwork statistical validation was implemented with the MATLAB® Statistical Toolbox (The Mathworks, Natick MA, USA).

12.3.7   INTERACTOME-TRANSCRIPTOME INTEGRATION ALGORITHM

The interactome-transcriptome integration (ITI) algorithm works by simultaneously examining interaction data and expression data to detect differentially expressed subnetworks, that is, interactome regions whose expression is globally differentially expressed among two experimental conditions. To accelerate detection and lower computing costs, the detection was parallelized by separating input datasets onto subsets

---

[5]http://ww.graphviz.org

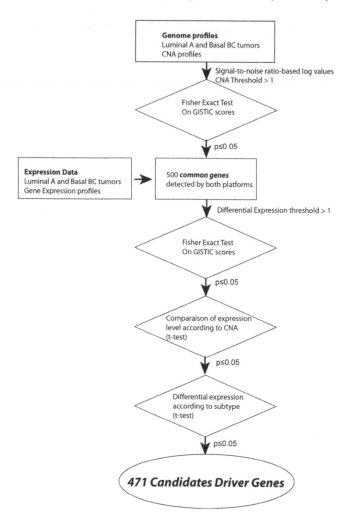

FIGURE 12.2 Initial candidate driver gene detection by comparative genomic hybridization and gene expression analysis. Protocol adapted from Bekhouche et al. [1].

on a Beowulf cluster. ITI works in two main steps, an initial subnetwork detection and a statistical validation step of subnetworks. To detect differentially expressed subnetworks, correlation between clinical or phenotypic status of the two conditions one wishes to analyze are computed. The set of interaction is then exhaustively searched for discriminative regions (Figure 12.3) by individually considering each node as a potential seed and recursively aggregating neighbors if they increase the subnetwork score. The score is calculated by Pearson correlation between average gene expression of all the genes belonging to the subnetwork and a vector representing the phenotype (see Equation 12.2). In [24], this was done with patient distant

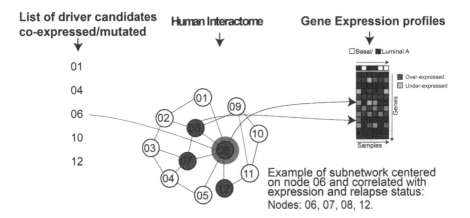

FIGURE 12.3   (SEE COLOR INSERT.) ITI algorithm and data organization. The 471 genes selected as seeds by the previous step are tested by ITI. Subnetworks are aggregated recursively around these seeds if their expression is correlated with the subtype.

metastasis free survival status (DMFS) clinical condition in order to predict relapse in BC. Neighbors of nodes are examined recursively and merged to the current subnetwork if their expression allows increasing the score by a minimal rate. The current minimal threshold score for a subnetwork to be retained before statistical validation is $S = 0.3$ and the minimal rate is $r = 0.03$. Once the score cannot be improved, no more nodes are considered for the current subnetwork.

The subnetwork score S of a subnetwork s on a dataset d is calculated by the score $S_{s,d}$ defined in Equation 12.2. This score measures the correlation (in the present case, this is not the absolute correlation, contrary to what was done in [24]) between the subnetwork average gene expression and the BC molecular subtype. A normalization term is calculated by using the total number of conditions in the subnetwork n_d. This term is not useful when analyzing a single dataset but allows us to scale scores appropriately when comparing multiple datasets. When analyzing multiple datasets simultaneously, the score S_s is computed by averaging individual scores over the datasets d (Equation 12.3) over the datasets list DS of size NS.

$$S_{s,d} = \frac{\sqrt{n_d}}{\sqrt{\max n_d(DS)}} \left| corr\left(\frac{1}{n} \sum_{g \in S} e(g,d), cc(d) \right) \right| \qquad (12.2)$$

$$S_s = \frac{1}{NS} \sum_{d \in DS} S(s,d) \qquad (12.3)$$

Subnetworks that are overlapping with subnetworks that have already been detected are not retained. Overlapping between two subnetworks A and B is calculated by maximum inclusion of B in A and A in B. The inclusion score is calculated by counting common genes included in subnetwork A to B and dividing by the total number of genes contained in subnetwork A. In practice, subnetworks overlapping by more than 50% are removed.

Once subnetworks are detected, they have to be statistically validated. This is done by assessing the statistical significance of the links between gene expression and protein-protein interaction. The statistical significance is measured by drawing a null distribution of scores by (i) selecting random subnetworks, and (ii) shuffling gene expression. This gives two null distributions of scores that are used to assign two distinct p-values, Pval1 and Pval2, to each subnetwork. We consider a subnetwork to be statistically significant if Pval1 and Pval2 are both under a defined significance threshold. The first random distribution assesses the pertinence of ITI. It is obtained by randomly selecting subnetworks, i.e., by replacing the recursive aggregation method detailed above by a random aggregation around a seed. The second null distribution assesses whether the link between gene expression and PPI is biologically sound and valid. This second distribution was obtained by shuffling luminal A and basal molecular subtypes. To keep random subnetworks comparable (in terms of size) to previously detected subnetworks, their distribution of size (modeled as Gaussian) was forced to be comparable to the subnetworks detected in the above step. After drawing the two distributions, they were modeled by a Gaussian mixture distribution. We then used this model to determine an appropriate threshold score and filter out subnetworks with scores that were considered not significant. Overlapping subnetworks are then clustered according to the inclusion score previously defined if they overlap by more than a threshold O_s specified as a percentage. For each cluster, the subnetwork having the largest score is kept.

12.4 ANALYSIS

In this section, we detail the analysis steps as well as analysis of driver genes obtained with the CNV-ITI pipeline.

12.4.1 DRIVER GENE DETECTION WITH INTERACTOME-TRANSCRIPTOME INTEGRATION

To separately detect subnetworks expressed in the luminal A and basal BC subtypes, we performed two separate ITI runs on the specific tumor sets. These two runs took as input the list of 471 candidate genes that met the initial filters and selected them as seeds to generate candidate subnetworks. To generate subnetworks in the whole human interactome at hand, two types of data were used. First, the interaction database previously assembled (Section 12.3.2) was parsed and used as a basis for network exploration. The expression data from the 148 tumors were superimposed to the network on a gene basis. Then, correlation between the BC molecular subtype and gene expression is computed and subnetworks were detected by using ITI with p-values fixed at $Pval_1 = 1.10^{-3}$ and $Pval_2 = 1.10^{-3}$ for the two null distributions of scores, respectively. After removal of overlapping networks (the overlapping score was fixed at $O_s = 60\%$), 123 subnetworks were detected to be differentially expressed and further retained for analysis, totaling 541 genes. Among these, 62 subnetworks

(279 genes) were detected to be expressed in basal subtype, and 61 subnetworks (262 genes) in luminal A subtype. A separate global analysis of gene expression and CNA gave, respectively, 5,000 genes differentially expressed and 1,000 genes with distinct frequency of alteration among the two studied molecular subtypes [44].

12.4.2 SUBNETWORK VISUALIZATION

Data visualization is a quite complex step as it involves data integration among PPI: gene expression on a large number of tumors encompassing two different subtypes, CNA information measured on the same tumor set, and a large body of annotation data for microarray probes, transcription factors, and genes (symbols and NCBI Entrez Gene accession number). To accommodate all these data types, we heavily modified the visualization routines of the original ITI pipeline [24], especially to superimpose the GISTIC values.

Figure 12.4 illustrates an example of subnetworks obtained with ITI and an integrated visualization (these data are accessible from the ITI Web site[6]). The subnetwork presented in Figure 12.4.A is expressed in luminal A subtype, while the one represented in Figure 12.4.B is expressed in basal subtype. For visualization, two additional figures of each subnetwork are generated for visualizing copy number variation (homozygous or heterozygous loss or gain). Sub-items 1, 2, and 3 of Figure 12.4 represent the ITI subnetwork score (blue = correlation with luminal A, red = correlation with basal subtype). GISTIC scores are directly represented on subnetworks with a red or green edge (red = gain, green = loss) for luminal A subtype (A.2 and B.2) and basal subtype (A.3 and B.3). The level of alteration or amplification is represented with a color code. The node is left white when the GISTIC score is not statistically significant. A screenshot of the complete CNV-ITI Web site is presented in Figure 12.5.

12.5 ANALYSIS AND RELEVANCE OF DETECTED SUBNETWORKS WITH RESPECT TO THE LITERATURE

The detected subnetworks are then analyzed for their biological relevance. Genes known to be specifically expressed in the two studied subtypes are found. In luminal A subtype, the gene that is the most frequently found in all subnetworks is without surprise ESR1 (found in nine subnetworks). FOXA1 is also frequently found by ITI (three subnetworks). ERG is found in one subnetwork. To the contrary, estrogen receptors or ERBB2 were not detected among basal tumors. Since these tumors are highly proliferative, genes related to cell cycle (cyclins in general, and CDKs) were found [4]. Also, among the 62 subnetworks expressed in basal, the Cyclin-dependent Kinase 6 was the most frequently found (in eight subnetworks). CDK6 is known to regulate tumor suppressor RB1. Genes coding for cyclin E1 (CCNE1) and the cyclin-dependent kinase 2 (CDK2) were also found in five and three subnetworks, respectively.

---

[6]http://iti.sourceforge.net/citi

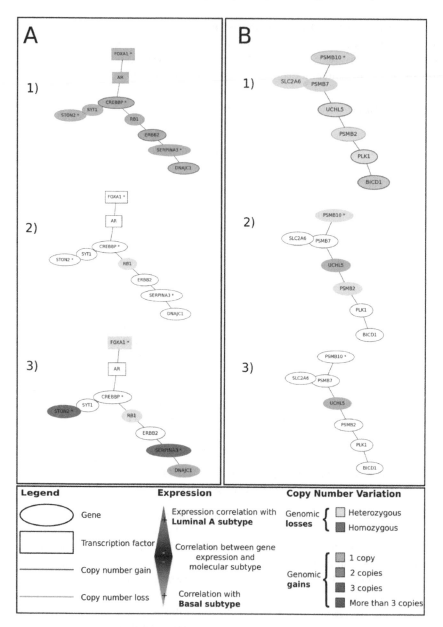

FIGURE 12.4 (SEE COLOR INSERT.) Examples of two subnetworks detected with ITI, expressed in the luminal A (A) and basal (B) subtypes. The two subnetworks are represented overlaid with three types of information related to gene expression and copy number variation. 1) Subnetwork with nodes colored as the gene expression correlation score with the phenotype (blue = correlated with luminal A subtype, red = correlated with basal subtype). 2) Subnetwork with nodes colored with CNA information (genomic loss or gain) with respect to luminal A subtype. 3) Subnetwork with nodes colored with CNA information (genomic loss or gain) with respect to basal subtype.

Subnetwork 12

Score

Dataset	Score	P-val 1	P-val 2	P-val 3
IPC-255	0.8978	1.603e-03	1.810e-04	8.053e-02

Subnetwork structure for each dataset

expression (LuminalA versus Basal)	IPC-255
aCGH-LuminalA	
aCGH-Basal	IPC-255

IPC255

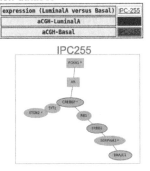

Score for each gene in subnetwork 12 in each dataset

Gene Symbol	Links	Frequency	Frequency Rank	Subnetwork score rank	Global rank	IPC-255
CREBBP		3	5	1	3	0.375
RB1		2	20	1	4	0.341
DNAJC1		1	47	65	69	0.580
SERPINA3		1	47	65	69	0.380
AR		14	1	1	1	0.847
ERBB2		2	20	1	4	0.336
SYT1		2	20	65	64	0.499
STON2		2	20	65	64	0.551
FOXA1		3	5	63	58	0.871

GO Enrichment output for subnetwork 12 in each dataset

Name	Accession Number	Link	P-val	Corrected P-val
prostate gland development	GO:0030850	GO	8.269E-09	2.02E-05
gland development	GO:0048732	GO	3.429E-07	4.189E-04
enucleate erythrocyte differentiation	GO:0043353	GO	5.998E-06	4.885E-03
N-terminal protein amino acid acetylation	GO:0006474	GO	8.394E-06	5.127E-03
regulation of lipid metabolic process	GO:0019216	GO	1.054E-05	5.152E-03
regulation of lipid kinase activity	GO:0043550	GO	1.119E-05	4.555E-03
regulation of T cell differentiation in the thymus	GO:0033081	GO	1.119E-05	3.905E-03
phosphoinositide 3-kinase cascade	GO:0014065	GO	1.119E-05	3.417E-03
urogenital system development	GO:0001655	GO	1.206E-05	3.273E-03
positive regulation of myeloid leukocyte differentiation	GO:0002763	GO	1.438E-05	3.513E-03
regulation of macrophage differentiation	GO:0045649	GO	1.438E-05	3.193E-03

FIGURE 12.5 ITI Visualization and organization. This shows visualization elements associated with subnetworks; subnetwork score and *p*-values; subnetwork graph structure and interaction, superimposed with gene expression data and CNA measurements; independent score measurement for each gene included in the subnetwork with links to annotation databases; GO enrichment information and *p*-values.

Bertucci et al. [4] constructed a list of discriminant genes in luminal A or basal subtypes from literature. We crossed the list of genes documented by Bertucci et al. [4] and the genes detected by ITI to see if CNV-ITI was able to retrieve genes with a known biological link with the phenotype. Among these, cyclin D1, MYB, a transcription factor that may play a role in tumorigenesis, and SMAD3, a major component of TGF signaling, were all present in luminal A. RUNX3 transcription factor was expressed in basal tumors. It is a tumor supressor gene that regulates carcinogenesis and that is silenced in breast cancer. The protein kinase LYN was also expressed in basal subtype. It encodes a tyrosine protein kinase that is known to be involved in basal signaling pathways [45]. CDK6 is known to be overexpressed in multiple cancer types [46]. As such, more than 5,000 genes were differentially expressed among the two subtypes [4], which makes the list of potential candidates for further analysis very large and potentially unpractical. This integrated analysis allows the reduction on a lower number of potential drivers ($n = 472$) and their interactors in the human interactome, which reduces the list of candidates, since only 114 genes were included in subnetworks. The complete gene list is presented in Figure 12.6.

12.5.1 IDENTIFICATION OF MUTATED GENES, TUMOR SUPPRESSORS, AND ONCOGENES

Figure 12.6 shows genes detected by the whole pipeline, with a detailed list of initial candidate genes and the list of genes found overexpressed in basal and overexpressed in luminal A subtypes. The GISTIC scores revealed many significant genomic losses in the two subtypes, particularly with basal ($n = 131$) tumors but also with luminal A tumors ($n = 78$). Genomic amplifications were also significant ($n = 78$ and $n = 56$, for basal and luminal A tumors, respectively). Although the chromosomal aberrations found are specific to each subtype, it is also possible to differentiate the two categories of tumors with the number of alterations, which are vastly superior in the basal populations. When we focused only on subnetwork seeds, 60% showed a genomic loss in the basal populations while only 21% in luminal A. Gains were less frequent and were observed for *CPB1*, *IL20RA*, *TNIK*, *EPB41L2*, and *MRC1* for basal tumors and for *CREBBP*, *TSC2*, *SPAG5*, and *TNFSRF17* for luminal A.

These losses have a significant impact on expression of corresponding transcripts and with genes with which they interact; this is why they are part of detected subnetworks (defined as groups of genes with significant expression changes throughout the interactome). These genes could be considered as drivers for basal ($n = 75$ drivers) and luminal A ($n = 30$ drivers). Genes associated with gains could also be considered as potential drivers, even though with a lower impact, as they also greatly influence gene expression in their neighborhood. To understand the role of these genes, their annotations were manually examined. Several genes were detected by CNV-ITI that had not been previously known to be involved in tumorigenesis. The carboxypeptidase B1 (*CPB1*), the *TRAF2* and *NCK* interacting kinase (*TNIK*), involved in *JNK* signaling, and *EPB41L2* have no known major role in carcinogenesis but were detected as differentially expressed.

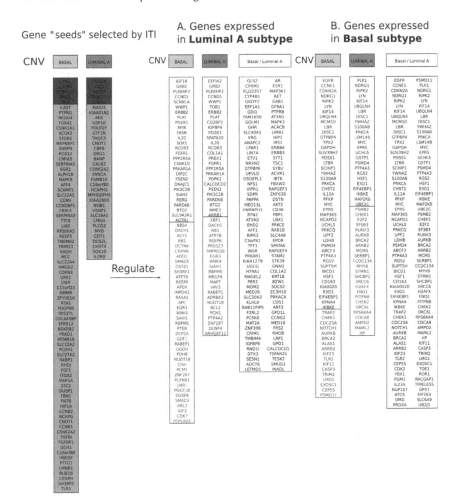

FIGURE 12.6 (SEE COLOR INSERT.) Detected driver genes with the CNV-ITI pipeline. On the left are represented the list of genes specifically expressed in luminal A and basal subtypes that have been retained as seeds by ITI (red = amplified, green = deleted). These are considered as putative driver genes, either tumor supressors or oncogenes. On the right, the list of genes regulated by these genes. In A, the list of genes found in subnetworks expressed in luminal A tumors. In B, the list of genes found in subnetworks expressed in basal tumors. For the A and B lists, we detailed the list of genes mutated in basal and luminal A tumors, or neither. In each of these sublists, genes marked as red are amplified and genes marked as green are deleted.

A total of 28 transcription factors were detected by CNV-ITI, while 8 of them were used as subnetwork seeds. These are *C16orf80*, *LMO4*, *GTF2B*, and *SOX10* (lost and under-expressed in luminal A tumors) and *FOXA1*, *EGR1*, *HIF1A*, *YBX1*, and *C16orf80* (lost and under-expressed in basal tumors). The early growth response 1 gene (*EGR1*) is known as a tumor suppressor gene. *FOXA1* is known to promote tumor growth in several types of cancer [47] and it has been established that SNP sites associated with breast cancer are enriched for *FOXA1* biding sites [48].

12.6 OTHER CNA-GENE EXPRESSION INTEGRATION METHODS FOR DETECTING DRIVER GENES IN CANCER

To illustrate the variety of approaches for detecting drivers in cancerology, we detail here two other methods, applied in the field of cancer for the detection of driver genes in melanoma (CONEXIC [14]), and in kidney carcinoma [19]. There were no methods previously developed for expression-CNV-PPI integration data, such as CNV-ITI, and hence these are considered to be the most relevant to our analysis.

12.6.1 CONEXIC METHOD BY AKAVIA ET AL.

The basic postulate by Akavia et al. [14] is that driving mutations are correlated with gene expression. A driver usually deregulates gene module expression that provokes tumorigenesis. A driver can also be deregulated without significant sequence alteration. Therefore, the method by Akavia et al. integrates gene expression and CNA to find signatures that are likely to contain drivers. The method is based on a previously published algorithm, module networks [49] that searches for regulated modules from gene expression data and precompiled lists of candidate control genes. COpy Number and Expression in Cancer (CONEXIC) extended Module Networks to make it suitable for driver detection in gene expression. CONEXIC uses a score-based search to detect gene modules that are differentially expressed and have the highest score within amplified or deleted regions. The output is a ranked list of modulators with high scores that correlate with differentially expressed modules in gene expression and that are located in significantly altered regions of the genome. The modules themselves are modulated or deleted, which indicates that they are likely to control gene expression of the corresponding modules. Because these alterations are recurrently found in a significant number of tumors, it is very likely that these modulators are driving tumorigenesis. This method was applied to melanoma samples as follows. First, a list of candidate drivers was generated using CNA profiles available from 101 melanoma samples by the GISTIC method. Then, CONEXIC was applied to 62 paired samples to select driver modulators. A total of 64 modulators were selected by the algorithm that explain the behavior of 7,869 genes. The top 30 were considered as drivers. To annotate the drivers and the genes they are modulating, an automated procedure was developed (LitVan) that connects genes to the complete text of published articles in the NCBI Pubmed Database.

12.6.2 METHOD BY BEROUKHIM ET AL.

Beroukhim et al. [19] performed an integrated analysis (CN changes and expression) on sporadic clear cell kidney carcinomas (ccRCC) and von-Hippel Lindau (VHL) disease. A total of 90 tumors were analyzed and searched for significant copy number changes. Amplified regions were then searched for consistent expression in gene expression data, which led to the identification of several relevant genes, including CDK2NA, CDK2NB, and MYC, among others. The analysis framework is articulated as follows. First, data were generated by extracting DNA and RNA from ccRCC tumors and VHL tumors. Significant regions were determined by the GISTIC method. Then, DNA from VHL and ccRCC tumors were profiled on CGH arrays. RNA was hybridized on DNA microarrays. Using GISTIC, seven amplified regions and seven deleted regions were identified in ccRCC with a q-value < 0.25. To take into account the fact that some peak regions may have been displaced by passengers events, more robust regions (called wide peak regions) were identified with boundaries that resisted a leave-one-out analysis that is robust to the iterative removal of one tumor. The first assumption of the authors is that the oncogene targets driven by CNAs are activated by overexpression. The search was done to prioritize genes in the peak amplified regions under the form of an integrated analysis. Specifically, genes being significantly expressed in amplified tumors (p-values were drawn by tumor label permutation) were selected, using a signal-to-noise ratio. Among the genes in the peak regions for which probes exist on the expression arrays, 23 were significant. MYC was consistently overexpressed in tumors with 8q24 amplification. A similar approach to detect under-expressed genes in detected regions yielded several tumor‑ suppressor genes (TSG). This approach identified CDKN2A, a known TSG, among others.

12.7 DISCUSSION

BC is a vastly heterogeneous disease, with tumors characterized by genomic events that are both common and unique, and associated with very heterogeneous gene expression changes. This makes the detection of relevant markers and predictive genes very difficult. We are specifically interested in the study of two subtypes that present opposite features, luminal A (differentiated) and basal (proliferative), and that are extremely different at the genomic, transcriptomic, and clinical levels.

To tackle the disease heterogeneity, integrative methods must be applied to take into account the information available at multiple biological levels and separate *driver* (genes that are the origin of the disease) from *passenger* (genes deregulated or altered as a collateral change to driver) genes. Methods integrating CNA information (measured by array-CGH) and gene expression were developed for various types of cancer. They were able to detect markers that are either expressed and amplified or under-expressed and deleted. In particular, the method from Beroukhim et al. [19] allowed the detection of new oncogenes in ccRCC tumors. The method by Akavia et al. [14], CONEXIC, made use of regulatory information to bring the detection to the level of modules, and separate driver from passenger events. Application to melanoma patients resulted in the identification of new candidate drivers as well as previously known oncogenes.

Applied to luminal A and basal BC subtypes, these methods still yield hundreds of potential candidates. Additional biological information must be taken into account by the detection algorithm to distinguish drivers and associated gene modules that are acting on the system—this has to be done separately for each subtype. We combined the integration method by Bekhouche et al. [1] and the ITI algorithm to build the CNV-ITI pipeline. A set of 471 candidate genes were first selected by a CNA-expression integration. These genes were submitted to ITI as subnetwork seeds to determine if they drive modules of interacting genes that are differentially expressed. ITI gave a list of differentially expressed subnetworks after a complete interactome parsing and score-based statistical validation. Only 24% of the 471 initially identified genes were finally retained in the 123 validated subnetworks,. The method was able to focus the analysis on relevant sets of genes. A total of 61 subnetworks expressed in the luminal A subtype and 62 expressed in the basal subtype were identified. Known markers were detected (*ESR1* for luminal A, cyclins, and kinases for basal), and new potential oncogenes and TSGs were identified. Finally, the number of candidate drivers was significantly reduced, which increased their statistical power as markers.

12.8 CONCLUSION

In this chapter, we describe the comprehensive genomic hybridization-interactome-transcriptome integration (CNV-ITI) pipeline for for the detection of driver genes in cancerology. This pipeline works in two steps. First, it detects candidate markers by overlapping CNA profiles and gene expression profiles. Then, these candidates are submitted to the interactome-transcriptome integration (ITI) pipeline for validation by searching for differentially expressed subnetworks in the human interactome. Retained drivers are the ones confirmed for being involved in differentially expressed subnetworks, since they are selected for modules that are driving the disease. All data is then stored in a bioinformatics resource for visualization and further analysis. We also improved previously published ITI visualization by superimposing deletion and amplification information in addition to the gene expression on subnetworks. As an illustrative example, we performed an analysis of specific subtypes in breast cancer with CNV-ITI. Two specific BC subtypes (luminal A (80 tumors) and basal (68 tumors)) were searched for driver and passenger genes. The CNV-ITI pipeline could potentially be applied to obtain an integrated view of the massive amount of data generated by international cancer consortia (The Cancer Genome Atlas, TCGA [50], the International Cancer Genome Consortium, ICGC, [51] or others). This would help obtain reliable markers that are not only significant, but also with the complete biological information and background under the form of gene interaction that would prove that these are, hence, drivers of the disease. These integrated analyses are a necessary step towards understanding the mechanisms that favor tumorigenesis in cancerology. It can be extended to integrate other levels of information such as point mutations and miRNA expression.

12.9 ACKNOWLEDGMENTS

M. Garia and R. Millat-Carus contributed equally to this work. We would like to thank our funding sources. This research was funded by the Institut National du

Cancer, the Ligue Nationale Contre le Cancer (label DB), and the Institut National de la Santé et de la Recherche Médicale. Support for the computational infrastructure was obtained from a Fondation pour la Recherche Médicale grant. Maxime Garcia is funded by the Institut National de la Santé et de la Recherche Médicale—Région Provence-Alpes Cote d'Azur Fellowship. Support for Raphaële Millat-Carus was obtained from the Institut National du Cancer Grant. Thanks to Sabrina Carpentier (Ipsogen, Marseille, France) for helpful discussions on the original ITI method, and Wahiba Gherraby for proofreading the manuscript.

REFERENCES

1. I. Bekhouche, P. Finetti, J. Adelaide, et al., "High-resolution comparative genomic hybridization of inflammatory breast cancer and identification of candidate genes," *PLoS One*, vol. 6, no. 2, p. e16950, 2011.

2. P. J. van Diest, J. A. Belien, and J. P. Baak, "An expert system for histological typing and grading of invasive breast cancer. first set up," *Pathology—Research and Practice*, vol. 188, pp. 405–409, Jun 1992.

3. F. Bertucci, P. Finetti, N. Cervera, D. Maraninchi, P. Viens, and D. Birnbaum, "Gene expression profiling and clinical outcome in breast cancer," *Omics: A Journal of Integrative Biology*, vol. 10, no. 4, pp. 429–443, 2006.

4. F. Bertucci, P. Finetti, N. Cervera, et al., "How different are luminal A and basal breast cancers?," *International Journal of Cancer*, vol. 124, pp. 1338–1348, Mar 2009.

5. C. M. Perou, T. Sorlie, M. B. Eisen, et al., "Molecular portraits of human breast tumours," *Nature*, vol. 406, pp. 747–752, Aug 2000.

6. T. Sorlie, C. M. Perou, R. Tibshirani, et al., "Gene expression patterns of breast carcinomas distinguish tumor subclasses with clinical implications," *Proceedings of the National Academy of Sciences, USA*, vol. 98, pp. 10869–10874, Sep 2001.

7. E. Charafe-Jauffret, C. Ginestier, F. Monville, et al., "How to best classify breast cancer: conventional and novel classifications (review)," *International Journal of Oncology*, vol. 27, pp. 1307–1313, Nov 2005.

8. M. Guedj, L. Marisa, A. de Reynies, et al., "A refined molecular taxonomy of breast cancer," *Oncogene*, vol. 31, pp. 1196–1206, Mar 2012.

9. C. Curtis, S. P. Shah, S.-F. Chin, et al., "The genomic and transcriptomic architecture of 2,000 breast tumours reveals novel subgroups," *Nature*, vol. 486, pp. 346–352, Jun 2012.

10. A. Prat, J. S. Parker, O. Karginova, et al., "Phenotypic and molecular characterization of the claudin-low intrinsic subtype of breast cancer," *Breast Cancer Research*, vol. 12, no. 5, p. R68, 2010.

11. C. Sotiriou, S.-Y. Neo, L. M. McShane, et al., "Breast cancer classification and prognosis based on gene expression profiles from a population-based study," *Proceedings of the National Academy of Sciences, USA*, vol. 100, pp. 10393–10398, Sept. 2003.

12. L. J. van 't Veer, H. Dai, M. J. van de Vijver, et al., "Expression profiling predicts outcome in breast cancer," *Breast Cancer Research*, vol. 5, no. 1, pp. 57–58, 2003.

13. Z. Hu, C. Fan, D. S. Oh, et al., "The molecular portraits of breast tumors are conserved across microarray platforms," *BMC Genomics*, vol. 7, p. 96, 2006.

14. U. D. Akavia, O. Litvin, J. Kim, et al., "An integrated approach to uncover drivers of cancer," *Cell*, vol. 143, pp. 1005–1017, Dec 2010.

15. G. Curigliano, "New drugs for breast cancer subtypes: targeting driver pathways to overcome resistance," *Cancer Treatment Reviews*, vol. 38, pp. 303–310, Jun 2012.

16. D. S. P. Tan and J. S. Reis-Filho, "Comparative genomic hybridisation arrays: high-throughput tools to determine targeted therapy in breast cancer," *Pathobiology*, vol. 75, no. 2, pp. 63–74, 2008.

17. A. Bergamaschi, Y. H. Kim, P. Wang, et al., "Distinct patterns of DNA copy number alteration are associated with different clinicopathological features and gene-expression subtypes of breast cancer," *Genes, Chromosomes and Cancer*, vol. 45, pp. 1033–1040, Nov 2006.

18. R. Beroukhim, G. Getz, L. Nghiemphu, et al., "Assessing the significance of chromosomal aberrations in cancer: methodology and application to glioma," *Proceedings of the National Academy of Sciences, USA*, vol. 104, pp. 20007–20012, Dec. 2007.

19. R. Beroukhim, J.-P. Brunet, A. Di Napoli, et al., "Patterns of gene expression and copy-number alterations in Von-Hippel Lindau disease-associated and sporadic clear cell carcinoma of the kidney," *Cancer Research*, vol. 69, pp. 4674–4681, Jun 2009.

20. W. M. Lin, A. C. Baker, R. Beroukhim, et al., "Modeling genomic diversity and tumor dependency in malignant melanoma," *Cancer Research*, vol. 68, pp. 664–673, Feb 2008.

21. H.-Y. Chuang, E. Lee, Y.-T. Liu, D. Lee, and T. Ideker, "Network-based classification of breast cancer metastasis," *Molecular Systems Biology*, vol. 3, p. 140, 2007.

22. Y. Wang, J. G. M. Klijn, Y. Zhang, et al., "Gene-expression profiles to predict distant metastasis of lymph-node-negative primary breast cancer," *The Lancet*, vol. 365, pp. 671–679, Feb. 2005.

23. M. J. van de Vijver, Y. D. He, L. J. van't Veer, et al., "A gene-expression signature as a predictor of survival in breast cancer," *The New England Journal of Medicine*, vol. 347, pp. 1999–2009, Dec. 2002.

24. M. Garcia, R. Millat-Carus, F. Bertucci, P. Finetti, D. Birnbaum, and G. Bidaut, "Interactome-transcriptome integration for predicting distant metastasis in breast cancer," *Bioinformatics*, vol. 28, pp. 672–678, Mar 2012.

25. M. Garcia, O. Stahl, P. Finetti, D. Birnbaum, F. Bertucci, and G. Bidaut, "Linking interactome to disease: A network-based analysis of metastatic relapse in breast cancer," in *Handbook of Research on Computational and Systems Biology: Interdisciplinary Applications*, 2011.

26. M. Ashburner, C. A. Ball, J. A. Blake, et al., "Gene ontology: tool for the unification of biology. The Gene Ontology Consortium," *Nature Genetics*, vol. 25, pp. 25–29, May 2000.

27. S. Tavazoie, J. D. Hughes, M. J. Campbell, R. J. Cho, and G. M. Church, "Systematic determination of genetic network architecture," *Nature Genetics*, vol. 22, pp. 281–285, Jul 1999.

28. E. W. Sayers, T. Barrett, D. A. Benson, et al., "Database resources of the National Center for Biotechnology Information," *Nucleic Acids Research*, vol. 40, pp. D13–D25, Jan 2012.

29. I. Rivals, L. Personnaz, L. Taing, and M.-C. Potier, "Enrichment or depletion of a GO category within a class of genes: which test?," *Bioinformatics*, vol. 23, pp. 401–407, Feb 2007.

30. K. H. Young, "Yeast two-hybrid: so many interactions, (in) so little time...," *Biology of Reproduction*, vol. 58, pp. 302–311, Feb. 1998.

31. E. N. Brody, L. Gold, R. M. Lawn, J. J. Walker, and D. Zichi, "High-content affinity-based proteomics: unlocking protein biomarker discovery," *Expert Review of Molecular*

Diagnostics, vol. 10, pp. 1013–1022, Nov 2010.

32. C. M. Deane, L. Salwinski, I. Xenarios, and D. Eisenberg, "Protein interactions: Two methods for assessment of the reliability of high throughput observations," *Molecular & Cellular Proteomics*, vol. 1, pp. 349–356, May 2002.

33. T. S. Keshava Prasad, R. Goel, K. Kandasamy, et al., "Human Protein Reference Database-2009 update," *Nucleic Acids Research*, vol. 37, pp. D767–D772, Jan 2009.

34. B. Aranda, P. Achuthan, Y. Alam-Faruque, et al., "The IntAct molecular interaction database in 2010," *Nucleic Acids Research*, vol. 38, pp. D525–D531, Jan 2010.

35. A. Ceol, A. Chatr Aryamontri, L. Licata, et al., "MINT, the molecular interaction database: 2009 update," *Nucleic Acids Research*, vol. 38, pp. D532–D539, Jan 2010.

36. L. Salwinski, C. S. Miller, A. J. Smith, F. K. Pettit, J. U. Bowie, and D. Eisenberg, "The Database of Interacting Proteins: 2004 update," *Nucleic Acids Research*, vol. 32, pp. D449–D451, Jan 2004.

37. A. K. Ramani, R. C. Bunescu, R. J. Mooney, and E. M. Marcotte, "Consolidating the set of known human protein-protein interactions in preparation for large-scale mapping of the human interactome," *Genome Biology*, vol. 6, no. 5, p. R40, 2005.

38. J. Tsai, R. Sultana, Y. Lee, et al., "RESOURCERER: a database for annotating and linking microarray resources within and across species," *Genome Biology*, vol. 2, no. 11, p. SOFTWARE0002, 2001.

39. F. Reyal, N. Stransky, I. Bernard-Pierrot, et al., "Visualizing chromosomes as transcriptome correlation maps: evidence of chromosomal domains containing co-expressed genes–a study of 130 invasive ductal breast carcinomas," *Cancer Research*, vol. 65, pp. 1376–1383, Feb 2005.

40. R. Sabatier, P. Finetti, N. Cervera, et al., "A gene expression signature identifies two prognostic subgroups of basal breast cancer," *Breast Cancer Research and Treatment*, vol. 126, pp. 407–420, Apr. 2011.

41. A. B. Olshen, E. S. Venkatraman, R. Lucito, and M. Wigler, "Circular binary segmentation for the analysis of array-based DNA copy number data," *Biostatistics (Oxford, England)*, vol. 5, pp. 557–572, Oct. 2004.

42. H. K. Lee, W. Braynen, K. Keshav, and P. Pavlidis, "ErmineJ: tool for functional analysis of gene expression data sets," *BMC Bioinformatics*, vol. 6, p. 269, 2005.

43. V. Matys, O.V. Kel-Margoulis, E. Fricke, et al., "TRANSFAC and its module TRANSCompel: transcriptional gene regulation in eukaryotes," *Nucleic Acids Research*, vol. 34, pp. D108–D110, Jan 2006.

44. J. Adelaide, P. Finetti, I. Bekhouche, et al., "Integrated profiling of basal and luminal breast cancers," *Cancer Research*, vol. 67, pp. 11565–11575, Dec 2007.

45. D. R. Croucher, F. Hochgrafe, L. Zhang, et al., "Involvement of Lyn and the atypical kinase SgK269/PEAK1 in a basal breast cancer signaling pathway," *Cancer Research*, vol. 73, pp. 1969–1980, Mar 2013.

46. P. J. Roberts, J. E. Bisi, J. C. Strum, et al., "Multiple roles of cyclin-dependent kinase 4/6 inhibitors in cancer therapy," *Journal of the National Cancer Institute*, vol. 104, pp. 476–487, Mar 2012.

47. M. Katoh, M. Igarashi, H. Fukuda, H. Nakagama, and M. Katoh, "Cancer genetics and genomics of human FOX family genes," *Cancer Letters*, vol. 328, pp. 198–206, Jan 2013.

48. M. R. Katika and A. Hurtado, "A functional link between FOXA1 and breast cancer SNPs," *Breast Cancer Research*, vol. 15, p. 303, Feb 2013.

49. E. Segal, M. Shapira, A. Regev, et al., "Module networks: identifying regulatory mod-

ules and their condition-specific regulators from gene expression data," *Nature Genetics*, vol. 34, pp. 166–176, Jun 2003.

50. Cancer Genome Atlas Research Network, "Comprehensive genomic characterization defines human glioblastoma genes and core pathways," *Nature*, vol. 455, pp. 1061–1068, Oct 2008.

51. The International Cancer Genome Consortium, "International network of cancer genome projects," *Nature*, vol. 464, pp. 993–998, Apr 2010.

13 Mining Gene-Sample-Time Microarray Data

Yifeng Li and Alioune Ngom

CONTENTS

13.1 INTRODUCTION

With recent advances in microarray technology, the expression levels of genes with respect to samples can be monitored synchronically over a series of time points. Such microarray data have three types of variables: genes, samples, and time points. Thus, they can be represented as tensor data of order three and are termed *gene-sample-time* microarray data or GST data for short. In the literature, GST data are known as three-dimensional (3D) data. In order to avoid any confusion with the dimensionality of a vector in linear algebra and multivariate statistics, we follow the definition in tensor (or multilinear) algebra and refer to GST data as order three tensor data, or three-way data. We should clarify that dimensionality refers to the number of features of a sample rather than the number of orders of a single value in a data set. Therefore, the dimensionality of a sample in a GST data set is the number of genes times the number of time points. Figure 13.1 gives an example of a GST data set in tensor and heatmap representations.

Machine learning and data mining approaches are among the main tools for analyzing GST data. For example, by applying classification and gene selection methods, candidate marker genes are selected in [1] from the IFNβ GST data. They can help with discovering the pathway of multiple sclerosis. By applying biclustering and triclustering methods, one can identify subpatterns of genes and samples. It has been well known that subtypes of breast cancer can be identified by biclustering [2, 3]. Classification techniques can be applied on GST data for the diagnosis of diseases, and the prognosis of an ongoing treatment. If a computational method suggests that the patient would not respond well to the current treatment based on the time-series data, then the therapy must be revised or stopped, as many treatments have severe side effects.

Current approaches of analyzing two-way microarray data, such as time-series microarrays (gene-time data) or tissue microarrays (gene-sample data), include supervised and unsupervised learning and data mining methods. However, two-way approaches, such as clustering models, for instance, may not be suitable to describe the relationships between the three variables in GST data. They cannot be directly generalized or extended for the three-way GST data. Current GST data contain many missing values in all three directions. They also contain a large number of genes with very few numbers of samples and time-points, and thus they require efficient and effective methods for tackling the potential *curse of dimensionality* arising while learning a computational model. In general, if the number of parameters is much larger than the number of training samples, and grows rapidly even exponentially as the number of dimensions increases, the curse of dimensionality occurs. In this chapter, we first summarize the techniques of handling missing values for microarray data. Next, we discuss the coherent gene clustering methods and the triclustering approaches recently proposed for GST data. We then introduce dynamic graphical models (specifically hidden Markov models) for the direct classification of GST data samples. After that, we present a dimensionality reduction technique based on tensor factorization. Finally, we show that kernel methods including *support vector machine* (SVM), sparse coding, and dictionary learning (which are initially designed for vectorial samples) can be used for classifying GST data.

13.2 MISSING VALUE ESTIMATION

Microarray data often suffer from *missing values* due to many factors such as insufficient resolution, image corruption, artifacts, systematic errors, or incomplete experiments. Microarray data analysis methods based on machine learning (such as clustering, dimensionality reduction, and classification) often require complete gene expression data sets, in order to perform robustly and effectively. Therefore, incomplete data sets need to be pre-processed for these methods before analysis, or handled carefully during their process. GST microarray data sets suffer from even more severe missing values problems. For example, if a patient is not available to obtain a sample at a certain time point, a whole vector for this patient is thus missing in the

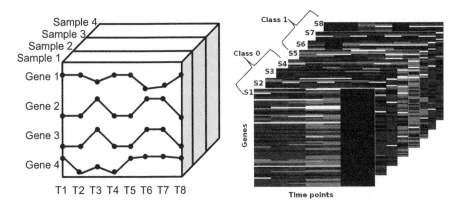

FIGURE 13.1 A 3-way tensor representation (left) and a heatmap representation (right) of a GST data set. The black columns in the heatmap representation correspond to missing time points.

data set.

There are three strategies for handling the issue of missing values in two-way data. First, we can remove the genes or samples (time points) with missing values. However, the main drawback is that the already small sample (time point) size becomes smaller and we may face the risk of having not enough data for learning a model. The second strategy is to impute (that is, to fill in) the missing values in the data [4, 5]. Missing values can be either imputed by a constant value (e.g., 0) or by feature averages, or alternatively they can be estimated by some statistical or machine learning methods. This strategy essentially completes an incomplete data set, and hence, avoids deleting genes or samples (or time points). The missing values can be estimated either before any analysis or during the learning of a model. The time complexity for making data complete before learning is usually much lower than that of combining missing value estimation with a learning model. The third strategy is that only the observed values are allowed to be used during the learning and prediction processes. Thus, it is usually called weighting strategy. For incomplete training data **D**, the objective of a weighted regression method for fitting a model can be expressed as

$$\min_{\mathbf{M}} \frac{1}{2} \|\mathbf{W} * (\mathbf{D} - \mathbf{M})\|_F^2, \tag{13.1}$$

where **M** contains the values estimated by the model, **W** is a weighting matrix with $w_{ij} = 1$ indicating that value d_{ij} is present in the data and $w_{ij} = 0$ indicating that d_{ij} is missing. Notation "$*$" is the Hadamard (element-wise) product operator. $\|\mathbf{A}\|_F$ is defined as the Frobenius norm of matrix **A**.

Some imputation methods have been proposed specifically for two-way microarray data sets since the work of [4, 5]. An imputation method that estimates a missing value by feature average value was investigated in [4, 5], and was shown to give the worst performance (with respect to normalized root mean squared error) among many imputation methods. In [4, 5], a k-nearest-neighbor based imputation method, or *KNNimpute* for short, has been proposed using the Euclidean, Pearson correlation, and variance minimization similarity metrics, and the Euclidean metric gave the best results. For each incomplete gene g_i with missing value g_{ij} in the j-th sample or time point, imputation is done by first finding its k nearest complete genes, and then taking the weighted average value at column j of those k genes as estimation of g_{ij}. The authors of [4, 5] also describes a singular-value-decomposition-based imputation method, *SVDimpute*. A set of eigen-genes is found by applying SVD on the complete genes only. Each incomplete gene is then represented as a linear combination of those eigen-genes. Linear regression using the *expectation maximization* (EM) algorithm is performed on those eigen-genes to estimate the missing values of given incomplete genes. *LLSimpute*, a local least squares method [6], represents an incomplete gene as a linear combination of its k nearest complete neighbors. Least square optimization is used to find the coefficients of the linear combination, which are then used for estimating the missing values of the incomplete gene. Other methods based on least squares regression are also introduced in [7, 8, 9]. The Bayesian principal component analysis method, *BPCAimpute* [10], applies PCA similarly to the SVD method. However, an EM-like Bayesian estimation algorithm is used to estimate the coefficients of the linear combinations for each incomplete gene to impute. In [11], genes are represented as cubic spline functions first, and then missing values for incomplete genes are estimated by resampling the continuous curves. An autoregressive-model-based missing value estimation method (*ARLSimpute*) [12], first applies auto-regression on a set of k similar genes (missing values are set to zero initially) to estimate their AR coefficients by means of a least squares method. Then, using the AR coefficients, missing values for all incomplete genes are imputed by means of another least squares error regression method. ARLSimpute is the only method devised specifically for time-series profiles; however, it works only on long stationary time-series data. It is also the only method that is able to impute a time point (entirely missing column of a gene-time microarray data). Except for the approach in [11], imputation methods for two-way microarray data are all based on similar principles: they either find the nearest complete genes to impute incomplete genes, or find eigen-genes to impute incomplete genes, or combine these two principles. Current methods, except the ARLSimpute method, are initially devised for static data, though they have been applied to time-series data. They do not work when an entire column of a two-way microarray data is missing.

Weighted non-negative matrix factorization (WNMF) is an implementation of the weighting strategy [13, 14]. Suppose the non-negative incomplete data $\mathbf{D}$ is to be decomposed into two non-negative matrices as $\mathbf{D} \approx \mathbf{AY}$. The weighted multiplicative update rules of WNMF are

$$\begin{cases} \mathbf{A} = \mathbf{A} * \dfrac{(\mathbf{W}*\mathbf{D})\mathbf{Y}^{\mathrm{T}}}{(\mathbf{W}*\mathbf{AY})\mathbf{Y}^{\mathrm{T}}} \\ \mathbf{Y} = \mathbf{Y} * \dfrac{\mathbf{A}^{\mathrm{T}}(\mathbf{W}*\mathbf{D})}{\mathbf{A}^{\mathrm{T}}(\mathbf{W}*\mathbf{AY})} \end{cases} . \tag{13.2}$$

It is even more important to handle missing values for GST data by imputation or weighting, because if the removal strategy is used and a value is missing in the three-way data set, the corresponding gene-slice, sample-slice, or time-slice should be deleted for completeness. If there are many missing values distributed randomly in a GST data set, by removing a slice for each missing value, there may not be complete data left for analysis. Furthermore, many data mining approaches, such as the triclustering methods introduced later, require complete data.

The missing value imputation issue for GST data was first investigated in [15], where KNNimpute and SVDimpute are extended for GST data and are called *3KN-Nimpute* and *3SVDimpute*, respectively. In both three-way methods, if a gene time-series has missing values, the missing values can be estimated by using KNNimpute or SVDimpute on either the same sample-slice or the same gene-slice. Using this flexibility, both three-way methods can impute the whole missing time point for a sample, while the original KNNimpute and SVDimpute for two-way data cannot estimate missing time points for a two-way time-series data set.

Similar to the EM-based SVD imputation method for two-way data, tensor factorization can be used to impute missing values for GST data. See Appendix 13.8 for a short introduction to tensor algebra. An iterative imputation method using PARAFAC decomposition is given in [16, 17]. The basic idea is the following. The missing values in the tensor data $\mathcal{D}$ are first initialized by random numbers. Next, $\mathcal{D}$ is factorized into $\mathcal{D} = [\![\mathbf{A}, \mathbf{B}, \mathbf{C}]\!]$ by PARAFAC. After that, the missing values are replaced with $d_{ijk} = \sum_{f=1}^{R} a_{if} b_{jf} c_{kf}$, where R is the tensor rank. The above steps are repeated until there is no change in the estimated values. The PARAFAC-alternating least squares with single imputation (PARAFAC-ALS-SI) [18] is an implementation of the above idea using the ALS algorithm. If $\mathcal{D}$ is non-negative, it is necessary that the factors $\mathbf{A}$, $\mathbf{B}$, and $\mathbf{C}$ must be restricted by non-negativity. In this situation, non-negative PARAFAC should be used [19, 20, 17].

The general weighted regression model for a three-way data set can be extended as

$$\min_{\mathcal{M}} \frac{1}{2} \|\mathcal{W} * (\mathcal{D} - \mathcal{M})\|_F^2. \tag{13.3}$$

The model to be learned may be constrained by non-negativity, orthogonality, or other constraints. The *incomplete data PARAFAC* algorithm (INDAFAC) [18] is an implementation to fit the PARAFAC model using a weighting strategy. The *sparse non-negative Tucker* algorithm (SN-TUCKER or HONMF) [21] is the extension of weighted sparse *non-negative matrix factorization* (NMF) into non-negative Tucker decomposition that can ignore missing values.

13.3 CLUSTERING

Clustering and biclustering are important techniques in two-way microarray data analysis for identifying biological patterns [2, 22, 23]. In this section, we survey the clustering methods for three-way GST data. In the context of clustering, a GST data set is usually represented as a *gene × sample × time* tensor.

13.3.1 BICLUSTERING

We first introduce the existing biclustering method for GST data. More details on bi-clustering can be found in Chapter 14. Jiang et al. proposed a pattern-based coherent clustering method in [24]. A *coherent gene cluster* contains a subset of genes and a subset of samples such that the genes behave coherently across the samples along the full time direction. Thus, this method is a biclustering method. Before introducing the mining algorithm, we give two definitions. First, for a gene g_i, Pearson's correlation coefficient is used to measure the correlation between the time-series of two samples. If for any pair of samples $s_{j_1}, s_{j_2} \in S$, $\rho(m_{ij_1}, m_{ij_2}) \geq \delta$, where $\rho(m_{ij_1}, m_{ij_2})$ is the Pearson's correlation coefficient between time series m_{ij_1} and m_{ij_2} and δ is a minimum coherence threshold, then gene g_i is said to be *coherent* on the subset of samples S. Second, for a subset of genes G, if any gene $g_i \in G$ is coherent on S, then the submatrix $G \times S$ is called a *coherent gene submatrix/cluster*.

In order to find the maximal subsets S where gene g_i is coherent, the correlation coefficients $\rho(m_{ij_1}, m_{ij_2})$ of any pair of samples need to be computed. Then, a graph is constructed, where an edge corresponds to $\rho(m_{ij_1}, m_{ij_2}) \geq \delta$. The maximal cliques in the graph are found by a recursive depth-first search on a set of enumeration trees, where a clique is a complete subgraph corresponding to a coherent sample subset.

To avoid costly computations, two search algorithms, sample-gene search and gene-sample search, are devised for mining coherent gene clusters. Both algorithms use depth-first search on the enumeration trees of samples and genes, respectively. In sample-gene search, for each subset of samples S, the maximal subsets of genes G are found such that $(G \times S)$ is a coherent gene cluster. $(G \times S)$ is then tested to see if it is a maximal coherent gene cluster. The gene-sample search algorithm works analogously. Both algorithms can discover identical complete sets of coherent gene clusters. However, the sample-gene search algorithm is more efficient, because the number of samples is much smaller than the number of genes.

The *extended dimension iterative signature algorithm* (EDISA) [25] is another biclustering algorithm that extends the *iterative signature algorithm* (ISA). ISA was proposed in [26] for clustering microarray data. ISA can be treated as a generalization of SVD. EDISA is a probabilistic biclustering approach for GST data. There are two phases in EDISA. First, EDISA is initialized by various gene expression *modules* (clusters), including coherent, independent response, and single response modules, which are defined mathematically. Second, the modules are refined by removing genes and samples until the corresponding module scores are satisfied. Third, the modules are merged into maximal ones.

13.3.2 TRICLUSTERING

One issue of the coherent gene clustering discovery approach is that only the full time space is considered. When a subinterval of time points T is considered, the resulting three-way subtensor $G \times S \times T$ called a coherent *tricluster*, where G is a subset of genes, S is a subset of samples, and T is a consecutive subinterval of time points. The main existing triclustering methods are introduced below.

TRICLUSTER [27] is the first triclustering method that can mine overlapping triclusters. The types of clusters that TRICLUSTER can identify include constant, scaling (multiplicative), and shifting (additive) patterns. TRICLUSTER is a graph-based method that has the following four phases. First, valid ratio-ranges for all pairs of samples are found for each gene-sample slice of a time point, and a weighted and directed range multigraph is constructed. Second, for each gene-sample slice, all maximal biclusters are obtained by a depth-first search on the range multigraph. Third, the maximal triclusters are obtained by enumerating the biclusters of all time points and extending a tricluster along the time direction. Finally, some triclusters are optionally merged and pruned if they are overlapping substantially. There exist some improvements of TRICLUSTER in either the aspect of computation or precision in literature. The *ParTriCluster* algorithm, proposed in [28], is a parallel version of TRICLUSTER. Because parallel computation is beyond the scope of this book, details of ParTriCluster are omitted.

Due to the strict constraints of TRICLUSTER, some patterns may be missed. In order to overcome this weakness, a general triclustering model (gTRICLUSTER) was proposed in [29]. By replacing the symmetry property of TRICLUSTER with *Spearman rank correlation* (SRC) in order to evaluate the local similarity of a pair of samples, gTRICLUSTER can capture more patterns than TRICLUSTER. Both TRICLUSTER and gTRICLUSTER consider only the similarity between a pair of samples.

The *order preserving triclustering* (OPTricluster) algorithm, proposed in [30], can measure both similarity and difference between a pair of samples. OPTricluster is designed for clustering three-way *short* time-series data. This algorithm is a fusion of combinatorial approach on the sample dimension and the order preserving concept on the time dimension. This algorithm has five steps. First, the gene expression data are discretized. Second, for each gene and each sample, the discrete values are ranked in the time axis. Third, distinct patterns are identified for each subset of the samples. Fourth, conserved and divergent clusters are identified. Finally, the resulting triclusters are evaluated by *p*-values and gene ontology. OPTricluster has two potential limitations for large GST data. First, it becomes inefficient for a large number of samples, since all possible subsets of samples are generated to identify distinct patterns. Second, subintervals of time points are not considered.

13.4 NON-KERNEL METHODS FOR CLASSIFICATION

Classification techniques for GST data can be used for diagnosis and prognosis. In this section, we shall introduce non-kernel classification and non-kernel feature extraction approaches for GST data including dynamic graphical models and tensor factorizations. In the next section, we introduce kernel approaches. In the context of classification, a GST data set is usually represented by a *Gene* × *Time* × *Sample* tensor. Most of the three-way classification and feature extraction methods have been performed on the IFNβ data [1]. *Interferon beta* (IFNβ) is a protein used for treating patients afflicted with multiple sclerosis (MS), among other diseases. Some MS patients who received IFNβ therapy do not respond well to the drug and the reasons are still not clear [31]. Medical researchers are seeking for genomic reasons via high-throughput data analysis. Baranzini et al. [1], among others researchers, applied Bayesian learning methods on a clinical microarray data set to determine pairs or triplets of genes that can discriminate between bad and good IFNβ responders. This data set is available online as a supplementary material [1]. The initial data set is a GST data sampled from 53 MS patients who were initially treated with equal doses of IFNβ over a time period. This initial data set contains the expression measurements for 76 genes at 7 time points (0, 3, 6, 9, 12, 18 and 24 months) for each patient, with 31 patients responding well and the remaining 22 responding badly to the treatment. This data set contains genes with missing expression measurements at some time points. Those genes and corresponding samples were removed from our analysis, and hence, the resulting "complete" data contains 53 genes and 27 samples (18 good responders and 9 bad responders). In this chapter, we denote this data set as IFNβ, and use it as the working data set for all the classification methods discussed below.

13.4.1 DIRECT CLASSIFICATION

The authors of [1] proposed an *integrated Bayesian inference system* (IBIS) to select triplets of genes for classifying IFNβ samples but using only the first time point, and thus did not benefit from (nor consider) the full GST data. In [32], *generative hidden Markov models* (GenHMMs) and *discriminative HMMs* (DiscHMMs) approaches were devised for classifying IFNβ samples. Samples from the same class are used to train a GenHMM, whereas samples from all classes are used to train a DiscHMM; then a test sample is assigned to a class based on the maximum conditional likelihood. The Baum-Welch algorithm is used to estimate the parameters of the models. For DiscHMMs, backward gene selection is first performed to find a small number of discriminative genes before training the models.

The authors of [33] proposed a robust constrained mixture estimation approach to classify the IFNβ data. This approach combines the constrained clustering method with a mixture estimation classification framework. Subdivision of classes and mislabeled samples can be investigated by this approach. During training, negative constraints were restricted to pairs of samples. The constrained mixture model, with linear HMMs, as components, is optimized by an EM algorithm. The supervised version of this approach (*HMMConst*) only uses the training set in the estimation of parameters, while the semi-supervised version (*HMMConstAll*) uses all the data. The emission probability for each state is modeled by a mixture of multivariate Gaussians for patient expression values, noise, and missing values, respectively. In order to select genes contributing to classification, an HMM-based gene ranking method is used. Each component of the mixture model is assigned to a class. When testing, a test sample is assigned to a class according to the maximum entry in their posterior distribution.

The classification performance of GenHMMs and DiscHMMs are compared on IFNβ data. While the implementations of HMMConst and HMMConstAll [33] are not available, we thus cannot investigate their performance. Nine-fold cross-

TABLE 13.1

Classification performance of dynamic graphical models on complete IFNβ data.

Method	Parameter	Specificity	Sensitivity	Accuracy
GenHMMs	-	0.8611 ± 0.036	0.5556 ± 0.000	0.7593 ± 0.044
DiscHMMs	-	0.8611 ± 0.036	0.5556 ± 0.000	0.7593 ± 0.044
GenHMMs	7	0.8611 ± 0.063	0.5611 ± 0.008	0.7611 ± 0.047
DiscHMMs	7	0.8611 ± 0.063	0.5611 ± 0.008	0.7611 ± 0.047

validation was employed to split the whole data into training sets and test sets. It was rerun 20 times, and the means and standard deviations of specificity, sensitivity, and accuracy are given in Table 13.1. The accuracy is defined as the ratio of the number of correctly predicted test samples to the total number of test samples. Good responders are treated as "negative," while bad responders are "positive." Therefore, specificity is the prediction accuracy of the good responders, that is, the ratio of the number of corrected predicted good responders to the total number good responders, while sensitivity is that of the bad responders. The parameter for GenHMMs and DiscHMMs is the number of selected genes; absence of such a parameter means gene selection was not used. First of all, we can see that both methods have low sensitivity and high specificity. This is consistent with the clinical result that the bad response to IFNβ is difficult to predict. Second, the average prediction accuracy of both methods are promising. Third, by selecting genes, the accuracy can be slightly improved.

13.4.2 FEATURE EXTRACTION

The following is an alternative to the direct classification using dynamic graphical models. First, we can conduct feature extraction on the GST data. After that, we can apply any standard classifiers for vectorial samples. A few new features can be extracted by *linear dimension reduction* (LDR) [34] methods, in order to capture useful information for classification or clustering. Each of the new features is a linear combination of the original features. A transformation matrix projects the original samples into a new space, termed *feature space*. A sample in the feature space is a *representation* of the corresponding original sample. Taking NMF, for example, a non-negative training set $\mathbf{X}^{\text{train}}$ with m genes and n samples of a gene-sample data can be decomposed into a non-negative basis matrix $\mathbf{A}^{\text{train}}$ and a non-negative coefficient matrix $\mathbf{Y}^{\text{train}}$, that is

$$\mathbf{X}_{m\times n}^{\text{train}} \approx \mathbf{A}_{m\times r}^{\text{train}}\mathbf{Y}_{r\times n}^{\text{train}}, \quad \mathbf{X}^{\text{train}}, \mathbf{A}^{\text{train}}, \mathbf{Y}^{\text{train}} \geq 0. \tag{13.4}$$

Each column of $\mathbf{Y}^{\text{train}}$ is a representation of the corresponding original sample in the feature space spanned by the columns of $\mathbf{A}^{\text{train}}$. In the feature space, a new feature is a linear combination of the original n genes. A sample in the original space

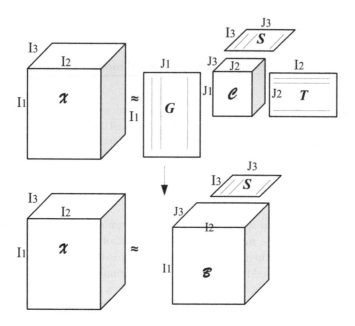

FIGURE 13.2 From Tucker3 decomposition (top) to Tucker1 decomposition (bottom).

can be mapped into the feature space by the transformation matrix $(\mathbf{A}^{\text{train}})^T$. The dimension of the feature space is much lower than that of the original space. LDR methods extend into *multilinear dimension reduction* (MLDR) methods in tensor algebra. The reader is referred to Appendix 13.8 for an introduction to tensor algebra, including PARAFAC decomposition and Tucker decomposition. Tensor decomposition has been applied to the analysis of high-order microarray data in [35] and [36]. We focus on the MLDR method for GST data below. Let $\mathcal{X}$ be a training set, from a GST data set, with I_1 genes, I_2 time points, and I_3 samples. Factorizing the GST data $\mathcal{X}$ by Tucker3 decomposition as in Equation (13.48), we can obtain

$$\mathcal{X} \approx \mathcal{B} \times_3 \mathbf{S} = [\![\mathcal{B}; \mathbf{I}_G, \mathbf{I}_T, \mathbf{S}]\!], \tag{13.5}$$

where $\mathcal{B} = \mathcal{C} \times_1 \mathbf{G} \times_2 \mathbf{T}$, $\mathbf{I}_G$ and $\mathbf{I}_T$ are identity matrices of sizes $I_1 \times I_1$ and $I_2 \times I_2$, respectively, $\mathbf{S}$ is of size $I_3 \times J_3$. The Tucker3 and Tucker1 decompositions are illustrated in Figure 13.2.

Making use of multilinear operations, we have

$$\begin{aligned}
\mathbf{X}_{(1)} &\approx \mathbf{I}_G \mathbf{B}_{(1)} (\mathbf{S} \otimes \mathbf{I}_T)^T \\
&= \mathbf{I}_G [\mathbf{B}_1, \mathbf{B}_2, \cdots, \mathbf{B}_{J_3}][\mathbf{s}_1 \otimes \mathbf{I}_T, \mathbf{s}_2 \otimes \mathbf{I}_T, \cdots, \mathbf{s}_{J_3} \otimes \mathbf{I}_T]^T \\
&= \sum_{r=1}^{J_3} \mathbf{I}_G \mathbf{B}_r (\mathbf{s}_r \otimes \mathbf{I}_T)^T,
\end{aligned} \tag{13.6}$$

where $\mathbf{B}_r = \mathcal{B}(:,:,r)$ is the r-th frontal slice of $\mathcal{B}(:,:,r)$, and $\mathbf{s}_r$ is the r-th column vector of $\mathbf{S}$. Via tensorization, we have

$$\mathcal{X} \approx \sum_{r=1}^{J_3} \mathbf{B}_r \times_3 \mathbf{s}_r, \tag{13.7}$$

which approximates the GST data, $\mathcal{X}$, by the summation of J_3 tensors. We can see it more clearly by the matrix formulation as follows:

$$\mathbf{X}_{(1)} \approx \mathbf{I}_G \mathbf{B}_{(1)} (\mathbf{S} \otimes \mathbf{I}_T)^T$$
$$= [\mathbf{B}_1, \mathbf{B}_2, \cdots, \mathbf{B}_{J_3}] \begin{bmatrix} s_{11}\mathbf{I}_T & \cdots & s_{I_31}\mathbf{I}_T \\ \vdots & \vdots & \vdots \\ s_{1J_3}\mathbf{I}_T & \cdots & s_{I_3J_3}\mathbf{I}_T \end{bmatrix}. \tag{13.8}$$

Thus, the k-th frontal slice of $\mathcal{X}$, that is, the k-th sample, can be fitted by the summation of the frontal slices of $\mathcal{B}$:

$$\mathbf{X}_{(1)k} \approx \sum_{r=1}^{J_3} \mathbf{B}_r s_{kr}, \tag{13.9}$$

where the coefficients are in the k-th row (denoted by $\mathbf{s}_k$) of $\mathbf{S}$.

We can see that $\mathcal{B}$ is the basis tensor for the samples and $\mathbf{S}$ is the encoding matrix. We can define the matrix space spanned by $\mathcal{B}$ as *feature space*, and $\mathbf{s}_k$ as the representation of the k-th sample in the feature space. In the sense of feature extraction, these matrix slices of $\mathcal{B}$ are the *features*. This reduces the original sample slice to a vector $\mathbf{s}_k$ in the feature space. Figure 13.3 illustrates the idea of tensor-factorization based feature extraction. Additionally, it is noted that $\mathcal{C} \times_2 \mathbf{T} \times_3 \mathbf{S}$ and $\mathcal{C} \times_1 \mathbf{G} \times_3 \mathbf{S}$ are the basis tensors for genes and time points, respectively. If the training set is decomposed by HONMF, the extracted non-negative features would be interpretable, and a sample will be an additive summation of the features.

In the test phase, each test sample $\mathbf{Y}_l$ is projected onto the feature space. $\mathbf{Y}_l$ is a linear combination of the basis matrices in $\mathcal{B}$:

$$\mathbf{Y}_l = \sum_{r=1}^{J_3} \mathbf{B}_r \alpha_r, \tag{13.10}$$

where $\alpha = [\alpha_1, \alpha_2, \cdots, \alpha_{J_3}]^T$ is the representation of $\mathbf{Y}_l$ in the feature space. Finding α is equivalent to solving the following generalized least squares problem:

$$\min_{\alpha} \| \mathbf{Y}_l - \sum_{r=1}^{J_3} \mathbf{B}_r \alpha_r \|_F^2. \tag{13.11}$$

The general solution to this problem is $\alpha_r = \frac{<\mathbf{Y}_l, \mathbf{B}_r>}{<\mathbf{B}_r, \mathbf{B}_r>}$ (see [37]), where $< \bullet, \bullet >$ is the inner product of two matrices. For different test samples, we place the α's in the corresponding rows of a coefficient matrix $\mathbf{A}$. We employed three Tucker models including HOSVD, HOOI, and HONMF. The unsupervised MLDR methods above based

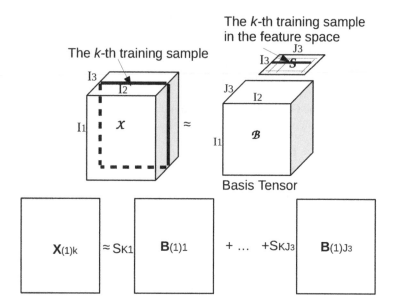

FIGURE 13.3 Tensor-factorization based feature extraction.

on these three Tucker models are denoted by uHOSVDls, uHOOIls, and uHONMFls, respectively.

Alternatively, given the test samples $\mathcal{Y}$, we can fix $\mathcal{C}$, $\mathbf{G}$, and $\mathbf{T}$ to calculate the coefficient matrix $\mathbf{A}$ of $\mathcal{Y}$. We need to find $\mathbf{A}$ that satisfies

$$\mathcal{Y} \approx \mathcal{C} \times_1 \mathbf{G} \times_2 \mathbf{T} \times_3 \mathbf{A}. \tag{13.12}$$

For HOSVD and HOOI, the mode matrices are orthogonal and $\mathbf{A}$ is the J_3 leading left singular vectors of $\mathbf{Z}_{(3)}$. $\mathbf{Z}_{(3)}$ is matricized from $\mathcal{Z}$, which is calculated by the following equation:

$$\mathcal{Z} = \mathcal{Y} \times_1 \mathbf{G}^T \times_2 \mathbf{T}^T. \tag{13.13}$$

For HONMF, the constraint on the mode matrices is non-negativity rather than orthogonality. Instead of solving the non-negativity constrained equation similar to Equation (13.13), $\mathbf{A}$ can be rapidly obtained using the update rules of the HONMF algorithm. We can iteratively update $\mathbf{A}$ only, while keeping $\mathcal{C}$, $\mathbf{G}$, and $\mathbf{T}$ constant. If this method is used for HONMF and Equation (13.13) is used for HOSVD and HOOI, then the resulting algorithms are denoted by uHONMFtf, uHOSVDtf, and uHOOItf, respectively.

Once $\mathbf{A}$ is obtained, we do not need to learn on the training samples and classify the test samples represented by the matrices. Instead, any classifier can be trained on the rows of $\mathbf{S}$ and classify the rows of $\mathbf{A}$. That is, the classification is conducted in the feature space.

TABLE 13.2

Classification performance of tensor factorization on complete IFNβ data.

Method	Parameters	Specificity	Sensitivity	Accuracy
uHOSVDls	7,3,3	0.8389±0.039	0.5944±0.020	0.7574±0.050
uHOOIls	4,3,10	0.9000±0.031	0.5000±0.012	0.7667±0.035
uHONMFls	3,5,3	0.8972±0.079	0.3056±0.034	0.7000±0.052
uHOSVDtf	4,2,3	0.7639±0.053	0.5500±0.041	0.6926±0.046
uHOOItf	3,7,3	0.8111±0.048	0.6611±0.055	0.7611±0.050
uHONMFtf	3,5,3	0.7889±0.029	0.8667±0.154	0.8148±0.040
sHOSVD	4,3,8	0.8306±0.054	0.6333±0.012	0.7648±0.044
sHOOI	3,4,4	0.7611±0.045	0.6667±0.000	0.7296±0.039
sHONMF	3,4,6	0.9583±0.110	0.0056±0.069	0.6407±0.075

Although the decomposition methods described above are unsupervised dimensionality reduction techniques, they can be modified to perform in a supervised manner, i.e., such that class information is taken into account during decomposition. Let m be the number of distinct class labels in the data. The idea is to first partition the training set into m subsets $\mathcal{X}^1, \mathcal{X}^2, \cdots, \mathcal{X}^m$, where each subset $\mathcal{X}^i$ contains only samples of class i. Next, m core tensors $\mathcal{B}^1, \mathcal{B}^2, \cdots, \mathcal{B}^m$ are obtained through decomposition using Equation (13.5). The resulting basis matrices are then normalized using the Frobenius norm. For a normalized test sample, we fit it using these basis tensors, respectively, through Equation (13.11). This sample is assigned to the class that obtains the minimal fitting residual. For simplicity, we denote the supervised version of HOSVD, HOOI, and HONMF based classification methods by sHOSVD, sHOOI, sHONMF. This supervised decomposition approach is described in [38] for handwritten recognition using HOSVD.

We implemented the above tensor-based approaches using MATLAB®. Our implementation is based on *The N-way Toolbox for MATLAB®* [39] and *Algorithms for SN-TUCKER (Higher-Order Non-Negative Matrix Factorization)* [21]. We used the k-nearest neighbor classifier with the Euclidean distance in the classification phase of our unsupervised methods. The parameter of the tensor decomposition based approaches are rank-(J_1, J_2, J_3), and grid search is performed to find the values of J_1, J_2, J_3 that yield the best classification performance. The parameters of all models were selected by grid search. The classification performance of 20 runs of 9-fold cross-validation is given in Table 13.2.

As shown in Table 13.2, uHONMFtf obtains the highest mean prediction accuracy (0.8148). uHOSVDls, uHOOIls, and uHOOItf obtain similar accuracies. This means that the tensor-decomposition-based unsupervised methods can capture discriminative information. uHONMFtf outperforms the HOSVD and HOOI based methods perhaps due to non-negativity. The reasons why uHONMFls and uHOOItf do not perform well needs further investigation. Good performance is also achieved by the supervised sHOSVD.

TABLE 13.3

Comparison of running times on complete IFNβ data.

Method	DiscHMMs	uHOSVDls	uHOOIls	uHONMFtf
Time (seconds)	2.117×10^3	1.321	1.057	1.662×10^3

From Table 13.2, we can conclude that the multi-dimensional reduction techniques are able to dramatically reduce the dimension of the original tensor data and can transform the sample matrices into new "equivalent" short vectors, which are used for classification. In uHONMFtf, for example, a 53 by 7 test sample can be represented by a vector of length 3 in the new feature space, thus reducing the data by 99.19% while preserving discriminative information.

The computing times (in seconds) of the tensor-factorization methods are compared with that of the HMM method in Table 13.3. The number of selected genes is set to 7 for DiscHMMs. The tensor-decomposition based approaches use the same parameter $(3,5,3)$, which are the numbers of factors in the three axes. In comparison with Table 13.1, it can be seen that the tensor-factorization based methods, HOSVD and HOOI, are much faster than the HMM based method while giving at least comparable classification results. uHONMFtf also took less time than DiscHMMs. If other optimization algorithms, such as active-set non-negative least squares algorithm, are used, we believe that the time-complexity of HONMF can be dramatically improved.

Finally, we should remind the readers willing to use NMF (or HONMF) to analyze their microarray data that the non-negativity of the data must be examined before using it. If the data have negative values, the non-negativity constraint should be enforced only on the coefficient matrix (or mode matrices of HONMF).

13.5   KERNEL METHODS FOR CLASSIFICATION

In the last section, graphical models and tensor factorization techniques are discussed for the classification of GST data. The graphical models make direct use of the temporal information in the GST data, while tensor factorization methods separate the dimension reduction from the classification phases. In this section, we shall show that the similarity between a pair of samples can be measured by a dynamical systems kernel, and many kernel classification and kernel dimensionality reduction methods can be used by taking the kernel matrices (rather than the two-way original samples) as inputs. That is, the corresponding classification and dimensionality reduction is *dimension-free*. Therefore, one may not need to propose new classifiers and dimensionality reduction techniques for GST data. In the following, we discuss three methods. The first method is to use the SVM classifier. The second is a new sparse coding method. The last one is the kernel sparse-representation-based dimensionality reduction method.

13.5.1 SUPPORT VECTOR MACHINE

The authors of [40] used an SVM classifier based on *dynamical systems kernel* to classify GST samples. Since each GST sample is represented by a time-series matrix, it is not appropriate to use the kernels, for example *radial basis function* (RBF), which take vectorial inputs, because the temporal structure would be deteriorated by vectorization. The dynamical systems kernel accepts matrix inputs and takes the temporal information into account. We define the dynamical systems kernel as follows. Two time-series matrix samples, say $\mathbf{X}$ and $\mathbf{X}'$, can be modeled by two separate *linear time invariant* (LTI) dynamical systems $\mathbf{X} = (\mathbf{P}, \mathbf{Q}, \mathbf{R}, \mathbf{S}, \mathbf{x}_0)$ (where $\mathbf{x}_0$ is a vector, and $\mathbf{P}$, $\mathbf{Q}$, $\mathbf{R}$, and $\mathbf{S}$ are matrices estimated by a SVD based approach) and $\mathbf{X}' = (\mathbf{P}', \mathbf{Q}', \mathbf{R}', \mathbf{S}', \mathbf{x}_0')$. The dynamical systems kernel between $\mathbf{X}$ and $\mathbf{X}'$ is defined as

$$k(\mathbf{X}, \mathbf{X}') = \mathbf{x}_0^{\mathsf{T}} \mathbf{M}_1 \mathbf{x}_0' + \frac{1}{e^{\lambda} - 1} [\text{trace}(\mathbf{SM}_2) + \text{trace}(\mathbf{R})], \qquad (13.14)$$

where $\mathbf{M}_1$ and $\mathbf{M}_2$ satisfy the Sylvester equation [40], and λ is a positive parameter of the kernel.

We now introduce the SVM on two-way data, and then show that only inner products of samples are needed. The SVM was inspired by statistical learning theory, in which the hard-margin SVM was first proposed in 1979 [41], which was followed by the soft-margin SVM in 1995 [42]. The fundamental principle motivating it is that it implements the *structural risk minimization* (SRM) inductive principle [43], which states that the actual risk is upper bounded by the trade-off between the empirical risk and model complexity. The SVM is a basis-expanded linear model that can be formulated as:

$$f(\mathbf{x}) = \mathbf{w}^{\mathsf{T}} \mathbf{x} + b, \qquad (13.15)$$

where $\mathbf{w}$ is normal vector to the hyperplane, and b is the bias. The decision function is the indicator:

$$d(\mathbf{x}) = \text{sign}[f(\mathbf{x}|\mathbf{w}^*, b^*)], \qquad (13.16)$$

with $\{\mathbf{w}^*, b^*\}$ being the optimal parameter with respect to some criteria. The geometric interpretation of the standard SVM is that the margin between two classes is maximized while keeping the samples of the same class on one side of the margin.

Suppose a two-way training set is represented by a matrix $\mathbf{X} \in \mathbb{R}^{m \times n}$, where each column corresponds to a training sample, and the class labels are in the column vector $\mathbf{y} \in \{-1, +1\}^n$. We define $\mathbf{Z}$ to be the sign-changed training samples with its i-th column defined as the element-wise multiplication of the class label and the input vector of the i-th training sample, that is, $\mathbf{z}_i = y_i * \mathbf{x}_i$. The optimization of the soft-margin C-SVM can be formulated as:

$$\min_{\mathbf{w}, b, \xi} \frac{1}{2} \|\mathbf{w}\|_2^2 + \mathbf{C}^{\mathsf{T}} \xi \qquad (13.17)$$

$$\text{s.t. } \mathbf{Z}^{\mathsf{T}} \mathbf{w} + b\mathbf{y} \geq \mathbf{1} - \xi$$

$$\xi \geq 0,$$

where $\mathbf{C} = \{C\}^m$ controls the tradeoff between the regularization term and the loss term, and ξ is a slack vectorial variable.

By considering the corresponding Lagrange function and *Karush-Kuhn-Tucker* (KKT) conditions, we can obtain the dual form of the optimization:

$$\min_{\mu} \frac{1}{2}\mu^T \mathbf{Z}^T \mathbf{Z}\mu - \mu^T \mathbf{1} \tag{13.18}$$

$$\text{s.t. } \mu^T \mathbf{y} = 0$$

$$0 \le \mu \le \mathbf{C},$$

where μ is the vector of Lagrangian multipliers (μ is a sparse vector). The nonzero multipliers correspond to the *support vectors*, which are crucial for classification. The training samples corresponding to zero multipliers can be ignored in further computations. The relation between the dual variable μ and the primal variable $\mathbf{w}$ is $\mathbf{w} = \mathbf{X}(\mu * \mathbf{y}) = \mathbf{X}_S(\mu_S * \mathbf{y}_S)$, where S is the set of indices of the non-zero multipliers. The bias b can be computed by $b = \frac{y_B - \mathbf{X}_B^T \mathbf{w}}{|B|} = \frac{y_B - \mathbf{X}_B^T \mathbf{X}_S(\mu_S * \mathbf{y}_S)}{|B|}$, where B is the index of the nonzero and unbounded multipliers corresponding to the support vectors on the margin border. From the formulation of the optimal $\mathbf{w}^*$ and b^*, the linear function in the decision function can be computed as:

$$f(\mathbf{x}) = \mathbf{w}^{*T}\mathbf{x} + b^* = \mathbf{x}^T \mathbf{X}_S(\mu_S^* * \mathbf{y}_S) + \frac{y_B - \mathbf{X}_B^T \mathbf{X}_S(\mu_S^* * \mathbf{y}_S)}{|B|}. \tag{13.19}$$

From Equations (13.18) and (13.19), we can see that the optimization and decision making of SVM only need the inner products of the training samples. By replacing the inner products by appropriate kernel functions, we can classify any data. Therefore, SVM with kernels is dimension-free. By using dynamical systems kernel, we can use SVM to classify time-series matrix samples. We compared the performance of SVM using dynamical systems kernel with that using RBF kernel on IFNβ data. Both methods are denoted by dsSVM and rbfSVM, respectively. The same experimental setting as in the previous section is used here. The comparison of both methods is given in Table 13.4. The parameter of rbfSVM is the parameter of the RBF function. The first parameter of dsSVM is the number of hidden states, and the second one is the parameter of the dynamical systems kernel function. We can see that rbfSVM fails to identify any positive sample, while dsSVM obtains a sensitivity of 0.422. The overall accuracy of dsSVM is 0.789, which is much higher than rbfSVM. Thus, we can conclude that the classification performance can be improved by considering structural information in GST data. Furthermore, the SVM method is more efficient than the graphical models (as shown in Table 13.3).

13.5.2 SPARSE CODING

Sparse coding classification methods have been used to classify GST data in [44]. The main idea of sparse coding methods is the following. First, all training samples are placed in a dictionary, and an unknown sample is approximated by a sparse linear

TABLE 13.4

Comparison of classification performance of SVM on complete IFNβ data.

Method	Parameters	Specificity	Sensitivity	Accuracy	Time
rbfSVM	1	1.000 ± 0.000	0.000 ± 0.000	0.667 ± 0.000	-
dsSVM	1,5	0.972 ± 0.082	0.422 ± 0.013	0.789 ± 0.023	93.474

combination of these training samples. Second, a sparse interpreter is used to predict the class label of the unknown sample based on the sparse coefficients. For vectorial samples, the sparse coding can be formulated as:

$$\mathbf{b} \approx \sum_{i=1}^{n} x_i \mathbf{a}_i = \mathbf{Ax}, \tag{13.20}$$

where $\mathbf{b}$ is an unknown sample, $\mathbf{A}$ is the training set with each training sample in a column, and $\mathbf{x}$ is the sparse coefficient.

The spare coefficient can be obtained by the following l_1-*regularized least squares* (l_1LS) model:

$$\min_{\mathbf{x}} \frac{1}{2}\|\mathbf{b} - \mathbf{Ax}\|_2^2 + \lambda\|\mathbf{x}\|_1, \tag{13.21}$$

where λ is a non-negative parameter to control the tradeoff of reconstructive error and sparsity. The l_1-norm is used to induce sparsity. This model is called l_1LS sparse coding. The sparse coefficient vector $\mathbf{x}$ may contain negative values, which may not be meaningful for classification. Furthermore, the optimization of l_1LS is non-smooth. Therefore, it may be slow for large numbers of training samples. In order to overcome these two issues, a non-negative constraint is used to induce sparsity. This leads to the following *non-negative least squares* (NNLS) model:

$$\min_{\mathbf{x}} \frac{1}{2}\|\mathbf{b} - \mathbf{Ax}\|_2^2, \tag{13.22}$$

$$\text{s.t. } \mathbf{x} \geq 0.$$

The non-negative coefficient vector is more meaningful, because for a vector $\mathbf{b}$, the model seeks for a conical hull spanned by using a small number of training samples. The optimization of the NNLS model is also easier to solve than the l_1LS model.

Once the sparse coefficient is obtained, we need to predict the class label of $\mathbf{b}$. Here we give two rules—the weighted k-*nearest neighbor* rule and the *nearest subspace* rule. Both rules are introduced below.

Suppose a k-length vector $\bar{\mathbf{x}}$ accommodates the k-largest coefficients from $\mathbf{x}$, and $\bar{\mathbf{c}}$ has the corresponding k class labels. The class label of $\mathbf{b}$ can be designated as $l = \arg\max_{i=1,\cdots,C} s_i$ where $s_i = \text{sum}(\delta_i(\bar{\mathbf{x}}))$ and C is the number of classes. $\delta_i(\bar{\mathbf{x}})$ is a k-length vector and is defined as

$$(\delta_i(\bar{\mathbf{x}}))_j = \begin{cases} \bar{x}_j & \text{if } \bar{c}_j = i, \\ 0 & \text{otherwise}. \end{cases} \tag{13.23}$$

This rule is called weighted *k-nearest neighbor* (*k*-NN), and was proposed in [45]. The maximum value of k can be n, the number of dictionary atoms. In this case, k is in fact the number of all non-zeros in $\mathbf{x}$. Alternatively, the *nearest subspace* (NS) rule, proposed in [46], can be used to interpret the sparse coding. The NS rule takes advantage of the property of discrimination of sparse coefficients. It assigns the class with the minimum regression residual to $\mathbf{b}$. More formally, it is expressed as $j = \min_{1 \le i \le C} r_i(\mathbf{b})$ where $r_i(\mathbf{b})$ is the regression residual corresponding to the i-th class and is computed as

$$r_i(\mathbf{b}) = \|\mathbf{b} - \mathbf{A}\delta_i(\mathbf{x})\|_2^2, \tag{13.24}$$

where $\delta_i(\mathbf{x})$ is defined analogously as in Equation (13.23).

Both the l_1LS model and the NNLS model can be kernelized as their optimization and decision only involves inner products of samples rather than the original samples. To see this, we can reformulate the NNLS model into a *non-negative quadratic programming* (NNQP) problem:

$$\min_{\mathbf{x}} \frac{1}{2}\mathbf{x}^T\mathbf{H}\mathbf{x} + \mathbf{g}^T\mathbf{x} \tag{13.25}$$
$$\text{s.t. } \mathbf{x} \ge 0,$$

where $\mathbf{H} = \mathbf{A}^T\mathbf{A}$, and $\mathbf{g} = -\mathbf{A}^T\mathbf{b}$. Furthermore, the k-NN rule only needs the class information of the training data, and the NS rule also uses inner products to calculate regression residuals. Therefore NNLS can be kernelized.

Other sparse coding classification methods can be found in [47], where sparse coding is conducted for each class separately and then a *nearest centroid* rule is used. Now, we extend the NNLS sparse coding method for tensor data. Without loss of generality, we suppose there are I_3 training samples represented by a three-way tensor $\mathcal{A}_{I_1 \times I_2 \times I_3}$. The third mode is the class axis. Each sample is a matrix of size $I_1 \times I_2$. Therefore, $\mathcal{A}(:,:,i)$ is the i-th training sample. Suppose we have P new samples in $\mathcal{B}_{I_1 \times I_2 \times P}$. Assume that each of such new samples can be regressed by a non-negative linear combination of the training samples. We then need to solve the following NNLS problem:

$$\min_{\mathbf{X}} \frac{1}{2}\|\mathcal{B} - \mathcal{A} \times_3 \mathbf{X}^T\|_F^2, \text{ s.t. } \mathbf{X} \ge 0, \tag{13.26}$$

where $\times_3$ is the mode-3 product [48] as defined in Appendix 13.8. Through matricizing tensors to matrices, we can convert the above optimization task into the equivalent formula:

$$\min_{\mathbf{X}} \frac{1}{2}\|\mathbf{B}_{(3)} - \mathbf{X}^T\mathbf{A}_{(3)}\|_F^2, \text{ s.t. } \mathbf{X} \ge 0, \tag{13.27}$$

TABLE 13.5

Comparison of classification performance of sparse coding on complete IFNβ data.

Method	Parameters	Specificity	Sensitivity	Accuracy	Time
l_1LS	2^{-4}	0.9000±0.0329	0.3750±0.0000	0.7444±0.0231	0.1152
NNLS	-	0.9000±0.0329	0.3750±0.0000	0.7444±0.0231	0.0573
Kl_1LS	2^{-14},[1,5]	0.7316±0.0468	0.7937±0.0596	0.7500±0.0386	4.2666
KNNLS	[1,5]	0.7895±0.0408	0.7375±0.0375	0.7741±0.0308	4.5199

where $\mathbf{A}_{(3)}$ is a matrix of $I_3 \times (I_1 \times I_2)$, unfolded from tensor $\mathcal{A}$ in mode 3. Using transposition, we have

$$\min_{\mathbf{X}} \frac{1}{2}\|\mathbf{B}_{(3)}^{\mathrm{T}} - \mathbf{A}_{(3)}^{\mathrm{T}}\mathbf{X}\|_F^2, \text{ s.t. } \mathbf{X} \geq 0. \qquad (13.28)$$

Now, Equation (13.28) can be solved by a two-way non-negative least squares algorithm, for example, the FC-NNLS in [45]. Our NNLS classifier can now be generalized for tensor data.

However, the drawback of this generalization is that the structural information within a sample is not considered. The objective (Equation (13.26)) uses the Euclidean distance, and hence the samples are actually vectorized in Equations (13.27) and (13.28). If we use other dissimilarity or similarity metrics that take the temporal information into account when classifying gene-sample-time data, the performance is expected to be increased. Since the dynamical systems kernel, defined in Equation (13.14), accepts matrix inputs and takes temporal information into account, we can thus apply it to our kernel NNLS method for GST data.

Using the same experimental setting as in previous sections, the classification performances of linear l_1LS, and NNLS and their kernel versions on GST data are compared in Table 13.5. We have the following observations. First of all, from the comparison between NNLS using linear kernel and its counterpart using the dynamical systems kernel, we can see that the latter yielded better results, because it considers temporal information within the samples. Thus, the structural information within the samples does contribute to the discrimination. Second, through comparing the computing time (in seconds), we can see that the sparse coding methods are very fast, compared with the graphical models, some tensor decomposition methods, and SVM (as shown in Tables 13.3 and 13.4).

13.5.3 KERNEL NMF AND DICTIONARY LEARNING

NMF has been applied to dimensionality reduction for gene-sample data in [49]. Also, we have shown, in Section 13.4, that the high-order NMF can be applied on GST data. In this section, we shall show that feature extraction can be conducted directly by kernel NMF and kernel dictionary learning in sparse representation.

Gaussian prior and uniform prior over the dictionary atoms have been used in [45] for kernel dictionary learning, respectively. Both priors aim to get rid of the arbitrary scale interchange between dictionary and coefficient. In the following, we first show the dictionary learning models and the generic algorithm of dictionary learning for two-way data. We then show that dictionary learning can be conveniently extended to kernel version. Using kernel dictionary learning, we can extract features from GST data.

Suppose matrix $\mathbf{D}_{m \times n}$ represents the data set of n training samples, and the dictionary $\mathbf{A}$ to be learned has k atoms. If the Gaussian prior is used on the dictionary atoms, our dictionary learning models of $l_1 LS$, NNLS, and l_1 NNLS are expressed as follows, respectively:

$$l_1 LS : \min_{\mathbf{A}, \mathbf{Y}} \frac{1}{2} \|\mathbf{D} - \mathbf{AY}\|_F^2 + \frac{\alpha}{2} \text{trace}(\mathbf{A}^T \mathbf{A}) + \lambda \sum_{i=1}^{n} \|\mathbf{y}_i\|_1, \tag{13.29}$$

$$NNLS : \min_{\mathbf{A}, \mathbf{Y}} \frac{1}{2} \|\mathbf{D} - \mathbf{AY}\|_F^2 + \frac{\alpha}{2} \text{trace}(\mathbf{A}^T \mathbf{A}) \tag{13.30}$$
$$\text{s.t. } \mathbf{Y} \geq 0,$$

and

$$l_1 NNLS : \min_{\mathbf{A}, \mathbf{Y}} \frac{1}{2} \|\mathbf{D} - \mathbf{AY}\|_F^2 + \frac{\alpha}{2} \text{trace}(\mathbf{A}^T \mathbf{A}) + \sum_{i=1}^{n} \lambda^T \mathbf{y}_i \tag{13.31}$$
$$\text{s.t. } \mathbf{Y} \geq 0,$$

where $\mathbf{Y}$ is the sparse coefficient matrix, α is a parameter to control the scale and smoothness of the dictionary atoms, λ is a parameter to control the sparseness of the coefficients, and $\lambda = \{\lambda\}^n$. One strength of the Gaussian prior based dictionary learning is that it is flexible to control the scales of dictionary atoms. However, it has two tradeoff parameters, which increase the model selection burden in practice.

Alternatively, in order to eliminate the parameter α, we can resort to a uniform prior over the dictionary. The corresponding dictionary learning models are given in the following equations, respectively:

$$l_1 LS : \min_{\mathbf{A}, \mathbf{Y}} \frac{1}{2} \|\mathbf{D} - \mathbf{AY}\|_F^2 + \lambda \sum_{i=1}^{n} \|\mathbf{y}_i\|_1 \tag{13.32}$$
$$\text{s.t. } \mathbf{a}_i^T \mathbf{a}_i = 1, \quad i = 1, \cdots, k,$$

$$NNLS : \min_{\mathbf{A}, \mathbf{Y}} \frac{1}{2} \|\mathbf{D} - \mathbf{AY}\|_F^2 \tag{13.33}$$
$$\text{s.t. } \mathbf{a}_i^T \mathbf{a}_i = 1, \quad i = 1, \cdots, k; \quad \mathbf{Y} \geq 0,$$

and

$$l_1 NNLS: \min_{\mathbf{A},\mathbf{Y}} \frac{1}{2}\|\mathbf{D} - \mathbf{AY}\|_F^2 + \sum_{i=1}^{n} \lambda^T \mathbf{y}_i \tag{13.34}$$

$$\text{s.t. } \mathbf{a}_i^T \mathbf{a}_i = 1, \quad i = 1, \cdots, k; \quad \mathbf{Y} \geq 0.$$

We devised block-coordinate-descent-based algorithms for the optimization of the above six models. The main idea is as follows: in the current step, $\mathbf{Y}$ is fixed, and the inner product $\mathbf{A}^T\mathbf{A}$, rather than $\mathbf{A}$ itself, is updated; in the next step, $\mathbf{Y}$ is updated while fixing $\mathbf{A}^T\mathbf{A}$ (a sparse coding procedure). The above procedure is repeated until the termination conditions are satisfied.

Now, we show that $\mathbf{A}$ can be analytically obtained. For normal prior over dictionary atoms, the optimization of finding $\mathbf{A}$ in Equations (13.29), (13.30), and (13.31) is to solve

$$\min_{\mathbf{A}} f(\mathbf{A}) = \frac{1}{2}\|\mathbf{D} - \mathbf{AY}\|_F^2 + \frac{\alpha}{2}\text{trace}(\mathbf{A}^T\mathbf{A}). \tag{13.35}$$

Taking the derivative with respect to $\mathbf{A}$ and setting it to zero, we have

$$\frac{\partial f(\mathbf{A})}{\partial \mathbf{A}} = \mathbf{AYY}^T - \mathbf{DY}^T + \alpha\mathbf{A} = 0. \tag{13.36}$$

We thus have

$$\mathbf{A} = \mathbf{DY}^{\ddagger}, \tag{13.37}$$

where $\mathbf{Y}^{\ddagger} = \mathbf{Y}^T(\mathbf{YY}^T + \alpha\mathbf{I})^{-1}$. The inner product $\mathbf{A}^T\mathbf{A}$ can thus be updated as:

$$\mathbf{R} = \mathbf{A}^T\mathbf{A} = (\mathbf{Y}^{\ddagger})^T\mathbf{D}^T\mathbf{DY}^{\ddagger}. \tag{13.38}$$

We can also compute $\mathbf{A}^T\mathbf{D}$ as follows:

$$\mathbf{A}^T\mathbf{D} = (\mathbf{Y}^{\ddagger})^T\mathbf{D}^T\mathbf{D}. \tag{13.39}$$

For the uniform prior, updating the unnormalized $\mathbf{A}$ while fixing $\mathbf{Y}$ in Equations (13.32), (13.33), and (13.34) is to solve the generalized least squares:

$$\min_{\mathbf{A}} f(\mathbf{A}) = \frac{1}{2}\|\mathbf{D} - \mathbf{AY}\|_F^2. \tag{13.40}$$

Taking derivative with respect to $\mathbf{A}$ and setting it to zero, we have

$$\mathbf{A} = \mathbf{DY}^{\dagger}, \tag{13.41}$$

where $\mathbf{Y}^{\dagger} = \mathbf{Y}^T(\mathbf{YY}^T)^{-1}$. The inner products of $\mathbf{R} = \mathbf{A}^T\mathbf{A}$ and $\mathbf{A}^T\mathbf{D}$ are computed similarly as for the Gaussian prior. The normalization of $\mathbf{R}$ is straightforward. We have $\mathbf{R} = \mathbf{R}./\sqrt{\text{diag}(\mathbf{R})\text{diag}(\mathbf{R})^T}$, where $./$ and $\sqrt{\bullet}$ are element-wise operators.

Algorithm 13.1 Dictionary Learning

Input: $\mathbf{K} = \mathbf{D}^T\mathbf{D}$, dictionary size k, α, λ
Output: $\mathbf{R} = \mathbf{A}^T\mathbf{A}$, $\mathbf{Y}$

Initialize $\mathbf{Y}$ and $\mathbf{R} = \mathbf{A}^T\mathbf{A}$ randomly;
$r_{prev} = Inf$; *% previous residual*

for $i = 1 : maxIter$ **do**
    Update $\mathbf{Y}$ by solving the active-set based l_1LS, NNLS, or l_1NNLS sparse coding algorithms;

    if Gaussian prior over $\mathbf{A}$ **then**
        update $\mathbf{R} = \mathbf{Y}^{\ddagger T}\mathbf{D}^T\mathbf{D}\mathbf{Y}^{\ddagger}$;
    end if

    if uniform prior over $\mathbf{A}$ **then**
        update $\mathbf{R} = \mathbf{Y}^{\dagger T}\mathbf{D}^T\mathbf{D}\mathbf{Y}^{\dagger}$;
        normalize $\mathbf{R}$ by $\mathbf{R} = \mathbf{R}./\sqrt{\text{diag}(\mathbf{R})\text{diag}(\mathbf{R})^T}$;
    end if

    if $i == maxIter$ or $i \mod l == 0$ **then**
        % check every l iterations
        $r_{cur} = f(\mathbf{A}, \mathbf{Y})$; *% current residual of a dictionary learning model*
        if $r_{prev} - r_{cur} \le \varepsilon$ or $r_{cur} \le \varepsilon$ **then**
            break;
        end if
        $r_{prev} = r_{cur}$;
    end if
end for

Learning the inner product $\mathbf{A}^T\mathbf{A}$ instead of $\mathbf{A}$ has the benefits of dimension-free computation and kernelization.

Fixing $\mathbf{A}$, $\mathbf{Y}$ can be obtained via our active-set algorithms. Recall that the sparse coding only requires the inner products $\mathbf{A}^T\mathbf{A}$ and $\mathbf{A}^T\mathbf{D}$. As shown above, we find that updating $\mathbf{Y}$ only needs its previous value and the inner product between training samples.

Due to the above derivation, we have the framework of solving our dictionary learning models as illustrated in Algorithm 13.1.

For a Gaussian dictionary prior, the l_1LS based kernel dictionary learning and sparse coding are expressed as follows, respectively:

$$\min_{\mathbf{A}_\phi, \mathbf{Y}} \frac{1}{2}\|\phi(\mathbf{D}) - \mathbf{A}_\phi\mathbf{Y}\|_F^2 + \frac{\alpha}{2}\text{trace}(\mathbf{A}_\phi^T\mathbf{A}_\phi) + \lambda\|\mathbf{Y}\|_1, \tag{13.42}$$

TABLE 13.6

Comparison of classification performance of kernel dictionary learning on complete IFNβ data.

Method	Prior	Specificity	Sensitivity	Accuracy	Time
KSR-l_1LS	Gaussian	0.9711±0.0352	0.1250±0.1118	0.7204±0.0320	39.9439
KSR-NNLS	Gaussian	0.9711±0.0310	0.1081±0.0000	0.7370±0.0329	13.3402
KSR-l_1NNLS	Gaussian	0.9579±0.0394	0.1437±0.0501	0.7167±0.0724	14.9027
KSR-l_1LS	uniform	0.9684±0.0349	0.1375±0.1111	0.7222±0.0321	29.7428
KSR-NNLS	uniform	0.9711±0.0310	0.1437±0.1066	0.7259±0.0429	7.4780
KSR-l_1NNLS	uniform	0.9763±0.0310	0.1563±0.1178	0.7333±0.0505	13.7721

$$\min_{\mathbf{X}} \frac{1}{2} \|\phi(\mathbf{B}) - \mathbf{A}_\phi \mathbf{X}\|_F^2 + \lambda \|\mathbf{X}\|_1,$$

where $\phi(\bullet)$ is a mapping function. Equations (13.30), (13.31), (13.32), (13.33), (13.34) and their sparse coding models can be kernelized analogously. Recall that the optimizations of the three pairs of linear models require only inner products of samples. Therefore, they can be easily extended to kernel versions by replacing these inner products with kernel matrices. In this paper, we use prefix "SR" before l_1LS, NNLS, and l_1NNLS to indicate that dictionary learning is involved in SR. We use prefix "KSR" before them to indicate the kernel versions.

Using a suitable kernel to measure the similarity between a pair of time-series samples, kernel dictionary learning can be applied on GST data for dimensionality reduction. A computational experiment is given in the following. We used the IFNβ data again. We used a dynamical systems kernel with the same parameters as in previous sections. The experimental results are shown in Table 13.6. We can see that the kernel sparse representation methods obtained lower sensitivity compared with the results of sparse coding methods in Table 13.5. This may be because the number of positive samples in the training set is very small, around eight, which may be insufficient for some dimensionality reduction techniques. However, it would be very interesting to investigate the performance of kernel sparse representation on a larger data set due to its computational flexibility. Finally, we can see that the kernel dictionary learning methods are very efficient for GST data as well.

13.6　CONCLUSIONS

In this chapter, we reviewed different machine learning approaches for analyzing GST data for various purposes. These techniques include missing value estimation, biclustering and triclustering, dimension reduction, and classification. We discuss three techniques for handling missing values in GST data, including removal, imputation, and weighting. The existing biclustering and triclustering methods are surveyed in this chapter as well. Direct time-series classification approaches are also discussed. We show the basic technique of dimension reduction using tensor factorizations. Kernel classification approaches, including SVM and kernel sparse coding, can be directly used for classifying GST data given an appropriate kernel, which can measure similarity between a pair of time-series samples. We also investigate kernel sparse representation models for kernel feature extraction.

There are still many challenges in each category. The current weighting strategies of handling missing values are very slow. It is therefore necessary to design a fast weighting method. NMF has been applied as a biclustering method for two-way time-series. It is interesting to investigate the tensor-factorization based triclustering method. The current HONMF is implemented by multiplicative update rules. Using other optimization techniques, for example active-set algorithms, may dramatically improve the efficiency. The current dynamical systems kernel is not symmetric and is unbounded, and hence a better kernel is needed to overcome these shortcomings. The most popular data set in this area is IFNβ, which has small numbers of genes, samples, and time points. More GST data sets need to be obtained from *Gene Expression Omnibus* [50] in order to investigate the performance of the methods reviewed in this chapter.

13.7 ACKNOWLEDGMENTS

We thank the anonymous reviewers for their comments to improve this chapter. This research has been partially supported by IEEE CIS Walter Karplus Summer Research Grant 2010, Ontario Graduate Scholarship 2011-2013, and Canadian NSERC Grants #RGPIN228117-2011.

13.8 APPENDIX: AN INTRODUCTION TO TENSOR ALGEBRA

In this appendix, we briefly introduce the main concepts in *tensor algebra*, which is necessary to help readers understand the methodologies presented in this chapter. Hereafter, we use the following notations unless otherwise noted:

1. A matrix is denoted by a bold capital letter, e.g., $\mathbf{A}$.
2. A (column) vector is denoted by a bold lowercase letter, e.g., $\mathbf{a}$.
3. A bold lowercase letter with a subscript $\mathbf{a}_i$ denotes the i-th column vector in matrix $\mathbf{A}$.
4. The italic lowercase letter with two subscripts a_{ij} is the (i, j)-th scalar element of matrix $\mathbf{A}$.
5. A boldface Euler script, e.g., $\mathcal{X}$, denotes a three (or higher)-way tensor, for example, $\mathcal{X} \in \mathbb{R}^{I_1 \times I_2 \times I_3}$.
6. $\mathbf{X}_{(1)p}$ denotes the p-th frontal slice of $\mathcal{X}$, of size $I_1 \times I_2$.
7. $\mathbf{X}_{(n)}$ denotes the matrix obtained through the mode-n matricization of the tensor $\mathcal{X}$. Columns of $\mathbf{X}_{(n)}$ are the mode-n fibers of tensor $\mathcal{X}$. A mode-n fiber is a vector defined through fixing every index but the n-th index. This is the extension of matrix row and column in tensor algebra. $\mathbf{X}_{(1)}$ therefore denotes the matrix of size $I_1 \times I_2 I_3$, unfolded in mode-1 of $\mathcal{X}$, that is

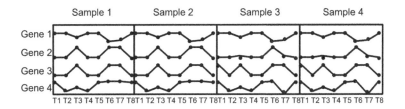

FIGURE 13.4 The mode-1 matricization of a three-way tensor.

$\mathbf{X}_{(1)} = [\mathbf{X}_{(1)1}, \mathbf{X}_{(1)2}, \cdots, \mathbf{X}_{(1)I_3}]$. In Figure 13.4, each column is a mode-1 fiber obtained via fixing the sample and time axes. Each such fiber represents the gene profiles of a specific sample at a specific time point. The matrix in Figure 13.4 is obtained by placing all mode-1 fibers together in a proper order.

8. The (i_1, i_2, i_3)-th scalar element of $\mathcal{X}$ is denoted by $x_{i_1 i_2 i_3}$.

Suppose that $\mathbf{a} \in \mathbb{R}^m$, $\mathbf{b} \in \mathbb{R}^n$, and $\mathbf{c} \in \mathbb{R}^r$. The operator "$\circ$" in $\mathbf{a} \circ \mathbf{b} = \mathbf{M} \in \mathbb{R}^{m \times n}$ is vector outer product. $\mathbf{M}$ is a rank-one matrix. Its elements can be computed as $m_{ij} = a_i b_j$. Similarly, the operation $\mathbf{a} \circ \mathbf{b} \circ c = \mathcal{M} \in \mathbb{R}^{m \times n \times r}$, where $\mathcal{M}$ is a rank-one tensor and $m_{ijk} = a_i b_j c_k$.

Suppose that $\mathbf{A} \in \mathbb{R}^{m_1 \times n_1}$ and $\mathbf{B} \in \mathbb{R}^{m_2 \times n_2}$, their Kronecker product is defined as

$$\mathbf{A} \otimes \mathbf{B} = \begin{bmatrix} a_{11}\mathbf{B} & a_{12}\mathbf{B} & \cdots & a_{1n_1}\mathbf{B} \\ a_{21}\mathbf{B} & a_{22}\mathbf{B} & \cdots & a_{2n_1}\mathbf{B} \\ \vdots & \vdots & \ddots & \vdots \\ a_{m_1 1}\mathbf{B} & a_{m_1 2}\mathbf{B} & \cdots & a_{m_1 n_1}\mathbf{B} \end{bmatrix}$$
$$= [\mathbf{a}_1 \otimes \mathbf{b}_1, \mathbf{a}_1 \otimes \mathbf{b}_2, \cdots, \mathbf{a}_{n_1} \otimes \mathbf{b}_{n_2-1}, \mathbf{a}_{n_1} \otimes \mathbf{b}_{n_2}]. \tag{13.43}$$

We have that $\mathbf{A} \otimes \mathbf{B} \in \mathbb{R}^{(m_1 m_2) \times (n_1 n_2)}$.

Suppose $\mathbf{A} \in \mathbb{R}^{m_1 \times n}$ and $\mathbf{B} \in \mathbb{R}^{m_2 \times n}$, their Khatri-Rao product is defined as

$$\mathbf{A} \odot \mathbf{B} = [\mathbf{a}_1 \otimes \mathbf{b}_1, \mathbf{a}_2 \otimes \mathbf{b}_2, \cdots, \mathbf{a}_{n-1} \otimes \mathbf{b}_{n-1}, \mathbf{a}_n \otimes \mathbf{b}_n]. \tag{13.44}$$

Therefore $\mathbf{A} \odot \mathbf{B}$ is of size $(m_1 m_2) \times n$.

The Hadamard product is also called element-wise product, which is denoted as $\mathbf{A} * \mathbf{B}$.

The n-mode product of a tensor $\mathcal{X}$ and a matrix $\mathbf{A}$, written as $\mathcal{X} \times_n \mathbf{A}$, is:

$$[\mathcal{X} \times_n \mathbf{A}]_{i_1 \times \cdots i_{n-1} \times j \times i_{n+1} \times \cdots \times i_N} = \sum_{i_n=1}^{I_n} x_{i_1 i_2 \cdots i_N} a_{j i_n}, \tag{13.45}$$

where $\mathcal{X} \in \mathbb{R}^{I_1 \times I_2 \times \cdots \times I_N}$ and $\mathbf{A} \in \mathbb{R}^{J \times I_n}$. This results in a tensor $\mathcal{Y} \in \mathbb{R}^{I_1 \times \cdots I_{n-1} \times J \times I_{n+1} \times \cdots \times I_N}$.

$\mathcal{X}$ can be matricized into matrices in different modes. For example, $\mathbf{X}_{(1)} = [\mathbf{X}_{(1)1}, \mathbf{X}_{(1)2}, \cdots, \mathbf{X}_{(1)I_3}]$ is matricized in the first mode (see Figure 13.4).

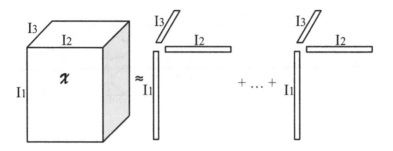

FIGURE 13.5 PARAFAC decomposition.

Tensor decomposition methods mainly include PARAFAC and Tucker decompositions [48]. PARAFAC is the abbreviation of *parallel factors*. It is also called *canonical decomposition* (CANDECOMP). It factorizes a tensor into a summation of rank-one tensors. Supposing that $\mathcal{X} \in \mathbb{R}^{I_1 \times I_2 \times I_3}$, PARAFAC decomposes $\mathcal{X}$ into

$$\mathcal{X} \approx \sum_{r=1}^{R} \mathbf{g}_r \circ \mathbf{t}_r \circ \mathbf{s}_r, \qquad (13.46)$$

where R is the rank of the reconstructed tensor, $\mathbf{g}_r \in \mathbb{R}^{I_1}$, $\mathbf{t}_r \in \mathbb{R}^{I_2}$, and $\mathbf{s}_r \in \mathbb{R}^{I_3}$ are columns of $\mathbf{G}$, $\mathbf{T}$, and $\mathbf{S}$, respectively. This factorization can be concisely written as $\mathcal{X} \approx [\![\mathbf{G}, \mathbf{T}, \mathbf{S}]\!]$, where $\mathbf{G}, \mathbf{T}, \mathbf{S}$ are called *factor matrices*. $\mathcal{X} \approx [\![\mathbf{G}, \mathbf{T}, \mathbf{S}]\!]$ can be matricized into different modes:

$$\begin{cases} \mathbf{X}_{(1)} \approx \mathbf{G}(\mathbf{S} \odot \mathbf{T})^{\mathrm{T}} \\ \mathbf{X}_{(2)} \approx \mathbf{T}(\mathbf{S} \odot \mathbf{G})^{\mathrm{T}} \\ \mathbf{X}_{(3)} \approx \mathbf{S}(\mathbf{T} \odot \mathbf{G})^{\mathrm{T}} \end{cases} . \qquad (13.47)$$

Figure 13.5 is an illustration of PARAFAC decomposition.

PARAFAC can be optimized by the *alternating least squares* (ALS) algorithm [17]. If non-negativity is constrained on all factor matrices, the *alternating nonnegative least squares* (ANLS) algorithm can be used [17].

The Tucker3 model of Tucker decomposition factorizes a tensor $\mathcal{X}$ into a core tensor $\mathcal{C}$ and three mode matrices $\mathbf{G}$, $\mathbf{T}$, and $\mathbf{S}$ as follows:

$$\mathcal{X} \approx \mathcal{C} \times_1 \mathbf{G} \times_2 \mathbf{T} \times_3 \mathbf{S} = \sum_{j_1=1}^{J_1} \sum_{j_2=1}^{J_2} \sum_{j_3=1}^{J_3} c_{j_1 j_2 j_3} \mathbf{g}_{j_1} \circ \mathbf{t}_{j_2} \circ \mathbf{s}_{j_3} = [\![\mathcal{C}; \mathbf{G}, \mathbf{T}, \mathbf{S}]\!], \quad (13.48)$$

where $\mathcal{C}$ is a core tensor and $\mathbf{G}, \mathbf{T}, \mathbf{S}$ are called mode matrices. The decomposition is illustrated in Figure 13.6. In light of Equation (13.48), it is clear that an element of core tensor $\mathcal{C}$ indicates the degree of interaction among the corresponding mode vectors from different mode matrices. For instance, $c_{j_1 j_2 j_3}$ reflects the interaction

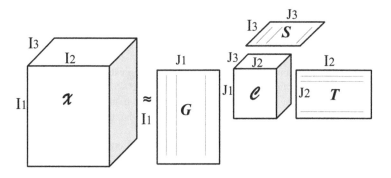

FIGURE 13.6   Tucker3 decomposition.

between $\mathbf{g}_{j_1}$, $\mathbf{t}_{j_2}$, and $\mathbf{s}_{j_3}$. $\mathcal{X} = [\![\mathcal{C}; \mathbf{G}, \mathbf{T}, \mathbf{S}]\!]$ can be matricized into

$$
\begin{cases}
\mathbf{X}_{(1)} \approx \mathbf{G}\mathbf{C}_{(1)}(\mathbf{S} \otimes \mathbf{T})^{\mathrm{T}} \\
\mathbf{X}_{(2)} \approx \mathbf{T}\mathbf{C}_{(2)}(\mathbf{S} \otimes \mathbf{G})^{\mathrm{T}} \\
\mathbf{X}_{(3)} \approx \mathbf{S}\mathbf{C}_{(3)}(\mathbf{T} \otimes \mathbf{G})^{\mathrm{T}}
\end{cases} . \tag{13.49}
$$

Generally speaking, there are no constraints on the core tensor and mode matrices in Tucker decomposition. However, constraints such as orthogonality, non-negativity, and non-Gaussianity can be enforced in a decomposition algorithm. For instance, HOSVD enforces the orthogonality constraints on the mode matrices and is among the most popular Tucker algorithms. It calculates the left singular matrices for different matrices in different modes as factors. The core tensor is obtained through $\mathcal{C} = \mathcal{X} \times_1 \mathbf{G}^T \times_2 \mathbf{T}^T \times_3 \mathbf{S}^T$. Interested readers are referred to [51] for more details. *Higher-order orthogonal iterations* (HOOI) is an alternating least squares (ALS) algorithm initialized by HOSVD, which gives better decomposition than HOSVD itself (see [52] and [48] for details). HOOI also generates orthogonal mode matrices. HONMF imposes non-negativity constraints on the core tensor and mode matrices. Multiplicative updates rules [53] corresponding to core and mode matrices have been extended in [21] for HONMF. The core tensor and mode matrices are alternatingly updated until the convergence criteria are met. The authors of [53] have observed that good interpretation and learning performance can be benefited by adding non-negativity and sparsity constraints to matrix factorization. Even though it can be imposed and controlled, sparsity is sometimes a by-product of non-negativity constrained matrix (maybe tensor also) factorization without explicit sparsity constraint.

REFERENCES

1. S. Baranzini, P. Mousavi, J. Rio, S. Caillier, A. Stillman, and P. Villoslada, "Transcription-based prediction of response to IFNβ using supervised computational methods," *PLOS Biology*, vol. 3, no. 1, p. e2, 2005.

2. C. Perou, T. Sorlie, M. Eisen, et al., "Molecular portraits of human breast tumours," *Nature*, vol. 406, pp. 747–752, 2000.

3. C. G. A. Network, "Comprehensive molecular portraits of human breast tumours," *Nature*, vol. 490, pp. 61–70, 2012.

4. O. Troyanskaya, M. Cantor, G. Sherlock, et al., "Missing value estimation methods for DNA microarrays," *Bioinformatics*, vol. 17, no. 6, pp. 520–525, 2001.

5. T. Hastie, R. Tibshirani, G. Sherlock, M. Eisen, P. Brown, and D. Botstein, "Imputing missing data for gene expression arrays," tech. rep., Division of Biostatistics, Stanford University, 1999.

6. H. Kim, G. H. Golub, and H. Park, "Missing value estimation for DNA microarray gene expression data: local least squares imputation," *Bioinformatics*, vol. 21, no. 2, pp. 187–198, 2005.

7. T. Bo, B. Dysvik, and I. Jonassen, "LSimpute accurate estimation of missing values in microarray data with least squares methods," *Nucleic Acids Research*, vol. 32, no. 3, p. e34, 2004.

8. D. Nguyen, N. Wang, and R. Carroll, "Evaluation of missing value estimation for microarray data," *Journal of Data Science*, vol. 2, pp. 347–370, 2004.

9. G. Brock, J. Shaffer, R. Blakesley, M. Lotz, and G. Tseng, "Which missing value imputation method to use in expression profiles: a comparative study and two selection schemes," *BMC Bioinformatics*, vol. 9, p. 12, 2008.

10. S. Oba, M. Sato, I. Takemasa, M. Monden, K. Matsubara, and S. Ishii, "Bayesian missing value estimation method for gene expression profile data," *Bioinformatics*, vol. 19, no. 16, pp. 2088–2096, 2003.

11. Z. Bar-Joseph, G. Gerber, and D. Gifford, "Continuous representations of time-series gene expression data," *Journal of Computational Biology*, vol. 10, no. 3-4, pp. 341–356, 2003.

12. M. Choong, M. Charbit, and H. Yan, "Autoregressive model based missing value estimation for DNA microarray time series data," *IEEE Transactions on Information Technology in Biomedicine*, vol. 13, no. 1, pp. 131–137, 2009.

13. S. Zhang, W. Wang, J. Ford, and F. Makedon, "Learning from incomplete ratings using non-negative matrix factorization," in *SDM*, (Philadelphia, PA), pp. 548–552, SIAM, 2006.

14. N. Ho, *Nonnegative Matrix Factorization Algorithms and Applications*. PhD thesis, Department of Mathematical Engineering, Louvain-la-Neuve, Belgium, 2008.

15. Y. Li, A. Ngom, and L. Rueda, "Missing value imputation methods for gene-sample-time microarray data analysis," in *IEEE Symposium on Computational Intelligence in Bioinformatics and Computational Biology*, Piscataway, NJ, USA, pp. 183–189, IEEE, IEEE Press, May 2010.

16. R. Bro, "PARAFAC. tutorial and applications," *Chemometric and Intelligent Laboratory Systems*, vol. 38, pp. 149–171, 1997.

17. R. Bro, *Multi-way Analysis in the Food Industry: Models, Algorithms, and Applications*. PhD thesis, Department of Dairy and Food Science, Royal Veterinary and Agricultural University, Denmark, 1998.

18. G. Tomasi and R. Bro, "PARAFAC and missing values," *Chemometric and Intelligent Laboratory Systems*, vol. 75, pp. 163–180, 2005.

19. P. Paatero, "A weighted non-negative least squares algorithm for three-way PARAFAC factor analysis," *Chemometric and Intelligent Laboratory Systems*, vol. 38, pp. 223–242, 1997.

20. R. Bro and S. D. Jong, "A fast non-negative constrained least squares algorithm," *Journal of Chemometrics*, vol. 11, pp. 393–401, 1997.

21. M. Morup, L. Hansen, and S. Arnfred, "Algorithms for sparse nonnegative Tucker decompositions," *Neural Computation*, vol. 20, no. 8, pp. 2112–2131, 2008.

22. Y. Cheng and G. Church, "Biclustering of expression data," in *Intelligent Systems for Molecular Biology*, pp. 93–103, 2000.

23. S. Madeira and A. Oliveira, "Biclustering algorithms for biological data analysis: A survey," *IEEE/ACM Transactions on Computational Biology and Bioinformatics*, vol. 1, no. 1, pp. 24–45, 2004.

24. D. Jiang, J. Pei, M. Ramanathan, C. Lin, C. Tang, and A. Zhang, "Mining gene-sample-time microarray data: a coherent gene cluster discovery approach," *Knowledge and Information Systems*, vol. 13, pp. 305–335, 2007.

25. J. Supper, M. Strauch, D. Wanke, K. Harter, and A. Zell, "EDISA: extracting biclusters from multiple time-series of gene expression profiles," *BMC Bioinformatics*, vol. 8, p. 334, 2007.

26. S. Bergmann, J. Ihmels, and N. Barkai, "Iterative signature algorithm for the analysis of large-scale gene expression data," *Physical Review E*, vol. 67, p. 031902, 2003.

27. L. Zhao and M. Zaki, "TRICLUSTER: an effective algorithm for mining coherent clusters in 3D microarray data," in *Proceedings of the ACM SIGMOD International Conference on Management of Data*, New York, NY, pp. 694–705, ACM, 2005.

28. R. Araujo, G. Ferreira, G. Orair, et al., "The ParTriCluster algorithm for gene expression analysis," *International Journal of Parallel Programming*, vol. 36, pp. 226–249, 2008.

29. H. Jiang, S. Zhou, J. Guan, and Y. Zheng, "gTRICLUSTER: a more general and effective 3D clustering algorithm for gene-sample-time microarray data," in *International Conference on Data Mining for Biomedical Applications*, (Berlin, Heidelberg), pp. 48–59, Springer-Verlag, 2006.

30. A. Tchagang, S. Phan, F. Famili, et al., "Mining biological information from 3D short time-series gene expression data: the OPTricluster algorithm," *BMC Bioinformatics*, vol. 13, p. 54, 2012.

31. B. Weinstock-Guttman, D. Badgett, K. Patrick, et al., "Genomic effects of IFN-β in multiple sclerosis patients," *The Journal of Immunology*, vol. 171, no. 5, pp. 2694–2702, 2003.

32. T. Lin, N. Kaminski, and Z. Bar-Joseph, "Alignment and classification of time series gene expression in clinical studies," *Bioinformatics*, vol. 24, no. ISMB 2008, pp. i147–i155, 2008.

33. I. Costa, A. Schonhuth, C. Hafemeister, and A. Schliep, "Constrained mixture estimation for analysis and robust classification of clinical time series," *Bioinformatics*, vol. 24, no. ISMB 2009, pp. i6–i14, 2009.

34. L. Rueda and M. Herrera, "Linear dimensionality reduction by maximizing the Chernoff distance in the transformed space," *Pattern Recognition*, vol. 41, pp. 3138–3152, 2008.

35. L. Omberg, G. Golub, and O. Alter, "A tensor higher-order singular value decomposition for integrative analysis of DNA microarray data from different studies," *Proceedings of the National Academy of Sciences*, vol. 104, no. 47, pp. 18371–18376, 2007.

36. S. Ponnapalli, M. Saunders, C. V. Loan, and O. Alter, "A higher-order generalized singular value decomposition for comparison of global mRNA expression from multiple organisms," *PLoS ONE*, vol. 6, no. 12, p. e28072, 2011.

37. B. Savas, "Analyses and tests of handwritten digit recognition algorithms," Master's

thesis, Dept. Mathathmatics Scientific Computing, Linköping University, Sweden, 2003.

38. B. Savas and L. Elden, "Handwritten digit classification using higher order singular value decomposition," *Pattern Recongtion*, vol. 40, pp. 993–1003, 2007.

39. C. Andersson and R. Bro, "The N-way toolbox for MATLAB," *Chemometrics and Intelligent Laboratory Systems*, vol. 52, pp. 1–4, 2000.

40. K. Borgwardt, S. Vishwanathan, and H. Kriegel, "Class prediction from time series gene expression profiles using dynamical systems kernels," in *Pacific Symposium on Biocomputing*, pp. 547–558, World Scientific Press, 2006.

41. V. Vapnik, *Estimation of Dependences Based on Empirical Data*. New York: Springer-Verlag, 1982.

42. C. Cortes and V. Vapnik, "Support vector networks," *Machine Learning*, vol. 20, pp. 273–297, 1995.

43. V. Vapnik, *Statistical Learning Theory*. New York: Wiley-IEEE Press, 1998.

44. Y. Li and A. Ngom, "Classification approach based on non-negative least squares," *Neurocomputing*, vol. 118, pp. 41–57, 2013.

45. Y. Li and A. Ngom, "Sparse representation approaches for the classification of high-dimensional biological data," *BMC Systems Biology*, 2013. In Press.

46. J. Wright, A. Yang, A. Ganesh, S. S. Sastry, and Y. Ma, "Robust face recognition via sparse representation," *IEEE Transactions on Pattern Analysis and Machine Intelligence*, vol. 31, no. 2, pp. 210–227, 2009.

47. J. Yang, L. Zhang, Y. Xu, and J.-Y. Yang, "Beyond sparsity: the role of L_1-optimizer in pattern recognition," *Pattern Recognition*, vol. 45, pp. 1104–1118, 2012.

48. T. Kolda and B. Bader, "Tensor decompositions and applications," *SIAM Review*, vol. 51, no. 3, pp. 455–500, 2009.

49. Y. Li and A. Ngom, "Non-negative matrix and tensor factorization based classification of clinical microarray gene expression data," in *Proceedings of the IEEE International Conference on Bioinformatics & Biomedicine*, Piscataway, NJ, pp. 438–443, IEEE, IEEE Press, Dec. 2010.

50. R. Edgar, M. Domrachev, and A. Lash, "Gene Expression Omnibus: NCBI gene expression and hybridization array data repository," *Nucleic Acids Research*, vol. 30, no. 1, pp. 207–210, 2002.

51. L. Lathauwer, B. De Moor, and J. Vandewalle, "A multilinear singular value decomposition," *SIAM Journal on Matrix Analysis and Applications*, vol. 21, no. 4, pp. 1253–1278, 2000.

52. C. Andersson and R. Bro, "Improving the speed of multi-way algorithms: Part I. Tucker3," *Chemometrics and Intelligent Laboratory Systems*, vol. 42, pp. 93–103, 1998.

53. D. D. Lee and S. Seung, "Learning the parts of objects by non-negative matrix factorization," *Nature*, vol. 401, pp. 788–791, 1999.

14 Systematic and Stochastic Biclustering Algorithms for Microarray Data Analysis

Wassim Ayadi, Mourad Elloumi, and Jin-Kao Hao

CONTENTS

14.1   INTRODUCTION

Microarray data measure the expression levels of a large number of genes, potentially all genes of an organism, in a number of different experimental conditions [1]. The conditions may come from different organs, from cancerous or healthy tissues, or even from different individuals. By observing these data, also called gene expression data, we find that the extraction of biological knowledge is even more difficult [2]. Usually, microarray data are represented by a data matrix M (Table 14.1), where the ith gene corresponds to the ith row and the jth condition to the jth column. One element m_{ij} of this data matrix represents the expression level of the ith gene under the jth condition, and is represented by a real number, which is usually the logarithm of the relative abundance of the mRNA (or ratio) of the gene under the specific condition.

Microarray data analysis is an essential step for the applications mentioned above in order to extract relevant biological knowledge embedded in these large datasets. However, the process of knowledge extraction is not a trivial task and data mining techniques are often used for this purpose. Among the techniques used, we can cite *clustering* [3] where we assume that the genes in a group may have similar behavior under all experimental conditions. Furthermore, clustering tries to find non-overlapping groups of genes [4]. Another interesting technique, called *biclustering* [5, 6], allows the identification of groups of genes that exhibit coherent expression forms across specific groups of experimental conditions. Biclustering relates to a different group of clustering algorithms that perform simultaneous row and column clustering. Biclustering algorithms have also been used in the literature to refer to other application fields with different names such as co-clustering, bi-dimensional clustering, and subspace clustering [6].

A biclustering cluster is called *bicluster*. Hence, a bicluster of genes (respectively conditions) is defined with respect to only a subset of conditions (respectively genes). In other words, a bicluster is a subset of genes showing similar behavior un-

---

TABLE 14.1

Gene expression data matrix.

	$cond_1$	...	$cond_j$	...	$cond_m$
$Gene_1$	m_{11}	...	m_{1j}	...	m_{1m}
...	...	...	...	...	...
$Gene_i$	m_{i1}	...	m_{ij}	...	m_{im}
...	...	...	...	...	...
$Gene_n$	m_{n1}	...	m_{nj}	...	m_{nm}

der a subset of conditions of the original expression data matrix. Let us note that a gene/condition can belong to more than one bicluster or none.

Formally, a *bicluster* can be defined as follows. Let $I = \{1, 2, \ldots, n\}$ be a set of indices of n genes, $J = \{1, 2, \ldots, m\}$ be a set of indices of m conditions and $M(I, J)$ be a data matrix associated with I and J. A *bicluster* associated with the data matrix $M(I, J)$ is a pair (I', J') such that $I' \subseteq I$ and $J' \subseteq J$.

The biclustering problem can be formulated as follows. Given a data matrix M, construct a group of biclusters B_{opt} associated with M such that:

$$f(B_{opt}) = \max_{B \in BC(M)} f(B) \tag{14.1}$$

where f is an objective function measuring the *quality*, i.e., degree of coherence, of a group of biclusters and $BC(M)$ is the set of all the possible groups of biclusters associated with M.

Clearly, biclustering is a highly combinatorial problem with a search space size $O(2^{|I|+|J|})$. In its general case, biclustering is NP-hard [5, 6].

In this chapter, we review two biclustering approaches for microarray data analysis: systematic and stochastic biclustering. The studied algorithms are evaluated on three popular microarray datasets. We show the relative performance of these algorithms to extract relevant biclusters. Biological significance of the selected genes within the biclusters is assessed with publicly available tools.

14.2 SYSTEMATIC BICLUSTERING ALGORITHMS

As we mentioned in the introduction of this chapter, the biclustering problem is NP-hard [5, 6]. Heuristic algorithms are usually used to find approximate biclusters. These algorithms belong to one of two main approaches: *systematic search* and *stochastic search*, based on *metaheuristic algorithms*.

In this section, we describe three types of systematic biclustering algorithms, such as the *enumeration algorithm*, the *greedy algorithm*, and the *divide-and-conquer algorithm*.

14.2.1 THE ENUMERATION ALGORITHM

The *enumeration algorithm* enumerates (explicitly or implicitly) all the solutions for the original problem. The enumeration process is generally represented by a search tree. By applying this approach to the biclustering problem, we identify all possible groups of biclusters in order to keep the *best* one. This approach has the advantage of being able to obtain the best solutions. Its disadvantage is that it is costly in computing time and in memory space. We describe two enumeration algorithms: *BiMine* [7] and *BiMine+* [8].

14.2.1.1 The *BiMine* Algorithm

BiMine [7] is an enumeration algorithm for biclustering of microarray data. It is based on a new evaluation function called average Spearman's rho (ASR) and a new tree structure, called bicluster enumeration tree (BET), to conveniently represent the different biclusters discovered during the enumeration process. In fact, an evaluation function is an indicator of the performance of a biclustering algorithm [9, 10].

Like any search algorithm, *BiMine* needs an *evaluation function* to assess the quality of a candidate bicluster. An *evaluation function* is an indicator of the coherence degree of a bicluster in a data matrix. There are several evaluation functions of a bicluster [5, 11, 12, 13, 14, 15]. The proposed evaluation function, called *average Spearman's rho* (ASR), is based on Spearman's rank correlation [16]. It is defined by the following equation:

$$ASR(I',J') = 2\max\left\{ \frac{\sum_{i\in I'}\sum_{j\in I',j\geq i+1}\rho_{ij}}{|I'|(|I'|-1)} \, , \, \frac{\sum_{k\in J'}\sum_{l\in J',l\geq k+1}\rho_{kl}}{|J'|(|J'|-1)} \right\} \qquad (14.2)$$

where:

ρ_{ij} ($i \neq j$) is the Spearman's rank correlation associated with the row indices i and j in the bicluster (I',J') [16], and

ρ_{kl} ($k \neq l$) is the Spearman's rank correlation associated with the column indices k and l in the bicluster (I',J').

Let us note that the values of Spearman's rank correlation belong to $[-1,1]$. A high (respectively low) Spearman's rank correlation value, i.e., *close* to 1 (respectively, *close* to -1), indicates that the two vectors are strongly (respectively weakly) coherent [16]. Thus, the values of the *ASR* function also belong to $[-1,1]$. Hence, a high (respectively, low) *ASR* value, i.e., *close* to 1 (respectively *close* to -1), indicates that the bicluster is strongly (respectively, weakly) coherent. Furthermore, it has been shown that the Spearman's rank correlation is less sensitive to the presence of noise in the data [17]. Since the *ASR* function is based on the Spearman's rank correlation, *ASR* is also less sensitive to the presence of noise in the data.

BiMine operates in two steps: construction of the *bicluster enumeration tree* (BET) and extraction of the best biclusters.

BiMine begins the construction of BET that represents every possible bicluster (node) that can be made from the data matrix M by using the following strategy: The ith child of a node is made up, on the one hand, from the union of the genes of the parent node and the genes of the ith uncle node, starting from the right side of the parent. On the other hand, it is made up of the intersection of the conditions of the parent and those of the ith uncle starting from the right side of the parent. Indeed, the BET permits us to represent the maximum number of significant biclusters and the links that exist between them.

Since the number of possible biclusters (nodes of the BET) increases exponentially, *BiMine* employs parametric rules to help the enumeration process to close (or cut) some hopeless tree nodes. If the ASR value associated with the ith child is smaller than or equal to the given threshold, then this child will be ignored. Notice that this parametric pruning rule based on a quality threshold is fully justified in this context. Indeed, if the current bicluster is not good enough, then it is useless to keep it because expanding such a bicluster certainly leads to biclusters of worse quality.

From this point of view, the pruning rule shares similar principles largely applied in optimization methods like dynamic programming. In addition, this pruning rule is essential in reducing the tree size and remains indispensable for handling large datasets.

Finally, the union of the leaves of the constructed BET that are not included in other leaves represents the best biclusters. *BiMine* is detailed in Algorithm 14.1. The algorithm calls *InitBET* and *ConstructBET*, which are depicted in Algorithms 14.2 and 14.3, respectively. We define in the following the needed notation:

M: data matrix,

T_n: sub-tree composed by the node n and his children,

n, n': nodes of BET,

$gene_n$: genes of the node n,

$cond_n$: conditions of the node n,

$Bc = (Ic, Jc)$: current bicluster,

β: threshold quality of a bicluster,

$\mathscr{B}$: biclusters results.

Algorithm 14.1 *BiMine*

Input: M, β
Output: $\mathscr{B}$
$BET \leftarrow InitBET(M)$
$\mathscr{B} \leftarrow ConstructBET(BET)$
return $\mathscr{B}$

Algorithm 14.2 *InitBET*

Input: M
Output: BET // Sub-tree composed by the empty node and its children
$T_0 \leftarrow$ empty node
for each *gene* $\in M$ **do**
    Extend T_0 by $Bc=(Ic,Jc)$ as a child of the root T_0
    where $Ic = gene$ and $Jc =$ conditions of this *gene*
end for
$BET \leftarrow T_0$
return BET

14.2.1.2 The *BiMine+* Algorithm

BiMine+ [8] is an enumerative heuristic algorithm that is based on the use of a new tree structure, called *modified bicluster enumeration tree* (MBET) (see Algorithms 14.4–14.7). The pruning rule used by *BiMine+* allows the algorithm to avoid both trivial biclusters and combinatorial explosion of the search tree.

The aim of *BiMine+* is to extract coherent and maximal size biclusters. The algorithm uses MBET to represent the identified biclusters, where each node of MBET

Algorithm 14.3 *ConstructBET*

Input: β, *BET* //*BET* current
Output: $\mathcal{B}$
$\mathcal{B} \leftarrow \emptyset$
for each node n in *BET* **do**
    for each node n' (sibling of n) not processed **do**
        $Ic \leftarrow gene_n \cup gene_{n'}$; $Jc \leftarrow cond_n \cap cond_{n'}$;
        if $ASR(Ic, Jc) \geq \beta$ **then**
            $Bc \leftarrow (Ic, Jc)$
            Insert Bc as a child of n
            if Bc is a vertex of the current sub-tree belonging to the first
            level **then**
                $\mathcal{B} \leftarrow \mathcal{B} \cup \{Bc\}$
            end if
        end if
    end for
    $T_n \leftarrow$ sub-tree composed by the node n and its children
    ConstructBET(T_n)
end for
return $\mathcal{B}$

contains the gene profile shape of a bicluster. The profile shape of a gene is defined as the behavior of this gene, i.e., up, down or no change, over the conditions of the bicluster to which this gene belongs. Indeed, according to the authors of [18, 19, 20], in microarray data analysis, genes are considered to be in the same cluster if their trajectory patterns of expression levels are similar across a set of conditions. To limit the size of MBET, *BiMine+* employs a pruning rule to eliminate any bicluster that has a number of conditions lower than a given threshold. Finally, *BiMine+* uses the ASR evaluation function to provide a final assessment of each extracted bicluster.

The *BiMine+* algorithm operates in three steps. First, we discretize the data matrix M to obtain M'. Second, we construct from M' MBET that represents every possible maximal bicluster with a low-level overlap. Finally, we select among the extracted biclusters those that have an ASR value equal to or greater than a fixed threshold.

To discretize the data matrix $M(I, J)$ into a data matrix M' we use the following equation:

$$M'[i,l] = \begin{cases} 1 & \text{if } M[i,l] < M[i,l+1] \\ -1 & \text{if } M[i,l] > M[i,l+1] \\ 0 & \text{if } M[i,l] = M[i,l+1] \end{cases} \tag{14.3}$$

with $i \in [1, \ldots, n]$ and $l \in [1, \ldots, m-1]$.

In the second step, we construct the MBET in a way similar to building BET except that nodes of MBET contain discrete values.

Since the number of the possible biclusters (nodes of MBET) increases exponentially, we employ a parametric rule to progressively prune some nodes. Indeed, a node is pruned if it has a number of conditions lower than a fixed threshold.

During the construction step, we extract the largest leaf (bicluster) from each subtree rooted by a node of the first level. In fact, the leaves of each subtree have a high-level overlap because they share, most of the time, the same genes. Hence, for each subtree we extract only the largest bicluster with a low-level overlap.

Algorithm 14.4 *BiMine+*

Input: M, δ, β
Output: $\mathscr{B}$
Discretize M by using Equation (14.3) to obtain M'
$MBET \leftarrow InitMBET(M')$
$\mathscr{B} \leftarrow BuildMBET(MBET)$
$\mathscr{B} \leftarrow SelectBiclusters(\mathscr{B})$
return $\mathscr{B}$

Algorithm 14.5 *InitMBET*

Input: M'
Output: *MBET* // Subtree made up by the empty node and its children
$T_0 \leftarrow$ empty node
for each *gene* $\in M'$ **do**
　　　Extend T_0 by $Bc=(Ic,Jc)$ as a child of the root,
　　　where $Ic = gene$ and $Jc =$ conditions of *gene*
end for
$MBET \leftarrow T_0$
return *MBET*

14.2.2 THE GREEDY ALGORITHM

The *greedy algorithms* construct a solution in a step-by-step way using a given quality criterion. Decisions made at each step are based on information at hand without worrying about the effect these decisions may have in the future. Moreover, once a decision is made, it becomes irreversible and is never reconsidered. By applying this approach to the biclustering problem, at each iteration, we construct submatrices of the data matrix by adding/removing a row/column to/from the current submatrix that maximizes/minimizes a certain function. We reiterate this process until no other row/column can be added/removed to/from any submatrix. The advantage of this approach is that it is fast. However, it may ignore good biclusters by partitioning them before identifying them.

In this subsection, we describe two greedy algorithms: *BicFinder* [21] and RMSBE [22].

Algorithm 14.6 *BuildMBET*

Input: *MBET* //Current *MBET*
Output: $\mathscr{B}$
$\mathscr{B} \leftarrow \emptyset$
for each node n in *MBET* **do**
    for each unprocessed brother n' of n **do**
        $Ic \leftarrow gene_n \cup gene_{n'}$; $Jc \leftarrow cond_n \cap cond_{n'}$;
        if $|Jc| \geq \delta$ **then** // δ: threshold of condition number
            $Bc \leftarrow (Ic, Jc)$
            Insert Bc as a child of n
            if Bc has a maximum size leaf in the current subtree, rooted at
            level 1 **then**
                $\mathscr{B} \leftarrow \mathscr{B} \cup \{Bc\}$
            end if
        end if
    end for
    $T_n \leftarrow$ subtree made up by n and its children
    $BuildMBET(T_n)$
end for
return $\mathscr{B}$

Algorithm 14.7 *SelectBiclusters*

Input: $\mathscr{B}$
Output: $\mathscr{B}$
for each bicluster (I', J') in $\mathscr{B}$ **do**
    if $ASR(I', J') < \beta$ **then**
        $\mathscr{B} \leftarrow \mathscr{B} \setminus \{(I', J')\}$
    end if
end for
return $\mathscr{B}$

14.2.2.1 The *BicFinder* Algorithm

BicFinder relies on two evaluation functions: the ASR function and the proposed average correspondence similarity index (ACSI), and utilizes a directed acyclic graph to construct its biclusters. Like *BiMine+*, *BicFinder* starts with the discretization of the initial data matrix $M(I, J)$ (see Equation (14.3)).

In order to compute ACSI, we define the following terms. Let the *correspondence similarity list* between genes g_i and g_j $(i < j)$, denoted by $CSL_{i,j}$, be the list. Each element of this list is represented by $T(M'[i, l] = M'[j, l])$ where $T(Func)$ is true, if and only if $Func$ is true, and $T(Func)$ is false otherwise. Let $NumCSL_{i,j}$ be the number of times we have a true value in $CSL_{i,j}$ and $MaxCSL_i = \max$ $\{NumCSL_{i,i+1}, NumCSL_{i,i+2}, \ldots, NumCSL_{i,n}\}$. We define the *correspondence simi-*

larity index as follows:

$$CSI(i,j,k) = \frac{\sum_{l=1}^{m-1} T(M'[i,l] = M'[j,l] = M'[k,l])}{MaxCSL_i} \tag{14.4}$$

with $i \in [1..n-2]$, $j \in [2..n-1]$, $k \in [3..n]$, $l \in [1..m-1]$ and $i < j < k$. Finally, for the whole bicluster, we define the *average correspondence similarity index* (ACSI) for the row i ($i \in I'$ and $i < j < k$):

$$ACSI_i(I',J') = 2\frac{\sum_{j\in I';j\geq i+1}\sum_{k\in I';k\geq j+1} CSI(i,j,k)}{(|I'|-1)(|I'|-2)} \tag{14.5}$$

The second step of *BicFinder* is to build a directed acyclic graph (DAG). A DAG associated with a data matrix M' is represented as follows: a node n_i represents a gene g_i and an arc connects a node n_i to a node n_j if and only if $i < j$. We assign $CSL_{i,j}$ to each arc (n_i, n_j).

The next step of *BicFinder* is to extract coherent biclusters. For each node n_i, *BicFinder* first initializes the associated bicluster $B_i=(I_i',J_i')$ to $(\emptyset,\emptyset)$. Then, we sort the arcs leaving n_i in a decreasing order based on the number of true values associated with each arc. *BicFinder* then considers the sorted arcs successively. Let (n_i,n_k) be the current arc; if the evaluation function ACSI associated with the bicluster $(I_i' \cup \{g_i,g_k\},J_i'\cup\{c_l,c_{l+1}$ such that $T(M'[i,l] = M'[k,l]) = true\})$ is greater than or equal to a fixed threshold α, then *BicFinder* sets $B_i=(I_i' \cup \{g_i,g_k\},J_i'\cup\{c_l,c_{l+1}$ such that $T(M'[i,l] = M'[k,l]) = true\})$. This process is repeated until all the arcs leaving n_i are processed.

Finally, *BicFinder* considers only the obtained biclusters for which the ASR evaluation function is greater than or equal to another fixed threshold β. The set of such biclusters represents a solution to our problem. *BicFinder* is detailed in Algorithm 14.8.

14.2.2.2   The MSB Algorithm

Liu and Wang proposed the maximum similarity biclustering algorithm (MSB) [22]. To measure the quality of the biclusters, they used a specific similarity scoring function to detect coherent sub-matrices. MSB does not employ a discretization procedure and performs well for overlapping and additive biclusters.

The MSB algorithm starts with the whole matrix as its initial bicluster. Then it iteratively removes the row or column in the bicluster with the worst similarity score, until there is one element in the current bicluster. During this process, $(n+m-1)$ submatrices are obtained, where n and m refer to the number of rows and columns in the input matrix, respectively. MSB only outputs one bicluster with the maximum similarity score. An extension of the MSB algorithm, called randomized MSB extension algorithm RMSBE, is also presented. RMSBE uses the average similarity scores between some pairs of genes in the bicluster, as well as of random selection to choose the reference genes.

Algorithm 14.8 The *BicFinder* Algorithm

Input: M, α (quality thresholds of ACSI), β

Output: $\mathscr{B}$

Discretize M using Equation (14.3) to obtain M' // Discretization step

Construct the DAG associated with M' // Construction step

$\mathscr{B} \leftarrow \varnothing$ // Extraction step

for each n_i in the DAG **do**

     $I'_i \leftarrow \varnothing; J'_i \leftarrow \varnothing$ // $B_i \leftarrow (I'_i, J'_i)$

     Sort the arcs leaving n_i in a decreasing way by considering the number of true's

     for each arc (n_i, n_k) **do**

         $Ic \leftarrow I'_i \cup \{g_i, g_k\}; Jc \leftarrow J'_i \cup \{c_l, c_{l+1}$ such that $T(M'[i,l] = M'[k,l])$ $= true)\};$ // Ic, Jc: current subset of genes and current subset of conditions

         if $ACSI_i(Ic, Jc) \geq \alpha$ **then**

             $B_i \leftarrow (Ic, Jc)$

         end if

     end for

     $\mathscr{B} \leftarrow \mathscr{B} \cup B_i$

end for

for each bicluster $B_i = (I'_i, J'_i)$ in $\mathscr{B}$ **do**   // Selection step

     if $ASR(I'_i, J'_i) < \beta$ **then**

         $\mathscr{B} \leftarrow \mathscr{B} \backslash B_i$

     end if

end for

return $\mathscr{B}$

14.2.3 THE DIVIDE-AND-CONQUER ALGORITHM

Generally, this method repeatedly divides the problem into smaller subproblems with similar structures to the original problem, until these subproblems become smaller enough to be solved directly. The solutions to the subproblems are then combined to create a solution to the original problem. With this approach, we start with a bicluster representing the whole data matrix, and then partition this matrix into two submatrices to obtain two biclusters. We recursively reiterate this process until we obtain a certain number of biclusters verifying a specific set of properties. This approach presents the similar advantages and disadvantages to the greedy algorithms.

One representative example of this approach is Bimax, which was proposed by Prelic et al. [23]. Bimax partitions the data matrix M' (M' is a discretization of a data matrix M, which contains only binary values where a cell m_{ij} contains 1 if $gene_i$ is expressed under $condition_j$ and 0 otherwise) into three submatrices, one of which contains only 0-cells. The algorithm is then recursively applied to the remaining two submatrices, and ends if the current matrix represents a bicluster that contains only 1's.

14.3   STOCHASTIC BICLUSTERING ALGORITHMS

In this section, we describe three types of stochastic biclustering algorithms that are based on *local search*, *evolutionary search*, and the *hybrid approach*.

14.3.1   THE ITERATED LOCAL SEARCH ALGORITHM

The *iterated local search algorithm* (ILS) can be described by a simple computing model [24]. A fundamental principle of ILS is to exploit the tradeoff between intensification and diversification. Intensification focuses on optimizing the objective function as far as possible within a limited search region while diversification aims to drive the search to explore new promising regions of the search space. The diversification mechanism of ILS—perturbation operator—has two aims: one is to jump out of the local optimum trap; the other is to lead the search procedure to a new promising region.

From the operational point of view, an ILS algorithm starts with an initial solution and performs local search until a local optimum is found. Then, the current local optimum solution is perturbed and another round of local search is performed with the perturbed solution. The advantage of this approach lies in the ability to explore large search spaces. This approach also offers the possibility of trade-off between solution quality and running time. Indeed, while the quality of a solution tends to improve gradually over time, the user can stop the execution at a chosen cut-off time. The disadvantage of this approach is that the search leads to sub-optimal solutions (local maxima). Below, we describe two ILS algorithms: BILS [25] and PDNS [26].

14.3.1.1   The BILS Algorithm

BILS is a stochastic neighborhood search algorithm. It uses a local search procedure and a perturbation-based diversification strategy. BILS uses a preprocessing step to transform the input data matrix M to a *Behavior Matrix* M'. This preprocessing step aims to highlight the trajectory patterns of the genes. Indeed, genes are considered to be in the same cluster if their trajectory patterns of expression levels are similar across a set of conditions. In our case, each column of M' represents the trajectory of genes between a pair of conditions in the data matrix M. The whole matrix M' provides useful information for the identification of related biclusters and the definition of a meaningful neighborhood and perturbation strategy.

The behavior matrix M' is defined as follows:

$$M'[i,l] = \begin{cases} 1 & \text{if } M[i,k] < M[i,q] \\ -1 & \text{if } M[i,k] > M[i,q] \\ 0 & \text{if } M[i,k] = M[i,q] \end{cases} \quad (14.6)$$

where $i \in [1..n]$, $l \in [1..J'']$ (with $J'' = m(m-1)/2$), $k \in [1..m-1]$, $q \in [2..m]$ and $q \geq k+1$.

Starting from an initial solution (which can be provided by a fast greedy algorithm to rapidly obtain a bicluster of reasonable quality), BILS tries to iteratively find biclusters of better and better quality. Basically, the improvement is carried out by removing "bad" genes from the current bicluster and adding one or more other "better" genes. Each application of this dual drop/add operation generates a new bicluster from the current bicluster. The way of identifying the possible genes to drop and to add defines the so-called neighborhood.

The general BILS procedure is given in Algorithm 14.9. Starting from an initial bicluster, called solution s, BILS uses the descent strategy to explore the neighborhood. At each iteration, we move to an improving neighboring solution $s' \in N(s)$ by considering the evaluation function $\mathbb{S}(s)$. This descent strategy based intensification phase stops when no improving neighbor can be found in the neighborhood. Thus, the last solution is the best solution found and corresponds to a local optimum. At this point, BILS triggers a diversification phase by perturbing the best solution to generate a new starting point for the next round of the search. The perturbation operator changes the best local optimum by randomly deleting 10% of genes of the best solution and adding 10% of genes among the best genes that are not included in the best solution. This perturbed solution is used by BILS as its new starting point.

The whole BILS algorithm stops when the best bicluster reaches a fixed quality or when the best solution found is not updated for a fixed number of perturbations.

Algorithm 14.9 The BILS Algorithm

Input: An initial bicluster s_0, quality threshold λ
Output: The best bicluster
Create the Behavior Matrix M'
$s \leftarrow s_0$ // current solution
repeat
  repeat
    Choose a pair of genes (g_i, g_j) belonging to s such that (g_i, g_j) have good quality
    Choose a pair of genes (g_j, g_r) belonging to s such that (g_j, g_r) have bad quality
    Identify all genes g_v, $v \notin I'$ such that (g_r, g_v) have good quality
    Generate neighbor s' by dropping g_i from s and adding all g_v
    if $((s')$ is better than $(s))$ **then**
      $s \leftarrow s'$
    end if
  until (no improving neighbor can be found in $N(s)$)
  Generate a new solution s by perturbing randomly 10% of the best solution
until (stop condition is verified)
return s

14.3.1.2 The PDNS Algorithm

Pattern-driven neighborhood search (PDNS) follows the general ILS scheme, called PDNS, for the biclustering problem (see Algorithm 14.10). PDNS, like BILS, is based on a solution representation encoded as a behavior matrix. It alternates between two basic components: a descent-based improvement procedure and a perturbation operator. PDNS uses the descent procedure to discover locally optimal solutions and the perturbation operator to displace the search to a new starting point in an unexplored search region. It also employs fast greedy algorithms to generate various initial biclusters of reasonable quality and a randomized perturbation strategy.

The originality of PDNS consists of the use of a *bicluster pattern* both in its search space and neighborhood definition. The bicluster pattern, which is a representation of the bicluster behavior, is used to evaluate genes/conditions of bicluster. This representation is defined by the behavior matrix of the bicluster, i.e., the trajectory patterns of the genes under all combined conditions of the bicluster.

Starting from an initial bicluster, PDNS uses the descent strategy to explore the pattern-based neighborhood and moves to an improving neighboring solution at each iteration. By using the bicluster pattern, we define a set of rules that allow us to qualify the goodness (or badness) of a gene and condition. Using these rules, PDNS iteratively replaces within the current bicluster bad genes/conditions by good ones, and hence progressively improves the quality of the bicluster under consideration. This iterative improvement procedure stops when the last bicluster attains a fixed quality threshold by considering the ASR evaluation function or when a fixed number Y of iterations is reached. At this point, PDNS triggers a perturbation phase by replacing randomly 10% of genes and conditions of the recorded best bicluster found so far. This perturbed bicluster is used as a new starting point for the next round of the descent search. PDNS stops when the best bicluster is not updated for a fixed number Z of perturbations.

14.3.2 THE EVOLUTIONARY ALGORITHM

The evolutionary computation approach is based on the natural evolutionary process such as population, reproduction, mutation, recombination, and selection. Candidate solutions of the given problem are sampled by a set of individuals in a population. An evaluation mechanism (fitness evaluation) is established to assess the quality of each individual. Evolution operators eliminate some (less fit) individuals and produce new individuals from the selected individuals.

By applying this approach to the biclustering problem, we start from an initial population of solutions, i.e., clusters, biclusters, or the whole matrix. Then, we measure the quality of each solution of the population by the fitness function. We select a number of solutions to produce new solutions by recombination and mutation operators. This process ends when a prefixed stop condition is verified. This approach shares the similar advantages and disadvantages with the local search approach. Below, we present our evolutionary algorithm called EvoBic [27].

The proposed *Evolutianary biclustering* algorithm (EvoBic) is an evolutionary approach to obtain maximal high-quality biclusters of highly-correlated genes. One particularity of EvoBic remains in the fact that it makes use of three main evaluation functions, which are explained below.

Algorithm 14.10 The PDNS Algorithm

Input: Initial data matrix M, Initial bicluster B_0, quality thresholds: α, β, *threshold_ASR*, Maximum number of iterations Y, Z

Output: The best bicluster B^*

Create the Behavior Matrix M' from M

Create the Behavior sub-Matrix $\bar{M}'$ for B_0

$B^* \leftarrow B_0$ // Record the best bicluster found so far

$s \leftarrow \bar{M}'$ // Set the initial solution

repeat

  repeat

    Construct bicluster pattern P from s

    $s' \leftarrow s \oplus mv_g(\alpha)$ /* Apply the row (gene) move operator: Replace the bad genes of s by genes among the good genes using the bicluster pattern*/

    Updated pattern P

    $s \leftarrow s' \oplus mv_c(\beta)$ /* Apply the column (condition) move operator: Replace the bad conditions of s by random columns among the good columns using the bicluster pattern */

    Reconstruct bicluster B from s

    if $ASR(B) > ASR(B^*)$ **then**

      $B^* \leftarrow B$ // Update the best bicluster found so far

    end if

  until $(ASR(B^*) \geq threshold\_ASR$ or we reach the maximum number of iterations $Y)$

  Generate a new solution s by perturbing randomly 10% of the recorded best solution

until $(B^*$ is not updated after a number Z of perturbations)

return B^*

14.3.2.1 Size

Most biclustering algorithms define the size of a bicluster by its number of elements $|I| * |J|$ which gives more chance to the number of genes to be maximized since the total number of genes is higher than the number of conditions [28]. To give the same chance to the number of genes and the number of conditions to be maximized, we choose to define the size of the biclusters by the following function:

$$f1: \qquad S(B) = \frac{1}{2}\frac{|I'|}{|I|} + \frac{1}{2}\frac{|J'|}{|J|} \qquad\qquad (14.7)$$

14.3.2.2 Mean Squared Residue

Cheng and Church [5] proposed *mean squared residue* (MSR), which measures the correlation of a bicluster. This measure is defined as follows:

$$f2: \qquad MSR(B) = \frac{1}{|I'||J'|} \sum_{i \in I', j \in J'} r(i,j)^2 \qquad (14.8)$$

where :

$$r(i,j) = m_{ij} - m_{iJ'} - m_{I'j} + m_{I'J'}$$

$$m_{I'J'} = \frac{\sum\limits_{i \in I', j \in J'} (m_{ij})}{|I'| \cdot |J'|}$$

$$m_{I'j} = \frac{\sum\limits_{i \in I'} (m_{ij})}{|I'|}$$

$$m_{iJ'} = \frac{\sum\limits_{j \in J'} (m_{ij})}{|J'|}$$

A high value of MSR (greater than a certain threshold δ) indicates that the bicluster is weakly coherent, while a low value of MSR indicates that it is highly coherent.

14.3.2.3 Average Correlation

Nepomuceno et al. [29] proposed the average correlation function to evaluate the correlation between genes in a bicluster. The authors indicate that the proposed function can find biclusters that cannot be found by the algorithms based on MSR because these algorithms might not find scaling patterns when the variance of gene values is high. The average correlation of B is denoted by $\rho(B)$ and is defined as follows:

$$f3: \qquad \rho(B) = \frac{2}{I'(I'-1)} \sum_{i=1}^{I'} \sum_{j=i+1}^{I'} \left| v_{(G_i, G_j)} \right| \qquad (14.9)$$

where $0 \le \rho(B) \le 1$ and $v_{(G_i, G_j)}$ correspond to the correlation coefficient between gene G_i and gene G_j.

The correlation coefficient between two vectors $X = \{x_1, x_2, \cdots, x_n\}$ and $Y = \{y_1, y_2, \cdots, y_n\}$ is defined as follows:

$$v_{(X,Y)} = \frac{cov(X,Y)}{\sigma_X \sigma_Y} = \frac{\sum_i^n (x_i - \bar{x})(y_i - \bar{y})}{n \sigma_X \sigma_Y}$$

where $cov(X,Y)$ represents the covariance of the vectors X and Y, σ_X and σ_Y correspond to the standard deviations of X and Y, and $\bar{x}$ and $\bar{y}$ are the means of the values of X and Y, respectively.

This measure varies between -1 and 1. If $v_{(X,Y)} = -1$ or $v_{(X,Y)} = 1$, the variables are linearly dependent, respectively, with negative correlation (when the value of X decreases, the value of Y increases and vice versa) and positive correlation (the value of X and the value of Y increase and decrease together), and if $v_{(X,Y)} = 0$ the variables are linearly independent. Thus, if $\rho(B) = 1$ the genes are highly correlated.

EvoBic can be summarized by five steps: First, we generate an initial population in a random manner in order to ensure diversity of the population. Second, we construct the set of individuals (biclusters) SI by selecting the best individuals from the population based on the evaluation functions $f1$ (maximize the size of biclusters), $f2$ (minimize the *mean squared residue* function [5]) and $f3$ (maximize the correlation between genes). Third, we create the new individuals by combining all individuals selected from the set SI. This mechanism combines all pairs of biclusters to generate offspring biclusters. Fourth, we update the set SI by selecting the best individuals from the new and the combined individuals, until it becomes stable. Finally, we add the best individuals to the biclusters, based on the fitness functions. If the number of biclusters is not reached, we restart with the first step. EvoBic is detailed in Algorithm 14.11.

Algorithm 14.11 The EvoBic Algorithm

Input: Microarray M, size of set of individuals S, size of population K, number of biclusters to be found $NbBi$
Output: The set of Bic
$Nb \leftarrow 0$
$Biclusters \leftarrow \emptyset$
while $(Nb < NbBi)$ **do**
    Initialize population P with K individuals (biclusters)
    $Stable \leftarrow FALSE$
    while $(Stable = FALSE)$ **do**
        $R1 \leftarrow S/3$ best biclusters from P (considering $f3$)
        $R2 \leftarrow S/3$ best biclusters from $P \setminus R1$ (considering $f1$)
        $R3 \leftarrow S/3$ best biclusters from $P \setminus (R1 \cup R2)$ (considering $f2$)
        $SI \leftarrow (R1 \cup R2 \cup R3)$
        $P \leftarrow P \setminus SI$
        $OldSI \leftarrow SI$
        $CombSI \leftarrow$ CombinationMethod (SI)
        $R1' \leftarrow S/2$ best biclusters from $SI \cup CombSI$ (considering $f1$ and $f3$)
        $R2' \leftarrow S/2$ best biclusters from $SI \cup CombSI \setminus R1'$ (considering $f2$)
        $SI \leftarrow (R1' \cup R2')$
        $P \leftarrow P \cup SI$
        if $(OldSI = SI)$ **then**
            $Stable \leftarrow TRUE$
        end if
    end while
    $Bic \leftarrow$ the best bicluster from SI (considering $f3$)
    $Nb \leftarrow Nb + 1$
end while
return (Bic)

	0	1	2	3
0	7	13	9	1
1	4	0	4	5
2	0	2	7	5
3	9	8	3	7
4	11	1	12	9
5	8	2	10	6

(a)

	0	2	3
1	4	4	5
2	0	7	5
4	11	12	9
5	8	10	6

(b)

FIGURE 14.1　Example of a data matrix.

14.3.2.4　Encoding of Bicluster

Most existing evolutionary algorithms for solving the biclustering problem represent the individuals (biclusters) by a fixed size binary string built by a bit string for genes with another one for condition. The string position set to 1 reflects the gene or condition belonging to the bicluster. This representation leads to high consumption of time and memory space since it requires exploration of all genes and conditions of each bicluster. In order to reduce time and memory space, we represent the individuals as a string composed by the ordered gene and condition indices. Let Figure 14.1(a) be an example of a data matrix and Figure 14.1(b) the bicluster that is represented by the string Bi: 1 2 4 5 // 0 2 3.

14.3.2.5　Genetic Operators

A single point crossover is used as a combination method to create new biclusters. The crossover combines all pairs of biclusters in the reference set (parent biclusters) to generate $S * (S - 1)$ children biclusters and each part of the bicluster (part genes and part conditions) is treated separately. Let the parent biclusters be:

$$Bi_1: g_1\ g_2\ \ldots\ g_n\ //\ c_1\ c_2\ \ldots\ c_p$$
$$Bi_2: g_1'\ g_2'\ \ldots\ g_m'\ //\ c_1'\ c_2'\ \ldots\ c_q'$$

where $g_n \leq g_m'$ and $c_p \leq c_q'$

First, we generate two random integers Ψ_1 and Ψ_1' corresponding, respectively, to the crossover point in genes and conditions of Bi_1 such that $g_1 \leq \Psi_1 \leq g_n$ and $c_1 \leq \Psi_1' \leq c_p$.

Second, we generate $\Psi_2 = g_i'$ and $\Psi_2' = c_j'$, the crossover points, respectively, in part genes and part conditions in Bi_2 where $g_{i-1}' \leq \Psi_1 \leq g_i'$ and $c_{j-1}' \leq \Psi_1' \leq c_j'$.

For example, let us consider the following parents:

$$Bi_1: 2\ 3\ 6\ 7\ 9\ 11\ 12\ 16\ 20\ //\ 0\ 5\ 9\ 10$$
$$Bi_2: 0\ 3\ 4\ 8\ 10\ 14\ 21\ 25\ 26\ 28\ 30\ //\ 2\ 4\ 6\ 8\ 12$$

Suppose that $\Psi_1 = 11$ and $\Psi_1' = 5$ therefore $\Psi_2 = 14$ and $\Psi_2' = 6$.
Thus, the resulting children are:

$$C_1: 2\ 3\ 6\ 7\ 9\ 11\ 14\ 21\ 25\ 26\ 28\ 30\ //\ 0\ 5\ 6\ 8\ 12$$
$$C_2: 0\ 3\ 4\ 8\ 10\ 12\ 16\ 20\ //\ 2\ 4\ 9\ 10$$

14.3.3 THE HYBRID APPROACH

The hybrid approach, also called *memetic approach*, tries to combine the neighborhood search and the evolutionary approaches. This hybrid approach is known to be quite successful in solving many hard combinatorial search problems. The purpose of such an approach is to take advantage of the complementary nature of the evolutionary and neighborhood search methods. Indeed, it is generally believed that the evolutionary framework offers more facilities for exploration, while neighborhood search has more capability for exploitation. Combining them may offer a better balance between exploitation and exploration, which is highly desirable for an effective search [30].

Mitra and Banka [31] presented the *multi-objective evolutionary algorithm* (MOEA) based on the notion of Pareto dominance. The authors tried to find biclusters with maximum size and homogeneity by using a multi-objective genetic algorithm called *non-dominated sorting genetic algorithm* (NSGA-II) [32] in combination with a local search procedure. Gallo et al. [33] presented another hybrid algorithm based on MOEA combined with a local search strategy. They extracted biclusters with multiple criteria like maximum rows, columns, homogeneity, and row variance. A mechanism for re-orienting the search in terms of row variance and size is provided. The mutation operator is performed when the individual needs to be mutated with a given probability. The gene/condition of the bicluster is mutated at a random position. The crossover operator is applied over both the genes and the conditions. Hence, when both children are obtained by combining at the end and at the center each of the two parents, the individual to become the only descendant is the non-dominated one. If both are non-dominated, one of them is chosen at random. The authors apply the local search procedure, based on the Cheng and Church procedure [5], on all the individuals in the resulting population of each generation.

14.4 EXPERIMENTAL RESULTS

In this section, we discuss the relative performance of the presented biclustering algorithms on three real DNA microarray datasets. For this purpose, we compare the obtained results to those from a selection of the most popular biclustering algorithms in the literature. This experimental study also aims to assess the relevance of these algorithms based on statistical and biological criteria.

The microarray datasets used for our experiments are the following:

> *Yeast cell cycle*: This dataset was described by Tavazoie et al. [34] and preprocessed by Cheng and Church [5]. It contains the expression of 2,884

genes under 17 conditions selected by Cheng and Church. It is available at `http://arep.med.harvard.edu/biclustering/`.

Saccharomyces cerevisiae: This dataset was described by Gasch et al. [35] and corresponds to a selection of 2,993 genes and a collection of 173 various conditions of stress. It is available at `http://www.tik.ethz.ch/sop/bimax/`.

Human B-cell Lymphoma: This dataset was described by Alizadeh et al. [36]. It contains the expression of 4,026 genes and 96 conditions. It is available at `http://arep.med.harvard.edu/biclustering/`.

These datasets are frequently used in the literature to assess the biclustering algorithms.

14.4.1 COMPARISON CRITERIA

To evaluate these algorithms on these microarray datasets, the following criteria are adopted:

(a) Statistical criteria:

1. Coverage: This criterion is defined as being the total number of cells of the matrix M covered by the biclusters.
2. p-value: To do a statistical validation, we compute the adjusted p-value, i.e., based on the Fisher's test, to measure the quality of the bicluster results. Indeed, the biclusters that have an adjusted p-value lower than 5% are considered as over-represented. This means that the majority of the genes of a bicluster have in common biological characteristics. The best biclusters have an adjusted p-value lower than 0.001%. This measure is computed by the tool FuncAssociate [37]. It is available at `http://llama.mshri.on.ca/funcassociate/`.

(b) Biological criteria: To biologically validate the obtained biclusters, we use GoTermFinder [38]. It is available at `http://db.yeastgenome.org/cgi-bin/GO/goTermFinder`. This tool allows us to identify three structures of ontologies describing the products of the genes in terms of biological process, molecular function, and cellular component. It also allows to identify the common functions between the genes.

14.4.2 PROTOCOL FOR EXPERIMENTS

The first series of experiments concern the statistical validation by computing the coverage for the datasets *Yeast cell cycle* and *Human B-cell Lymphoma*, and the adjusted p-value for the datasets *Yeast cell cycle* and *Saccharomyces cerevisiae*. The second series of experiments are applied to *Yeast cell cycle* with the aim of studying the biological significance of the extracted bigroups. The initial solutions of BILS and PDNS are obtained from two fast greedy algorithms: CC [5] and OPSM [39].

TABLE 14.2

Numbers of biclusters obtained by our biclustering algorithms.

Algorithm	Yeast cell cycle	Saccharomyces cerevisiae	Human B-cell Lymphoma
BiMine	18,021	20,582	24,317
BiMine+	883	1,441	2,266
BicFinder	708	1,163	727

14.4.3   OBTAINED BICLUSTERS

The *BiMine* algorithm extracts a very large number of biclusters. This implies a high number of biclusters with a high degree of overlapping. Indeed, the strategy used by *BiMine* allows us to extract all the leaves of BET. This strategy was given up by *BiMine+* by considering only the biclusters of maximal size for each sub-tree of MBET as shown in Table 14.2. Concerning BILS and PDNS, in each execution they improve a given input bicluster. EvoBic has as input the extracted number of biclusters.

After an execution, a step of post-processing was applied to decrease the number of biclusters to be analyzed. Indeed, the number of generated biclusters is very important (sometimes on the order of a few thousand), making their analysis difficult. We then followed the strategy of [22, 23, 40, 41] to keep the 100 largest and most significant biclusters.

14.4.4   STATISTICAL VALIDATION

14.4.4.1   Coverage

To estimate the performances of *BiMine*, *BiMine+*, *BicFinder* and EvoBic, we compute the total number of cells of the dataset covered by biclusters. We compare the results of our algorithms with those of reported results of CC [5], EA FRAMEWORK [42], SEBI [40], MOEA [31], SMOB [13], MOPSOB [43], CMOPSOB [44].

This test is not applied to the algorithms BILS and PDNS. Indeed, they improve separately a bicluster given in the input, and keep almost the same coverage of the input bicluster.

Table 14.3 shows the coverage (in percentage) of each algorithm for each dataset. We notice that most of the algorithms have closer rates. For example, for *Yeast cell cycle*, *BiMine+* covers 51.76% of the cells of the initial matrix while *BicFinder* covers 55.43%. However, *BiMine* has weak performance. This is understandable by the fact that *BiMine* extracts thousands of biclusters of small sizes. EvoBic has very high coverage because it extracts very large biclusters by using the evaluation functions $f1$ (see Section 14.3). MOEA covers 51.34% of the cells of the Yeast cell cycle, while CC (respectively, EA FRAMEWORK) covers 67.30% (50.99%) of the cells.

TABLE 14.3

The bicluster coverage for *Yeast cell cycle* and *Human B-cell Lymphoma* datasets.

Dataset	Algorithm	Total Coverage	Coverage of genes	Coverage of conditions
Yeast	*BiMine*	13.36%	32.84%	100%
cell cycle	*BiMine+*	51.76%	68.65%	100%
	BicFinder	55.43%	76.93%	100%
	EvoBic	44.21%	98.58%	70.59%
	MOEA	51.34%	-	-
	EA FRAMEWORK	50.99%	-	-
	CC	81.47%	97.12%	100%
	SEBI	38.14%	43.55%	100%
	SMOB	40.39%	47.02%	100%
	CMOPSOB	53.80%	75.20%	100%
	MOPSOB	52.40%	-	-
Human B-cell	*BiMine*	8.93%	26.15%	100%
Lymphoma	*BiMine+*	21.19%	46.26%	100%
	BicFinder	44.24%	55.89%	100%
	EvoBic	61.21%	73.53%	100%
	MOPSOB	36.90%	-	-
	MOEA	20.96%	-	-
	SEBI	34.07%	38.23%	100%
	SMOB	33.52%	45.05%	100%
	CC	36.81%	91.58%	100%
	CMOPSOB	38.30%	48.10%	100%

In Human B-cell Lymphoma data MOEA covers 20.96% of the cells, whereas an average of 36.81% of cells is covered by CC. The CC algorithm masks the extracted biclusters by random values [5]. Thus, it forbids the genes/conditions already discovered to be selected in the next search step. This type of mask leads to a high cover, while preventing the discovery of large biclusters.

14.4.4.2   P-Value

To estimate the statistical relevance of our algorithms, we compare the obtained results to those of CC, ISA, *Bimax*, and OPSM on *Yeast cell cycle* and *Saccharomyces cerevisiae* datasets. The aim is to compute the adjusted p-value for a set of genes. We use the tool *FuncAssociate* [37] to estimate the extracted bigroups.

Figure 14.2 shows the p-values ($p = 5\%$, 1%, 0.5%, 0.1%, and 0.001%) for each algorithm, the total percentage of extracted biclusters. Indeed, we notice that with

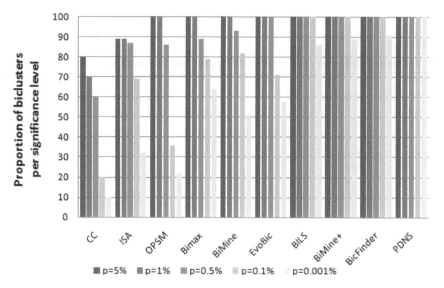

FIGURE 14.2 Proportions of biclusters significantly enriched by GO annotations on the *Yeast cell cycle* dataset.

PDNS (respectively BILS) 100% (respectively 86%) of biclusters are statistically significant with $p < 0.001\%$. This implies that these algorithms improve considerably the biclusters of CC and OPSM, which have, respectively, 10% and 22% of biclusters with $p < 0.001\%$. *BicFinder* (respectively *BiMine+*) has 91% (respectively 88%) of biclusters $p < 0.001\%$. *BiMine* has 100% of the tested biclusters with $p = 5\%$ and $p = 1\%$. Of the biclusters obtained by *BiMine*, 93% (respectively 82%) have $p = 0.5\%$ (respectively 0.1%). EvoBic has a performance comparable to *BiMine*. On the other hand, good results are obtained by *Bimax* where 89% (respectively 79%) of biclusters are extracted with $p = 0.5\%$ (respectively $p = 0.1\%$). Finally, 51% (respectively 64%) of extracted biclusters by *BiMine* (respectively *Bimax*) have a value of $p < 0.001\%$. The best results are achieved by PDNS and *BicFinder*.

Figure 14.3 presents the same experiment on the *Saccharomyces cerevisiae* dataset. All the results of the other algorithms, namely, CC, ISA, *Bimax*, and OPSM, were reported in [23]. We also compare the results of our algorithms to those of RMSBE [22] and Samba [45] which are reported in [22]. Figure 14.3 shows that *BicFinder*, PDNS, and RMSBE are the most successful approaches compared to the other algorithms. The results of *BicFinder*, PDNS, and RMSBE show that 100% (98% for RMSBE) of extracted biclusters are statistically significant and have a value of $p < 0.001\%$. BILS and *BiMine+* have similar performances. Also, we notice that *BiMine+* (respectively BILS) outperforms CC, Samba, ISA, *Bimax*, OPSM, and *BiMine* with 85% (respectively 89%) of biclusters having $p < 0.001\%$, whereas *BiMine* has comparable performances with *Bimax*, it outperforms CC and ISA. *BiMine* has good performances for $p = 1\%$ and $p = 0.5\%$ compared to OPSM. However, OPSM outperforms *BiMine* for $p = 5\%$, $p = 0.1\%$ and $p = 0.001\%$.

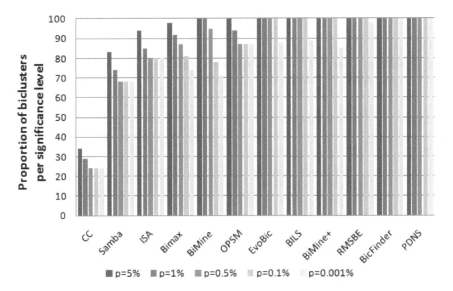

FIGURE 14.3 Proportions of biclusters significantly enriched by GO annotations on the *Saccharomyces cerevisiae* dataset.

14.4.5 BIOLOGICAL VALIDATION

To estimate biologically the extracted biclusters we used GOTermFinder [38], which is designed to search for the significant shared gene ontology (GO) [46] terms of the groups of genes and provides the user with the means to identify the characteristics that the genes may have in common.

We present the significant shared GO terms (or parent of GO terms) used to describe two selected biclusters for each algorithm on the *Yeast cell cycle* dataset. These biclusters are randomly chosen among the over-represented biclusters (having an adjusted *p*-value lower than 0.001%).

Table 14.4 presents the significant shared GO terms for two biclusters extracted by *BiMine* with (11 genes; 11 conditions) and (12 genes; 13 conditions) in each bicluster with ASR equal to 0.8690 and 0.8873, respectively, for biological process, molecular function, and cellular component. As in [47], we report the most significant GO terms shared by these biclusters. By analyzing the first bicluster, genes (YDL003W, YDL164C, YDR097C, YDR440W, YKL113C, YLL002W, YLR183C, YNL102W) are particularly involved in the process of cellular response to DNA damage stimulus, response to DNA damage stimulus, cellular response to stress, cellular response to stimulus, response to stress, and response to stimulus.

The values within parentheses after each GO term, such as (66.7%, 1.87e-08) in the first bicluster, indicate the cluster frequency and the statistical significance. The cluster frequency (66.7%) shows that out of 12 genes in the first bicluster, 8 belong to this process, and the statistical significance is provided by a *p*-value of 1.87e-08 (highly significant).

TABLE 14.4

Most significant shared GO terms (process, function, component) for two biclusters extracted by *BiMine* on *Yeast cell cycle*.

Biclusters	Biological Process	Molecular Function	Cellular Component
12 genes; 13 conditions	*cellular response to* *DNA damage stimulus* $(66.7\%, 1.87 * 10^{-8})$ *response to DNA* *damage stimulus* $(66.7\%, 6.30 * 10^{-8})$ *cellular response* *to stress* (66.7%, $2.12 * 10^{-7})$ *cellular response* *to stimulus* (66.7%, $3.25 * 10^{-7})$ *DNA repair* $(50\%, 2.58 * 10^{-5})$ *response to stress* $(66.7\%, 2.98 * 10^{-5})$	*chromatin* *binding* (25%, 0.00037)	microtubule organizing *center part* (16.7%, 0.00742)
11 genes; 11 conditions	*cell cycle process* $(63.6\%, 2.93 * 10^{-5})$ *cell cycle* $(63.6\%, 6.85 * 10^{-5})$	*GTPase activator* *activity* (18.2%, 0.00994)	*microtubule* *cytoskeleton* $(45.5\%, 6.33 * 10^{-6})$ *microtubule* *organizing* *center* (36.4%, $4.97 * 10^{-5})$ *spindle pole body* $(36.4\%, 4.97 * 10^{-5})$ *spindle pole* $(36.4\%, 6.77 * 10^{-5})$

Table 14.5 presents the significant shared GO terms used to describe two selected sets of genes (extracted by *BiMine+* on yeast cell-cycle dataset) with (136 genes; 6 conditions) and (131 genes; 7 conditions) in each bicluster. These two biclusters have, respectively, an ASR value equal to 0.24 and 0.73. By analyzing the first bicluster (Table 14.5), genes (YAL059W, YBL072C, YBR048W, YBR181C, YBR189W, YCR031C, YDL083C, YDL208W, YDR025W, YDR064W, YDR418W,

TABLE 14.5

Most significant shared GO terms (process, function, component) for two biclusters extracted by *BiMine+* on *Yeast cell cycle*.

Biclusters	Biological Process	Molecular Function	Cellular Component
136 genes; 6 conditions	*translation* (43.4%, $5.05 * 10^{-23}$) *maturation of SSU-rRNA* (14.7%, $1.73 * 10^{-13}$) *ribosome biogenesis* (25.7%, $1.11 * 10^{-12}$) *maturation of SSU-rRNA from tricistronic rRNA transcript (SSU-rRNA, 5.8S rRNA, LSU-rRNA)* (14.0%, $1.78 * 10^{-12}$) *ribonucleoprotein complex biogenesis* (25.7%, $6.44 * 10^{-11}$)	*structural constituent of ribosome* (36.8%, $1.84 * 10^{-39}$) *structural molecule activity* (36.8%, $6.09 * 10^{-30}$)	*cytosolic ribosome* (39.0%, $5.98 * 10^{-50}$) *cytosolic part* (37.5%, $8.78 * 10^{-43}$) *ribosome* (42.6%, $5.99 * 10^{-39}$) *ribosomal subunit* (36.8%, $3.79 * 10^{-38}$) *cytosolic small ribosomal subunit* (19.9%, $9.20 * 10^{-29}$)
131 genes; 7 conditions	*DNA-dependent DNA replication* (12.2%, $2.42 * 10^{-9}$) *DNA strand elongation* (8.4% , $4.03 * 10^{-9}$) *DNA strand elongation during DNA replication* (8.4%, $4.03 * 10^{-9}$) *lagging strand elongation* (6.9%, $9.90 * 10^{-9}$) *DNA metabolic process* (22.1%, $2.74 * 10^{-8}$) *DNA replication* (13.0%, $7.38 * 10^{-8}$)	*double-stranded DNA binding* (5.3%, 0.00035) *structure-specific DNA binding* (6.1%, 0.00276) *structural constituent of ribosome* (10.7%, 0.00612)	*replication fork* (11.5%, $1.65 * 10^{-12}$) *nuclear replication fork* (9.9%, $4.04 * 10^{-11}$) *non-membrane-bounded organelle* (40.5%, $8.17 * 10^{-10}$) *intracellular non-membrane-bounded organelle* (40.5%, $8.17 * 10^{-10}$)

YDR447C, YDR450W, YER074W, YER131W, YGR214W, YJR123W, YLR048W, YLR068W, YLR167W, YLR192C, YLR441C, YML026C, YMR143W, YMR230W, YMR269W, YNL096C, YNL112W, YNL302C, YOL040C, YOL127W, YOR056C, YOR293W, YPL090C, YPR102C) are particularly involved in the ribosome biogenesis and ribonucleoprotein complex biogenesis.

TABLE 14.6

Most significant shared GO terms (process, function, component) for two biclusters extracted by *BicFinder* on *Yeast cell cycle*.

Biclusters	Biological Process	Molecular Function	Cellular Component
92 genes; 16 conditions	*translation* (59.7%, $4.33 * 10^{-30}$) *cellular protein metabolic process* (65.2%, $1.48 * 10^{-18}$) *protein metabolic process* (65.2%, $1.43 * 10^{-17}$)	*structural constituent* (52.1%, $3.67 * 10^{-48}$) *of ribosome structural molecule activity* (53.2%, $3.76 * 10^{-40}$)	*cytosolic ribosome* (53.2%, $6.22 * 10^{-56}$) *cytosolic part* (54.3%, $8.15 * 10^{-52}$)
50 genes; 17 conditions	*cellular response to DNA damage stimulus* (37.3%, $3.24 * 10^{-14}$) *response to DNA damage stimulus* (37.3%, $5.27 * 10^{-13}$)	*double-stranded DNA binding* (9.8%, 0.00026) *cyclin-dependent protein kinase regulator activity* (7.8%, 0.00124)	*replication fork* (21.6%, $3.36 * 10^{-12}$) *chromosome* (35.3%, $1.94 * 10^{-10}$) *chromosomal part* (31.4%, $5.28 * 10^{-9}$)

Table 14.6 presents the significant shared GO terms used to describe the two selected sets of genes (extracted by *BicFinder*) with (92 genes; 16 conditions) and (50 genes; 17 conditions) in each bicluster with ASR equal to 0.4543 and 0.7844, respectively, for biological process, molecular function, and cellular component. By analyzing the first bicluster (Table 14.6), genes (YEL034W, YER074W, YER117W, YER131W, YGR214W, YHL001W, YIL069C, YJL111W, YJL136C, YJL177W, YJL189W, YJL190C, YJR123W, YKL056C, YKL156W, YKL180W, YKR057W, YKR094C, YLR029C, YLR048W, YLR075W, YLR150W, YLR167W, YLR185W, YLR248W, YLR249W, YLR325C, YLR344W, YLR367W, YLR380W, YLR406C, YLR441C, YLR448W, YML026C, YML063W, YML073C, YMR143W, YMR225C, YNL030W, YNL067W, YNL096C, YNL162W, YNL301C, YNL302C, YOL039W, YOL040C, YOL127W, YOL139C, YOR167C, YOR234C, YOR293W, YOR312C, YOR369C, YPL037C, YPL081W, YPL090C, YPL143W, YPR043W, YPR102C, YPR163C) are particularly involved in the cellular protein metabolic process and protein metabolic process.

Tables 14.7, 14.8, and 14.9 present the significant shared GO terms used to describe the two selected biclusters obtained by BILS, PDNS, and EvoBic, respectively.

TABLE 14.7

Most significant shared GO terms (biological process, molecular function, cellular component) of BILS for two biclusters on *Yeast cell-cycle*.

Biclusters	Biological Process	Molecular Function	Cellular Component
42 genes;	translation	structural constituent of ribosome	cytosolic ribosome
4 conditions	(67.95%, 2.86e-35)	(45.21%, 2.50e-70)	(53.01%, 1.05e-76)
	cellular protein	Structural molecule activity	ribosomal subunit
	metabolic process	(21.96%, 6.06e-54)	(35.71%, 1.08e-68)
	(47.82%, 2.59e-16)	translation factor activity, nucleic acid binding (34.61%, 0.00445)	cytosolic part
	cellular macromolecule		(65.34%, 1.01e-66)
	biosynthetic process		
	(41.27%, 1.74e-15)		
36 genes;	response to stimulus	structural constituent of ribosome (23.19%, 9.19e-24)	cytosolic ribosome
5 conditions	(57.28%, 0.00092)		(39.48%, 1.09e-23)
	response to stress	structural molecule activity (51.39%, 3.78e-12)	ribosomal subunit
	(11.63%, 0.00454)		(17.82%, 3.28e-23)
		oxidoreductase activity (42.58%, 2.36e-05)	cytosolic part
			(37.61%, 7.35e-22)

14.5 CONCLUSION

In this chapter, we have presented an overview of systematic [7, 8, 21, 22, 23] and stochastic biclustering algorithms [25, 26, 27, 31, 33]. First, we have given a brief description of microarrays and the biclustering problem. Afterwards, we have presented in detail several works on biclustering of microarray data. Three types of systematic biclustering algorithms adopting *enumeration*, *greedy*, and *divide-and-conquer* algorithms, and three types of stochastic algorithms adopting *iterated local search*, *evolutionary*, and *hybrid* approaches are introduced.

TABLE 14.8

Most significant shared GO terms (process, function, component) PDNS for biclusters on *Yeast Cell-Cycle* dataset.

Biclusters	Biological Process	Molecular function	Cellular component
78 genes;	translation	structural constituent of ribosome	cytosolic ribosome
9 conditions	(58.1%, 8.71e-37)	(51.3%, 4.48e-59)	(53.00%, 5.97e-70)
83 genes;	nucleic acid metabolic	phosphatase regulator	nucleus
11 conditions	process (34.0%, 2.45e-11)	activity (1.7%, 0.00041)	(44.8%, 3.46e-15)

The performance of the presented algorithms are evaluated on three DNA microarray datasets. We have tested the biological significance using a gene annotation tool to show that these works are able to produce biologically relevant biclusters.

Given the diversity of microarray data and evaluation criteria, it seems that no single biclustering approach dominates all the other approaches. The existing biclustering tools have specific advantages and limitations and could be jointly applied. Moreover, given that large and complex microarray data become available, it is useful to devise more powerful bicluster methods. One way to achieve this is to make the search method more intelligent by integrating various specific knowledge into the search mechanisms and operators.

REFERENCES

1. P. Baldi and G. W. Hatfield, "DNA microarrays and gene expression," *From Experiments to Data Analysis and Modelling*. Cambridge: Cambridge University Press, 2002.
2. L. Lazzeroni and A. Owen, "Plaid models for gene expression data," tech. rep., Stanford University, 2000.
3. J. Quackenbush, "Microarray analysis and tumors classification," *The New England Journal of Medicine*, vol. 354, no. 23, pp. 2463–2472, 2006.
4. T. Yun and G. Yi, "Biclustering for the comprehensive search of correlated gene expression patterns using clustered seed expansion," *BMC Genomics*, vol. 14, no. 144, 2013.
5. Y. Cheng and G. M. Church, "Biclustering of expression data," in *Proceedings of the Eighth International Conference on Intelligent Systems for Molecular Biology*, pp. 93–103, 2000.
6. S. C. Madeira and A. L. Oliveira, "Biclustering algorithms for biological data analysis: A survey," *IEEE/ACM Transactions on Computational Biology and Bioinformatics*, vol. 1,

TABLE 14.9

Most significant shared GO terms (process, function, component) EvoBic for biclusters on *Yeast Cell-Cycle*.

Biclusters	Biological Process	Molecular Function	Cellular Component
2527 genes;	*cytoplasmic translation*	*structural molecule*	*replication fork*
6 conditions	(2.57%; 2.10e-4)	*activity*	(1.60%; 4.67e-7)
	response to stimulus	(6.60%; 9.7e-3)	*nuclear replication fork*
	(11.40%; 4.90e-6)		(1.40%; 9.10e-7)
	cellular response to stimulus		*non-membrane-bounded organelle*
	(9.66%; 2.22e-5)		(20.60%; 1.20e-6)
20 genes;	*nucleoside transport*	*nucleobase-containing*	
3 conditions	(20%; 6.30e-4)	*compound transmembrane transporter activity*	
		(30%; 4.44e-6)	
	purine-containing compound transmembrane transport	*nucleobase transmembrane transporter activity*	
	(20%; 1.51e-3)	(20%; 1.40e-4)	

no. 1, pp. 24–45, 2004.

7. W. Ayadi, M. Elloumi, and J. Hao, "A biclustering algorithm based on a bicluster enumeration tree: application to DNA microarray data," *BioData Mining*, vol. 2, December 2009.

8. W. Ayadi, M. Elloumi, and J. Hao, "Bimine+: An efficient algorithm for discovering relevant biclusters of DNA microarray data," *Knowledge-Based Systems*, Elsevier, vol. 35, pp. 224–234, November 2012.

9. P. Orzechowski, "Proximity measures and results validation in biclustering—a survey,"

in *Proceedings of the 12th International Conference on Artificial Intelligence and Soft Computing*, Zakolpane, Poland, Lecture Notes in Computer Science, Volume 7895, pp. 206–217, Springer, 2013.

10. F. Divina, B. Pontes, R. Giráldez, et al., "An effective measure for assessing the quality of biclusters," *Computers in Biology and Medicine*, vol. 42, no. 2, pp. 245–256, 2012.

11. F. Angiulli, E. Cesario, and C. Pizzuti, "Random walk biclustering for microarray data," *Journal of Information Sciences*, pp. 1479–1497, 2008.

12. K. O. Cheng, N. F. Law, W. C. Siu, and A. W. Liew, "Identification of coherent patterns in gene expression data using an efficient biclustering algorithm and parallel coordinate visualization," *BMC Bioinformatics*, vol. 210, no. 9, pp. 1282–1283, 2008.

13. F. Divina and J. S. Aguilar-Ruiz, "A multi-objective approach to discover biclusters in microarray data," in *Proceedings of the 9th Annual Conference on Genetic and evolutionary computation*, (New York, NY, USA), pp. 385–392, 2007.

14. J. A. Hartigan, "Direct clustering of a data matrix," *Journal of the American Statistical Association*, vol. 337, no. 67, pp. 123–129, 1972.

15. L. Teng and L. Chan, "Discovering biclusters by iteratively sorting with weighted correlation coefficient in gene expression data," *Journal of Signal Processing Systems*, vol. 50, no. 3, pp. 267–280, 2008.

16. E. L. Lehmann and H. J. M. D. Abrera, "Nonparametrics: Statistical methods based on ranks," rev. ed., Englewood Cliffs, NJ: Prentice-Hall, pp. 292–323, 1998.

17. R. Balasubramaniyan, H. llermeier, E. Weskamp, and J. Kamper, "Clustering of gene expression data using a local shape-based similarity measure," *Bioinformatics*, vol. 21, pp. 1069–1077, 2005.

18. Y. Luan and H. Li, "Clustering of time-course gene expression data using a mixed-effects model with B-splines," *Bioinformatics*, vol. 19, no. 4, pp. 474–482, 2003.

19. S. Peddada, E. Lobenhofer, L. Li, C. Afshari, C. Weinberg, and D. Umbach, "Gene selection and clustering for time-course and dose-response microarray experiments using order-restricted inference," *Bioinformatics*, vol. 19, no. 7, pp. 834–841, 2003.

20. A. Schliep, A. Schonhuth, and C. Steinhoff, "Using hidden Markov models to analyze gene expression time course data," *Bioinformatics*, vol. 19, no. Suppl 1, pp. i255–i263, 2003.

21. W. Ayadi, M. Elloumi, and J. Hao, "BicFinder: a biclustering algorithm for microarray data analysis," *Knowledge and Information Systems*, Springer, vol. 30, pp. 341–358, February 2012.

22. X. Liu and L. Wang, "Computing the maximum similarity bi-clusters of gene expression data," *Bioinformatics*, vol. 23(1), pp. 50–56, 2007.

23. A. Prelic, S. Bleuler, P. Zimmermann, et al., "A systematic comparison and evaluation of biclustering methods for gene expression data," *Bioinformatics*, vol. 22, no. 9, pp. 1122–1129, 2006.

24. H. Lourenco, O. Martin, and T. Stutzle, *Handbook of Meta-heuristics*, ch. Iterated Local Search: Framework and Applications, pp. 321–353. Heidelberg: Springer, second edition, 2010.

25. W. Ayadi, M. Elloumi, and J. Hao, "Iterated local search for biclustering of microarray data," in *Proceedings of the 5th annual conference on Pattern Recognition in Bioinformatics*, Lecture Notes in Computer Science, Volume 6282, pp. 219–229, 2010.

26. W. Ayadi, M. Elloumi, and J. Hao, "Pattern-driven neighborhood search for biclustering of microarray data," *BMC Bioinformatics*, vol. 13, p. S11, May 2012.

27. W. Ayadi, O. Maatouk, and H. Bouziri, "Evolutionary biclustering algorithm of gene

expression data," in *Proceedings of the 23rd International Workshop on Database and Expert Systems Applications (DEXA), IEEE*, pp. 206–210, 2012.

28. K. Seridi, L. Jourdan, and G. Talbi, "Multi-objective evolutionary algorithm for biclustering in microarrays data," in *Proceedings of the IEEE Congress of Evolutionary Computation*, New Orleans, LA, USA, pp. 2593–2599, 2011.

29. J. A. Nepomuceno, A. Troncoso, and J. S. Aguilar-Ruiz, "Evolutionary metaheuristic for biclustering based on linear correlations among genes," in *Proceedings of the 2010 ACM Symposium on Applied Computing (SAC 2010)*, Sierre, Switzerland, pp. 22–26, 2010.

30. J.-K. Hao, *Memetic Algorithms in Discrete Optimization*, ch. Handbook of Memetic Algorithms. Studies in Computational Intelligence, pp. 73–94. Heidelberg: Springer, 2012.

31. S. Mitra and H. Banka, "Multi-objective evolutionary biclustering of gene expression data," *Journal of Pattern Recognition*, pp. 2464–2477, 2006.

32. C. A. C. Coello, G. B. Lamont, and D. A. V. Veldhuizen, *Evolutionary Algorithms for Solving Multi-Objective Problems (Genetic and Evolutionary Computation)*, Secaucus, NJ, USA:Springer, 2nd edition, 2002.

33. C. Gallo, J. Carballido, and I. Ponzoni, "Microarray biclustering: A novel memetic approach based on the PISA platform," in *EvoBIO '09: Proceedings of the 7th European Conference on Evolutionary Computation, Machine Learning and Data Mining in Bioinformatics, LNCS*, (Berlin, Heidelberg), pp. 44–55, Springer-Verlag, 2009.

34. S. Tavazoie, J. D. Hughes, M. J. Campbell, R. J. Cho, and G. M. Church, "Systematic determination of genetic network architecture," *Nature Genetics*, vol. 22, pp. 281–285, 1999.

35. A. Gasch, P. Spellman, C. Kao, et al., "Genomic expression programs in the response of yeast cells to environmental changes," in *Molecular Biology of the Cell, 11(12)*, pp. 4241–4257, 2000.

36. A. Alizadeh, M. Eisen, R. Davis, et al., "Distinct types of diffuse large (b)-cell lymphoma identified by gene expression profiling," *Nature*, vol. 403, pp. 503–511, 2000.

37. G. Berriz, J. Beaver, C. Cenik, M. Tasan, and F. Roth, "Next generation software for functional trend analysis," *Bioinformatics*, vol. 25, no. 22, pp. 3043–3044, 2009.

38. E. Boyle, S. Weng, J. Gollub, et al., "GOTermFinder–open source software for accessing gene ontology information and finding significantly enriched gene ontology terms associated with a list of genes," *Bioinformatics*, vol. 20, no. 18, pp. 3710–3715, 2004.

39. A. Ben-Dor, B. Chor, R. Karp, and Z. Yakhini, "Discovering local structure in gene expression data: the order-preserving submatrix problem," in *Proceedings of the Sixth Annual International Conference on Computational Biology (RECOMB 2002)*, New York, NY, USA, pp. 49–57, 2002.

40. F. Divina and J. S. Aguilar-Ruiz, "Biclustering of expression data with evolutionary computation," *IEEE Transactions on Knowledge & Data Engineering*, vol. 18, no. 5, pp. 590–602, 2006.

41. Y. Christinat, B. Wachmann, and L. Zhang, "Gene expression data analysis using a novel approach to biclustering combining discrete and continuous data," *IEEE/ACM Transactions on Computational Biology and Bioinformatics*, vol. 5, no. 4, pp. 583–593, 2008.

42. S. Bleuler, A. Prelic, and E. Zitzler, "An EA framework for biclustering of gene expression data," in *Proceedings of Congress on Evolutionary Computation*, pp. 166–173, 2004.

43. J. Liu, Z. Li, F. Liu, and Y. Chen, "Multi-objective particle swarm optimization biclus-

tering of microarray data," in *Proceedings of the IEEE International Conference on Bioinformatics and Biomedicine (BIBM 2008)*, Washington, DC, USA, pp. 363–366, IEEE Computer Society, 2008.

44. J. Liu, Z. Li, X. Hu, and Y. Chen, "Biclustering of microarray data with MOSPO based on crowding distance," *BMC Bioinformatics*, vol. 10, no. S-4, 2009.

45. A. Tanay, R. Sharan, and R. Shamir, "Discovering statistically significant biclusters in gene expression data," *Bioinformatics*, vol. 18, pp. S136–S144, 2002.

46. G. O. Consortium, "Gene ontology: tool for the unification of biology," *Nature Genetics*, vol. 25, pp. 25–29, 2000.

47. U. Maulik, A. Mukhopadhyay, and S. Bandyopadhyay, "Combining pareto-optimal clusters using supervised learning for identifying co-expressed genes," *BMC Bioinformatics*, vol. 10, no. 27, 2009.

15 Reconstruction of Regulatory Networks from Microarray Data

Yiqian Zhou, Rehman Qureshi, Francis Bell, and Ahmet Sacan

CONTENTS

15.1   PROBLEM DESCRIPTION

15.1.1   THE IMPORTANCE OF BIOCHEMICAL INTERACTION NETWORKS

Life can be regarded as a complex system in which genes, gene products, and other metabolites interact with each other. It is an essential step in system biology to uncover these biochemical interactions and organize them into a regulatory network. By inferring a regulatory network, which may contain a large number of components, scientists obtain a wider view of the biological system and a better understanding of its dynamic nature. With the availability of regulatory networks, scientists are able to answer questions such as: "how does a specific biological system respond to external stimulus or treatment," "what is the stable state of a cellular process under certain conditions," and "how will a biological process behave if some portion of the system were abnormal?" With the insights gained from regulatory network reconstruction, scientists have the ability to control and optimize biological systems, which leads to many practical applications in biotechnology and medicine.

15.1.2   ADVENT OF MICROARRAY TECHNOLOGIES AND DATA-DRIVEN MODELING

The development of microarray technologies has enabled a high throughput evaluation of gene expression, providing large scale quantitative data of cellular activity at the molecular level (see Chapter 1).

The availability of large scale gene expression data has shifted the starting point of the knowledge generation cycle of individual biological regulatory networks and spawned the development of data-driven modeling of biological systems [1]. Traditionally, a hypothesis was constructed from background knowledge and tested by experiments, and then the hypothesis was either verified or modified for further experimental testing. This cycle was repeated until the solution of the problem was satisfactory (Figure 15.1). However, when there are many possible testable hypotheses and their detailed experimental validation is not feasible, it is more practical and resource effective to generate and prioritize candidate hypotheses from the prior data using computer based inductive reasoning. Inference of biological regulatory networks is a complex task and the search space of possible interactions is far too large. On the other hand, there is a great amount of gene expression data accumulated from microarray studies; thus the question of how to utilize these data to generate sound hypotheses is of great interest. Data-driven inference of a network, which generates a reasonable hypothesis, is a key step in understanding the biological systems.

15.1.3   GENE REGULATORY NETWORKS

In a typical biochemical regulatory network (Figure 15.2), there may be many different kinds of biochemical interactions and these regulations may take place at different levels such as at the DNA level, the transcript level, or the protein level. Thus reconstruction of the real regulatory network describing these physical interactions requires different kinds of knowledge and experimental data, and the integration of

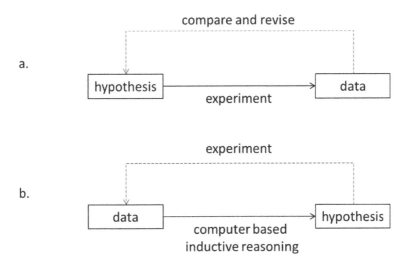

FIGURE 15.1 The availability of large scale gene expression data has shifted the starting point of the cycle of knowledge generation. (a) Traditionally, a hypothesis is generated based on background knowledge and then tested by experiments. (b) With the large amount of available gene expression data and little suitable background knowledge, the hypothesis is generated based on the analysis of data.

the information is a complex process. Feist et al. [2] described a detailed framework in which various experimental data can be integrated to reconstruct biochemical networks in microorganisms.

A gene regulatory network is an "abstract" representation of the complex gene regulatory system (15.2.b). Based on expression data of thousands of genes from microarrays, it is possible to infer how the expression of a gene relates to the expression of other genes; even in the absence of direct data on the concentrations of protein products or metabolites. A model describing "gene-gene" interactions can be called a gene regulatory network (GRN) [3]. GRN models usually do not necessarily describe the real physical interactions, but include indirect relationships between genes via proteins, non-coding RNAs, and other metabolites. A GRN can be represented as a graph, in which nodes represent genes and edges represent direct or indirect interactions between genes (Figure 15.2).

There are many models and approaches that have been proposed to infer the GRN based on gene expression data from microarrays. Based on the type of network being inferred, these models and approaches can be categorized as Boolean networks, Bayesian networks, co-expression networks, and differential equation models (Figure 15.3). In the remainder of this chapter, we describe each of these categories in more detail.

a.

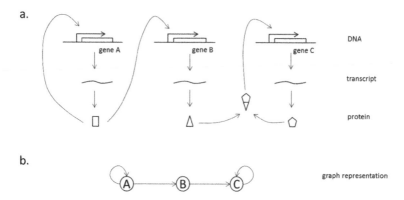

b.

FIGURE 15.2   Gene regulatory network (GRN) can be regarded as an "abstract" representation of a real biochemical regulatory network. (a) The real regulation of gene expression may happen at different levels, and there are different types of interactions. (b) A graph model of gene regulatory network (GRN) is an abstract representation describing how genes influence each other.

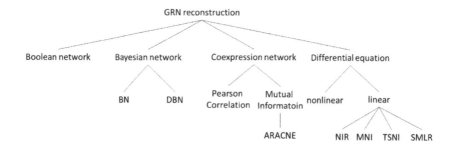

FIGURE 15.3   Gene regulatory network reconstruction methods described in this chapter.

15.2   MODELS OF GENE REGULATORY MODELS

15.2.1   BOOLEAN NETWORKS

Boolean network models represent the earliest attempts at GRN modeling. They were first proposed by Kauffman in 1969 [4]. Boolean networks attempt to track gene expression at discrete time steps allowing a gene to have only two possible states, expressed (a value of 1) or not expressed (a value of 0).

Boolean values define the current state of genes in the network. These values are applied to Boolean functions to determine the gene states at the next discrete time step. Let n be the number of genes in the network, and k be the maximum number of inputs to a Boolean function. In order to reduce the computational complexity and generate stable results, the number of inputs to a function is usually kept small ($k \leq 20$). Networks are defined by the number of nodes and the maximum number

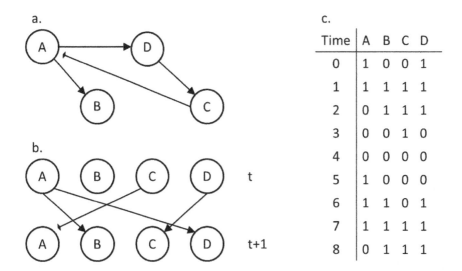

FIGURE 15.4 A $n=4\,k=1$ Boolean network (a) as a directed graph, (b) as a wiring diagram, and (c) as a value table of gene states at discrete time steps. In the network, Gene A is expressed unless Gene C was expressed in the previous time step. When Genes A and D are expressed and Genes B and C are repressed, the network engages in a six-state attractor. The values for times 7 through 13 will be the same as times 1 through 7. In (a, b), arrows represent stimulation and diamonds are suppression.

of inputs to functions used to construct them, e.g., a $n = 1,000$, $k = 3$ network. The formal definition of a Boolean network, $G(V,F)$, with a set of nodes, V, and functions, F, is given as:

$$V = \{X_1, X_2, , X_n\} \quad F = \{f_1, f_2, , f_n\} \qquad (15.1)$$

where each X_i represents a gene and f_i is a Boolean function that can have a maximum of k inputs. The number of inputs to each function can vary. This notation is consistent throughout the literature.

There are several approaches for visualizing a Boolean network. Directed graphs and wiring diagrams are the most common. A directed graph is constructed from a Boolean network with genes as nodes and edges representing Boolean functions. A wiring diagram is constructed with two levels of nodes representing discrete time steps and connections between layers indicating Boolean functions. It can also be helpful to show trajectory tables for networks. Trajectory tables provide the output of the network given various inputs. In trajectory tables, usually all possible inputs are enumerated. There are often many possible inputs for a network. Value tables listing the sequence of output states can be used to simplify trajectory tables when the initial condition is known. Figure 15.4 illustrates a simple $n = 4$, $k = 1$ Boolean network.

The simplest Boolean network models are called synchronous, meaning that all states are updated simultaneously. Future states of the network are determined from previous states. As a result, once a particular state is reached, the following states are always the same as dictated by the Boolean functions. For every Boolean network, a cycle will eventually emerge. For a Boolean network with n nodes, there are a maximum of 2^n different states. After 2^n time steps, at least one state will cycle back to itself. These cycles are called attractors. Attractors represent steady states of the model. Point attractors contain one state for steady state conditions. Dynamic attractors contain multiple states for a limit-cycle. For example, the phase changing of the cell-division cycle is a dynamic cycle.

Attractors have been widely studied. They represent stable phenotypic structures [5]. They provide information about the system being modeled. A useful tool for identifying and studying attractors is DDLab (Discrete Dynamics Lab) [6, 7], a publicly available graphic software for studying discrete dynamical systems. DDLab is capable of finding all attractors of a Boolean network. Before attractors can be identified, the network has to be created.

Before a Boolean network can be created, the expression data must be discretized. Once the data has been discretized, a network that explains the data can be generated. There is no standard method to translate discrete data into a network. The simplest methods involve some form of clustering. Clustering emphasizes positive correlations that identify similar regulations, positive and negative correlations that identify antagonistic regulations, and correlations of mutual information that detect complex relations [8]. The clustering techniques used are chosen to best fulfill the intended use of the Boolean network; each technique provides unique insights. When clustering is used, networks are often created by hand. These networks are very subjective. To objectively create networks, reverse engineering is used. Reverse engineering creates networks that are defined by a set of rules and are reproducible.

Many algorithms for reverse engineering Boolean networks have been proposed and research is still ongoing. The Reverse Engineering Algorithm (REVEAL) is an early algorithm that accomplishes this task. REVEAL utilizes the Shannon entropy of binary systems to compute a score for mutual information between sets of genes [9]. By trying to find the smallest set of input genes that most accurately explain the state of an output gene, REVEAL is able to construct binary networks from expression data. The algorithm was proven and refined by Akutsu et al. [10, 11]. The limitation of the refined algorithm is its inefficiency when the maximum number of inputs to a function is greater than or equal to three. Lahdesmaki et al. improved upon the algorithm of Akutsu et al., achieving efficiency for any number of inputs, assuming the number of inputs was pre-defined [12]. This method is known as the Best-Fit method. The Best-Fit method allowed many Boolean networks to be constructed from time series expression data. Examples of such networks include DNA damage repair in *E. Coli* [13], cell cycle in *S. pombe* [14], and IL-2 stimulated T cell responses in *M. musculus* [15].

Despite their practicality, simple Boolean networks are limited as models for genetic networks. They do not represent the genetic mechanisms exactly as they appear in living cells. Gene expression is not switched on or off as is a Boolean toggle. Gene expression is often varied on a continuous scale. Fortunately, this scale is often sigmoidal in shape. As Kauffman explains, sigmoidal patterns can be modeled as Boolean switches without large errors [5]. Because of the almost step-wise nature of sigmoidal functions, it is acceptable to model sigmoidal functions as Boolean switches. Only a small error is found. Other modeling techniques such as Bayesian networks and differential equation networks are used to more accurately simulate genetic mechanisms with less error. These techniques are more computationally complex and require more time to implement. The specific nature of an experiment and the sensitivity at which interactions are modeled dictate which technique to use.

Szallasi and Liang attempted to address the issue of biological relevance of Boolean networks. They proposed realistic Boolean networks. Realistic Boolean networks account for regulatory factors such as concentrations of antagonistic proteins by incorporating additional variables and Boolean functions into the model [16]. Szallasi and Liang used realistic Boolean networks to better understand the phenomenon of carcinogenesis. Although they showed promising results, realistic Boolean networks have not found widespread use. The additional time and space constraints imposed by the method act as deterrents for its use.

Another problem with simple Boolean networks is the synchronous nature of the model. In the model, all genes are subjected to changes in the system simultaneously at discrete time steps. This is not a realistic model of genes in cells. Gene expression is varied at different time points for each gene. The simple Boolean network is appropriate for showing expression changes over periods of time, but it cannot depict immediate interactions between genes. Updating the network asynchronously has the possibility of solving this problem. Limited work with asynchronous Boolean networks has been performed [17, 18, 19, 20], but the results suggest that their complexity makes asynchronous Boolean networks impractical for genome wide experiments.

The most serious limitation of simple Boolean networks for modeling GRN based on gene expression data is the inability to compensate for noise without an extensive prior knowledge. This limitation is exacerbated by the large amount of noise inherent in high throughput gene expression technologies and the incomplete knowledge of genetic interactions. Random Boolean networks (RBN) were the first method to address this directly. RBN are a class of Boolean networks where inputs and Boolean functions are randomly assigned. The number of inputs is kept constant to reduce complexity and to allow computation. In the case where the number of inputs is the same as the number of nodes, $k = n$, the network is informally called *a grand ensemble* [5]. In this network, every gene serves as an input for every gene. This network is extremely chaotic and varies greatly for different initial conditions [5, 21]. It is rarely used to model actual systems. When the number of inputs is small ($k \leq 20$), the RBN is less chaotic but does not accurately depict the RGN being studied. Although the RGN is not accurately depicted, the overall trends of the RBN are similar to the overall trends of the genetic network. For this reason, RBN are appropriate for initial studies of RGN [22]. For more detailed studies of RGN to capture behavior

a.   b.

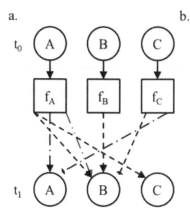

Function	Probability	Style
$f_A^{(1)}$	0.45	- - -
$f_A^{(2)}$	0.20	- · -
$f_A^{(3)}$	0.35	- ·· -
$f_B^{(1)}$	0.30	- - -
$f_B^{(2)}$	0.70	
$f_C^{(1)}$	0.42	- - -
$f_C^{(2)}$	0.58	- · -

FIGURE 15.5 Visualization of a simple PBN with $n = 3$, $k = 1$, and $N = 3$. (a) Wire diagram. Boxes represent sets of predictors for each gene. The dashed edges show the result if a predictor is selected. (b) Function table. The style of edge for each function appears in the table. $F_2^{(B)}$ is a function that does not change gene expression. It has no edges associated with it.

beyond the general trends, other methods are more appropriate.

Probabilistic Boolean networks (PBN) account for noise associated with gene expression data when an unknown RGN is encountered, without increasing complexity significantly. PBNs were first proposed in 2002 [23]. Since then, they have been well studied and improved. In PBN, each node is assigned a set of Boolean functions, called predictors, that determine the next state of the network. A set of Boolean functions can have a variable number of predictors with a maximum of N. Each predictor in a set of functions has an associated probability of occurring given the network's current state. The sum of the probabilities of predictors in a set of functions must equal unity at all times. A predictor is randomly selected for each node at each time step in accordance with predictor probabilities. The formal definition of a Boolean network is changed to account for many possible functions for each node. A PBN, $G(V, F)$, is defined similar to Eqn. 15.1, but each function, f_i, is a set of predictors, defined as:

$$f_i = \{f_i^{(1)}, f_i^{(2)}, ..., f_i^{(\lambda)}\} \tag{15.2}$$

where $f_i^{(j)}$ is a predictor and λ is the number of predictors for node i. λ cannot be greater than N. When only one predictor is used for each node ($N = 1$), the model functions as a simple Boolean network. Because PBNs are often complex with many nodes and several predictors for each node, wire diagrams become difficult to interpret and are rarely used. Instead, state transition diagrams and function tables are utilized. Figure 15.5 shows a wire diagram and function table of a simple PBN with $n = 3$, $k = 1$, and $N = 3$.

Most PBNs use predictors that do not alter the probabilities of the next predictors to be selected. When PBNs utilize dependent predictors, determining the probability of a predictor being selected is very complicated [23]. It often requires solving complex differential equations and is usually very time consuming. To reduce the computational complexity of PBNs, most models avoid dependent predictors. Reducing complexity is important because PBNs are more complex than simple Boolean networks, due to the selection of random functions. This is a compromise made to account for noise and uncertainty.

Although PBNs have a finite number of possible states, the model does not necessarily enter a steady state. This is because PBNs are not deterministic. The same state of the model can produce different states. If a PBN does enter into a single steady state, the state is called absorbing. If the PBN enters into a stable set of states, the set is called an irreducible set. Absorbing states and irreducible sets of states are similar to point and dynamic attractors, respectively. Steady states only occur in PBN models that allow for them and often require specific initial conditions. For some PBN models, steady-states are needed. A PBN model of cancer in which cells continue to reproduce regardless of external stimuli is an example of such a model, where an irreducible set of states would be appropriate. PBN models of GRN inferred from microarray data may contain steady states. These states should be analyzed for relevance.

The following example illustrates how a PBN model of a GRN can be inferred from time-series microarray data. First, experiments must be run or data retrieved from experiments testing a condition at different time points. Public microarray data repositories include such datasets. For this example, we used the data from a study performed by Tirosh et al. comparing genome-wide transcript profiles of four related yeast species under a variety of stresses at time points up to 90 minutes [24]. The data from the study is available on the Gene Expression Omnibus (GEO) as accession GDS2910 [25, 26]. For this example, the data is limited to samples of *S. paradoxus*, from the first experimental strain, subjected to environments at a temperature of $37^{o}C$. This subset is selected because multiple samples are tested at each time point and heat shock response is an extensively studied phenomenon in yeast. The entire genome can be modeled, but the model would be extremely complex and inappropriate for this example. For simplicity, this example focuses on the Slt2 MAPK signaling pathway. It has been shown by several studies that this pathway is activated in response to heat [27, 28, 29]. In the pathway, cell wall integrity proteins (Wsc1, Wsc2, Wsc3, Mid2) are stimulated by changes in the cell wall caused from the increased temperature. These proteins start a signal cascade of a Ras homolog protein (Rho1), a protein kinase C (Pkc1), kinases of mitogen-activated protein (Bck1, Mkk1, Mkk2, Slt2), and regulatory proteins (Swi4, Swi6, Rlm1) resulting in expression of an enzyme for cell wall remodeling (Fks2). The complete pathway as defined by KEGG [30] is shown in Figure 15.6.

Expression values from genes involved in the Slt2 MAPK pathway present on the microarray technology used are isolated and processed. The only gene of the pathway not measured in the microarray experiment was *Fks2*. If data for a gene from any sample contain missing values, the gene was excluded from the example.

FIGURE 15.6   KEGG pathway of Slt2 MAPK in yeast.

Several genes contained missing data, and only seven genes did not contain missing data: *Wsc2, Rho1, Pkc1, Slt2, Mlp1, Swi4, Swi6*. The average expression value
of each gene at each time step is calculated. A student's *t*-test is performed to determine the probability that an expression value of a gene at a given time point is
different than the expression value of the gene at the initial time point. The *p-values*
from the *t*-test are used to cluster the data into two groups by *k*-means clustering.
The group designations are used with the *best-fit* method of Lahdesmaki et al. [12]
with a maximum input value of 3 to create a PBN model of the pathway using Bool-
Net, a package developed for the R programing language for the analysis of Boolean
networks [31]. The network wiring diagram is shown in Figure 15.7. The inaccuracies of the model can be attributed to: the noise inherent in microarrays, the limited
number of samples taken at each time point, and the large amount of time relative
to cellular activity between time points. Despite its imperfections, the model captures key aspects of the regulatory network. Activation of *Wsc2* causes expression
changes in *Rho1, Pkc1, Slt2*, and *Mlp1* resulting in the increased expression of *Swi4*
and *Swi6*. The model also shows that the proteins in the middle of the pathway interact with each other, although the exact relationships are unclear. If the pathway was
not known, the relationships among *Rho1, Pkc1, Slt2*, and *Mlp1* could be discovered
using more targeted experimental approaches.

15.2.2   BAYESIAN NETWORKS

Bayesian networks (BN) can model gene expression by treating each gene as a random variable governed by a probability distribution function, determined by the
product of conditional probabilities [3]. Bayesian networks represent the dependence
between the levels of expression of different genes. BNs are ideal for describing
processes where the value of each component is dependent upon the values of a
small number of other components [32]. The networks produced by BNs are directed
acyclic graphs (DAG). Each gene in the network is dependent on the probabilities of
a set of other genes; this set of genes is assumed to regulate the first gene in some
way and is referred to as that gene's *parents*. In a Bayesian Network the probability
distribution function for each gene is a product of the conditional probabilities of all
of the "ancestor" genes in the network. One of the major issues facing Bayesian network models is the large number of possible networks that can be constructed from a
particular set of genes. A particular directed acyclic graph that best fits the data must
be determined.

   In order to understand the mechanics of Bayesian networks, consider a finite set

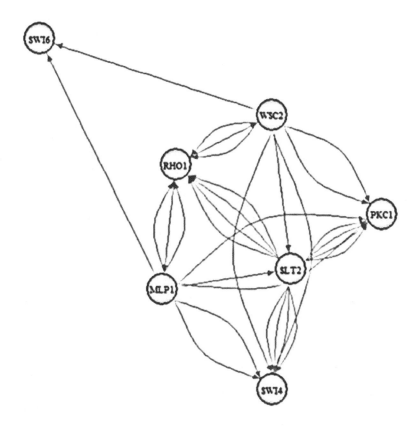

FIGURE 15.7 PBN of Slt2 pathway inferred from microarray data. Each edge represents one predictor. Edges connecting the same nodes represent predictors with different probabilities of occurrence.

of random variables, S, as:

$$S = \{X_1, X_2, X_3, ..., X_n\} \tag{15.3}$$

Two components make up a Bayesian network: the directed acyclic graph, G, where each vertex corresponds to a variable in the set S and the conditional probability distribution of each variable based on its parents in G, which is denoted as Θ. The Markov assumption, which is visualized by the directed acyclic graph, states that given a set of parent vertices, each vertex in G is conditionally independent of the vertices that are not its descendants. Using this conditional independence, the joint probability distribution for the entire network can be expressed as:

$$P(X_1, X_2, X_3, ..., X_n) = \prod_{i=1}^{n} P(X_i | Pt(X_i)) \tag{15.4}$$

where $Pt(X_i)$ represents the set of parents of X_i. The conditional probability distributions can arise from discrete or continuous variables. The parents of a variable are expressed as:

$$Pt(X_i) = \{A_i, ..., A_k\} \tag{15.5}$$

where k is the number of parents. If each value in the previous set possesses a discrete value from a set of finite size, then the probability of X_i given its parents can be represented as a table that specifies the value of X_i for each value in the set of parents. If there are m possible values for each discrete variable then the table will specify m^k possible distributions. When continuous real valued variables, such as those present in gene expression datasets, are used there are infinite possible distributions. In such cases, linear Gaussian conditional probability densities can be used, where each variable X_i is assumed to follow a normal distribution with a mean that is dependent on its parents' values. However, the variance of this distribution will be independent of the parents. The resulting joint distribution for the network will be a multivariate Gaussian distribution [32].

The large search space of possible directed acyclic graphs for a given set of nodes makes identifying a network that is ideal for a particular dataset difficult. The most common solution to this problem is the use of a scoring function to evaluate the potential graphs. The two most common scoring functions are the Bayesian information criteria (BIC) and the Bayesian Dirichlet equivalence (BDe). These scoring methods incorporate penalties to prevent overfitting the dataset [33]. However, finding the graph with the maximum score out of all possible graphs is known to be an NP-hard problem [34]. *A priori* biological information can also be utilized in order to restrict the number of possible graphs, for example, Ong et al. utilized the fact that certain *E. Coli* genes are co-transcribed, and thus co-regulated, to identify edges between these genes in their networks [35]. Due to the exponential size of the search space for possible networks, heuristic search methods, such as greedy-hill climbing, Markov Chain Monte Carlo, and simulated annealing are employed. Furthermore, model averaging or bootstrapping techniques can be utilized to select the ideal network from several highly scoring networks identified by the selected search heuristics, and can also be employed to determine confidence intervals for the interactions [33]. Information theory-based scoring functions such as mutual information can also be employed. It is important to note that the edges derived from Bayesian network methods do not constitute direct causal relationships, but rather a correlation between the expressions of the genes.

Bayesian networks can handle incomplete or noisy data, combine heterogeneous data types, and avoid over-fitting. However, Bayesian networks are unable to model feedback loops, because they cannot generate graphs with cycles. Dynamic Bayesian networks (DBNs) were created to overcome this limitation. Unlike the regular Bayesian networks, dynamic Bayesian networks require time-series gene expression data instead of static gene expression data. Dynamic Bayesian networks model the gene regulatory network as a graph with two layers of nodes. The first layer of nodes represents the expression of the genes at time $t - 1$, and the second layer represents the gene expression at time t. The expression of each gene is dependent on the ex-

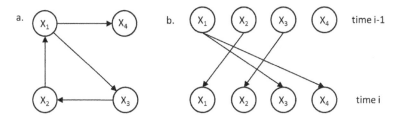

FIGURE 15.8 On the left is an example network. On the right is how the network would be represented by a dynamic Bayesian network. The dynamic Bayesian framework is able to overcome the limitation of Bayesian networks and can model the cyclic behavior of the gene regulatory network using two layers of nodes. Each row of nodes in the dynamic Bayesian network represents a consecutive time point.

pression of its parents at the previous time point. This arrangement of nodes allows for the representation of a network containing cycles with an acyclic graph. Figure 15.8 illustrates how a gene regulatory network can be represented by a dynamic Bayesian network.

A time-series microarray dataset with n samples, each corresponding to a unique time point, and p genes, can be expressed as an n-by-p matrix, where each row represents the gene expression at a discrete time point, and each column corresponds to a particular gene. We can express the joint probability as:

$$P(x_{1,1}, \cdots, x_{p,n}) = P(X_1)P(X_2|X_2) \cdots P(X_n|X_{n-1}) \qquad (15.6)$$

where $X_{i,t}$ on the left-hand side represents the expression of gene i at time t, and X_t on the right-hand side represents the column vector of the expression of all genes at time t. It is assumed that the structure of the network does not change between time points. (There are DBN studies about time-varying network structure that we will introduce later.) The conditional probability of one time point based on the previous time point can be expressed as the product of the conditional probabilities of each gene given its set of parent genes:

$$P(X_t|X_{t-1}) = P(x_{1,t}|P_{1,t-1}) \cdots P(x_{p-1,t-1}|P_{p,t-1}) \qquad (15.7)$$

where $P_{i,t-1}$ is the vector of parents of the ith gene at time $t-1$ [36, 37]. This equation is analogous to the joint probability distribution of the standard Bayesian network, but it differs because the expression of a gene is dependent on its parents' expressions at the previous time point.

Like standard Bayesian networks, a scoring function is needed to evaluate the possible network topologies. Similar scoring functions and search algorithms and techniques can be applied to dynamic Bayesian networks. Bayesian and dynamic Bayesian networks have been widely applied to the gene regulatory network reconstruction problem [38, 39, 40].

Recently, there are studies of DBN on time-varying GRN structures. In those studies, the topology of GRN were no longer assumed to be static, but varying during the time course. Song et al. [41] proposed a formalism in which $P(X_t|X_{t-1})$ is time dependent. They decompose the problem by finding the neighbor of each gene separately. To learn the neighborhood, they assume the network is sparse and varies smoothly, and then transform the problem to a $l1$-regularized square linear regression problem. Lébre et al. [42] proposed a method called auto regressive time varying models (ARTIVA) to learn time varying GRN from time series expression data. For each gene, regression models are learned for distinct phases separated by change points.

15.2.3 CO-EXPRESSION NETWORKS

Like clustering analysis [43], the construction of a co-expression network is based on the measure of similarity between gene expression profiles. The rationale behind co-expression networks is straightforward: if two genes have similar gene expression profiles, they are likely to interact with each other. Thus, if a metric can be established to evaluate the similarity, a GRN can be constructed by connecting genes that have similarity over a certain cutoff threshold.

Pearson correlation is one of the simplest metrics for similarity, and it is suitable for large scale networks because of its computational cost efficiency. In 2003, Stuart et al. [44] utilized the Pearson correlation in their network reconstruction study of 3182 DNA microarrays from 4 different organisms (humans, flies, worms, yeasts). They first constructed a set of "metagenes" across multiple organisms and then utilized the Pearson correlation to identify pairs of genes that had significantly correlated expression values. They obtained a correlation rank of all pairs of genes and calculated the probability of observing a particular configuration of ranks by chance. Finally, they obtained a co-expression network that contained 3,416 genes and 22,163 interactions.

Despite its low data requirement and simplicity, Pearson correlation cannot handle non-linear similarity. As the number of available microarray datasets has steadily been increasing, researchers have begun to use other metrics that utilize larger sample sizes. Mutual information, based on information theory, is a popular alternative that is capable of capturing the general similarity between two variables. The definition of mutual information is based on Shannon entropy [45]. For a discrete random variable X with n outcomes $\{x_1, x_2, \cdots, x_n\}$, the entropy $H(X)$, is defined as:

$$H(X) = -\sum_{i=1}^{n} p(X_i) \log p(X_i) \qquad (15.8)$$

where $p(x_i)$ is the corresponding probability of outcome x_i.

Entropy $H(X)$ measures the uncertainty of variable X. The conditional entropy of two variables X and Y taking values x_i and y_i is defined as:

$$H(X|Y) = -\sum_{i=1}^{n} p(x_i, y_i) \log \frac{p(x_i, y_i)}{p(y_i)} \qquad (15.9)$$

where $p(x_i, y_i)$ is the probability that $X = x_i$ and $Y = y_i$.

Conditional entropy $H(X|Y)$ can be regarded as the uncertainty in the random variable X given Y. The mutual information can be expressed in terms of entropy and mutual entropy as:

$$I(X,Y) = H(X) - H(X|Y) \qquad (15.10)$$

which measures the average amount of information Y conveys for X, or the reduction of uncertainty about X if Y is given. For gene expressions, which are continuous random variables, mutual information can be estimated by approaches based on discretization or approaches based on kernel density estimation [46]. Mutual information provides a metric for the general similarity between variables.

Butte et al. [47] first proposed a methodology, termed relevance networks (RelNet), that computes pairwise mutual information for all genes. In the study, they used 79 microarrays containing 2,467 genes in yeast, and calculated pairwise mutual information 3,041,811 times in total. Pairs with mutual information higher than the threshold were considered to represent interacting genes.

Later in 2005, Margolin et al. [48] developed the model called ARACNE (algorithm for the reconstruction of accurate cellular networks), which is also based on pairwise mutual information. ARACNE aimed to improve the inference performance by eliminating the majority of indirect interactions. The property called data processing inequality in information theory is applied. The idea is that if gene $g1$ and $g3$ interact through a third gene, $g2$, then the data processing inequality (DPI), shown in Eqn. 15.11, must be satisfied:

$$I(g1,g3) \leq \min\{I(g1,g2),I(g2,g3)\} \qquad (15.11)$$

That is, if three genes are connected, the edge with the smallest mutual information would be removed. The ARACNE model was successfully applied to the study of microarray expression profiles of human B cells [49]. While ARACNE infers GRN from steady-state data, Zoppoli et al. [50] proposed a method called TimeDelay-ARACNE that works with time-series data. TimeDelay-ARACNE tries to extract dependencies between genes at different time delays with a stationary Markov random field.

Because of its low computational cost and low data requirement, co-expression networks are usually used to infer global properties of large-scale of regulatory networks. However, the drawback of using co-expression networks is that since the similarity metric only accounts for a pairwise relationship, the models do not consider interactions including multiple genes.

15.2.4 DIFFERENTIAL EQUATION MODELS

Differential equations are widely used for modeling dynamic systems in engineering. As for gene regulatory networks, a system of ordinary differential equations (ODE) can be used to describe the change rate of gene expression as a function of expressions of other genes and external perturbation. These systems are described by the following differential equation:

$$\frac{dx_i}{dt} = f_i(\vec{x}, \vec{u}, \vec{\theta}) \quad i = (1, 2, ..., n) \tag{15.12}$$

where variable x_i represents the expression level of gene i, vector $\vec{x} = (x_1, x_2, \cdots, x_n)$ represents the expressions of other genes in the system, vector $\vec{u}$ represents the external perturbations, like gene knockouts or chemical treatments, and vector $\vec{\theta}$ is the set of parameters. By estimation of the function $f_i, (i = 1, 2, \cdots, n)$, the structure of a GRN can be established.

Unlike co-expression models, differential equation models are capable of describing the dynamic behavior of GRN quantitatively, and thus they can be used not only to study the topology of the GRN but also to simulate the GRN dynamics. Depending on the type of function f, the differential equation models for a GRN can be nonlinear or linear. Generally non-linear models are more complex and require more data to estimate the parameters. In addition, non-linear models usually require prior knowledge about the system to choose proper form of function. On the other hand, linear models require less data and usually no prior knowledge for parameters learning. Given the fact that microarray data are noisy and under-sampled, linear models are more suitable for GRN modeling. As described in the following section, there are several well developed methods based on linear differential equation models.

15.2.4.1 Nonlinear Differential Equation Models

In general, the actual biochemical regulatory systems are complex and nonlinear [51] and numerous nonlinear ODE models have been proposed to describe the regulatory networks. However, due to the complexity of the models, the estimation of the parameters requires a large amount of data. Thus, nonlinear ODE models are often only suitable for small-scale networks. For example, Sakamoto et al. [52] used genetic programming to infer the right-hand side of Eqn. 15.12, which can be of arbitrary form. They limited the number of genes to around three. For larger networks, further assumptions and preprocessing are needed.

One of the most well-studied nonlinear ODE models is the S system, which can be regarded as the canonical form of general non-linear differential equations [53]. The model for a network containing n genes is a system of nonlinear differential equations of the form:

$$\frac{dx_i}{dt} = \alpha_i \prod_{j=1}^{n} x_j^{g_{i,j}} + \beta_i \prod_{j=1}^{n} x_j^{h_{i,j}} \quad i = (1, 2, ..., n) \tag{15.13}$$

where x_i is expression level of gene i. Parameters α_i and β_i are called rate constants, and $g_{(i,j)}$ and $h_{(i,j)}$ are called kinetic orders. Since the number of parameters in the model is proportional to the square of number of genes, it is a challenge to reconstruct large-scale differential equation networks. Kimura [54] proposed a method for inferring the large-scale network with the S system model. The method is based on a problem decomposition strategy of dividing a problem into sub-problems.

Another strategy for implementing a nonlinear model is to restrict the nonlinear function f to specific types. However, picking a suitable function requires prior knowledge and experience. As a result, many data-driven methods based on linear models have been proposed. For more information on nonlinear genetic regulatory system modeling, please refer to Section 6 of the review by de Jong [20].

15.2.4.2 Linear Differential Equation Models

Linear models generally do not require extensive prior knowledge about the regulatory network, and they are suitable for larger scale networks because of their relative simplicity. As mentioned above, the behavior of gene expression can be modeled by nonlinear differential equations. Within the small neighborhood of a particular point of interest, this nonlinear system can be approximated to the first order by a system of linear equations. Consider a system containing n genes and p perturbations. Then, for each gene i, the rate of change of gene expression is described as a function of expression of other genes and external perturbation:

$$\frac{dx_i}{dt} = \sum_{j=1}^{n} a_{i,j} x_j + \sum_{l=1}^{p} b_{i,l} u_l \quad i = (1, 2, ..., n) \tag{15.14}$$

where x_i is the mRNA concentration of gene i, $a_{i,j}$ can be regarded as the influence of gene j on gene i, u_l is the lth external perturbation, and $b_{i,l}$ can be regarded as the influence of perturbation l on gene i in this experiment.

For compactness, Eqn. 15.14 can be rewritten in matrix form as follows:

$$\frac{d\vec{x}}{dt} = A\vec{X} + B\vec{u} \tag{15.15}$$

where $\vec{x} = (x_1, x_2, \cdots, x_n)^T$, $\vec{u} = (u_1, u_2, \cdots, u_p)^T$ and A is an n-by-n matrix containing coefficients $a_{(i,j)}$, B is a n-by-p matrix containing coefficients $b_{i,l}$.

If there are m experiments, then there are m equations like Eqn. 15.15; combining them together in matrix form results in the matrix form of the linear equation:

$$\frac{dX}{dt} = AX + BU \tag{15.16}$$

where $X = [\vec{x}_1 | \vec{x}_2 | \cdots | \vec{x}_m]$, in which $\vec{x}_k$ is a column vector containing gene expression level of n genes in the kth experiment. Thus X is an n-by-m matrix representing the expression level of all n genes in m experiments, and similarly $U = [\vec{x}_1 | \vec{x}_2 | \cdots | \vec{x}_m]$ is a p-by-m matrix representing the external perturbation in m experiments.

Notice that matrix A describes how genes interact with each other, and thus by estimating matrix A we can infer the GRN. However, in most cases, $m \ll n$, such that matrix A cannot be calculated directly. Furthermore, due to limitations of experimental measurement techniques, the expression data are noisy, making inference of A more challenging. In recent years, several methods have been proposed to solve this problem, including singular value decomposition (SVD), multiple regression, mode-of-action by network identification, time series network identification, and stepwise multiple linear regression. Each of these methods are described in more detail below.

15.2.4.3 GRN Inference Using Singular Value Decomposition

Yeung et al. [55] proposed an approach to reconstructing GRNs based on singular value decomposition (SVD) and robust regression. The goal is to deduce matrix A in Eqn. 15.16 by using the measured data of X, BU, and $\frac{dx}{dt}$. The approach is a two-step procedure. First, they use singular value decomposition to solve the system of ODEs, Eqn. 15.16, and obtain a family of matrices A. The family of solutions represents networks that are consistent with the measured data. Next, the best candidate can be chosen according to the prior knowledge of the biological network. If no prior knowledge is available, inspired by the observation that GRNs are sparse, they pick the network among candidates by maximizing the number of zero entries in matrix A using robust regression based on the L_1 norm.

More specifically, decomposing X^T by using SVD results in:

$$X^T = U\Sigma V^T \tag{15.17}$$

where U and V are orthogonal, meaning that $UU^T = I$ and $VV^T = I$. Substituting Eqn. 15.17 into Eqn. 15.16 and rearranging results in Eqn. 15.18:

$$AV\Sigma U^T = \frac{dX}{dt} - BU \tag{15.18}$$

One particular solution to Eqn. 15.19 is:

$$A_0 = (\frac{dX}{dt} - BU)U\Sigma^{-1}V^T \tag{15.19}$$

The family of solutions consistent with the measurements is given by:

$$A = A_0 + CV^T \tag{15.20}$$

where C is an arbitrary scalar coefficient matrix.

The next step is to choose C such that A has maximum number of zero entries. Yeung et al. [55] proposed to set $A = 0$ in Eqn. 15.20 and obtained an over-determined equation $CV^T = -A_0$ and then found the exact fit plane passing through as many points as possible. In order to do this, they used L_1 norm regression that minimizes the sum of absolute values of errors to calculate C. This SVD-based approach was tested and validated in numerous experiments on model gene networks in their study [55].

15.2.4.4 Network Identification by Multiple Regression

One potential drawback of the approach above is that it requires data to estimate the time derivative $\frac{dX}{dt}$. Gardner et al. [56] proposed a method called network identification by multiple regression (NIR), which used only steady-state expression ($\frac{dX}{dt} = 0$) measurements.

If only one gene is perturbed in each experiment, Eqn. 15.14 and Eqn. 15.15 can be written as Eqn. 15.21 and Eqn. 15.22, respectively:

$$\frac{dx_i}{dt} = \sum_{j=1}^{n} a_{i,j} x_j + u_i \quad i = (1, 2, ..., n) \tag{15.21}$$

$$\frac{d\vec{x}}{dt} = A\vec{x} + \vec{u} \tag{15.22}$$

If there are m such perturbation experiments, similarly to Eqn. 15.16, we have:

$$\frac{dX}{dt} = AX + U \tag{15.23}$$

Since the concentration of n genes are at steady state, $\frac{d\vec{x}}{dt} = 0$, Eqn. 15.23 can be simplified as:

$$AX = -U \tag{15.24}$$

Similarly, by retrieving the connectivity matrix A, we can describe the network. Thus, the remaining problem is to solve the linear system. Since sample size is usually limited, $m \ll n$, and measurement of gene expression is noisy, it is preferred to have an over-determined system (more equations than unknowns) and use statistical regression. NIR solves this problem by assuming that for each gene, there is a maximum number of regulators, k. Then, the solutions of all possible combinations are calculated by least-square regression. The best solution is chosen based on the significance of the regression calculated by the F-test. NIR was applied to reconstruct regulatory networks containing nine genes in the SOS pathway in $E.$ $coli$. Each perturbation was accomplished by overexpressing one of the nine genes with arabinose-controlled episomal expression plasmid.

15.2.4.5 Mode-of-Action by Network Identification

Since NIR requires a well-designed perturbation experiment, it is not applicable to many datasets. Di Bernardo et al. [57] proposed a method called mode-of-action by network identification (MNI) to find the solution of the system without information about permutation u. Thus, MNI is suitable for the analysis of a wider range of microarray data. The idea of MNI is based on the assumption that any external stimuli acts on only a small number of genes, and thus most coefficients, $a_{i,j}$, in Eqn. 15.21 will be zero.

For each gene i, if there are m steady states experiments, we have:

$$\sum_{j=1}^{n} a_{i,j} x_{j,l} = -u_{i,l} \quad l = (1, 2, ..., m) \tag{15.25}$$

Extracting the experiments in which $-u_{i,l'} = 0$, we obtain

$$\sum_{j=1}^{n} a_{i,j} x_{j,l'} = 0 \quad l' = (1, 2, ..., m') \tag{15.26}$$

When implementing MNI, only experiments in which the perturbation, $-u_{i,l}$, is zero are considered. Determining these experiments is not trivial. Di Bernardo et al. [57] proposed a recursive method starting with an initial estimate of $\hat{a}_{i,j}$. The external influence, $\hat{u}_{i,j}$, is calculated using Eqn. 15.25. Any experiment with an external influence greater than a pre-determined threshold is removed for further calculations. Eqn. 15.26 is used to obtain a new estimate of $\hat{a}_{i,j}$. The method is repeated using the new estimate and continues until $\hat{a}_{i,j}$ and $\hat{u}_{i,l}$ converge.

As in NIR, Eqn. 15.26 is underdetermined. Thus, additional constraints are needed to find a reliable solution. Unlike NIR, which uses subset regression to identify a small set of non-zero coefficients, MNI uses the fact that expression profiles of many genes are similar. Thus, gene expressions can be represented by a reduced set of "characteristic" or "meta" genes by using SVD. The original space of gene expression with dimension n is first mapped into the space of metagenes with reduced dimensionality. The recursive procedure described above is then used to identify a network for the metagenes. The estimated perturbation is then mapped back into n gene space.

MNI was proposed for the application of finding target genes of a particular treatment. Di Bernardo et al. [57] applied MNI to the analysis 515 whole-genome yeast gene expression datasets resulting from different perturbation experiments and correctly enriched the target gene and pathway for most compounds.

15.2.4.6 Time Series Network Identification

Bansal et al. [13] proposed a method based on time series expression experiments, called time series network identification (TSNI). Similar to NIR, at a particular time point, the rate of synthesis of a transcript is represented as a function of the expression of the other genes and the external perturbation (See Eqn. 15.21, Eqn. 15.22, and Eqn. 15.24). Eqn. 15.27 is obtained when Eqn. 15.24 is converted to the corresponding discrete form:

$$X_{t+1} = A_d X_t + B_d U_t \qquad (15.27)$$

where X_{t+1} is the gene expression at time point $t+1$ and X_t at time point t. This equation states that the gene expression level at one time point depends on the expression profile and the external perturbation at the previous time point.

Similar to MNI, TSNI applies SVD to decompose matrix Y and solve Eqn. 15.27 with reduced dimensions and then maps the obtained solution back into the original space to obtain A_d and B_d. TSNI is suitable for the reconstruction of GRN containing genes of interest by analysis of time series gene expression data resulting from a particular perturbation. Bansal et al. [13] applied TSNI to recover a nine-gene subnetwork, part of the DNA-damage response pathway in E. coli, using experimental data obtained by treatment with Norfloxacin.

15.2.4.7 Stepwise Multiple Linear Regression

We have recently proposed an approach to reconstruct GRN based on time series gene expression data assuming no knowledge of perturbation [58]. The expression level of each gene is modeled as a linear function of expression levels of genes in the preceding time step. Eqn. 15.27 can be simplified to yield:

$$X_{t+1} = AX_t \tag{15.28}$$

This model can be generalized by utilizing prior τ time points as in Eqn. 15.29, where A_τ is a $n \times (n\tau)$ matrix:

$$X_{t+1} = A_\tau[X_{t-2}, X_{t-1}, ..., X_{t-\tau}]^T \tag{15.29}$$

The coefficients $a_{i,j}$ are identified using stepwise multiple linear regression (SMLR) with a forward selection strategy. In each forward selection step, individual predictor variables are considered for addition based on their statistical significance in the regression fitting. The p-value of an F-statistic for each variable is calculated to test the model including and excluding that variable using the null hypothesis whose weight coefficient is zero. SMLR provides a locally optimal set of predictors. For smaller networks, one may apply brute force and test all combinations of predictor variables to obtain the globally optimal set of predictors. In order to find a balance between capturing the underlying expression pattern and over-fitting, a proper p-value and a suitable number of previous time points to be used in the model should be chosen. Using four-fold cross-validation experiments on a set of cell-cycle datasets, we showed that using the previous two time points, and setting a p-value such that for each gene there were on average two to three predictors, would give the optimal performance. SMLR was tested in the reconstruction of yeast cell cycle GRN [58]. In an analysis of a sub-network containing 14 genes, the edge prediction precision exceeded 50.

15.3   INTEGRATING OTHER KNOWLEDGE

The complexity of a GRN model is exponential to the size of the network. Due to the abundance of parameters to be estimated and the insufficiency of available microarray gene expression data, it is necessary to impose some constraints in order to limit the search space. One widely used strategy is utilizing the sparseness of gene networks and thus minimizing the number of edges in the graph. By setting up a maximum number of regulators for each gene, the total number of candidate models to consider is significantly reduced. As mentioned earlier, the SVD method by Yeung et al. [55] also uses the sparseness of gene networks to select a proper GRN graph.

A study by Jeong et al. [59] indicates that biological networks are usually scale-free, which means the probability of a node having k edges, $P(k)$, follows a power-law: $P(k) \sim k^{-\lambda}$, where λ is a constant. Most of the genes are sparsely connected and only a few genes, called hubs, have many connections. The structure of the network is mainly determined by those highly connected nodes. Utilizing this property as a constraint, Chen et al. [60] modified the co-expression GRN model proposed by Agrawal [61]. The modification was shown to have better performance than the original method.

Another strategy to handle the problem of insufficiency of expression data is to identify gene modules or to cluster genes into "metagenes" or "prototypical genes" [62] based on expression patterns. There are several studies conducted in GRN reconstruction that identify modules of co-regulated genes [39, 63]. Guthke et al. [64] proposed a method that first clusters genes and chooses a cluster-representative gene for each cluster in accordance with physiological knowledge. Then, a hypothetical network is constructed with those cluster-representative genes.

There is also knowledge about GRN from the literature or databases, and this knowledge can be used for more accurate GRN inference if it is suitable for the particular problem. Bayesian networks provide a flexible scheme for integrating prior knowledge. Given prior knowledge ζ, the probability of model θ, $p(\theta|\zeta)$ is called the *prior* and $p(\theta|D,\zeta)$ is called *posterior*. $p(D|\theta,\zeta)$ is then the probability of observing the data given the model and prior knowledge, and thus describing how the model fits the data. According to Bayes' theorem, $p(\theta|D,\zeta) \propto (\theta|\zeta)p(D|\theta,\zeta)$, which means posterior is proportional to the product of prior probability of model and the fitness. Jensen et al. [65] proposed an approach to impose a prior distribution based on the prior information from ChIP binding and promoter sequence data and also weighting methodology for balancing the two prior information types. Nariai et al. [66] added knowledge of protein-protein interactions to estimation under a Bayesian statistical framework. Hartemink et al. [67] incorporated genomic location data in yeast to add constraints in model construction. However, whether the prior information can improve performance of the BN model mainly depends on the quality of information and how the gene expression data relates to the prior knowledge. Geier et al. [68] investigated the influence of integrating prior knowledge in BN and found that using prior knowledge gives a benefit when the size of the data is small in BN. Prior knowledge can also be integrated in differential equation models, where constraint from prior knowledge can be imposed directly, e.g., on coefficients $a_{i,j}$ in matrix A in Eqn. 15.15.

15.4   VERIFYING THE MODEL

As described in the knowledge cycle in Figure 15.1, any reconstructed hypothetical GRN should be verified and then modified in accordance with experimental data, which is considered the gold standard. The steps are repeated until the GRN can be considered to be true. However, before the experimental data is available, there are still several strategies to verify the reconstructed GRN and evaluate the performance of the modeling algorithm.

With only the gene expression data from microarrays, the performance of the model is usually tested by resampling. The model can be evaluated by k-fold cross validation, in which the available expression data set is divided into k sub-samples, and at each time, $k-1$ sub-samples are used as training data. Another approach is to randomize the input data to obtain a statistical confidence value for the predictions. For example, the time points of time-series gene expression data can be randomly permutated and GRN models are constructed repeatedly with the permutated data in order to evaluate the performance.

The proposed model can also be tested by *in silico* experimental data. Mendes et al. [69] proposed an artificial gene network based on defined topology and kinetics. Then, the synthetic data from the artificial gene network can be used to test and compare GRN inference algorithms. Geier et al. [68] used data generated from the set of synthetic networks. Yeung et al. [55] also tested their model with three synthetic data experiments. One well-known assessment of reconstruction methods is called DREAM [70], *Dialogue on Reverse-Engineering Assessment and Methods*, which attempted to provide "gold standard" data and metrics that could be used to compare the performance of various methods. The DREAM challenge invites participants to provide solutions to its reverse engineering question. Information about the newest DREAM challenge (DREAM8) can be found on the Web site http://www.the-dream-project.org/.

Another way to evaluate the performance of a model is to search the literature and databases for relevant information. There are various databases containing information for different pathways. For example, the *Kyoto Encyclopedia of Genes and Genomes* (KEGG) [60] is an on-line database that records networks of molecular interactions in cells for various organisms. The constructed GRN can be compared with the known network archived in KEGG. For more databases of regulatory networks and pathways, see the meta-database, Pathguide [71] (http://www.pathguide.org/), which, to date, provides information about more than 500 Web-accessible databases and resources.

15.5 CONLUSION AND FUTURE WORK

Inferring the gene regulatory network is a major step in systems biology. It not only helps scientists organize knowledge, but also assists scientists in generating hypotheses for further experiments, as a part of the iterative process of knowledge generation. Although the goal of GRN inference is to uncover the real biochemical interactions and its dynamics, due to insufficiency of data, GRN inference usually focuses on "lower resolution," that is, instead of real biochemical interactions and their reaction rate, scientists usually focus on finding the "influential network," which describes how expression of one gene interacts with the expression of other genes.

Microarray is a well-developed high-throughput technique, and gene expression data from microarrays are most commonly used for GRN inference. This chapter reviewed several mathematical schemes for GRN inference based on microarray gene expression data; including Boolean networks, Bayesian networks, co-expression networks, and differential equation models. Each model has its own advantages and drawbacks. The selection of a model is dependent on the purpose and application of the study. For example, when the prediction is qualitative, a logical model such as Boolean networks is preferable; when we want the real-valued dynamics of GRN, differential equation models are more suitable. The selection of the model also depends on the scale of the problem and availability of data; for smaller networks, Boolean networks and dynamic Boolean networks are usually good candidate models for reconstructing the topology of GRN. For larger networks, differential equation and co-expression networks are often used. Despite advances in quantitative assessment and internal validation of the networks produced, it remains an art to choose the best model. Since simpler models are easier to interpret and less prone to over-fitting, when two models provide equally good performances for explaining the observations, it is recommended to use the simpler model. "Everything should be made as simple as possible, but not simpler" (Albert Einstein).

So far no single model is error-free and no single inference method has optimal performance for all data sets [72]. It is a natural idea to combine different models. Marbach et al. [72] showed that the integrated results of multiple inference methods have robust performance across different data sets. Vignes et al. [73] proposed a meta-analysis that integrates the inferred results of three different methods. Li et al. [74] proposed a method called differential equation-based local dynamic Bayesian network (DELDBN), which integrates differential equations and the DBN model. They solve the differential equations and infer dependence by identifying the Markov blanket of target variables.

With the rapid accumulation of microarray data and biological knowledge, as well as the development of other high-throughput techniques, integrating data and knowledge from different sources is important to improve inference accuracy. Wang et al. [75] proposed a method based on linear programming and SVD to find the most consistent network structure with respect to multiple microarray data sets and the sparseness of the network. Chen et al. [76] integrated Bayesian prior from the epigenetic data of histone modification, and applied DBN to infer GRN from time series gene expression data. Wang et al. [77] proposed a GRN inference method using genome-wide fitness data from knockout libraries, which is an emerging high-throughput data type.

In conclusion, to generate better and useful hypotheses, scientists need a GRN with finer resolution. So far, no single inference method can achieve this goal and there is always lack of data. Combining the advantages of different inference models and integrating data and knowledge from different sources are promising directions for future studies.

15.6 FURTHER READING

For more information, please refer to survey papers on gene regulatory network reconstruction [3, 20, 33, 78, 79, 80, 81, 82, 83].

REFERENCES

1. R. Goodacre, S. Vaidyanathan, W.B. Dunn, G.G. Harrigan, and D.B. Kell, "Metabolomics by numbers: acquiring and understanding global metabolite data," *Trends in Biotechnology*, vol. 22, no. 5, pp. 245–252, 2004.
2. A.M. Feist, M.J. Herrgard, I. Thiele, J. L. Reed, and B.O. Palsson, "Reconstruction of biochemical networks in microorganisms," *Nature Reviews Microbiology*, vol. 7, no. 2, pp. 129–143, 2009.
3. T.S. Gardner and J.J. Faith, "Reverse-engineering transcription control networks," *Physics of Life Reviews*, vol. 2, no. 1, pp. 65–88, 2005.

4. S.A. Kauffman, "Metabolic stability and epigenesis in randomly constructed genetic nets," *Journal of Theoretical Biology*, vol. 22, no. 3, pp. 437–467, 1969.

5. S.A. Kauffman, *The Origins of Order: Self-Organization and Selection in Evolution.* New York: Oxford University Press, 1993.

6. A. Wuensche, "Genomic regulation modeled as a network with basins of attraction," in *Proceedings of the Pacific Symposium on Biocomputing*, vol. 3, pp. 89–102, 1998.

7. A. Wuensche, "Discrete dynamics lab," in *Artificial Life Models in Software*, pp. 215–258, Springer, 2009.

8. P. D'haeseleer, S. Liang, and R. Somogyi, "Genetic network inference: from co-expression clustering to reverse engineering," *Bioinformatics*, vol. 16, no. 8, pp. 707–726, 2000.

9. S. Liang, S. Fuhrman, R. Somogyi, et al., "REVEAL, a general reverse engineering algorithm for inference of genetic network architectures," in *Proceedings of the Pacific Symposium on Biocomputing*, vol. 3, p. 2, 1998.

10. T. Akutsu, S. Miyano, and S. Kuhara. "Identification of genetic networks from a small number of gene expression patterns under the Boolean network model," in *Proceedings of the Pacific Symposium on Biocomputing*, vol. 4, pp. 17–28, 1999.

11. T. Akutsu, S. Miyano, and S. Kuhara, "Algorithms for identifying boolean networks and related biological networks based on matrix multiplication and fingerprint function," *Journal of Computational Biology*, vol. 7, no. 3-4, pp. 331–343, 2000.

12. H. Lähdesmäki, I. Shmulevich, and O. Yli-Harja, "On learning gene regulatory networks under the boolean network model," *Machine Learning*, vol. 52, no. 1-2, pp. 147–167, 2003.

13. M. Bansal, G. Della Gatta, and D. Di Bernardo, "Inference of gene regulatory networks and compound mode of action from time course gene expression profiles," *Bioinformatics*, vol. 22, no. 7, pp. 815–822, 2006.

14. M. I. Davidich and S. Bornholdt, "Boolean network model predicts cell cycle sequence of fission yeast," *PLoS ONE*, vol. 3, no. 2, 2008.

15. S. Martin, Z. Zhang, A. Martino, and J.L. Faulon, "Boolean dynamics of genetic regulatory networks inferred from microarray time series data," *Bioinformatics*, vol. 23, pp. 866–874, Mar. 2007.

16. Z. Szallasi and S. Liang, "Modeling the normal and neoplastic cell cycle with realistic boolean genetic networks: Their application for understanding carcinogenesis and assessing therapeutic strategies," in *Proceedings of the Pacific Symposium on Biocomputing*, vol. 3, pp. 66–76, 1998.

17. R. Thomas, D. Thieffry, and M. Kaufman, "Dynamical behaviour of biological regulatory networksi. biological role of feedback loops and practical use of the concept of the loop-characteristic state," *Bulletin of Mathematical Biology*, vol. 57, no. 2, pp. 247–276, 1995.

18. F. Greil and B. Drossel, "Dynamics of critical kauffman networks under asynchronous stochastic update," *Physical Review Letters*, vol. 95, p. 048701, July 2005.

19. X. Deng, H. Geng, and M. Matache, "Dynamics of asynchronous random boolean networks with asynchrony generated by stochastic processes," *Biosystems*, vol. 88, Mar. 2007.

20. H. De Jong, "Modeling and simulation of genetic regulatory systems: a literature review," *Journal of Computational Biology*, vol. 9, no. 1, pp. 67–103, 2002.

21. L. Glass and C. Hill, "Ordered and disordered dynamics in random networks," *EPL (Europhysics Letters)*, vol. 41, no. 6, p. 599, 1998.

22. S. Kauffman, C. Peterson, B. Samuelsson, and C. Troein, "Random boolean network models and the yeast transcriptional network," *Proceedings of the National Academy of Sciences*, vol. 100, pp. 14796–14799, Dec. 2003.

23. I. Shmulevich, E.R. Dougherty, S. Kim, and W. Zhang, "Probabilistic boolean networks: a rule-based uncertainty model for gene regulatory networks," *Bioinformatics*, vol. 18, pp. 261–274, Feb. 2002.

24. I. Tirosh, A. Weinberger, M. Carmi, and N. Barkai, "A genetic signature of interspecies variations in gene expression," *Nature Genetics*, vol. 38, no. 7, pp. 830–834, 2006.

25. R. Edgar, M. Domrachev, and A.E. Lash, "Gene Expression Omnibus: NCBI gene expression and hybridization array data repository," *Nucleic Acids Research*, vol. 30, no. 1, pp. 207–210, 2002.

26. T. Barrett, D.B. Troup, S.E. Wilhite, et al., "NCBI GEO: archive for functional genomics data sets—10 years on," *Nucleic Acids Research*, vol. 39, no. suppl 1, pp. D1005–D1010, 2011.

27. Y. Kamada, U.S. Jung, J. Piotrowski, and D.E. Levin, "The protein kinase C-activated MAP kinase pathway of Saccharomyces cerevisiae mediates a novel aspect of the heat shock response," *Genes & Development*, vol. 9, no. 13, pp. 1559–1571, 1995.

28. M.C. Gustin, J. Albertyn, M. Alexander, and K. Davenport, "Map kinase pathways in the yeast Saccharomyces cerevisiae," *Microbiology and Molecular Biology Reviews*, vol. 62, no. 4, pp. 1264–1300, 1998.

29. R. Serrano, H. Martín, A. Casamayor, and J. Ariño, "Signaling alkaline pH stress in the yeast saccharomyces cerevisiae through the wsc1 cell surface sensor and the Slt2 MAPK pathway," *Journal of Biological Chemistry*, vol. 281, no. 52, pp. 39785–39795, 2006.

30. M. Kanehisa, S. Goto, M. Hattori, et al., "From genomics to chemical genomics: new developments in KEGG," *Nucleic Acids Research*, vol. 34, no. suppl 1, pp. D354–D357, 2006.

31. C. Müssel, M. Hopfensitz, and H.A. Kestler, "BoolNetan R package for generation, reconstruction and analysis of boolean networks," *Bioinformatics*, vol. 26, no. 10, pp. 1378–1380, 2010.

32. N. Friedman, M. Linial, I. Nachman, and D. Pe'er, "Using Bayesian networks to analyze expression data," *Journal of Computational Biology*, vol. 7, no. 3-4, pp. 601–620, 2000.

33. M. Bansal, V. Belcastro, A. Ambesi-Impiombato, and D. Di Bernardo, "How to infer gene networks from expression profiles," *Molecular Systems Biology*, vol. 3, no. 1, 2007.

34. S. Imoto, K. Sunyong, T. Goto, et al., "Bayesian network and nonparametric heteroscedastic regression for nonlinear modeling of genetic network," in *Proceedings of the IEEE Computer Society Bioinformatics Conference*, Stanford, CA, USA, pp. 219–227, IEEE, 2002.

35. I.M. Ong, J.D. Glasner, and D. Page, "Modelling regulatory pathways in E. coli from time series expression profiles," *Bioinformatics*, vol. 18, no. suppl 1, pp. S241–S248, 2002.

36. S.Y. Kim, S. Imoto, and S. Miyano, "Inferring gene networks from time series microarray data using dynamic Bayesian networks," *Briefings in Bioinformatics*, vol. 4, no. 3, pp. 228–235, 2003.

37. M. Zou and S.D. Conzen, "A new dynamic Bayesian network (DBN) approach for identifying gene regulatory networks from time course microarray data," *Bioinformatics*, vol. 21, no. 1, pp. 71–79, 2005.

38. A.J. Hartemink, D.K. Gifford, T.S. Jaakkola, et al., "Using graphical models and genomic expression data to statistically validate models of genetic regulatory networks,"

in *Pac. Symp. Biocomput*, vol. 6, pp. 422–433, 2001.

39. E. Segal, M. Shapira, A. Regev, et al., "Module networks: identifying regulatory modules and their condition-specific regulators from gene expression data," *Nature Genetics*, vol. 34, no. 2, pp. 166–176, 2003.

40. I. Nachman, A. Regev, and N. Friedman, "Inferring quantitative models of regulatory networks from expression data," *Bioinformatics*, vol. 20, no. suppl 1, pp. i248–i256, 2004.

41. L. Song, M. Kolar, and E.P. Xing, "Time-varying dynamic Bayesian networks," in *Advances in Neural Information Processing Systems*, pp. 1732–1740, 2009.

42. S. Lebre, J. Becq, F. Devaux, M. Stumpf, and G. Lelandais, "Statistical inference of the time-varying structure of gene-regulation networks," *BMC Systems Biology*, vol. 4, no. 1, p. 130, 2010.

43. M.B. Eisen, P.T. Spellman, P.O. Brown, and D. Botstein, "Cluster analysis and display of genome-wide expression patterns," *Proceedings of the National Academy of Sciences*, vol. 95, no. 25, pp. 14863–14868, 1998.

44. J.M. Stuart, E. Segal, D. Koller, and S.K. Kim, "A gene-coexpression network for global discovery of conserved genetic modules," *Science*, vol. 302, no. 5643, pp. 249–255, 2003.

45. C.E. Shannon and W. Weaver, "A mathematical theory of communication," *Bell System Technical Journal*, vol. 27, pp. 379–423 and 623–656, 1948.

46. R. Steuer, J. Kurths, C.O. Daub, J. Weise, and J. Selbig, "The mutual information: detecting and evaluating dependencies between variables," *Bioinformatics*, vol. 18, no. suppl 2, pp. S231–S240, 2002.

47. A.J. Butte and I.S. Kohane, "Mutual information relevance networks: functional genomic clustering using pairwise entropy measurements," in *Proceedings of the Pacific Symposium on Biocomputing*, vol. 5, pp. 418–429, 2000.

48. A.A. Margolin, I. Nemenman, K. Basso, et al., "Aracne: an algorithm for the reconstruction of gene regulatory networks in a mammalian cellular context," *BMC Bioinformatics*, vol. 7, no. Suppl 1, p. S7, 2006.

49. K. Basso, A.A. Margolin, G. Stolovitzky, U. Klein, R. Dalla-Favera, and A. Califano, "Reverse engineering of regulatory networks in human B cells," *Nature Genetics*, vol. 37, no. 4, pp. 382–390, 2005.

50. P. Zoppoli, S. Morganella, and M. Ceccarelli, "Timedelay-aracne: Reverse engineering of gene networks from time-course data by an information theoretic approach," *BMC Bioinformatics*, vol. 11, no. 1, p. 154, 2010.

51. E. Klipp, R. Herwig, A. Kowald, C. Wierling, and H. Lehrach, *Systems biology in practice: concepts, implementation and application*. WeinHeim: Wiley-Blackwell, 2008.

52. E. Sakamoto and H. Iba, "Inferring a system of differential equations for a gene regulatory network by using genetic programming," in *Proceedings of the 2001 Congress onEvolutionary Computation*, Seoul, South Korea, vol. 1, pp. 720–726, IEEE, 2001.

53. M.A. Savageau and E.O. Voit, "Recasting nonlinear differential equations as S-systems: a canonical nonlinear form," *Mathematical biosciences*, vol. 87, no. 1, pp. 83–115, 1987.

54. S. Kimura, K. Ide, A. Kashihara, et al., "Inference of s-system models of genetic networks using a cooperative coevolutionary algorithm," *Bioinformatics*, vol. 21, no. 7, pp. 1154–1163, 2005.

55. M.S. Yeung, J. Tegnér, and J.J. Collins, "Reverse engineering gene networks using singular value decomposition and robust regression," *Proceedings of the National Academy of Sciences*, vol. 99, no. 9, pp. 6163–6168, 2002.

56. T.S. Gardner, D. Di Bernardo, D. Lorenz, and J.J. Collins, "Inferring genetic networks and identifying compound mode of action via expression profiling," *Science Signaling*, vol. 301, no. 5629, p. 102, 2003.

57. D. di Bernardo, M.J. Thompson, T.S. Gardner, et al., "Chemogenomic profiling on a genome-wide scale using reverse-engineered gene networks," *Nature Biotechnology*, vol. 23, no. 3, pp. 377–383, 2005.

58. Y. Zhou, R. Qureshi, and A. Sacan, "Data simulation and regulatory network reconstruction from time-series microarray data using stepwise multiple linear regression," *Network Modeling Analysis in Health Informatics and Bioinformatics*, vol. 1, no. 1-2, pp. 3–17, 2012.

59. H. Jeong, B. Tombor, R. Albert, Z.N. Oltvai, and A.L. Barabási, "The large-scale organization of metabolic networks," *Nature*, vol. 407, no. 6804, pp. 651–654, 2000.

60. G. Chen, P. Larsen, E. Almasri, and Y. Dai, "Rank-based edge reconstruction for scale-free genetic regulatory networks," *BMC Bioinformatics*, vol. 9, no. 1, p. 75, 2008.

61. H. Agrawal, "Extreme self-organization in networks constructed from gene expression data," *Physical review letters*, vol. 89, no. 26, p. 268702, 2002.

62. E.P. van Someren, L.F. Wessels, and M.J. Reinders, "Linear modeling of genetic networks from experimental data," in *Proceedings of the 8th International Conference on Intelligent Systems for Molecular Biology*, La Jolla, CA, USA, vol. 8, p. 155, 2000.

63. Z. Bar-Joseph, G.K. Gerber, T.I. Lee, et al., "Computational discovery of gene modules and regulatory networks," *Nature Biotechnology*, vol. 21, no. 11, pp. 1337–1342, 2003.

64. R. Guthke, U. Möller, M. Hoffmann, F. Thies, and S. Töpfer, "Dynamic network reconstruction from gene expression data applied to immune response during bacterial infection," *Bioinformatics*, vol. 21, no. 8, pp. 1626–1634, 2005.

65. S.T. Jensen, G. Chen, and C.J. Stoeckert, Jr, "Bayesian variable selection and data integration for biological regulatory networks," *The Annals of Applied Statistics*, Institute of Mathematical Statistics, pp. 612–633, 2007.

66. N. Nariai, S. Kim, S. Imoto, and S. Miyano, "Using protein-protein interactions for refining gene networks estimated from microarray data by Bayesian networks," in *Proceedings of the Pacific Symposium on Biocomputing*, pp. 336–347, 2003.

67. A.J. Hartemink, D.K. Gifford, T.S. Jaakkola, and R.A. Young, "Combining location and expression data for principled discovery of genetic regulatory network models," in *Proceedings of the Pacific Symposium on Biocomputing*, pp. 437–449, 2002.

68. F. Geier, J. Timmer, and C. Fleck, "Reconstructing gene-regulatory networks from time series, knock-out data, and prior knowledge," *BMC Systems Biology*, vol. 1, no. 1, p. 11, 2007.

69. P. Mendes, W. Sha, and K. Ye, "Artificial gene networks for objective comparison of analysis algorithms," *Bioinformatics*, vol. 19, no. suppl 2, pp. ii122–ii129, 2003.

70. G. Stolovitzky, D. Monroe, and A. Califano, "Dialogue on reverse-engineering assessment and methods," *Annals of the New York Academy of Sciences*, vol. 1115, no. 1, pp. 1–22, 2007.

71. G.D. Bader, M.P. Cary, and C. Sander, "Pathguide: a pathway resource list," *Nucleic Acids Research*, vol. 34, no. suppl 1, pp. D504–D506, 2006.

72. D. Marbach, J.C. Costello, R. Kuffner, et al., "Wisdom of crowds for robust gene network inference," *Nature Methods*, vol. 9, no. 8, pp. 796–804, 2012.

73. M. Vignes, J. Vandel, D. Allouche, et al., "Gene regulatory network reconstruction using Bayesian networks, the Dantzig selector, the Lasso and their meta-analysis," *PLoS ONE*, vol. 6, no. 12, p. e29165, 2011.

74. Z. Li, P. Li, A. Krishnan, and J. Liu, "Large-scale dynamic gene regulatory network inference combining differential equation models with local dynamic Bayesian network analysis," *Bioinformatics*, 2011.

75. Y. Wang, T. Joshi, X.S. Zhang, D. Xu, and L. Chen, "Inferring gene regulatory networks from multiple microarray datasets," *Bioinformatics*, vol. 22, no. 19, pp. 2413–2420, 2006.

76. H. Chen, D. Maduranga, P. Mundra, and J. Zheng, "Integrating epigenetic prior in dynamic Bayesian network for gene regulatory network inference," in *Proceedings of the IEEE Symposium on Computational Intelligence in Bioinformatics and Computational Biology*, Singapore, pp. 76-82, 2013.

77. L. Wang, X. Wang, A.P. Arkin, and M.S. Samoilov, "Inference of gene regulatory networks from genome-wide knockout fitness data," *Bioinformatics*, vol. 29, no. 3, pp. 338–46, 2013.

78. M. Hecker, S. Lambeck, S. Toepfer, E. van Someren, and R. Guthke, "Gene regulatory network inference: Data integration in dynamic models – a review," *Biosystems*, vol. 96, no. 1, pp. 86 – 103, 2009.

79. G. Karlebach and R. Shamir, "Modelling and analysis of gene regulatory networks," *Nature Reviews Molecular Cell Biology*, vol. 9, no. 10, pp. 770–780, 2008.

80. F. Markowetz and R. Spang, "Inferring cellular networks–a review," *BMC Bioinformatics*, vol. 8, no. Suppl 6, p. S5, 2007.

81. K.H. Cho, S.M. Choo, S. Jung, J.R. Kim, H.S. Choi, and J. Kim, "Reverse engineering of gene regulatory networks," *Systems Biology*, vol. 1, no. 3, pp. 149–163, 2007.

82. E. van Someren, L. Wessels, E. Backer, and M. Reinders, "Genetic network modeling," *Pharmacogenomics*, vol. 3, no. 4, pp. 507–525, 2002.

83. N. A. van Riel, "Dynamic modelling and analysis of biochemical networks: mechanism-based models and model-based experiments," *Briefings in Bioinformatics*, vol. 7, no. 4, pp. 364–374, 2006.

16 Multidimensional Visualization of Microarray Data

Urška Cvek and Marjan Trutschl

CONTENTS

16.1 INTRODUCTION

Large amounts of multidimensional data resulting from the high-throughput technologies are not completely served by the biostatistical techniques alone and are complemented with visual, knowledge discovery, and other techniques that enable data exploration. We first formalize the record and dimension representation for microarray data into the microarray expression space. We review some of the classic information visualization techniques, followed by more novel, multidimensional visualization techniques that are capable of displaying larger, higher-dimensional data sets.

Classic visualization techniques range from scatterplots, histograms, and Venn diagrams, to box plots and many statistical plots that are still in heavy use today. Modern visualization techniques are targeted towards high-dimensional data and span from parallel coordinates and self-organizing map visualizations, to principal components analysis and other methods. As Saraiya et al. pointed out in 2004 [6], there is an abundance of tools available to biologists that includes many modern visualization techniques, although their utility has been mostly reserved to visualization experts that work hand in hand with domain experts.

Information visualization deals with presentation of information in a visually meaningful manner. Every visualization should reveal information inherent in the data. We use color, shape, size, and distance in order to uncover this information and reveal patterns. The complexity and richness of techniques available benefit from interactivity that ranges from brushing to filtering to parameter adjustment. For a more general review of visualization, we recommend Bertin [7], Cleveland [8] and Tufte [9, 10].

16.1.1 GENES, EXPERIMENTS AND THE EXPRESSION SPACE

We define a gene as an *"expression vector"* with each experiment as a separate, distinct axis. We assume that the experiments have been preprocessed and signal values obtained. For example, if our microarray data has three experiments, then our expression vectors have three dimensions and can be represented in the x-y-z space. Two genes with similar expression values for each experiment will be spatially near in this expression space while two genes that have very different values will be spatially far from each other. Dimensionality grows to n dimensions where n is the number of experiments, each gene (expression vector) represented as a single point in the n-dimensional space.

16.1.2 SAMPLE DATA: TRANSITIONAL CELL CARCINOMA OF THE BLADDER

In this chapter, we utilize a data set of the transitional cell carcinoma (TCC) of the bladder generated at the Clifford lab at LSU Health Sciences Center in Shreveport [1]. TCC ranks fourth in incidence of all cancers in the developed world, yet the mechanisms of its origin and progression remain poorly understood and there are few useful diagnostic or prognostic biomarkers for this disease. Transgenic mice carrying a low copy number of the SV40 large T (SV40T) oncogene express specific murine uroplakin II promoters [11]. The transgenic mice, UPII-SV40T, develop a condition closely resembling human carcinoma *in situ* (CIS) starting as early as 6 weeks of age, progressing to invasive TCC from 6 months of age onward. Affymetrix's Mouse GeneChip (Mouse Genome 430 2.0) was used to determine a relative expression level of over 39,000 mouse transcripts (45,101 probe sets), representing the majority of the transcribed mouse genome in a given mRNA sample. Duplicates were performed for SV40T and non-transgenic littermates (WT) at each time point, yielding a set of arrays for two factors: mouse genotype (WT or SV40T) and week (3, 6, 20 and 30), creating eight targets or an eight-dimensional expression space. The WT line at the 6-week time point was an exception with only one quality array due to RNA degradation.

16.2   CLASSIC VISUALIZATIONS

Classic visualization techniques typically used on microarray data range from scatterplots, histograms, line and box plots to heatmaps and dendrograms. Many variations on the theme exist today, and we showcase some of the most common approaches using a sample TCC data set.

We can separate the techniques into two categories: *point-based*, where one marker is used per record (gene or experiment) and *aggregate*, where one marker represents a group or a number of records (genes or experiments). Scatterplots are a familiar point-based technique, where a few pixels are used to visually represent the record in a two-dimensional or three-dimensional layout. Heatmaps are utilized similarly, although a record is typically represented as a cell in a matrix. Histograms are an example of aggregate displays; each histogram bar represents a group of records that fits a certain criteria. Aggregate visualization techniques are used to visually present summary statistics; they pictorially and concisely convey a large amount of information.

16.2.1   SCATTERPLOT AND MATRIX OF SCATTERPLOTS

A scatterplot is a point projection of the data into a two-dimensional or three-dimensional space represented on the screen in the classic *x-y* or *x-y-z* formats. This is the most commonly utilized data visualization today (no matter the application field and with microarray data being no exception), where it is commonly used to display measured experiment intensities. Scatterplots can be drawn with almost any visualization package available today and are useful for comparing two samples, channels from the same ratio (e.g., Cy3, Cy5) data or duplicate spots of the sample. We can display individual genes or experiments and use the color, size, shape, texture, motion, and even sound attributes of the records to map data to. In Figure 16.1 we observe the highest congruency of the 3-week distributed and log-normalized values of replicates of the SV40 arrays. The least agreement between any of the four pairs is detected in the 30 week arrays, which is suspected to be due to the length of the treatment. Week six and week twenty show a similar agreement between the pairs

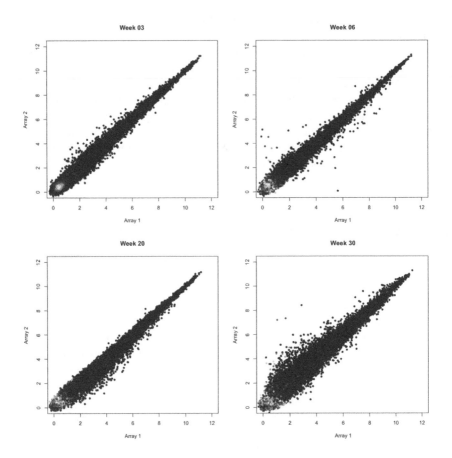

FIGURE 16.1 Scatterplot matrix of SV40 probe intensity (log-based intensity of background adjusted, normalized and summarized) values of the arrays. Each point represents one probe set with density of points (the number of probes drawn at the same location) mapped to the linearized grey color scale; higher intensity color (white) indicates high point density and lower intensity color (black) indicates low point density.

of arrays. Local density is mapped to the linearized grey color scale [12], with highest density mapped to the white color and lowest density mapped to the black color. Microarrays typically have a larger number of genes that map to smaller values and it is not unusual to see data that is heavily biased towards the low values, such as in these samples. These scatterplots are used in the preprocessing stages of microarray analysis in order to evaluate the quality of arrays and potentially determining which arrays might have to be excluded from further data analysis.

Scatterplots are most typically used to show trends in the data, correlate dimensions of the data, or select a set of records that fit a profile [13, 14]. In their most general form, scatterplots are related to iconographic and pixel displays. When the number of records displayed is large, the points become very dense and form black (or colored) shapes that cannot be visually distinguished. Binning has been used to overcome this issue, and variations on this theme utilize the record density to provide clarity for high-volume scatterplots. Scatterplots are also used to display data that has been transformed from its original dimensions, such as using multidimensional scaling, principal component analysis, independent component analysis, and other techniques, as shown in a later section of this chapter.

A matrix of scatterplots has been long in use before its publication [15] and is an array of scatterplots displaying all possible pairwise combinations of dimensions or coordinates. Scatterplot matrices have become increasingly common in general purpose statistical software programs; we generated them with the R project and its accompanying tools [4]. For n-dimensional data this yields $\frac{n(n-1)}{2}$ scatterplots with shared scales, although most often n^2 scatterplots are displayed. The concept is simple and powerful and extends to other plot formats, such as the quantile-quantile plot. A matrix of scatterplots can also be arranged in a non-array format, such as circular and hexagonal layout, visually linking features of related scatterplots.

16.2.2 HEATMAP AND DENDROGRAM

A heatmap is an array of cells where each cell is colored based on some data value or function on the data. This method is a generalization of a scatterplot where the points are grid cells and each cell is colored. There are many named variants, such as a clustered image map or patchgrid. The most common visual representation of microarray data represents a column as an experiment, and a row as an expression vector for a particular gene, coloring each of the array cells. This sort of two-dimensional clustering was originally used for analysis of gene expression array data by Hastie et al. [16] and was first used in transcriptomics by Eisen et al. in their milestone publication [17]. Its prevalent use continues through today [18, 19, 20].

For a set of genes and experiments the expression matrix typically appears random, until clustering or some other rearranging approach reorders the rows, columns, or both, making the patterns visually apparent. This gives a visual overview of the experiments and genes, most commonly using black color when the log (ratio) value is close to zero, red color when positive (larger than zero), and green when negative (color scheme progresses from red through black to green). The color scale thus indicates the differential expression, with the brighter elements more highly expressed. A variety of color schemes can be used to illustrate the heatmap, with perceptual advantages and disadvantages for each [12]. While the red-to-green colormaps are often used in heatmaps, this should be discouraged due to a number of perceptual issues this scheme brings.

Dendrograms typically accompany heatmaps when clustering is used to reorder the records [21, 22]. A dendrogram is a branching tree visualization that presents the results of a top-down or bottom-up hierarchical clustering algorithm. Dendrograms typically display other degree of similarity measures or the number of shared characteristics (such as biological taxonomy). Each dendrogram stem (node) represents a cluster of records and its children sub-clusters (stems). Dendrograms are typically displaying the correlation coefficients or Euclidean distances among the records. Dendrograms do not necessarily expose the structure that exists in the data, and generally represents only the between-record distances as path lengths of the leaves of the dendrogram. Some of the drawbacks of the technique include the inability to display negative interactions, only connecting each record (gene or experiment) to the tree with a single stem. In reality, we commonly encounter negative interactions (such as tumor suppressor genes), and transcription factors that are responsible for the regulation of multiple other genes.

We applied hierarchical clustering to expression vectors over the time course of the TCC data (3, 6, 20, and 30 weeks) and we used the R project to draw the heatmap and dendrogram [4]. One of the most commonly used tools to visualize the biological heatmaps and dendrograms is TIBCO Spotfire [23]. Hierarchical clustering is a partitioning algorithm (working either in an agglomerative or divisive fashion) that provides us with single element clusters all the way through one cluster of all records. In order to obtain the desired number of clusters, we make a cut in the dendrogram at the desired level. The heatmap we present in Figure 16.2 uses red, green, and black to represent genes that are upregulated, downregulated, or unchanged, respectively, in the UPII-SV40 mice when compared to the WT littermates over the time course. Each column represents one of the fifteen arrays of the experiment, and each row represents a gene. The last cluster shown at the bottom in the right heatmap has been enlarged and rescaled into the left heatmap and represents genes that are highly upregulated at all four time points (more strongly expressed in WT bladders). The analysis focused on the most strongly upregulated and downregulated groups of genes that could contain candidate biomarkers for both, premalignant and later stage TCC.

16.2.3   HISTOGRAM

A histogram is a representation of univariate data's (measurements of a single quantitative variable) frequency distribution by means of rectangles whose width represents class intervals and whose areas are proportional to the corresponding frequencies [24]. It is an aggregate visualization technique, as it uses one marker to represent a number of records (genes or experiments, for example). The histogram is a widely used graphical method that is at least 100 years old [8] and is included as part of the R tools [4]. As applied to microarrays, they are most commonly used to analyze probe and signal intensities of microarray data, or visualizing the density of each of the arrays (smoothed histograms of raw and log scale intensities) [25, 26]. They can also compare counts of genes that fit a certain profile, such as up or down regulation.

For the TCC data set, we show the frequency of raw perfect match (PM) and mismatch (MM) probe log-intensities of our sample data set of WT and SV40 arrays (Figure 16.3). The distributions have similar shapes, with a higher frequency of a narrower intensity window present in the MM set.

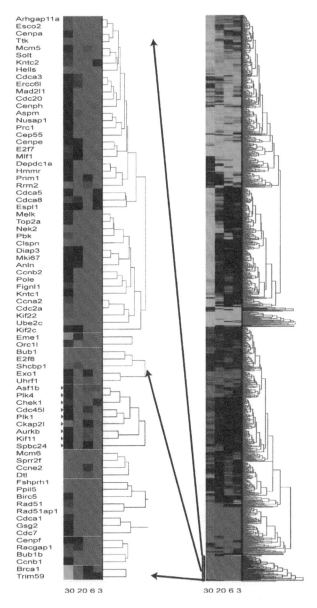

FIGURE 16.2   (SEE COLOR INSERT.) Heatmap and dendrogram of the hierarchical clustering of the four time points (3, 6, 20, 30 weeks). Genes that are upregulated, downregulated or unchanged, respectively, in the UPII-SV40 mice when compared with the WT littermates are shown. The right image contains approximately 1,900 genes, and its last cluster of genes (more strongly expressed in WT bladders) has been enlarged and rescaled (genes highly up-regulated at all four time points [1]). Dendrograms on the right of the heatmaps show the hierarchical clustering of the genes.

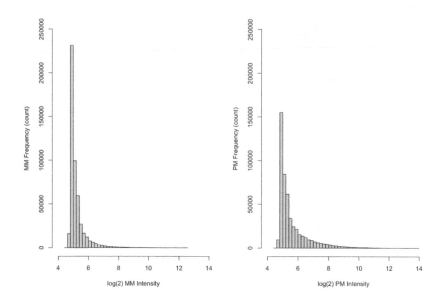

FIGURE 16.3 Histograms of the raw perfect match (PM) and mismatch (MM) probe log-intensities of our sample data set. Both sets have similarly shaped distributions.

16.2.4 BOX PLOT

A *box plot* is an aggregate technique that is typically used in the preprocessing and quality assessment stage, displaying the unprocessed log scale and raw probe intensities of the experiments (of microarray experiments). A box plot is a standard technique to present the *five-number summary*, which includes the smallest and largest observations, the 1st, 2nd (median) and 3rd quartiles [27]. Typically, the ends of the box are at the quartiles, so that the box length is the interquartile range, with the median marked by a line within the box. Two lines ("whiskers") outside of a box extend to the smallest and largest record values. Box plots are utilized to compare probe intensity levels among the arrays of a data set. Box plots are frequently used in the quality assessment of the probe intensity levels of microarray experiments [28]. For example, a box for an array that has spatial effects or quality problems may have a greater spread or may not be centered near the mean for the other arrays.

The original fifteen array signals of the TCC data set are shown in Figure 16.4, first of the raw log probe intensities, followed by preprocessed log probe intensities. This gives a simple summary of the distribution of the probes. We can see in Figure 16.4 that all of the arrays have similar probe intensity readings, which gives us further assurance of the good quality of the arrays in the TCC experiment. We utilized the R project [4] and Bioconductor tools [29] to generate these images.

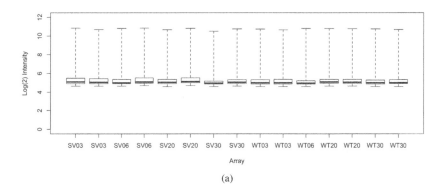

(a)

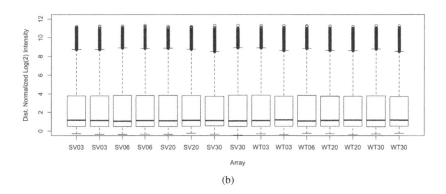

(b)

FIGURE 16.4 Boxplots of the 15 arrays from the TCC data set. (a) Unprocessed log scale probe-level intensities, (b) Log scale probe intensities after RMA preprocessing was applied. RMA background corrects, normalizes, and summarizes the probe level data [2].

16.3 MODERN MULTIDIMENSIONAL VISUALIZATIONS

Modern information visualization techniques encompass multidimensional and multivariate datasets such as microarray experiments. They can also be divided into point-based techniques and aggregate techniques, just like the classic visualizations. The common thread to multidimensional modern techniques is the need for interaction, such as filtering, selection, zooming and rotation, and integration of advanced computational techniques in order to better understand the underlying patterns in the data. Some of the techniques available today include tools developed specifically for microarray datasets and provide additional insights into the biology of individual records of values (genes or experiments). Some techniques are based on data mining algorithms, such as principal component analysis (PCA), independent component analysis (ICA), multidimensional scaling (MDS), and the self-organizing map algorithm (an unsupervised neural network technique).

16.3.1 PARALLEL COORDINATES

Parallel coordinates [30, 31] are a geometric projection visualization technique, representing dimensions on horizontally arranged parallel common axes. This is a point-based technique, where each record corresponds to a polyline intersecting the axes at the record's dimensional values. Although parallel coordinates can display a large number of dimensions in a dataset, the number of records that can be visualized is rather limited. Visualizing a medium or a large number of records results in over-plotting and clutter of featureless blobs that hide the underlying structure of the data. Dimension reordering and summarization, such as clustering, sampling, or filtering, are attempts at overcoming this issue.

A number of parallel-coordinate-based techniques organize the records first and then visualize the cluster centroids [32, 33, 34]. Users can navigate and filter the data and select a desired focus region and level of detail. Parallel coordinates are commonly augmented by additional information displays or interactive tools, such as histograms, frequency or density information, and coupled visualizations. For example, visual data mining displays utilize cluster centroids placed on top of the parallel coordinates and tracked statistical measures displayed as static or animated glyphs in a separate coordinate system such as those shown in Figure 16.10 (right) and described by Ericson [35].

In Figure 16.5, we show the over-plotting or occlusion that occurs when we visualize the set of 585 unique TCC genes that are differentially expressed in SV40 line at week 6 and/or week 20 using a proprietary program. We display all of the 15 experiments of the dataset and order them by first listing the SV40 lines followed by the WT lines. No significant differences across the dimensions are noticeable, although we can detect that the majority of the records have low to mid values. There are only a few records whose signal values are relatively high. In the center of the display we notice a dip in the expression, which is due to the switch from the SV40 to the WT and is not a true pattern in the data.

16.3.2 VISUALIZATIONS BASED ON DATA MINING ALGORITHMS

A number of different visualization techniques are based on a point-representation of a record, whose dimensions are derived from a data mining algorithm. Principal component analysis (PCA), multidimensional scaling (MDS) and independent component analysis (ICA), together with the self-organizing map (SOM) are frequently used to visually represent the gene expression data. All of these techniques are available as part of the R project core and extended tools [4]. We cover the SOM technique in more detail in section 16.3.3 and expand its utility beyond two-dimensional projections.

Introduced by Pearson [36], principal component analysis (PCA) is a dimension reduction technique applied in microarray analysis and exploratory data analysis in general. PCA reduces the data set to a number of uncorrelated variables with largest possible variance preserved [37]. The output of a PCA algorithm is a series of principal components (PCs) with each successive principal component being orthogonal to

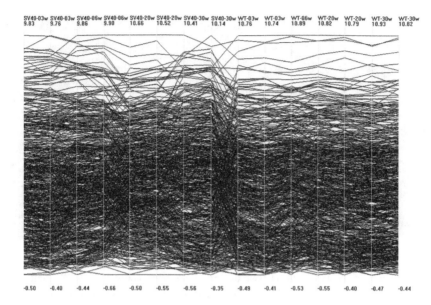

FIGURE 16.5 Parallel coordinates of the 15-dimensional dataset related to the 585 early changing genes of the TCC data set [3]. This otherwise dimensionally capable visualization turns into an incomprehensible collection of polylines when presented with such a dataset. 3D parallel coordinates address this shortcoming by utilizing a modified version of a self-organizing map algorithm (see 16.3.5.3).

the previous ($PC_n \perp PC_{n+1}$). The first principal component (PC_1) exhibits the largest variability in the dataset. Successive principal components represent the largest remaining variability in the dataset [38]. PCA can use the correlation or covariance matrix for its basis. When variables scales are similar, the covariance matrix is usually used, and the correlation matrix standardizes the data when variables are on different scales. Typically gene expression data have been normalized, and we can use the covariance matrix when the data are gene expression patterns from the same platform with similar range and scale.

To visualize computed principal components for a dataset, either a 2D or 3D scatter plot, possibly a scatter plot matrix, is used. Additionally, the first three principal components reveal the largest insight into the dataset and are considered the most useful in detecting patterns, trends, and/or clusters. If there exist patterns, trends and/or clusters, this method will most likely identify them. If no obvious patterns exist, the co-located records are highly correlated. Figure 16.6 shows a two and a three-dimensional representation of the first three principal components of a subset of the TCC genes. Figure 16.7(a) is the PCA of the fifteen arrays of the TCC data set. It shows that the replicates are fairly similar to each other and that the two sets (WT and SV40) can be distinctively separated.

PCA and multidimensional scaling (MDS) are both linear dimensionality reduc-

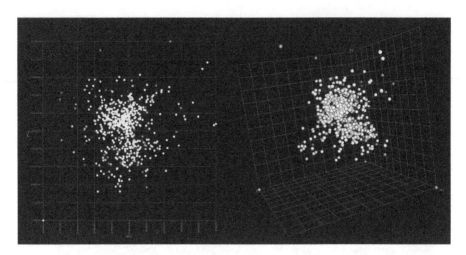

FIGURE 16.6 (SEE COLOR INSERT.) Principal component analysis of the 585 early changing genes of the TCC data set. Two-dimensional scatter plot of PC_1 vs. PC_2 (left) and three-dimensional scatter plot of PC_1 vs. PC_2 vs. PC_3 (right) show relationships among the records in a high-dimensional dataset, revealing relationships that usually remain hidden when visualized with classic visualization techniques.

tion schemes. MDS visualizes the level of similarity of individual cases of a data set. It refers to the set of related ordination techniques used in information visualization, to display the information contained in a distance matrix. An MDS aims to place each object in an n-dimensional space, such that the between-object distances are preserved as well as possible. Unlike PCA (where most of the variance in the data is captured in the first axis and each subsequent axis captures progressively less variance in the data), axes in MDS are arbitrary and distance units along each axis do not reflect equal quantitative distances at other sections of the same axis [39]. Unsupervised MDS of the time series of the TCC data set in Figure 16.7(b) reveals two distinct and homogeneous clusters: one comprised of the WT and the other of the SV40 arrays. MDS has recently been used to analyze endometrial cancer [40], and complemented the PCA analysis by Boedigheimer et al. [41].

Independent component analysis (ICA) defines a generative model for the observed multivariate data, which is typically a large set of samples. In the model, the data variables are assumed to be linear mixtures of some unknown latent variables (assumed mutually independent and non-Gaussian), which are called independent components of the data. These independent components, also called factors, can be found by the ICA. While PCA is based on the information given by the second order statistics, ICA goes up to high order statistics, and its result is assumed to be more meaningful than the one gained by PCA. ICA is often perceived as an extension of PCA and it works better on the data that have already been preprocessed by PCA [42].

Additional data mining techniques that are used for classification of data include

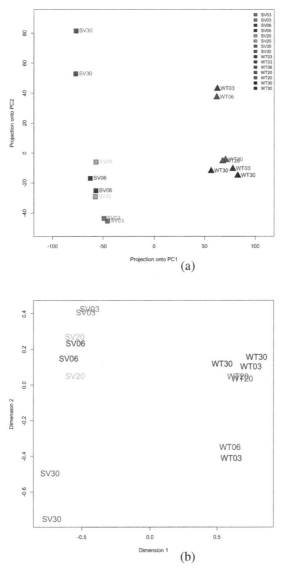

FIGURE 16.7 (SEE COLOR INSERT.) PCA and MDS plots of the 15 arrays of the TCC data set using the complete set of genes on the array generated using the R project [4]. (a) Triangles represent the WT arrays while squares represent the SV40 arrays, with a clear separation of the two sets. The majority of the replicates are positioned in close proximity to each other, indicating good replication of the data. (b) Similar observations can be made about the MDS plot of the arrays based on the top 5000 genes, which shows the between-object distances of the arrays. We can conclude that the WT are most similar to each other, and similar relationships between the individual time points can be observed, just as in the PCA plot.

linear discriminant analysis (LDA), also known as the Fisher discriminant analysis, singular value decomposition (SVD), canonical correlation analysis (CCA), latent semantic analysis (LSA), and others. PCA is probably one of the best known of the techniques of multivariate analysis in statistics and its computation is very simple and the definition straightforward. Thus, it has been extensively applied to microarray data in its original form and with variations.

Visualizations such as those in Figure 16.6 benefit from linked views where selection of gene-related records is performed in one view (i.e., scatter plot) and related data displayed in another view (i.e., list, table, etc.) [43, 44, 45]. This way, users can interactively explore features of interest and form new or confirm existing hypotheses.

16.3.3   SELF-ORGANIZING MAP

A self-organizing map (SOM) is an unsupervised neural network that facilitates mapping of a set of n-dimensional vectors (records of data) to a two-dimensional scatter-plot-like topographic map [46, 47, 48]. Training of an unsupervised neural network is completely data-driven, without a target condition that would have to be satisfied (as in a supervised neural network). Therefore, the output of an SOM algorithm represents relationships among the input vectors. The SOM combines an analytic and graphical technique of grouping the data with intent to reduce its size. In other words, the SOM is a summarization technique that attempts to reduce the complexity of the dataset by displaying clusters of the data in a grid-like layout. Therefore the input vectors with similar properties map to the same or neighboring output nodes. This turns out to be ideal for multidimensional microarray datasets.

However, the results produced by the SOM algorithm may be misinterpreted, if taken out of context. For example, the distance between the neighboring weight vectors does not correspond to the physical location of these vectors on the matrix of output nodes as described by Ultsch [49]. Additionally, the output often shows a scatter-plot-like visualization represented by a grid of output nodes ($Nodes_{width} \times Nodes_{height}$). It is important to note that the output nodes are usually not visualized and that there are no two dimensions in the original dataset that map to x and y-coordinates. Therefore, the data should be interpreted with caution.

Nevertheless, the widespread use of the SOM algorithm is attributed to its simplicity. Besides gene expression analysis, the SOM algorithm has been utilized in a wide spectrum of applications from machine vision and image analysis to neurophysiological research, and visualization [50] and [51]. We created the SOM and related visualizations with our proprietary package. Other tools that can generate a SOM include the R project package *som* [52], and a visualization package *somplot* [53].

16.3.4   RADIAL VISUALIZATION OR RADVIZ

Radial display technique places dimensional anchors (dimensions) around the perimeter of a unit circle and utilizes spring constants to represent relational values among points is the dimension-reduction technique known as Radviz (radial visualization) [54].

As shown in Figure 16.8, Radviz utilizes spring constants to represent dimensional values among the records. A record is represented as a vector $(x_{i_1}, ..., x_{i_n})$ of n dimensions. Its position is determined by the pull of the position vectors $(\bar{S}_1, ..., \bar{S}_n)$. The record in our example consists of eight dimensions ($n = 8$), ordered on a unit circle in counter-clockwise, equidistant fashion. Each position vector points from the center of a unit circle to the corresponding dimensional anchor on the perimeter of a unit circle. The values of each dimension are usually normalized to $[0..1]$ to eliminate the effects associated with the variable min/max values. Each data point marker is displayed at the location where the sum of all spring forces equals zero and the stiffness of each spring is proportional to the value of the corresponding dimension. Therefore, the point representing a record of dimensional values ends up at the position where the spring forces are in equilibrium. Additionally, the genes with similar expression patterns are plotted in proximity to each other, forming clusters. However, such clustering may result in overplotting of records with gene expression profiles with scaled signal values. For example, vectors $[1, 5, 2, 10]$, $[10, 50, 20, 100]$, and $[8, 40, 16, 80]$ map to the same location although the latter two are "closer" to each other. 3D Radviz addresses this issue by pulling the records into the third dimension based on all (or a selected set of) dimensions (see *3D Radial Visualization or 3D Radviz* in 16.3.5.2).

The position of the data point marker depends largely on the arrangement of dimensions around the unit circle. However, data with similar values are always placed in proximity to each other. The technique has been complemented by dimension ordering approaches, where the dimension order is determined by the structure of the data or the inherent class separation [55, 56, 57, 58, 59, 60].

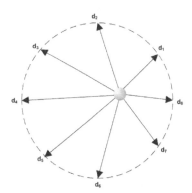

FIGURE 16.8 Dimensions from a dataset are ordered along the unit circle in Radviz. The location of the record marker is determined based on Hooke's law—at the equilibrium of forces with each position vector pulling from the center of a unit circle to the corresponding dimensional anchor on the perimeter of a unit circle.

16.3.5 INNFOVIS: A NEW BREED OF DIMENSIONALLY CAPABLE VISUALIZATIONS

iNNfovis encompasses several widely utilized visualizations such as 2D and 3D scatterplots, parallel coordinates, Radviz, and histogram. These are so-called hybrid visualizations and rely on a modified Kohonen's self-organizing map algorithm. They have been successfully utilized in the analysis of microarray and other high-dimensional datasets. We describe a couple of iNNfovis algorithms below.

16.3.5.1 SmartJitter: Self-Organized Scatter Plot

SmartJitter is a variation of Kohonen's SOM that performs competitive learning while preserving the general shape of a scatter plot, a well-understood and heavily utilized classic visualization technique. SmartJitter falls within a family of algorithms designed to preserve the topological features of a visualization technique while increasing the intrinsic dimensionality (see Section 16.4) of a classic SOM. This makes it suitable for gene expression analysis where two dimensions (i.e., control expressions) in a dataset are mapped to x and y-axes and then pulled into the third dimension based on additional dimensional values (i.e., experiment-related expression).

Although SmartJitter is derived from Kohonen's concept of an SOM, it is different from Kohonen's SOM in that competition is restricted to a subset of the map [61, 62], removing the full-connectivity condition present in Kohonen's SOM. As part of this algorithm, each input vector is mapped to a primary mapping that associates it to a subset (a.k.a., cell) of secondary output nodes (secondary mapping) placed orthogonally to the primary mapping. When the primary mapping is 2D (primary grid) and the secondary mapping is 3D, primary cells are called "stacks," and the visualization is referred to as SmartJitter Stacks [63]. The winning output node is selected from this subset of output nodes based on the shortest distance between the input vector and the weight vectors associated with each output node (secondary mapping) as shown in Figure 16.9.

In the case of a SmartJitter, the purpose of the primary mapping is to preserve the topology of the scatter plot and at the same time provide a good approximation of the dataset. Kohonen permits all nodes to be candidates for selection during competition [64]. However, SmartJitter only allows a stack of output nodes associated with the primary grid cell identified as part of primary mapping to be considered [61, 62]. This results in constrained clustering, enhancing widely utilized classic visualizations such as scatter plot. Such visualizations provide additional insight into the multidimensional dataset and at the same time address issues related to occlusion or overplotting observed in a classic scatter plot visualization. Moreover, records representing similar data (such as genes with similar expression profiles) are plotted in proximity to each other. Clusters contain records with similar profiles, thus revealing new insights as shown in Figure 16.10 . Brushing randomly-chosen clusters of co-located records shows close association among them. Dimensional values mapped to the primary grid (x and y-coordinates) can be selected arbitrarily. However, it is generally the best practice to leave selection of dimensions to a domain expert.

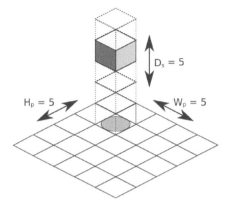

FIGURE 16.9 The winning node for a record constrained to a stack of output nodes associated with its primary mapping location (represented by the filled oval). The winning node is a node such that the distance between the output node's weight vector and the input vector is minimal. Note that H_p (height—primary grid), W_p (width—primary grid), and D_s (depth—secondary grid) can be set to virtually any positive, nonzero integer (i.e., they are independent variables) keeping in mind that large values associated with these variables could lead to memory-intensive topologies that can potentially exhaust the system resources.

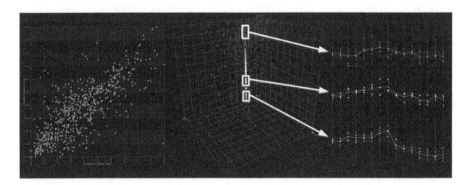

FIGURE 16.10 (SEE COLOR INSERT.) A comparison of a traditional scatter plot (left) and a self-organizing map enhanced SmartJitter Stacks visualizations with records related to the 585 early changing genes in the TCC data set [3] being pulled into the third dimension (middle). This technique alleviates the issue of overplotting in a classic scatter plot visualization. Brushing three randomly chosen clusters within an arbitrary stack (middle) reveals the properties of co-located records—they show close association of co-located records colored by the z-value (right). White lines denote related cluster centroids (right). Dimensional values mapped to x and y-coordinates were selected arbitrarily (left and middle image) to demonstrate the power of this technique. The z-values in the middle image were calculated using a SmartJitter algorithm. Separation of otherwise overlapping records is obvious.

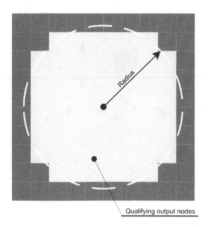

FIGURE 16.11   Primary grid of qualifying output nodes in a 3D Radviz. The number of rows and columns in a primary grid is assigned arbitrarily, keeping in mind that finer grids of output nodes result in higher computational complexity and increased memory footprint.

16.3.5.2   3D Radial Visualization or 3D Radviz

Primary mapping is based on the Radviz visualization and establishes primary mapping on the two-dimensional surface (Figure 16.11). Secondary mapping is similar to the one described in SmartJitter Section 16.3.5.1, where an output node in a primary grid is associated with a stack of secondary output nodes that qualify for mapping in the third dimension (Figure 16.12). The output of this algorithm and related visualization is shown in Figure 16.13.

16.3.5.3   3D Parallel Coordinates

The dimensionally capable nature of a parallel coordinates plot lends itself to visualization of high-dimensional datasets [65]. Such visualizations tend to result in unwanted inter-dimensional blobs consisting of polylines representing the records in the dataset. While such plotting may uncover previously unknown inter-dimensional relationships, it may also hide them [47, 66]. 3D parallel coordinates address the problem of over-plotted or occluded polylines with the integration of the SOM algorithm into the classic parallel coordinates algorithm.

Self-organized parallel coordinates (SOPC) are constructed by augmenting common axes associated with dimensional values in a dataset with an SOM-like grid (see Figure 16.14). This is where the records are self-organized based on multiple, if not all, dimensional values of interest. While the two-dimensional view of a self-organized parallel coordinates plot matches the original parallel coordinates plot, the plot also provides self-organized dimensional information based on the records' dimensional data along the z-axis. Just like in the previously described SOM-enhanced algorithms, the core algorithm consists of two steps: primary and secondary mapping or training, extending the original parallel coordinates into self-organized parallel co-

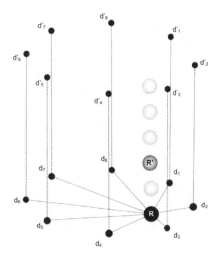

FIGURE 16.12   Three-dimensional version of Radviz utilizes a similar approach as exhibited in Radviz 2D. Primary mapping locates the winning output node on a two-dimensional primary grid of output nodes (16.11). The record is then pulled into the third dimension based on all or, if so desired, a subset of dimensional values in the dataset. Each output node in the primary grid is associated with a stack of secondary output nodes used for mapping of records in the third dimension.

ordinates. Final output of the SOPC algorithm yields informative and dimensionally capable visualization that aids in the analysis of microarray experiments (see Figure 16.15).

16.3.6   BIASOGRAM

Gene expression measurements are susceptible to technical bias caused by variation in extraneous factors such as RNA quality and array hybridization conditions, which increase the likelihood of identification of false positive genes. This is an important issue in large clinical trials, which can be affected by batch effects and varying specimen quality, causing non-random measurement error and leading to spurious correlation to the technical factor [67]. A Biasogram [68] is an orthogonal projection of the expression matrix of all genes onto a plane defined by a clinical variable and a technical nuisance variable. They can be created using the R project package *biasogram* [69]. The resulting plot indicates the extent to which each gene is correlated with the clinical variable or the technical variable and this should in principle refine existing gene signatures and eliminate genes correlated with the technical bias. Its input data consists of: 1) a gene expression matrix X with dimensions equal to the number of probes (or probe sets) and the number of specimens; 2) outcome vector Y indicating the outcome of each specimen; and 3) a bias vector B indicating the technical bias (or other nuisance variable) of each specimen. This method does assume linear relationships between gene expression, outcome, and bias, and that we are able

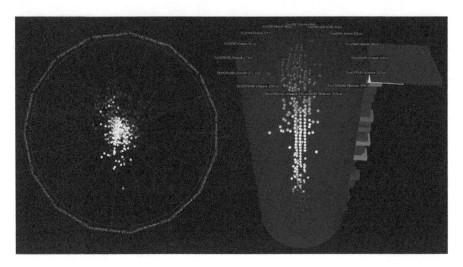

FIGURE 16.13 (SEE COLOR INSERT.) Two (left) and three-dimensional (right) Radviz visualization of dimensional values of the 585 early changing genes in the TCC data set [3]. Three-dimensional version of Radviz utilizes a similar approach as exhibited in Radviz 2D. Primary mapping locates the winning output node on a two-dimensional surface (left). The record is then pulled into the third dimension based on all or a subset of dimensional values in a dataset. This technique alleviates heavy overplotting noticeable in a two-dimensional Radviz (left).

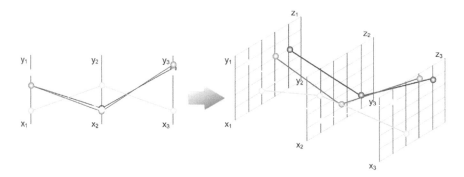

FIGURE 16.14 Classic parallel coordinates (left) and the corresponding three-dimensional self-organized parallel coordinates (right). Records are pulled into the third dimension based on dimensional values resulting in records exhibiting similar properties being plotted in proximity to each other.

to identify a likely source of technical bias, which may not be readily apparent in all experiments. The example in Figure 16.16 shows that the technical bias (in this case the fraction of "present" calls), is uncorrelated with the outcome (the genotype— SV40 or WT). The graphs are colored by record count, and we can interpret that most of the records do not exhibit extreme technical bias.

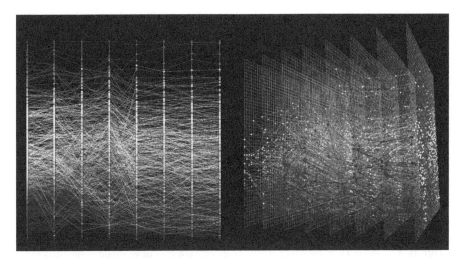

FIGURE 16.15 (SEE COLOR INSERT.) Two-dimensional parallel coordinates representation of gene expression data (left) and the corresponding three-dimensional self-organized parallel coordinates of dimensional values of the 585 early changing genes in the TCC data set [3] (right). Records are pulled into the third dimension based on dimensional values resulting with records exhibiting similar properties being plotted in proximity to each other.

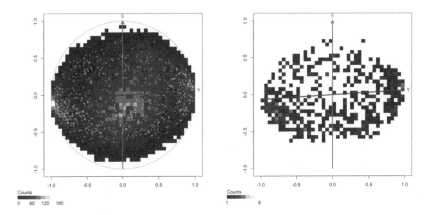

FIGURE 16.16 (SEE COLOR INSERT.) Biasograms of the 45,101 expression values with the 585 early changing genes (indicated by the circles and shown separately in the right image). We used the genotype as the outcome vector (Y) and the fraction of "present" calls as the source of technical bias (B).

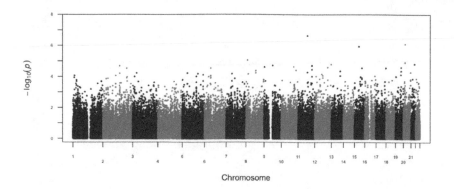

FIGURE 16.17 Basic Manhattan plot of all of the chromosomes of an experimental data set [5]. The x-axis shows the chromosome and the y-axis the negative logarithm of the association p-value for each SNP displayed.

16.3.7 MANHATTAN PLOT

In recent years, a lot of studies utilizing microarrays have been expanding to the system level. They include the use of a multitude of sources of data, and combine several patients and/or diseases. Genome-wide association study (GWAS) is an examination of many common genetic variants in different individuals with a goal to establish variants that are associated with the trait(s) [70]. Millions of genetic variants are examined from each individual sample of DNA read with the SNP arrays. GWAS typically focus on two groups of participants or organisms: individuals with a disease (samples) and similar individuals without (controls). This produces a large amount of data that needs to be analyzed and visualized in order to pursue the goal of identifying SNPs that are associated with the disease.

A Manhattan plot is a type of scatterplot that displays data with a large number of data points with a distribution of higher-magnitude values and many of non-zero amplitude (Figure 16.17), and is thus an insightful visualization for GWAS studies. In GWAS Manhattan plots the genomic coordinates are displayed along the x-axis, with the negative logarithm of the association p-value for each SNP displayed on the y-axis, color-coded by chromosome. Three tools that visualize Manhattan plots from GWAS data include the *manhattan* function [5] using the R project package *ggplot2* [71], the SNPEVG graphical tool [72], and Broad Institute's Integrative Genomics Viewer [73].

16.4 INTRINSIC DIMENSIONALITY OF VISUALIZATIONS

The goal of this section is to define metrics that identify how visualizations deal with high-dimensional data, such as data generated by the microarray experiments, when rendered on an output device. The main problems are that points may overlap and that dimensional data may be lost in the projection, which can lead to limited insight or in some cases misinterpretation of data.

16.4.1 INTRINSIC DIMENSION

Given an n-dimensional space, the intrinsic dimension of a visualization is defined to be the largest $k, k \leq n$, for which a set of k unit vectors can be uniquely perceived. For example, the intrinsic dimension of a 2D scatterplot is 2. This is because dimensional values mapped to x and y-coordinates can be uniquely perceived.

16.4.2 INTRINSIC RECORD RATIO

Given an n-dimensional space, the intrinsic record ratio of a visualization is defined to be $\frac{k}{n}$, where k is the largest value for which the set of 2^n binary vectors can be uniquely perceived in a visualization. In other words, the intrinsic record ratio represents the percentage of records that can be uniquely distinguished, if one had reasonably distributed records.

16.4.3 INTRINSIC COORDINATE DIMENSION

Given an n-dimensional space, the intrinsic coordinate dimension of a visualization is defined to be the largest $k, k \leq n$ for which k-dimensional values of any vector in the n-dimensional space can be uniquely identified in the visualization.

Therefore, the intrinsic dimension of a 2D scatterplot is 2 and its intrinsic coordinate dimension is 2. Additionally, the intrinsic dimension for a 3D scatterplot is 3 whereas its intrinsic coordinate dimension is 2. This is because the data is visualized on a 2D output device. Note that in many cases the intrinsic coordinate dimension is smaller than the intrinsic dimension. Using rigid transformations such as rotate, pan, and zoom, one can often increase the intrinsic coordinate dimension. However, this requires interaction with a visualization and it is heavily dependent on an analyst's skills.

There are a number of factors that can affect the outcome of a visual representation. Color, size, and shape of the points representing a data record will make a difference. Perception is dependent on the viewer as well as the environment. Screen resolution and size have a significant bearing on the evaluation of intrinsic dimensions since the metric involves perception of unique points or values.

16.5 CONCLUSION

The analysis of microarray and similar high-dimensional datasets requires advanced visualization techniques that are now more readily available in standard analysis tools of choice for most life scientists. While the above-described visualization techniques may help confirm and/or form new hypotheses, it is important to stress that other sources of data such as pathways, ontologies, and external and previously published data can augment primary microarray experiment data, making it more complex, but in many ways easier to comprehend. We should not neglect the domain expertise that can play a major role in microarray data analysis since only a domain expert may be able to determine whether identified patterns or lists of genes carry biological significance, and has the capacity to test them in a wet lab setting. In any case, microarrays proved to be an indispensable technology for studies related to the behavior of genes under various conditions. The techniques described above represent the basis for the visual analysis of microarray experiment data and are an important step of its general analysis. At the same time, they serve as a stepping stone for further utilization with advanced and novel visual techniques in the rapidly evolving field of biomedical visualization.

In the future, the field of biomedical visualization will continue to focus on the development of tools as well as integration of tools and data from various sources and modes under the umbrella known as "visual analytics" [74]. The principles of visual analytics will advance the field of biomedical visualization by integrating analytical reasoning, interaction, data transformation and representation into novel, yet to be developed tools [75]. Since the effective analysis of microarray and related data relies on collaboration of analysts and domain experts, these efforts will focus on the development of collaborative tools that will facilitate exploration of the data by entities that are potentially not co-located. This means that the tools of the future will make heavy use of hardware such as multi-core CPUs, cloud storage and computing resources and capable GPUs as well as networking and distributed databases, requiring well-defined mechanisms for integration of disparate tools and sources of data. Overall, the future of biomedical visualization is bright and will get even brighter as new technologies, requiring sophisticated tools, are introduced.

ACKNOWLEDGMENTS

We would like to thank Phillip C.S.R. Kilgore for his assistance and for proofreading this manuscript.

REFERENCES

1. R. Stone, II, A. L. Sabichi, J. Gill, et al., "Identification of genes correlated with early-stage bladder cancer progression," *Cancer Prevention Research*, vol. 3, pp. 776–786, 2010.

2. R. A. Irizarry, B. Hobbs, F. Collin, et al., "Exploration, normalization, and summaries of high density oligonucleotide array probe level data," *Biostatistics*, vol. 4, no. 2, pp. 249–264, 2003.

3. U. Cvek, M. Trutschl, R. Stone, II, Z. Syed, J. L. Clifford, and A. L. Sabichi, "Multidimensional visualization tools for analysis of expression data," in *International Conference on Bioinformatics and Computational Biology*, World Academy of Science, Engineering and Technology, 2009.

4. R Core Team, *R: A Language and Environment for Statistical Computing*. R Foundation for Statistical Computing, Vienna, Austria, 2013. ISBN 3-900051-07-0.

5. S. Turner, "Annotated Manhattan plots and QQ plots for GWAS using R, revisted." Last accessed: 08/01/2013. Online: `http://gettinggeneticsdone.blogspot.com/2011/04/annotated-manhattan-plots-and-qq-plots.html`.

6. P. Saraiya, C. North, and K. Duca, "An evaluation of microarray visualization tools for biological insight," in *Proceedings of the 10th IEEE Symposium on Information Visualization (InfoVis 2004)*, Austin, TX, USA, pp. 1–8, 2004.

7. J. Bertin, *Semiologie Graphique*. Berlin: Walter de Gruyter, Inc., 2 ed., 1973.

8. W. S. Cleveland, *Visualizing Data*. Summit, New Jersey: Hobart Press, 1993.

9. E. Tufte, *Envisioning Information*. Cheshire, CT: Graphics Press, 2 ed., 1990.

10. E. Tufte, *The Visual Display of Quantitative Information*. Cheshire: Graphics Press, 2 ed., 2001.

11. Z. T. Zhang, J. Pak, E. Shapiro, T. T. Sun, and X. Wu, "Urothelium-specific expression of an oncogene in transgenic mice induced the formation of carcinoma in situ and invasitve transitional cell carcinoma," *Cancer Research*, vol. 59, no. 14, pp. 3512–3517, 1999.

12. H. Levkowitz, *Color Theory and Modeling for Computer Graphics, Visualization, and Multimedia Applications*. Norwell, MA: Kluwer Academic Publishers, 1997.

13. T. R. Golub, D. K. Slonim, P. Tamayo, C. Huard, M. Gaasenbeek, J. P. Mesirov, H. Coller, M. L. Loh, J. R. Downing, M. A. Caligiuri, and C. D. Bloomfield, "Molecular classification of cancer: class discovery and class prediction by gene expression monitoring," *Science*, vol. 286, pp. 531–537, 1999.

14. A. Izzotti, G. A. Calin, V. E. Steele, et al., "Chemoprevention of cigarette smoke-induced alterations of microRNA in rat lungs," *Cancer Prevention Research*, vol. 3, no. 1, pp. 62–72, 2010.

15. D. F. Andrews, "Plots of high-dimensional data," *Biometrics*, vol. 29, pp. 125–136, 1972.

16. T. Hastie, R. Tibshirani, and R. Friedman, *The elements of statistical learning*, pp. 453–480. Springer-Verlag New York, Inc., first ed., 2001.

17. M. B. Eisen, P. T. Spellman, P. O. Brown, and D. Botstein, "Cluster analysis and display of genome-wide expression patterns," *Proceedings of the National Academy of Sciences*, vol. 95, no. 25, pp. 14863–14868, 1998.

18. W.-J. Fang, Y. Zheng, L.-M. Wu, et al., "Genome-wide analysis of aberrant DNA methylation for identification of potential biomarkers in colorectal cancer patients," *Asian Pacific Journal of Cancer Prevention*, vol. 13, pp. 1917–1921, 2012.

19. A. C. Culhane, M. S. Schroeder, R. Sultana, et al., "GeneSigDB: a manually curated database and resource for analysis of gene expression signatures," *Nucleic Acids Research*, vol. 40, pp. D1060–1066, January 2012.

20. N. Geblenborg, S. O'Donoghue, N. Baliga, et al., "Visualization of omics data for systems biology," *Nature Methods*, vol. 7, no. 3, pp. S56–S68.

21. M. S. Waterman and T. F. Smith, "On the similarity of dendrograms.," *Journal of Theoretical Biology*, vol. 73, pp. 789–800, Aug 1978.

22. N. Kim, H. Park, N. He, H. Y. Lee, and S. Yoon, "QCanvas: An advanced tool for data clustering and visualization of genomics data.," *Genomics Informatics*, vol. 10, pp. 263–265, Dec 2012.

23. TIBCO Software Inc., "http://spotfire.tibco.com/," 2013. Last accessed: 8/1/2013. Online: `http://spotfire.tibco.com/`.

24. http://www.merriam webster.com/dictionary/histogram, "histogram," 2013.

25. M. Marczyk, R. Jaksik, A. Polanski, and J. Polanska, "Adaptive filtering of microarray gene expression data based on Gaussian mixture decomposition," *BMC Bioinformatics*, vol. 14, no. 101, 2013.

26. O. G. Shaker, O. A. Hammam, and M. M. Wishahi, "Is there a correlation between HPV and urinary bladder carcinoma?," *Biomedicine and Pharmacotherapy*, vol. 67, no. 3, pp. 183–191, 2013.

27. M. E. Spear, *Charting Statistics*. New York, NY: McGraw-Hill Book Company, Inc., 1952.

28. C. Gillespie, G. Lei, R. Boys, A. Greenall, and D. Wilkinson, "Analysing time course microarray data using bioconductor: a case study using yeast2 affymetrix arrays," *BMC Research Notes*, vol. 3, no. 81, 2010.

29. R. C. Gentleman, V. J. Carey, D. M. Bates, et al., "Bioconductor: Open software development for computational biology and bioinformatics," *Genome Biology*, vol. 5, p. R80, 2004.

30. A. Inselberg, "The plane with parallel coordinates," *The Visual Computer*, vol. 1, pp. 69–91, 1985.

31. A. Inselberg and B. Dimsdale, "Parallel coordinates: a tool for visualizing multidimensional geometry," in *Proceedings of the First IEEE Conference on Visualization*, San Francisco, California, USA, pp. 361–378, 1990.

32. W. Peng, M. Ward, and E. Rundensteiner, "Clutter reduction in multi-dimensional data visualization using dimension reordering," in *Proceedings of the IEEE Symposium on Information Visualization (InfoVis 2004)*, Austin, TX, USA, pp. 89–96, 2004.

33. M. Ward, "XmdvTool: integrating multiple methods for visualizing multivariate data," in *Proceedings of the IEEE Conference on Visualization (InfoVis 1994)*, Austin, TX, USA, pp. 326–333, 1994.

34. J. Wang, W. Peng, M. Ward, and E. Rundensteiner, "Interactive hierarchical dimension ordering, spacing and filtering for exploration of high dimensional datasets," in *IEEE Symposium on Information Visualization (InfoVis 2003)*, Seattle, Washington, USA, pp. 105–112, 2003.

35. D. Ericson, J. Johansson, and M. Cooper, "Visual data analysis using tracked statistical measures within parallel coordinate representations," in *Proceedings of the Third International Conference on Coordinated and Multiple Views in Exploratory Visualization (CMV 2005)*, London, England, pp. 42–53, 2005.

36. K. Pearson, "On lines and planes of closest fit to systems of points in space," *Philosophical Magazine*, vol. 2, no. 6, pp. 559–572, 1901.

37. L. J. P. van der Maaten, E. O. Postma, and H. J. van den Herik, "Dimensionality Reduction: A Comparative Review," 2008.

38. D. Amaratunga and J. Cabrera, *Exploration and analysis of DNA microarray and protein array data*. Wiley Series in Probability and Statistics, Hoboken, NJ: John Wiley and Sons, Inc., 2003.

39. I. Borg and P. J. Groenen, *Modern multidimensional scaling*. No. XXII in Springer Series in Statistics, New York, NY: Springer, 2nd ed., 2005.

40. J. I. Risinger, J. Allard, U. Chandran, et al., "Gene expression analysis of early stage endometrial cancers reveals unique transcripts associated with grade and histology but not depth of invasion," *Frontiers in Oncology*, vol. 3, no. 139, 2013.

41. M. J. Boedigheimer, D. J. Freeman, P. Kiaei, M. A. Damore, and R. Radinsky, "Gene expression profiles can predict panitumumab monotherapy responsiveness in human tumor xenograft models," *Neoplasia*, vol. 15, no. 2, pp. 125–132, 2013.

42. S. Vaseghi and H. Jetelova, "Principal and independent component analysis in image processing," in *Proceedings of the 14th ACM International Conference on Mobile Computing and Networking*, Los Angeles, California, USA, pp. 1–5, 2006.

43. J. C. Roberts, "State of the art: Coordinated & multiple views in exploratory visualization," in *Proceedings of the Fifth International Conference on Coordinated and Multiple Views in Exploratory Visualization (CMV 2007)*, Washington, DC, USA, pp. 61–71, IEEE Computer Society, 2007.

44. M. Q. Wang Baldonado, A. Woodruff, and A. Kuchinsky, "Guidelines for using multiple views in information visualization," in *Proceedings of the Working Conference on Advanced Visual Interfaces (AVI 2000)*, New York, NY, USA, pp. 110–119, ACM, 2000.

45. C. North and B. Shneiderman, "Snap-together visualization: a user interface for coordinating visualizations via relational schemata," in *Proceedings of the Working Conference on Advanced Visual Interfaces (AVI 2000)*, New York, NY, USA, pp. 128–135, 2000.

46. T. Kohonen, "Neurocomputing: foundations of research," ch. Self-organized formation of topologically correct feature maps, pp. 509–521, Cambridge, MA: MIT Press, 1988.

47. T. Kohonen, "The self-organizing map," *Proceedings of the IEEE*, vol. 78, no. 9, pp. 1464–1480, 1990.

48. T. Kohonen, M. R. Schroeder, and T. S. Huang, eds., *Self-Organizing Maps*. Secaucus, NJ: Springer-Verlag New York, Inc., 3rd ed., 2001.

49. A. Ultsch and C. Vetter, "Self-organizing-feature-maps versus statistical clustering: A benchmark," Tech. Rep. 9, University of Marburg, Department of Mathematics, University of Marburg, Marburg, Germany, 1994.

50. T. Kohonen, ed., *Self-organizing maps*. Secaucus, NJ, USA: Springer-Verlag New York, Inc., 1997.

51. T. Kohonen, E. Oja, O. Simula, A. Visa, and J. Kangas, "Engineering applications of the self-organizing map," *Proceedings of the IEEE*, vol. 84, pp. 1358–1384, Oct. 1996.

52. J. Yan, *som: Self-Organizing Map*, 2010. R package version 0.3-5.

53. B. Schulz and A. Dominik, *somplot: Visualisation of hexagonal Kohonen maps*, 2013. R package version 1.6.4.

54. P. Hoffman, G. Grinstein, K. Marx, I. Grosse, and E. Stanley, "DNA visual and analytic data mining," in *Proceedings of the IEEE Conference on Information Visualization*, Phoenix, AZ, USA, pp. 437–442, 1997.

55. K. M. Daniels, G. G. Grinstein, A. Russell, and M. Glidden, "Properties of normalized radial visualizations," *Information Visualization*, vol. 11, no. 4, pp. 273–300, 2012.

56. L. Di Caro, V. Frias-Martinez, and E. Frias-Martinez, "Analyzing the role of dimension arrangement for data visualization in radviz," in *Proceedings of the 14th Pacific-Asia Conference on Advances in Knowledge Discovery and Data Mining - Volume Part II (PAKDD 2010)*, Berlin, Heidelberg, pp. 125–132, Springer-Verlag, 2010.

57. J. Sharko, G. Grinstein, and K. A. Marx, "Vectorized radviz and its application to multiple cluster datasets," *IEEE Transactions on Visualization and Computer Graphics*, vol. 14, pp. 1444–1427, Nov. 2008.

58. E. Bertini, L. Dell'Aquila, and G. Santucci, "Springview: cooperation of radviz and parallel coordinates for view optimization and clutter reduction," in *Proceedings of the Third International Conference on Coordinated and Multiple Views in Exploratory Visualization (CMV 2005)*, London, England, pp. 22–29, July.

59. G. Leban, I. Bratko, U. Petrovic, T. Curk, and B. Zupan, "Vizrank: finding informative data projections in functional genomics by machine learning," *Bioinformatics*, vol. 21, pp. 413–414, Feb. 2005.

60. P. Au, M. Carey, S. Sewraz, Y. Guo, and S. M. Rüger, "New paradigms in information visualization," in *Proceedings of the 23rd Annual International ACM SIGIR Conference on Research and Development in Information Retrieval*, Athens, Greece, pp. 307–309,

2000.

61. M. Trutschl, G. Grinstein, and U. Cvek, "Intelligently resolving point occlusion," in *Proceedings of the Ninth Annual IEEE Conference on Information Visualization (InfoVis 2003)*, Washington, DC, USA, pp. 131–136, IEEE Computer Society, 2003.

62. U. Cvek, M. Trutschl, J.C. Cannon, R.S. Scott, and R.E. Rhoads, "2d and 3d neural-network based visualization of high-dimensional biomedical data," in *IV '07: Proceedings of the 11th International Conference on Information Visualization*, Washington, DC, USA, pp. 545–550, IEEE Computer Society, 2007.

63. M. Trutschl, P. Kilgore, and U. Cvek, "High-performance visualization of multidimensional gene expression data," in *Proceedings of the Third International Conference on Networking and Computing (ICNC 2012)*, Hangzhou, China, pp. 76–84, 2012.

64. T. Kohonen, "Self-organized formation of topologically correct feature maps," *Biological Cybernetics*, vol. 43, pp. 59–69, 1982.

65. A. Inselberg and B. Dimsdale, "Parallel coordinates: a tool for visualizing multidimensional geometry," in *Proceedings of the 1st Conference on Visualization (VIS 1990)*, Los Alamitos, CA, USA, pp. 361–378, IEEE Computer Society Press, 1990.

66. E. Oja, "Simplified neuron model as a principal component analyzer," *Journal of Mathematical Biology*, vol. 15, pp. 267–273, 1982.

67. A. C. Eklund and Z. Szallasi, "Correction of technical bias in clinical microarray data improves concordance with known biological information," *Genome Biology*, vol. 9, no. 2, 2008.

68. M. Krzystanek, Z. Szallasi, and A. C. Eklund, "Bioasogram: Visualization of counfounding technical bias in gene expression data," *PLoS ONE*, vol. 8, no. 4, 2013.

69. M. Krzystanek, Z. Szallasi, and A. C. Elkund, "Biasogram," 2013. Last accessed: 8/22/2013. http://cbs.dtu.dk/biotools/biasogram/.

70. G. Gibson, "Hints of hidden heritability in GWAS," *Nature Genetics*, vol. 42, no. 7, pp. 558–560, 2010.

71. H. Wickham, *ggplot2: Elegant Graphics for Data Analysis*. New York: Springer, 2009.

72. S. Wang, D. Dvorkin, and Y. Da, "SNPEVG: a graphical tool for GWAS graphing with mouse clicks," *BMC Bioinformatics*, vol. 13, no. 319, pp. 1471–2105, 2012.

73. J. T. Robinson, H. Thorvaldsdóttir, W. Winckler, et al., "Integrative genomics viewer," *Nature Biotechnology*, vol. 29, no. 1, pp. 24–26, 2011.

74. J.J. Thomas and K.A. Cook, *Illuminating the Path: The Research and Development Agenda for Visual Analytics*. Los Alamitos, CA: National Visualization and Analytics Ctr, 2005.

75. J. Kielman, J. Thomas, and R. May, "Foundations and frontiers in visual analytics," *Information Visualization*, vol. 8, pp. 239–246, 2009.

17 Bioconductor Tools for Microarray Data Analysis

Simon Cockell, Matthew Bashton, and Colin S. Gillespie

CONTENTS

17.1 INTRODUCTION

17.1.1 WHAT IS BIOCONDUCTOR?

Bioconductor [1] is an open source software project for the R statistical computing language. The main aim of the project is to provide R packages to facilitate the analysis of DNA microarray, sequencing, SNP and other genomic data. In addition

459

to various packages for analysis of data, they also distribute meta-data packages that provide useful annotation. The stable non-development version of Bioconductor is normally released biannually.[1] In addition to providing packages the Bioconductor site (bioconductor.org) also provides documentation for each package, often a brief vignette and a more comprehensive user guide.

17.1.2 THE MICROARRAY ANALYSIS WORKFLOW

There are a number of microarray platforms for studying gene expression. The basic workflow used for deriving a list of differentially expressed genes follows the same basic plan, irrespective of the platform. When using Bioconductor, these steps are usually:

1. Data reading. Data is read from text or binary format files into R. Raw data in R is stored as an `ExpressionSet`.
2. Normalization. This process makes groups of arrays comparable to one another.
3. Quality control. This step ensures the data are of sufficient quality to enable effective analysis. This usually takes two forms: i) sample-level QC, where outlier samples and poor quality arrays are discarded; and ii) probe-level QC, where poor quality probes and non-expressed probes are removed from consideration.
4. Differential expression analysis. Groups of arrays are compared, and genes that vary significantly between them are identified.
5. Visualization. Making graphs and other visual representations of the data on a microarray can be helpful when identifying differentially expressed genes, or groups of interesting genes.
6. Functional analysis. Once a gene list has been generated, functional analysis provides useful insight into the biological processes underlying the changes observed in the microarray experiment.

Bioconductor contains packages that perform each of these steps for each of the main array platforms, with many steps; the same package will work with array data from any platform (see Table 17.1).

We have chosen to focus on Illumina BeadArrays™ in this chapter as the central example for analysis as from our experience they are currently the most popular platform in usage by molecular biologists. This is mainly due to their cost advantage over comparable arrays from Affymetrix. In addition, by having multiple copies of each probe present on many different beads, Illumina BeadArrays™ provides within-array technical replicates that variant stabilization transform (VST) can take advantage of to improve differential expression reporting and reduce false positives [2]. Also, there are known issues with Affymetrix GeneChips, which cause systematically incorrect values of intensity to be reported for particular probes and probe sets, which do not reflect the abundance of the intended target cRNA molecules [3, 4].

---

[1]http://www.bioconductor.org/news/bioc_2_11_release/

17.1.3 ILLUMINA MICROARRAYS

The Illumina BeadArray[TM] microarray platform makes use of BeadChips, which, as their name suggests, are composed of tiny beads. These beads are made of silica and are just 3μm in diameter, and are covered in hundreds of thousands of seventy-nine-nucleotide-long oligonucleotide sequences; individual beads each have a different probe sequence. A unique property of BeadChips, which distinguishes them from Affymetrix arrays, is that every array is unique, and the beads self assemble into microwells found on the slide. As a consequence of this, most probes will be present on the array $\sim$ 30 times [17]. An additional consequence is that each array has to be decoded during its manufacture so that the positions of each probe are known; the first twenty-nine nucleotides of the probe sequence are reserved for decoding the array during manufacture. An Illumina BeadChip can support the analysis of multiple samples having up to 12 copies of each array on a chip. The HumanHT-12 v4.0 Expression BeadChip has over 47,000 probes, which are based on RefSeq 38 and Unigene.[2] Being randomly assembled means that Illumina arrays do not suffer so much from spatially localized artifacts. Additionally, the presence of multiple copies of each bead can be used as technical replicate information for variance stabilizing transformation (VST) [18, 19].

17.1.4 LUMI

The lumi package [5], available in Bioconductor,[3] is specifically designed to process BeadArrray data providing VST and quality control steps specifically tuned for this technology. lumi assumes that data has been pre-background corrected in Illumina's BeadStudio or GenomeStudio, although, if it has not, they provide the lumiB function that mimics the background correction found therein. Should the user want to investigate processing the raw image files from the scanner directly, in order to have control over the background correction and image processing steps, then the Bioconductor package beadarray needs to be used. This, however, depends on the BeadScan Illumina microarray scanner software being set to output raw TIFF files, which is not a default setting, at the time the arrays are processed. Consequently we are focusing on analysis downstream of BeadStudio or GenomeStudio. Two key features of lumi that give it an advantage over other methods, which are largely based on Affymetrix data processing methods, are VST, which takes advantage of multiple beads per probe when transforming the data and outperforms $\log_2$ based transformations [18], and Robust Spline Normalization (RSN), which is designed for Illumina data and combines the advantageous features of both quantile normalization (fast, gene rank order preserving) and loess normalization, which is continuous.[4]

---

[2]http://www.illumina.com/Documents/products/datasheets/datasheet_gene_exp_analysis.pdf

[3]http://www.bioconductor.org/packages/release/bioc/html/lumi.html

[4]http://www.bioconductor.org/packages/2.11/bioc/vignettes/lumi/inst/doc/lumi.pdf

TABLE 17.1

Suggested Bioconductor packages for dealing with particular steps of the analysis workflow in different array platforms.

Workflow step	Array platform			
	Illumina	Affymetrix	Agilent	NimbleGen*
1. Data reading	lumi [5], beadarray (for RAW TIFFs and low level pre-processing)	affy [6], simpleaffy [7]	marray [8], agilp, limma	oligo [9], Ringo [10]
2. Normalization	lumi	affy, simpleaffy, gcrma (simpleaffy or gcrma are recommended for GCRMA normalization over RMA offered in affy)	marray, agilp, limma	oligo, Ringo
3. Quality Control	lumi, beadarray [11], affycoretools, arrayQualityMetrics [12]	affy, simpleaffy, affycoretools, arrayQualityMetrics	marray, limma	Ringo
4. Differential expression analysis	limma [13]	limma, simpleaffy	limma	limma
5. Visualization	gplots [14]	gplots	gplots	gplots
6. Functional Analysis	GOstats [15], RamiGO [16]	GOstats, RamiGO	GOstats, RamiGO	GOstats, RamiGO

Note: *nimbleGen^TM arrays were discontinued by Roche as of November 7, 2012.*

17.1.5   LIMMA

While `lumi` handles the transformation, normalization, and quality control of Illumina microarray data, in order to obtain a list of differentially expressed genes, the user needs to make use of the `limma` package. This has advantages over standard *t*-tests since it makes use of linear models and empirical Bayes methods in order to determine significant differentially expressed genes [13, 20]. The starting point for analysis in `limma` is a matrix of expression values from `lumi`, which is accessed via the `exprs` method provided by the core `Biobase` package of Bioconductor.

17.2   IMPORT AND PREPROCESSING

17.2.1   INSTALLING BIOCONDUCTOR PACKAGES

Provided that R is already installed, installing packages from the Bioconductor repository is straightforward. First, we source the installation script:

```
source("http://bioconductor.org/biocLite.R")
```

and then run the downloaded function, `biocLite`, to install the standard packages:

```
biocLite()
```

This will install the `Biobase`, `IRanges`, and `AnnotationDbi` packages (and their dependencies). Additional packages can be installed by directly specifying their name:

```
## From Bioconductor
biocLite(c("ArrayExpress", "arrayQualityMetrics", "GOstats",
           "GEOquery", "lumi", "lumiHumanAll.db",
           "lumiHumanIDMapping", "RamiGO"))
## From CRAN
install.packages("gplots")
```

Table 17.2 gives an overview of the packages used in this chapter. A text file of the R commands used can be found at

```
https://github.com/csgillespie/illumina-analysis/
```

17.2.2   LOADING AND NORMALIZING THE DATA

The data for this chapter can be downloaded (within R) directly from the ArrayExpress Web site. We simply load the `ArrayExpress` and `lumi` packages

TABLE 17.2

Packages (and versions) used in this chapter.

Package	Version	Package	Version
ArrayExpress	1.20.0	arrayQualityMetrics	3.16.0
Biobase	2.20.1	GEOquery	2.26.2
GOstats	2.26.0	gplots	2.11.3
limma	3.16.6	lumi	2.12.0
lumiHumanAll.db	1.20.0	lumiHumanIDMapping	1.10.0
RamiGO	1.6.0		

```
library("ArrayExpress")
library("lumi")
```

and use the getAE function to download the data file. Since the data is Illumina BeadArray, the standard ArrayExpress method will make some false assumptions, so we use getAE instead:

```
ae = getAE("E-MTAB-1593", type = "full")
raw_data = lumiR(ae)
```

The raw_data object contains all the information associated with this microarray experiment; for example, the raw data, information on the protocol, and details of the MIAME metadata. It is an S4 R LumiBatch object. This object is specifically used to contain and describe Illumina data within R. It extends ExpressionSet, one of the key Bioconductor classes. To investigate the LumiBatch object (or any R object), we can use the str command to obtain a detailed overview of its structure:

```
str(raw_data)
```

Alternatively, we can obtain a short summary using print(raw_data).

It is also desirable to use the plot() method to investigate some of the properties of the individual arrays. This quality control measure allows any obviously aberrant arrays to be detected and removed. There are a range of possible plots, including density plots of intensity (line graphs or boxplots), plots of pairwise correlation, and a plot of the coefficient of variance.

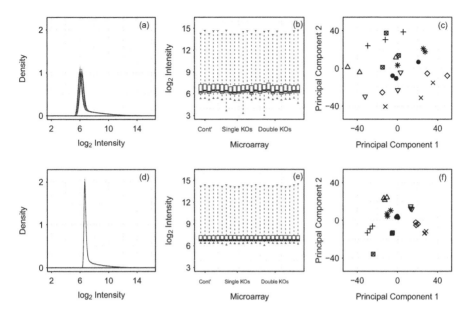

FIGURE 17.1 Top row: raw data. Bottom row: normalized. (a,d) Density plots. (b,e) Box-plots. (c,f) PCA plots.

```
## Substitute "density" for one of the other plot types:
## "boxplot", "pair", "MA" or "cv"
plot(raw_data, what="density")
```

Plots of the raw and normalized data are given in Figure 17.1. As the density plots show, the intensity across the microarrays in our experiment can be quite varied (Figure 17.1a-c). We can make a valid assumption here that this variance is experimental, and not biological. The result of this large amount of experimental variability is that it masks the biological variance in which we are interested. We therefore have to treat the data obtained from the arrays in such a way that this systematic variability is masked, while the important and interesting biological variability is maintained. Normalization is intended to achieve this purpose by ensuring all the samples in an experiment follow the same underlying statistical distribution, but the variances within observations of a particular probe should remain, and therefore be discoverable.

There is no *single* method for normalizing microarray data sets. In this chapter, we will use one of the standard methods. Normalization is carried out over two steps. First, a transformation is applied to stabilize the variance across probes [18]:

```
vst_data = lumiT(raw_data)
```

The variance stabilization transformation exploits the within-array technical replicates (i.e., the bead-level data) generated from Illumina microarrays to model the relationship between the mean and the variance. Using VST means the differential expression of more genes can be detected at the same time as reducing false positives [18]. Other (less sophisticated) methods are a $\log_2$ and cubic root transformation.

The second step is to normalize between chips

```
## Use robust spline normalization (rsn)
rsn_data = lumiN(vst_data, method = "rsn")
```

This normalization step uses the rsn method and forces the intensity values for different samples (microarrays) to have the same distribution [5].

When the raw data is loaded, the lumiQ function is automatically called to provide a list of summary data about the arrays that are stored in the QC slot of the LumiBatch object. Following normalization this function needs to be called again to update the data in the QC slot to reflect changes in the underlying data, viz.

```
qc_data = lumiQ(rsn_data)
```

Information held in the QC slot includes the mean, standard deviation, detectable probe ratios, and sample correlation data, in addition to control probe information for each array. The subsequent S4 plot methods used to render QC plots are also dependent on the data in the QC slot of the LumiBatch object.

The raw and transformed data can be easily compared via plotting:

```
plot(qc_data)
plot(raw_data)
```

17.2.3   QUALITY CONTROL

The arrayQualityMetrics package is a generic set of quality control routines that can be applied to many types of array data. It produces a report of quality metrics, which can be used to make assessments about overall array quality, and diagnose batch effects. To generate the report, we load the package

```
library("arrayQualityMetrics")
```

then run the arrayQualityMetrics function

```
arrayQualityMetrics(expressionset=qc_data, outdir="qc")
```

The user can view the output from the arrayQualityMetrics function at

https:
//github.com/csgillespie/illumina-analysis/blob/master/qc/

To render the page in a Web browser, we use the following link

http://goo.gl/bksFR

Examining the *distance between arrays* metric[5] and the principal component plot from the report (the latter figure is given in Figure 17.1) strongly suggests that array 23 (TF3TF4B) is an outlier—thus, we will remove this array from any further analysis. We have to do this removal in the raw data, since the removal of any whole sample will affect our between-array normalization, so this has to be repeated with the subset of 23 arrays that passed quality control:

```
raw_data_post = raw_data[, -23]
vst_data_post = lumiT(raw_data_post)
rsn_data_post = lumiN(vst_data_post, method = "rsn")
qc_data_post = lumiQ(rsn_data_post)
```

The *array intensity distributions*[6] (not shown) metric suggests that array 11 may also be an outlier, although in this case its variation from the other arrays is less convincing, and hence we will retain the array in the analysis.

Now that we can be confident the arrays we have retained are of sufficient quality, we want to ensure the probes we are analyzing on those arrays also pass a stringent quality check. We begin by extracting the data matrix from the LumiBatch object:

```
exprs_data = exprs(qc_data_post)
treatments = c("Ctrl", "TF1", "TF2", "TF3", "TF4",
               "TF1TF4", "TF2TF4", "TF3TF4")
array_names = rep(treatments, each=3)[1:23]
colnames(exprs_data) = array_names
```

Then we use the detectionCall to find the probes that are below a detection threshold.

```
present_count = detectionCall(raw_data_post)
select_data = exprs_data[present_count > 0, ]
```

This method exploits the detection *p*-value found in the raw data to determine whether or not a probe is detected above a threshold level in each of the samples

---

[5]https://raw.github.com/csgillespie/illumina-analysis/master/qc/outhm.png

[6]https://github.com/csgillespie/illumina-analysis/raw/master/qc/outbox.pdf

of our experiment. If the detection *p*-value is less than 0.01 (by default, this can be changed by passing the Th= parameter to `detectionCall`), then the probe is found to be detected. The filter that is applied at this stage removes a probe from all the samples if it is not detected on any of the 23 arrays. A probe that is detected on at least one array is retained.

Overall, this procedure has removed approximately 50% of probes:

```
nrow(select_data)/nrow(exprs_data)
```

```
## [1] 0.4709
```

17.3   DIFFERENTIALLY EXPRESSED GENES

17.3.1   ARRAY ANNOTATION

Oligonucleotide identifiers provided by array manufacturers are technology specific, proprietary, and mutable. It was felt that an external identifier would be beneficial to solve the problem of identifier permanency, and enable cross-platform data integration. nuIDs are one implementation of just such a strategy. Each nuID is a string of letters and numbers that encode, via a variation of Base64, a lossless compression of the probes sequence in addition to a checksum [21]. The probe names on our Illumina arrays can be mapped to nuIDs, and these can then be used to stably map the probes to other identifiers (in this case, gene symbols and gene names).

This annotation is useful for much of the downstream analysis, providing human readable identifiers for genes that can be included in the output of the differential gene detection to come.

```
library("lumiHumanAll.db")
library("annotate")
probe_list = rownames(select_data)
nuIDs = probeID2nuID(probe_list)[, "nuID"]
symbol = getSYMBOL(nuIDs, "lumiHumanAll.db")
name = unlist(lookUp(nuIDs, "lumiHumanAll.db", "GENENAME"))
```

To avoid mixing up the order of the IDs and symbols, we combine the vectors into a single data frame

```
anno_df = data.frame(ID = nuIDs, probe_list, symbol, name)
```

17.3.2   USING LIMMA TO DETECT DIFFERENTIALLY EXPRESSED GENES

The `limma` package has become the de-facto way of analyzing microarrays for differentially expressed genes. By using an empirical Bayes method to moderate the variance of observations across the microarrays, limma tends to give more robust gene-lists than those generated by standard Student's *t*-tests. We set our samples up by defining the design of our experiment. The in-built R method model.matrix can be used to do this; the resulting design matrix tells limma which samples belong in which groups. The lmFit method then fits a linear model for each gene in the group of arrays:

```
library("limma")
design = model.matrix(~0 +
                      factor(array_names, levels=treatments))
colnames(design) = treatments
num_parameters = ncol(design)
fit = lmFit(select_data, design)
```

Now we can set up the contrasts, or comparisons, that we want to analyze. In the case of this experiment we have eight sample groups: a control, four individual siRNA knock-downs and three combination knock-downs. We want to analyze the effect of each transcription factor knock-down separately and also detect the presence of any interaction effects. To detect main effect differences, we would have a contrast of the form

$$TF1 - Ctrl$$

To investigate interactions, the contrast would be slightly more complicated:

$$(TF1TF4 - Ctrl) - (TF1 - Ctrl) - (TF4 - Ctrl) = TF1TF4 - TF1 - TF4 + Ctrl$$

Using the makeContrasts method provided by limma, we set up the following contrast matrix:

```
cont_mat = makeContrasts(TF1-Ctrl, TF2-Ctrl, TF3-Ctrl, TF4-Ctrl,
               TF1TF4-TF1-TF4+Ctrl, TF2TF4-TF2-TF4+Ctrl,
               TF3TF4-TF3-TF4+Ctrl, levels=treatments)
fit2  = contrasts.fit(fit, contrasts=cont_mat)
```

We then fit these contrasts to the limma model of the data, and apply the empirical Bayes statistics to derive moderated *t*-statistics of differential expression for all the probes on the arrays:

```
fit2 = eBayes(fit2)
fit2$genes = anno_df
```

To enable other functions, notably topTable, to access the gene names we append the anno_df data frame to the fit2 list.

TABLE 17.3

Top five differentially expression genes for the contrast `TF1 - Ctrl.`

Rank	Gene symbol	Mean expression	$\log_2$ fold change	Adjusted p-value
1	NDRG1	9.51	-1.64	5.5×10^{-16}
2	SCARNA11	7.49	1.73	1.1×10^{-14}
3	ZDHHC7	10.53	-1.21	1.1×10^{-13}
4	PAICS	11.10	-1.25	2.3×10^{-13}
5	C1orf85	10.55	-1.59	2.5×10^{-13}

The final stage of the `limma` analysis is to apply `topTable` in order to produce a list of differentially expressed genes. `topTable` takes a number of arguments that allow for precise filtering of the gene list by a number of criteria, including applying a cut-off to the $\log_2$ fold change, and the p-value of the moderated t-tests performed (see Table 17.3). `topTable` also allows us to apply a multiple testing correction to the test statistics produced. This is important because utilizing "raw" p-values in over 20,000 independent statistical tests will lead to an unreasonably high error rate when predicting differentially expressed genes (by definition, a rate of 1 in 20, or around 1,000 false positives per contrast for our arrays). Applying a multiple testing correction allows this error rate to be controlled for, and a lower false positive rate to be guaranteed.

```
## Filter by fold change (1.5x) and p-value (0.05) cutoffs
## Adjusted using Benjimini-Hochberg false discovery rate
topTable(fit2, coef="TF1 - Ctrl", p.value=0.05, lfc=log2(1.5))
```

17.4 VISUALLY INVESTIGATING DIFFERENCES

When dealing with so many data points, it can be useful to obtain a visual overview to quickly identify particularly interesting genes. In this section, we consider three standard graphical methods.

17.4.1 VOLCANO PLOTS

When looking for interesting genes it can be helpful to restrict attention to differentially expressed genes that are both statistically significant and of potential biological interest. This objective can be achieved by considering only significant genes that show, say, at least a two-fold change in their expression level. A volcano plot can be used to summarize the fold change and p-value information produced by `limma`; the

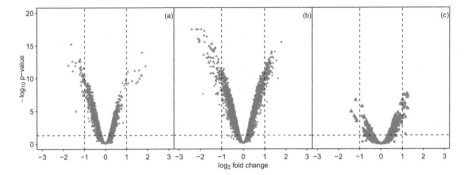

FIGURE 17.2 Volcano plots. The triangles indicate genes whose $\log_2$ absolute fold change is greater than one and has an adjust p-value greater than 0.05. The figures above are volcano plots for the contrasts (a) `TF1 - Ctrl` (b) `TF4 - Ctrl` (c) `TF1TF4 - TF1 - TF4 + Ctrl`.

log fold change is plotted on the x-axis and the negative $\log_{10} p$-value is plotted on the y-axis.

To construct the plot, the user should extract all genes, with associated p-value and fold change for a particular contrast:

```
gene_list = topTable(fit2, coef="TF1 - Ctrl",
                              number=nrow(anno_df))
```

The volcano plot is created using the standard plotting function

```
plot(gene_list$logFC, -log10(gene_list$adj.P.Val),
   col=1+(abs(gene_list$logFC) > 1 & gene_list$adj.P.Val < 0.05))
```

In this plot, we use the `col` argument to color points where the adjusted p-value was less than 0.05 and the absolute log fold change was above one. Figure 17.2 shows volcano plots for three particular contrasts.

17.4.2 VENN DIAGRAMS

Typically, when comparing interesting genes over different contrasts, we often are interested in the gene overlap. The function `classifyTestsF` from the `limma` package classifies each gene as being up, down, or not significant. In this example, we choose a p-value cut-off of 0.01

```
## p-values are adjusted for multiple testing
results = classifyTestsP(fit2, p.value = 0.01, method = "fdr")
```

When classifying the tests, we control for multiple testing using the FDR correction [22]. These results can then be visualized using a Venn diagram

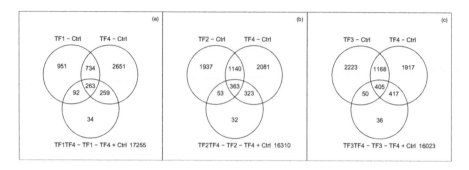

FIGURE 17.3　Venn diagrams showing the number of overlapping genes between contrasts.

```
vennDiagram(results[, c("TF1 - Ctrl", "TF4 - Ctrl",
                        "TF1TF4 - TF1 - TF4 + Ctrl")])
```

This command generates Figure 17.3a.

17.4.3　HEATMAPS

Lists of genes that change expression are of limited value taken on their own. What is needed to provide more value to these lists is context. We can look at how the genes that change in one of our contrasts is affected in all the others by constructing a heatmap of the fold changes found in each of the contrasts. This visual tool allows us to quickly identify genes that are similarly regulated across conditions, or those that have opposing regulation. Due to the probe-level clustering that is applied during heatmap construction, we can also spot groups of genes that are regulated in similar ways across the constrasts. Genes that share regulation in this way are often functionally related.

In order to construct this plot, we first need the full results of the four contrasts we are plotting.

```
tf1_table = topTable(fit2, coef="TF1 - Ctrl",
                     n=length(probe_list), sort.by="logFC")
tf2_table = topTable(fit2, coef="TF2 - Ctrl",
                     n=length(probe_list), sort.by="logFC")
tf3_table = topTable(fit2, coef="TF3 - Ctrl",
                     n=length(probe_list), sort.by="logFC")
tf4_table = topTable(fit2, coef="TF4 - Ctrl",
                     n=length(probe_list), sort.by="logFC")
```

Now we construct a list of the top ten genes from each contrast (sorted by fold change). From each gene table we take the top ten genes, then select the unique probes, since the same probe may occur in more than one list

```
all_signames = unique(c(rownames(tf1_table[1:10,]),
                        rownames(tf2_table[1:10,]),
                        rownames(tf3_table[1:10,]),
                        rownames(tf4_table[1:10,])))
```

Now that we have a list of genes, we extract these genes from each data frame

```
full = data.frame(tf1_table[all_signames,]$logFC,
                  tf2_table[all_signames,]$logFC,
                  tf3_table[all_signames,]$logFC,
                  tf4_table[all_signames,]$logFC,
                  row.names=all_signames)
colnames(full) = paste0("TF", 1:4)
```

Another couple of manipulations are necessary, to provide information necessary for the rendering of the heatmap. We need to know the smallest and largest fold change values, so the extent of the range of values represented can be passed to the heatmap function, and we can keep zero (i.e., no change) in the center of the color spectrum. We also filter out unannotated genes at this stage, since they are of limited interest.

```
row_names = as.character(tf1_table[rownames(full), ]$symbol)
## Filter out unannotated genes
hm_data = full[!is.na(row_names), ]
row_names = row_names[!is.na(row_names)]
```

Finally, the heatmap itself is plotted (Figure 17.4). The heatmap.2 function from the gplots package is responsible for a number of data manipulations while drawing the heatmap. It rearranges the data rows (i.e., the probes) based on the results of a clustering step (using a Euclidean distance matrix, by default), and it colors the cells of the heatmap based on the fold change value for that probe in the contrast that the column represents. The breaks argument defines the scope of the color gradient, ensuring the the most "blue" cell is the one with the biggest negative fold change, that the most "red" cell is the one with the largest positive fold change, and that white (the center of the color gradient) is set at zero.

```
library("gplots")
breaks = c(seq(min(full), 0, length.out=128),
           seq(0, max(full), length.out=128))
heatmap.2(as.matrix(hm_data), dendrogram="row", Colv=FALSE,
          col=bluered(255), key=TRUE, labRow=row_names,
          breaks=breaks, symkey=FALSE, density.info="none",
          trace="none", cexRow=0.5, cexCol=0.75)
```

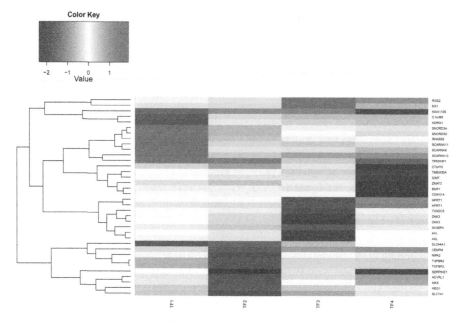

FIGURE 17.4 (SEE COLOR INSERT.) Clustering of the top ten differentially expressed genes from each main contrast. Red and blue correspond to up- and down-regulation, respectively.

The resulting heatmap is shown in Figure 17.4. It is worth noting that many heatmaps are constructed using a red-green color scheme. However, since 5%–10% of the male population are red-green color blind, this is a particularly poor choice of colors.

17.5 FUNCTIONAL INTERPRETATION OF DIFFERENTIALLY EXPRESSED GENES

The functional relationship of the genes in our lists can be studied more directly by using the GOstats package [23] to find Gene Ontology [24] terms that are statistically over-represented among the genes in the lists. To begin, we load the package

```
library("GOstats")
```

and then select the "top genes"

```
sig_values = tf1_table[tf1_table$adj.P.Val < 0.05 &
                        abs(tf1_table$logFC) > log2(1.5), ]
sig_probes = as.character(sig_values$probe_list)
```

Then we map the vector of significant probes to Entrez gene identifiers (as these are the identifiers GOstats uses to find GO terms)

```
## Map probe id to Entrez gene identifiers
entrez = unique(unlist(lookUp(nuIDs[sig_probes],
                              "lumiHumanAll.db", "ENTREZID")))
entrez = as.character(entrez[!is.na(entrez)])
```

We also need to define a "gene universe," that is, the global set of terms among which we want to look for over-representation. If a particular GO term is not present in the genes represented on our microarray, then there is no point in testing to see if it is over-represented. In this case, we construct our universe from the probes on the arrays that passed our probe-level quality control step 17.2.3.

```
## Determine the universe of possible entrez ids
entrez_universe = unique(unlist(
            lookUp(nuIDs, "lumiHumanAll.db", "ENTREZID")))
entrez_universe = as.character(
            entrez_universe[!is.na(entrez_universe)])
```

To test whether a particular group is over-represented, we use a hyper-geometric test; this is provided by the GOstats package. We create an S4 object that contains all the necessary information

```
params = new("GOHyperGParams",
    geneIds=entrez,
    universeGeneIds=entrez_universe,
    annotation="lumiHumanAll.db",
    ontology="BP",
    pvalueCutoff= 0.01,
    conditional=FALSE,
    testDirection="over")
```

The params object is then passed to the hyperGTest function

```
hyperg_result = hyperGTest(params)
```

which returns a GOHyperGResult object. Printing this object provides a summary of the results

```
print(hyperg_result)
```

```
## Gene to GO BP  test for over-representation
## 3124 GO BP ids tested (213 have p < 0.01)
## Selected gene set size: 216
##      Gene universe size: 9765
##      Annotation package: lumiHumanAll
```

TABLE 17.4

Top five gene ontology identifiers with a brief description.

GO Identifier	Description	Occurrences (Universe size)	p-value
GO:0022403	Cell cycle phase	62 (711)	1.9×10^{-18}
GO:0022402	Cell cycle process	68 (896)	1.4×10^{-17}
GO:0000278	Mitotic cell cycle	57 (667)	1.2×10^{-16}
GO:0007049	Cell cycle	76 (1176)	2.6×10^{-16}
GO:0051301	Cell division	38 (402)	9.8×10^{-12}

Since we have performed multiple tests, we use the adjusted p-values

```
## Adjust the p-values for multiple testing (FDR)
go_fdr = p.adjust(pvalues(hyperg_result), method="fdr")
```

We can then select the significant GO terms

```
## Select the Go terms with adjusted p-value less than 0.01
sig_go_id = names(go_fdr[go_fdr < 0.01])
## Retrieve significant GO terms for BP (Biological process)
sig_go_term = getGOTerm(sig_go_id)[["BP"]]
```

The top five gene ontology identifiers are given in Table 17.4.

Once we have a list of over-represented terms, we can use the RamiGO package to visualize them. RamiGO takes a list of terms and constructs a graph from them, and their parent terms.

```
## Visualize the enriched GO categories
library("RamiGO")
amigo_tree = getAmigoTree(sig_go_id)
```

The resulting graph can be examined for clusters of related enriched terms, which may be more important for the context of the experiment than isolated over-represented terms would be. The graph can be found on this chapter's Web site.

As we can see from the top GO terms for this contrast in Table 17.4 (and the biological process graph produced by RamiGO), knocking down the expression of TF1 has had a demonstrable effect on a number of processes related to the progression of the cell cycle. We may simply hypothesize, therefore, that TF1 is an important regulator of these processes. Similar analyses with the other contrasts we have performed allow us to draw similar conclusions as to the function of the other knocked-down transcription factors. The top GO terms for each of these other transcription factor knockdowns also contain a lot of terms indicating involvement of those transcription factors in the cell cycle, suggesting that all of the transcription factors being studied are involved with this important process.

GOstats can also be used to investigate statistical enrichment of other functional classifications, for example, the other categories of the Gene Ontology (Molecular Function and Cellular Component) or pathway data, taken from the Kyoto Encyclopaedia of Genes and Genomes (KEGG). It is therefore possible to build up a very rich picture of the functions of our lists of genes in this way.

17.6   SUMMARY

Bioconductor provides us with all the tools we need to go from raw data in a public repository to deriving novel conclusions about the function of groups of genes, via data quality control and detection of differentially expressed genes. We have only scratched the surface of the 671 packages available (at time of writing; see http://bioconductor.org for an up-to-date list). Import and processing of Affymetrix microarrays is provided by the affy and simpleaffy packages [25, 26, 27]. Other methods of differential expression analysis are available, in packages such as RankProd [28]. Batch effects, systematic errors that can cause many problems with microarray analysis, can be dealt with using the methods available in the sva package [29], especially ComBat [30]. Bioconductor also does not just provide packages for analyzing expression microarrays, but there is also extensive support for different types of arrays, such as methylation arrays, arrayCGH, SNP arrays, among others. Beyond microarrays, there are also many packages for the analysis of proteomics data and latterly next-generation sequencing analysis. Coupled with generic packages for inferring biological meaning, and visualizing large and complex datasets, Bioconductor is increasingly a one-stop solution for many bioinformatics analyses.

ACKNOWLEDGMENTS

We would like to thank Neil Perkins (Newcastle University) for the kind permission to use his microarray data for this chapter.

ADDITIONAL INFORMATION

The R code used in this book chapter can be downloaded from the book chapter's Web page:

https://github.com/csgillespie/illumina-analysis/

The chapter was constructed using R version 3.0.1, Bioconductor version 2.12, and the knitr package [31].

REFERENCES

1. R. C. Gentleman, V. J. Carey, D. M. Bates, et al., "Bioconductor: open software development for computational biology and bioinformatics," *Genome Biology*, vol. 5, no. 10, p. R80, 2004.

2. S. M. Lin, P. Du, W. Huber, and W. A. Kibbe, "Model-based variance-stabilizing transformation for Illumina microarray data," *Nucleic Acids Research*, vol. 36, no. 2, p. e11, 2008.

3. H. P. Shanahan, F. N. Memon, G. J. G. Upton, and A. P. Harrison, "Normalized Affymetrix expression data are biased by G-quadruplex formation," *Nucleic Acids Research*, vol. 40, no. 8, pp. 3307–3315, 2012.

4. G. J. G. Upton and A. P. Harrison, "Motif effects in Affymetrix GeneChips seriously affect probe intensities," *Nucleic Acids Research*, vol. 40, no. 19, pp. 9705–9716, 2012.

5. P. Du, W. A. Kibbe, and S. M. Lin, "lumi: a pipeline for processing Illumina microarray," *Bioinformatics*, vol. 24, no. 13, pp. 1547–1548, 2008.

6. L. Gautier, L. Cope, B. M. Bolstad, and R. A. Irizarry, "affy – analysis of Affymetrix GeneChip data at the probe level," *Bioinformatics*, vol. 20, no. 3, pp. 307–315, 2004.

7. C. L. Wilson and C. J. Miller, "Simpleaffy: a BioConductor package for Affymetrix quality control and data analysis," *Bioinformatics*, vol. 21, no. 18, pp. 3683–3685, 2005.

8. J. Wang, V. Nygaard, B. Smith-Sørensen, E. Hovig, and O. Myklebost, "MArray: analysing single, replicated or reversed microarray experiments," *Bioinformatics*, vol. 18, no. 8, pp. 1139–1140, 2002.

9. B. S. Carvalho and R. A. Irizarry, "A framework for oligonucleotide microarray preprocessing," *Bioinformatics*, vol. 26, no. 19, pp. 2363–2367, 2010.

10. J. Toedling, O. Skylar, O. Sklyar, et al., "Ringo – an R/Bioconductor package for analyzing ChIP-chip readouts," *BMC Bioinformatics*, vol. 8, p. 221, 2007.

11. M. J. Dunning, M. L. Smith, M. E. Ritchie, and S. Tavaré, "beadarray: R classes and methods for Illumina bead-based data," *Bioinformatics*, vol. 23, no. 16, pp. 2183–2184, 2007.

12. A. Kauffmann, R. Gentleman, and W. Huber, "arrayqualitymetrics – a Bioconductor package for quality assessment of microarray data," *Bioinformatics*, 2009.

13. G. K. Smyth, "Linear models and empirical Bayes methods for assessing differential expression in microarray experiments," *Statistical Applications in Genetics and Molecular Biology*, vol. 3, 2004.

14. G. R. Warnes, B. Bolker, L. Bonebakker, et al., *gplots: Various R programming tools for plotting data*, 2013. R package version 2.11.3.

15. S. Falcon and R. Gentleman, "Using GOstats to test gene lists for GO term association," *Bioinformatics*, vol. 23, no. 2, pp. 257–258, 2007.

16. M. S. Schröder, D. Gusenleitner, J. Quackenbush, A. C. Culhane, and B. Haibe-Kains, "RamiGO: an R/Bioconductor package providing an AmiGO visualize interface," *Bioinformatics*, vol. 29, no. 5, pp. 666–668, 2013.

17. K. L. Gunderson, "Decoding randomly ordered DNA arrays," *Genome Research*, vol. 14, no. 5, pp. 870–877, 2004.

18. S. M. Lin, P. Du, W. Huber, and W. A. Kibbe, "Model-based variance-stabilizing transformation for Illumina microarray data," *Nucleic Acids Research*, vol. 36, no. 2, p. e11, 2008.

19. K. Kuhn, "A novel, high-performance random array platform for quantitative gene expression profiling," *Genome Research*, vol. 14, no. 11, pp. 2347–2356, 2004.

20. G. K. Smyth, J. Michaud, and H. S. Scott, "Use of within-array replicate spots for assess-

ing differential expression in microarray experiments," *Bioinformatics*, vol. 21, no. 9, pp. 2067–2075, 2005.

21. P. Du, W. A. Kibbe, and S. M. Lin, "nuid: a universal naming scheme of oligonucleotides for Illumina, Affymetrix, and other microarrays," *Biology Direct*, vol. 2, p. 16, 2007.

22. Y. Benjamini and Y. Hochberg, "Controlling the false discovery rate: a practical and powerful approach to multiple testing," *Journal of the Royal Statistical Society, Series B*, pp. 289–300, 1995.

23. S. Falcon and R. Gentleman, "Using gostats to test gene lists for GO term association," *Bioinformatics*, vol. 23, no. 2, pp. 257–258, 2007.

24. M. Ashburner, C. A. Ball, J. A. Blake, et al., "Gene ontology: tool for the unification of biology. The Gene Ontology Consortium," *Nature Genetics*, vol. 25, no. 1, pp. 25–29, 2000.

25. L. Gautier, L. Cope, B. M. Bolstad, and R. A. Irizarry, "affy – analysis of Affymetrix GeneChip data at the probe level," *Bioinformatics*, vol. 20, no. 3, pp. 307–315, 2004.

26. C. L. Wilson and C. J. Miller, "Simpleaffy: a Bioconductor package for Affymetrix quality control and data analysis," *Bioinformatics*, vol. 21, no. 18, pp. 3683–3685, 2005.

27. C. S. Gillespie, G. Lei, R. J. Boys, A. Greenall, and D. J. Wilkinson, "Analysing time course microarray data using Bioconductor: a case study using yeast2 Affymetrix arrays," *BMC Research Notes*, vol. 3, no. 1, p. 81, 2010.

28. F. Hong, R. Breitling, C. W. McEntee, B. S. Wittner, J. L. Nemhauser, and J. Chory, "Rankprod: a Bioconductor package for detecting differentially expressed genes in meta-analysis," *Bioinformatics*, vol. 22, no. 22, pp. 2825–2827, 2006.

29. J. T. Leek, W. E. Johnson, H. S. Parker, A. E. Jaffe, and J. D. Storey, "The sva package for removing batch effects and other unwanted variation in high-throughput experiments," *Bioinformatics*, 2012.

30. W. E. Johnson, C. Li, and A. Rabinovic, "Adjusting batch effects in microarray expression data using empirical Bayes methods," *Biostatistics*, vol. 8, no. 1, pp. 118–127, 2007.

31. Y. Xie, *Dynamic Documents with R and knitr*. London: Chapman and Hall/CRC, 2013.

Index

Printed and bound by CPI Group (UK) Ltd, Croydon, CR0 4YY

21/10/2024

01777104-0002